COVID-19 AND THE GLOBAL PREDATORS: *WE ARE THE PREY*

Peter R Breggin MD
Ginger Ross Breggin

LAKE
EDGE
PRESS

COVID-19 AND THE GLOBAL PREDATORS: WE ARE THE PREY

Peter R Breggin MD
Ginger Ross Breggin

Printing History
Lake Edge Press / 2021

Cover photograph and book design
by Ginger R Breggin

ISBN: 978-0-9824560-6-4

Lake Edge Press
206A Dryden Road PMB 112
Ithaca, NY 14850

Selected Books by Peter R. Breggin MD and Ginger Ross Breggin

Electroshock: Its Brain-Disabling Effects, 1979

Psychiatric Drugs: Hazards to the Brain, 1983

Toxic Psychiatry, 1991

Beyond Conflict, 1992

Talking Back to Prozac, coauthor Ginger Ross Breggin, 1994

The War Against Children, coauthor Ginger Ross Breggin, 1994

Psychosocial Approaches to Deeply Disturbed Persons (senior editor), 1996

Brain-Disabling Treatments in Psychiatry: Drugs, Electroshock, and the Role of the FDA, 1997

The Heart of Being Helpful: Empathy and the Creation of a Healing Presence, 2006

Talking Back to Ritalin, 1998

The War Against Children of Color: Psychiatry Targets Inner City Youth, coauthor Ginger Ross Breggin, 1998 [update of *The War Against Children*]

Your Drug May Be Your Problem: How and Why to Stop Taking Psychiatric Medications, coauthor David Cohen, 1999

Reclaiming Our Children: A Healing Solution to a Nation in Crisis, 2000

The Antidepressant Fact Book, 2001

Dimensions of Empathic Therapy, coedited by Ginger Ross Breggin and Fred Bemak, 2002

The Ritalin Fact Book, 2002

Your Drug May Be Your Problem: How and Why to Stop Taking Psychiatric Medications, Second Edition, coauthor David Cohen, 2007

Brain-Disabling Treatments in Psychiatry: Drugs, Electroshock and the Psychopharmaceutical Complex, Second Edition, 2008

Medication Madness: The Role of Psychiatric Drugs in Cases of Violence, Suicide, and Crime, 2008

Wow, I'm an American! How to Live Like Our Nation's Heroic Founders, 2009

Psychiatric Drug Withdrawal: A Guide for Prescribers, Therapists, Patients and Their Families, 2013

Guilt, Shame and Anxiety: Understanding and Overcoming Negative Emotions, 2014

Suing for Freedom: The Legal Documents that Started the Ohio COVID-19 Freedom Case, with Attorney Thomas Renz (in press)

DEDICATION

W e dedicate our book to those courageous physicians who have defied their medical and scientific establishments, their governments, and other powerful authorities by offering patients inexpensive, safe, effective, and lifesaving treatments for COVID-19 and who have dared to speak out against ill-advised public health measures which have caused more misery and death than the pandemic itself.

We further dedicate this book to three exemplars of these honorable physicians:

Peter A. McCullough MD, MPH

Elizabeth Lee Vliet MD

Vladimir "Zev" Zelenko MD

And we thank them for the inspiring and informative introductions that they have generously written for our book.

IMPORTANT NOTICE

Our YouTube Channel is down and other platforms are threatened. To keep up with our work, sign up for our Free Frequent Alerts. Go to *www.breggin.com*

CONTENTS

PART ONE:
ANTHONY FAUCI AND THE MAKING OF COVID-19

PART TWO:
COVID-19 POLICIES AND THE END OF SCIENCE

PART THREE:
COVID-19 AND THE PREDATORY GLOBALISTS

PART FOUR:
RECOVERING OUR LIBERTY

HOW TO FIND LIFESAVING COVID-19 TREATMENT GUIDES AND DOCTORS

T reatment for COVID-19 requires specialized medical knowledge and should begin as early as possible, preferably at home before hospitalization is required, with guidance from a physician experienced with and dedicated to early treatment. Although almost entirely sparing young and healthy people, COVID-19 can be very serious and life-threatening, especially to older people and those with multiple medical vulnerabilities.

A regularly updated free *GUIDE TO COVID EARLY TREATMENT— Options to Stay Out of Hospital and Save Your Life!* is published on the Truth For Health Foundation website. Elizabeth Lee Vliet MD, the President and CEO, and Peter A. McCullough MD, MPH, Senior Medical Advisor, lead the foundation and are the authors of this new and extraordinarily valuable educational resource that also provides access to treating physicians. Here is the link:

https://www.truthforhealth.org/patientguide/patient-treatment-guide/

You can also access a similar booklet, *A Guide to Home-Based COVID-19 Treatment*, a free resource published by the Association of American Physicians and Surgeons (AAPS), that also provides access to treating physicians. Here is the link:

https://aapsonline.org/covidpatientguide/

For a recent scientific review, see Peter McCullough et al. (2020)[1] written by 57 authors with overwhelming support for early home treatment of COVID-19 in high-risk adults.[2] According to Dr. McCullough, "Powerful multidrug regimens *reduce hospitalization by 87% and death by 75%*."[3]

The two guidebooks do not provide individual medical advice or prescribe treatment. This is also true of our book, *COVID-19 and the Global Predators*.

Instead, we[4] provide an educational service for patients, their families, and their doctors to know what options are available.

The Pandemic Is the Opening Salvo

There have been lifetimes in the last few months. We have been writing as if to catch up with the future, instead of finishing the book. It is August 4, 2021, and we must be done. The book should be published in a matter of weeks. Hopefully, it will provide the world with a better understanding of what is unfolding before our eyes in a growing, at times blinding, blaze of oppression.

Our book, *COVID-19 and the Global Predators*, began with an inquiry into the source of the obviously irrational, contradictory, and harmful public health policies implemented in the name of COVID-19. Eventually, it grew so broad and deep in scope that it now requires an introductory summation of some of our major discoveries and conclusions.

A BASIC PUBLIC HEALTH PRINCIPLE

A basic public health principle is, "*The more society can function normally during a pandemic—the better the people will manage and survive.*" This policy was turned on its head into, "The more society can be frightened, isolated, controlled, and suppressed—the safer it will be from the virus."

Denying the Origin of the Virus
in American and Chinese Labs

From the start, the source of SARS-CoV-2 was denied, although it is the obvious product of collaborative U.S./Chinese gain-of-function research, making deadly SARS-CoV viruses in American and Chinese labs. The origin of SARS-CoV-2 is easily traced and documented in multiple scientific publications, most of them funded in part by the National Institutes of Health (NIH) and Anthony Fauci's National Institute of Allergy and Infectious Diseases (NIAID).

Some gain-of-function projects funded by NIAID and NIH involved American collaborations with China. Other NIAID and NIH funding continues to support individual American and Chinese Communist researchers creating dangerous viruses. Fauci lied under oath before a Senate committee when he said neither NIAID nor NIH funded gain-of-function research conducted in China or the U.S.

Ultimately, SARS-CoV-2 was released from the Wuhan Institute of Virology. Whether or not it was done intentionally is a difficult question we will fully address and then draw conclusions about in Chapter 4.

NIAID and NIH continue to fund these deadly gain-of-function experiments in American labs.[5] Other U.S. agencies have also contributed funding to gain-of-function research, ominously including the shadowy Defense Advanced Research Projects Agency (DARPA). DARPA is a part of the Department of Defense. Research into making deadly viruses and vaccines in the U.S. is inevitably connected to perpetrating and defending against biological attack, much as it is in other Western nations.

Early on, the question arose, "Why would the U.S. support the Chinese Communist biowarfare program by sharing advanced research and experimentation on making pandemic viruses?"

Eventually, we would find many of the basic explanations.

We Send Critical Information to
President Trump and He Takes Action

In mid-April 2020, we published a blog and video about Anthony Fauci and NIAID's support of the seemingly treasonous U.S./Chinese Communist collaboration making deadly viruses. We sent our blog and video up lines to President Donald Trump's inner circle, and two days later the President cancelled all funding for U.S./Chinese viral research funding, although we will find reason to believe that Fauci continues to find ways around this prohibition.

This early success gave us hope for having a continued impact, and we went to work on what became *COVID-19 and the Global Predators: We Are the Prey*. Both of us have been working hard for sixteen months, often up to ten hours a day, researching and gathering information for our book from innumerable colleagues and sources. We have also been sharing as much of our knowledge as we can with the rest of the world through written reports, media, and circulating the manuscript as it developed.

Suppressing the Home-based Early and Effective Treatment of COVID-19

We soon learned that most of the relevant identifiable institutions in the West—NIH, NIAID, the FDA, the CDC, WHO, the major and social media, the scientific and medical establishments, the tech companies, the universities, governments, and multiple billionaires—were suppressing and continue to suppress the early and very effective treatments for COVID-19.

Hundreds of honorable and heroic doctors were and continue to be threatened, risking the loss of their professional careers, as they successfully treat COVID-19 patients. They are saving lives with readily available, safe, and effective treatments for infection with SARS-CoV-2. All the treatments are inexpensive, including hydroxychloroquine, ivermectin, and assorted biological and nutraceutical approaches.

Peter A. McCullough MD, Elizabeth Lee Vliet MD, and Vladimir "Zev" Zelenko MD—each of these three medical doctors are leading the early, home-based treatment movement and helping to save millions of lives. Each of them has honored us by reviewing our manuscript and writing introductions to this book.

We asked ourselves, "Why do the global predators continue to violently suppress early home-based treatments, causing the unnecessary deaths of millions of people?"

Then we found out why. There is a catch in the federal legislation called the Emergency Use Authorization (EUA) that empowers the government to finance drug company costs for the development of vaccines and to rush them onto the market at breakneck speed without standard FDA approval as safe and effective. Under the EUA, the U.S. could not finance and rush through the experimental and potentially lethal vaccines *unless there were no safe and effective treatments already in place.*

Bill Gates, Fauci, and the drug companies with their predatory investors must continue to do everything in their power to discredit existing safe and effective treatments. At any time, if they fail to suppress all effective treatments, their enormous ambitions for money and power from the unwarranted experimental mass vaccinations will collapse like a flimsy house of cards.

Hydroxychloroquine-based home treatments reduce hospitalizations by 87% and deaths by 75%, while ivermectin-based treatments may be as good or better.

But the global predators suppress these life-saving therapies to swell their wealth and power by forcing unneeded, unproven, dangerous vaccinations on the entire world. As ghastly as it seems, many of the most wealthy and powerful people and organizations in the world would rather see people die without treatment than lose their much-anticipated pandemic opportunity to coerce all of us into being experimentally vaccinated while having the government cover their expenses and protect them from lawsuits.

After reading this preface, Peter A. McCullough MD sent this important analysis to us on June 28, 2021:

> We are more than six months into the largest mass vaccination program in U.S. history. There have been no press briefings to the public on overall vaccine safety by the CDC, FDA, NIH, or White House Task Force. The silence has been deafening despite 411,391 adverse event reports to the CDC with 54,606 urgent care visits, 23,257 hospitalizations, and 6,985 deaths as of June 25, 2021.[6] No analysis of these adverse event reports has been given to our terrified citizens who are being forced to gamble their health and their lives on unsafe, experimental vaccines or lose the right to exercise their Constitutional rights to travel or to go work, to school, or to public and private events.

If there were complete reporting of every vaccine related safety event, the already huge numbers of injured and dead from the vaccines would probably be tenfold or greater. At this rate, since the launch of the mass vaccination program, there will be more casualties and damage to the health of Americans attributable to the vaccines than to the COVID-19 respiratory illness during that same period.

A Well-Planned Long-Term Strategy: Vaccinating the World for Wealth and Power

The deplorable, wicked actions by those in charge of COVID-19 became much more understandable in light of the necessity of their preventing early, effective treatments for COVID-19, while simultaneously rushing through highly remunerative experimental vaccines under the Emergency Use Authorization. They were sacrificing millions of lives in order to preserve their ability to get taxpayer financing and a free pass from the FDA to flood the world with their vaccines at breakneck speed.

Learning that COVID-19 was about forcing humanity to submit to experimental, dangerous, and costly vaccines, and to use that justification for increasing totalitarian power, opened up a broader understanding. Then we could begin to unravel the overall scheme for exploiting humanity.

The Global Predator Network

We uncovered how a loose but coordinated array of billionaires, tech companies, public health schools and authorities, major worldwide corporations, and their allies had been planning for at least four years to make a financial killing on what they defined and repeatedly predicted as the inevitable and soon-to-arrive pandemic. Four years ahead of time, the globalists began making massive marketing expenditures to prepare the world for accepting dangerously rushed vaccination programs in the future. The propaganda splurge was stunning, and before 2017 was over, they were already predicting and preparing the world specifically for a SARS-CoV pandemic through which to impose their vaccines on humanity at the first opportunity.

From as early as January 2017, the global predators planned for the financial windfall to come specifically from making and marketing untried, highly dangerous mRNA and DNA vaccines. They already had programs called "platforms" in the early stages of development with Moderna and Pfizer that would, several years later, in 2020, be named and unveiled as Operation Warp Speed.

The globalists were also organizing and collecting a massive fusion of corporate, philanthropic, and government financing, which is the backbone of Klaus Schwab's Great Reset announced in 2020. They were already working with many world governments and agencies, and huge investors, without any involvement of actual legislative bodies, courts, or citizens' groups placing restraints on them. Wielding so much power, influence, and money, they met nothing but appreciation from people and organizations ambitious to have a piece of the action.

Connections to Communist China

From the start of COVID-19 in 2020, the collaborating globalists have included America's top tech corporations: Microsoft, Apple, Alphabet (Google and YouTube), Facebook, and Twitter. Almost every single American billionaire remains deep in the predatory morass: Jeff Bezos, Bill Gates, Warren Buffett (a partner with Gates until he recently resigned), Mark Zuckerberg, the Waltons, Steve Balmer, Larry Page, and Michael Bloomberg. All are deeply invested in China. The more recently crowned king of the billionaires, Elon Musk, has well-advertised ambitions for Tesla and other products in the Chinese market. As *Bloomberg Business* recently declared,[7] "Elon Musk Loves China, and China Loves Him Back—for Now."

Meanwhile, the coalition continues to include surrogates for the top predators. Two of the most powerful are Anthony Fauci (very close to Bill Gates) and Tedros of WHO (also very close to Gates and even closer to the Chinese Communists).

This lockstep support of global predation includes the two premier American schools of public health, Johns Hopkins and Harvard. The medical and scientific establishments are marching into the future with the globalist predatory plans, including the most highly respected medical journals, *The Lancet*, *The New England Journal of Medicine* (*NEJM*), and *The Journal of the American Medical Association*

(*JAMA*). All of them continue to fully cooperate with the globalist predatory goals of exonerating the Chinese Communists, preventing early treatment, and pushing experimental vaccines onto humanity. They participate in crushing the Western democratic republics with draconian public health measures that disrupt individual lives, society, and the economy.

All the old and supposedly "progressive" radio and TV media unflinchingly support the global predators, while protecting Communist China. They include ABC, NBC, and CBS; National Public Radio; and CNN and MSNBC. Although getting shaky about it, they continue to reject the Wuhan Institute origin of SARS-CoV-2. They reject the truth as racist and conspiratorial. They also back every new political oppression in the name of COVID-19, including the worst actions of extremist Democratic governors. The *Wall Street Journal*, *The Washington Post*, *The New York Times*, and *USA Today*—the old establishment print media—behave as one with their radio and TV counterparts.

Many, if not most, large American universities are also closely tied to global predators and, surprisingly so, to the Chinese Communist Party. Chinese businesses and scientific enterprises enrich the universities. The universities in return provide profits, patents, industrial and military secrets, and training for Communist scientists in high-tech, virology, and other critical areas for national security. The situation is so bad that some universities are under federal investigation for their ties to China and to the Chinese Communist Party.

Again, our book documents all that we are presenting in this preface.

Most of the world's largest corporations and banks, regardless of their national origin, are deeply involved in global exploitation. All of them salivate over current and future opportunities to invest heavily in Chinese markets. They show no hesitation about working with a predatory government that uses slave labor, and they are happy to make money from Chinese enterprises, such as those involved in surveillance and media censoring, that oppress the Chinese people. Nor do they display any qualms about strengthening the world's most threatening dictatorship, which is competing with America for dominance and developing weapons of mass destruction for use against America as the last, great bastion of individual and political liberty.

Also supporting the global predators, we find key American politicians. Among them are recent Republican presidents before Donald Trump and many top Republican establishment leaders; the anti-Trumpers; and the Koch family.

The entire progressive political establishment has also become entwined with the highest circles of worldwide industry and finance and the Chinese government. There can be no mistaking the intentions of the Chinese Communist Party—it has always intended to reshape the world into a Communist empire with itself as the supreme ruler. Any denial of the Communist goals by Western leadership is self-deluding and ultimately suicidal.

Even Pope Francis has become a Communist collaborator. The Pope is taking millions of dollars from China and in return giving the Communists the power to select Catholic bishops in China. This has never been done before. Simultaneously, the Pope is pushing the Church as far to the Left as possible.

Every global predator we could identify is financially wedded to and filled with admiration for, not the United States or other Western democracies, but for Communist China and totalitarianism. This ultimate realization from our research was totally dismaying.

COVID-19 policies have aimed at and caused a massive shutdown of Western economies. The intention of the globalists has been to harm and intimidate democratic republics such as the United States, Great Britain, Germany, and Australia. The ultimate goal is to vastly increase top-down government, authoritarianism, and totalitarianism throughout the world.

The global predators want to stop the U.S. and other Western nations from impeding their global predation. This menacing group includes most of the world's billionaires and the major institutions of the West, and it is backed by the Chinese Communist Party. This coalition of evil could ultimately result in a thousand-year reign of terror under the Chinese Communist Empire.

WE DISCOVER THE MASTER PLAN!

As we have already summarized, the entire plan for the fulfillment of these main predatory goals was firmly established by 2017. More recently, we found the master plan that was presented in July 2017. We found it shortly before publication of this book in the form of a 21-page, detailed PowerPoint presentation made by CEPI to the World Health Organization.

CEPI is the Coalition for Epidemic Preparedness Innovations. It was officially unveiled in early 2017 by Bill Gates, along with cofounder Klaus Schwab of the World Economic Forum (WEF). The giant philanthropic health fund,

Wellcome Trust, was also among the original founders. The CEPI master plan, presented to WHO in the summer of 2017, was essentially aimed at empowering each other to run the next pandemic and hence to govern much of the world's activities.

In the CEPI master plan, which WHO adopted, Bill Gates manages the vaccine production and profits from the coming pandemic while WHO manages the medical and scientific community. WHO's major role became managing the pandemic in the interest of the Chinese Communist Party and Bill Gates, two of the largest donors to WHO.

At its inception in January 2017, CEPI was already a massive organization that epitomized the yet-unnamed Great Reset strategy. CEPI takes funding from governments, philanthropic organizations, and billionaires, and then reinvests it into the pharmaceutical industry, supposedly for the benefit of humanity:[8]

> CEPI is an innovative partnership between public, private, philanthropic, and civil organisations. It was founded by the governments of India and Norway, the Bill & Melinda Gates Foundation, Wellcome and the World Economic Forum. Other partners include multinational pharmaceutical corporations, the World Health Organization, and NGOs.

> CEPI is supported by several leading pharmaceutical companies with strength in vaccines – GSK, Merck, Johnson & Johnson, Pfizer, Sanofi, and Takeda, plus the Biotechnology Innovation Organisation.

CEPI's complex partnerships, investments, and governances are transnational, autonomous, and probably beyond the reach of most regulatory agencies.[9] It goes into partnership with and funnels vast amounts of money into the pharmaceutical industry, including Chinese corporations.[10] CEPI recently boasted, "By the end of 2019, CEPI had a total of 19 vaccine candidates against five priority pathogens and three rapid response platforms in its portfolio."[11] Bill Gates and CEPI could not have been better positioned to make a fortune from COVID-19.

But Bill Gates wanted to guarantee that he would suffer no losses in the coming pandemic while collecting dazzling profits. In one of the most stunning financial boondoggles in history, Bill Gates' CEPI announced a fundamental "principle" of its working agreement with WHO: "No Loss: vaccine developers should be reimbursed for their direct and indirect costs." This goal would guarantee huge personal profits in the coming pandemic for Bill Gates' vast investments in the pharmaceutical industry. But that was not enough for the master manipulator. As another "principle" worked out with WHO, Gates' CEPI foundation would take any excess profits from the drug companies to "pay back CEPI funding." That way, the wealth of the Gates' controlled fund would also grow (Chapter 15).

The Desperate Need for an Ideology Results in the Great Reset

With exceptions—such as the Chinese Communist Party and aggressive Muslim organizations—most global predators have no strong ideological commitments. They are strictly driven by personal ambition and the acquisition of wealth and power. They are not patriotic toward any nation. They have no religion and most are agnostics or atheists, with many identifying themselves as humanists or transhumanists. Their ethics are relativist, situational, and tailored to meeting their greater goals. This vastly limits their ability to take over the world because people who lack a strongly motivating ideology also lack courage and the required determination to succeed or to die in the effort.

Many of the heroic founders of America believed they were fighting much more to fulfill God's purposes than for their own more petty ambitions. The signers of the Declaration of Independence knew they were putting their names on their death warrant, should King George get his hands on them. Their dedication inspired me to write a little book for families titled, *Wow, I'm an American! How to Live Like Our Nation's Heroic Founders.*

For some modern global predators—including Bill Gates, President Joe Biden, and his environmental czar John Kerry—their identified "intellectual" spokesperson is Klaus Schwab. Schwab promotes the concept of the Great Reset—essentially a combination of predatory capitalism and predatory progressivism working with predatory governments—while opposing the principles of individual and political liberty and the world's democratic republics. Schwab is a lackluster leader, and his philosophy is pitifully weak; but his Great Reset concept affords us the ability to see what they have in store for us should they triumph.

The Great Reset aims to further empower people of great wealth and influence to use docile top-down governments to exploit the world. In the process, it plans to distort human nature and contemporary societies to create more conformist people. The newly engineered human, this product of transhumanism, will happily live under their governance, despite crushing deprivations of personal freedom and quality of life. Population control will help enable the remaining people to have enough food and limited amenities, while preserving planet Earth by reducing the number of human consumers and polluters.

Schwab spelled out all of the above in his 2020 book, *COVID-19 and The Great Reset*. He acknowledged that the greatest threat to globalist ambitions are the patriotic, democratic republics, especially the United States with its love of individualism and freedom. He admitted that the nemesis of globalism was President Donald Trump.

In his book, Schwab warned that globalism was taking a beating from President Trump and his America First policies. America was finally refusing to further empower Communist China or to sacrifice American workers and businesses to the global economy and its politics. Schwab was hopeful, however, and boasted about using COVID-19 to further his globalist schemes.

The global predators are neither idealistic progressives nor freedom-loving capitalists. They are devoted neither to the common good nor to individual freedom. They are predators eager to collaborate with anyone who shares the primary goal of controlling, dominating, and exploiting humanity.

If the global predators succeed in further weakening America, until no nation remains to resist them, they will be in for a shock. The global predators will not thrive under the Chinese Communist Empire that they so avidly support. They will either slavishly serve the empire or be extinguished. Meanwhile, a few patriotic democracies, led by a reinspired America, are the only hope for the survival of individual and political freedom in the world.

President Trump's Critical Failures

In mid-April 2020, when President Donald Trump responded to our public disclosures by cancelling the several-year U.S. collaboration with the Chinese Communists in making deadly SARS-CoVs, we were vastly encouraged. Unfortunately, President Trump did not follow up by cancelling all gain-of-function research. And then our President let things get progressively and tragically worse.

Trump ultimately supported the most massively organized assault on humanity in history, one especially focused on America and the free world. It crushed societies and economies around the world. Churches, small businesses, and schools were shut down. Children learned to act like humbled prisoners, and adults became arrogant prison guards. Humanity itself became demoralized under the crushing weight, not of the pandemic itself, but of the draconian "public health measures" that were calculated to render humans more docile and amenable to control.

SARS-CoV-2 was less deadly to most of the population than the seasonal flu and most dangerously struck the very elderly and impaired, but few had to die. Yet millions died, and continue to die, because President Trump gave up on early home-based treatment. He could have asserted his presidential authority to issue an executive order to make highly successful home-based treatment widely available across America, including the release of 63 million doses of hydroxychloroquine from government stockpiles. The entire trajectory of the pandemic would have been blunted, resulting in few hospitalizations or deaths. He received information that could have provided him the legal basis for such actions. Instead, he enthusiastically supported Operation Warp Speed to produce the dangerous, experimental, and already proven-deadly vaccines. America, the last bastion of freedom, is falling before the spearhead of globalism in the form of a spike protein too small to see with the naked eye.

After heroically standing up to the global predators on political and economic issues, President Trump played into their hands by refusing to fight their totalitarian assault in the guise of treating COVID-19. He allowed the destruction of the wonderful economy he had built and did little or nothing to prevent the closing of schools, churches, and small businesses.

It is impossible to know if any human being could have stood up against such an unexpected, unprecedented onslaught of evil, but many of us had hoped for this President to lead us through the worst. When President Trump failed, we still hoped, when he was out of office and free of Deep State influences, that he would then realize he had been duped into betraying humanity in his handling of COVID-19.

Former President Trump continues to boast about "his" vaccine program and to urge people to take these gene-modifying, experimental shots. He ignores the coercion being applied to our citizens to make them submit to the jabs, and he fails to address the threats of mandates and vaccine passports that will expand globalist control over their lives. Former President Trump does not address the horrific prospect of imposing these experimental vaccines on younger and younger children. He also continues to betray humanity by failing to support early home-based treatments which, in the great majority of cases, prevent the need for hospitalization and eliminate most deaths.

All of this is sad, but true. However, it does not undermine a major theme of this book: President Trump's America First policies deeply threatened the globalist predators and brought them out like giant bats flying in black clouds from dark caves at the witch's beckoning. President Trump's pro-America, and hence anti-globalist policies, included closing the southern border to wholesale invasions, raising employment rates and real wages to new heights among minority and working-class people, bringing back industries to America, and defying the Koch family, the Bushes, and other Republican globalists.

Seeing a huge setback in their previously unfettered exploitations, the global predators struck back. The Chinese Communists, with the collaboration of WHO and a coalition led by Bill Gates and his surrogate Anthony Fauci, acted precipitously and inflicted COVID-19 on the U.S. and much of the world. These events, so previously impossible to contemplate—so beyond our own imaginations before we began our research—are thoroughly documented in this book.

All of President Trump's accomplishments were erased or set back by his submission to the totalitarian COVID-19 strategies of the global predators. However, his earlier monumental successes forced these exploiters of humanity out into the open, alerting millions of Americans, and many others throughout the world, to their identities and to the enormity of their threat.

During his campaign and in his administration until COVID-19, Trump contributed marvelously to the revival of the spirit of freedom in America. Now, increasing numbers are joining the movement to restore our nation's great principles of individual liberty and political freedom. We must each do everything we can to retake America as a nation devoted to freedom.

The oppressors of free people everywhere, both foreign and domestic, want us demoralized, discouraged, hopeless, and depressed. They want us to give up and give in to their orders and their New World Order.

We must not give in. We must never give up. We must set aside our anxieties and sadness, and our wishes that someone would save us—someone big and brash and brave enough to stand up to all the oppressors. There is no one great superhero to come to the rescue of the people of Earth.

Seekers of liberty and seekers of truth, rebels, individualists, explorers, cowboys, artists, nonconformists, intellectuals, mavericks, misfits, professionals, Divergents, free thinkers, and people of faith—even establishment members who now see the light—all of us must stand together. We have only ourselves. We must with true grit and respect and grace forge ahead and create a free tomorrow.

What Can We Do Now?

What can you do? Read our book to learn more details about the global predators who are threatening and weakening America while exploiting all humanity. Familiarize yourself with and support the freedom-loving patriotic individuals and organizations who are joining together to fight for the revival of America and other nations as democratic, constitutional republics. Get to know the great numbers of us who are already working together to restore our nation and to inspire other nations to stand up against the global predators and their major backer, the Chinese Communist Party.

You will find yourself no longer alone. You will become empowered. You will make wonderful new friends. Your life will take on new meaning when you realize you have a rare opportunity in history to fight for Western values, the Judeo-Christian traditions, and the existence of individual and political liberty on Earth.

INTRODUCTION

by Peter A. McCullough MD, MPH

Historians will write about the inexplicable medical response to the fatal SARS-CoV-2 pandemic. Doctors edged out of the circle of empathy responded to their acutely ill patients by saying, "I don't treat COVID-19," and were supported by the National Institutes of Health (NIH) and U.S. Food and Drug Administration (FDA), both of which actively obstructed any early treatment to sick Americans.

The media asked no questions and stuck to the narrative of reporting grim news and then blaming the victims for not wearing masks or complying with directives. Censorship went from covert to overt, and the "Trusted News Initiative" was announced as an explicit program to filter out any negative information on vaccine development. All the while, "stakeholders" had organized and were executing a highly profitable, controlling, and nefarious mass coronavirus vaccination program.

The public had to be held in fear and pay a heavy price with hospitalizations and deaths to fully accept vaccines that were not fully evaluated or safe, and were promoted coercively with full indemnification to the manufacturers and those administering the injections. The only threats to this plan were heroic doctors and journalists who risked it all to save as many lives as possible and expose the emerging truths about SARS-CoV-2, the treatable COVID-19 illness, and immunity enjoyed by survivors.

Breggin is an intrepid scholar and is assiduous and methodological as he assembles all the pieces to the puzzle. His research, carried out with his wife Ginger, is impeccable, and his incisive approach sears the neck of those whose aim it is to wield power, control, and instill fear among the world's wealthiest nations.

This book is a must-read-now to understand what is happening before our very eyes. Sit forward with both feet on the ground and be ready for the most gripping nonfiction that just rocketed into your library. What you will learn may save you and your family in the turbulent times ahead.

—Peter A. McCullough MD, MPH, FACC, February 2021

Dr. McCullough is Professor of Medicine, Texas A & M College of Medicine, Dallas, Texas. He is an internist, cardiologist, and epidemiologist, and the editor-in-chief of *Cardiorenal Medicine* and *Reviews in Cardiovascular Medicine*. He has authored over 600 cited works in the National Library of Medicine and lectured across the world on contemporary medical issues. In 2020, he testified before the U.S. Senate on the early treatment of COVID-19 and is senior author of the most extensive scientific publications outlining the pathophysiological basis and rationale for early ambulatory treatment of COVID-19. His methods reduced hospitalization and death by 85 percent, thus making the COVID-19 epidemic the largest avoidable source of suffering, hospitalization, and death in modern human history. Most of the drugs and concepts formulated by Dr. McCullough were available by early April 2020.[12]

INTRODUCTION

by Vladimir Zelenko MD

Mass Murder by COVID-19

The Breggins carefully examine the science and the facts of the tragic situation surrounding COVID-19. I want to place these mass murders brought about by improper treatment and withholding lifesaving treatment in their truly horrific context. The mismanagement of the COVID-19 pandemic is akin to mass murder and the genocide of the elderly and infirm. The root cause of this crime against humanity is the denial of man's divine origin. Man is made in "the image of G-d" and, therefore, his or her life has intrinsic value and natural rights. From this perspective, it is self-evident that each individual life has sanctity, intrinsic value, and must be preserved whenever possible. This is succinctly stated in the Declaration of Independence: "We hold these truths to be self-evident, that all men are created equal, that they are endowed by their Creator with certain unalienable Rights, that among these are Life, Liberty and the pursuit of Happiness."

In contradistinction to man's divine origin, Darwin's theory of evolution as well as Galton's ideas of eugenics touted the erroneous belief in the dominance hierarchy in human genetics. This discredited pseudoscience aimed to improve the genetic quality of the human population by excluding people and groups judged to be inferior and exalting those judged to be superior. This is the ideological basis of the Nazis that used it to justify their torture and murder of Jews, disabled people, and other minority groups.

Despite a plethora of scientific data, lifesaving information and access to vital medications are being suppressed for the majority of the human race. This has so far led to the tragic and preventable deaths of over three million people. The perpetrators of this historically heinous crime are motivated by the desire for power and control over the human race. These modern-day slave masters believe that they are übermensch (superhuman) with the right to decide who should live or die.

The denial that the human race is made in "the image of G-d" invariably leads to the slippery slope of moral relativism, which is governed by the principle of "the survival of the fittest." Therefore, the self-proclaimed übermensch of this generation believe in their right to effectuate "The Great Reset" and "New World Order." In order to facilitate "The Great Reset," the World Economic Forum publicly stated their goals for the year 2030:

- You will own nothing and you will be happy.

- Whatever you want you can rent and it will be delivered by drones.

- The United States will not be the world's leading superpower.

- A handful of countries will dominate.

- You'll eat much less meat. An occasional treat and not a staple, for the good of the environment and your health.

- A billion people will be displaced by climate change.

- We will have to do a better job at welcoming and integrating refugees.

- There will be a global price on carbon. This will help make fossil fuels history.

- Western values will have been tested to the breaking point.

In the last year, I have personally witnessed the tragic and preventable deaths of hundreds of people from COVID-19. Despite a Samsonian effort to advocate for effective, safe, and affordable prehospital treatments, my colleagues and I have been threatened and censored.

It is my supposition that this suppression of lifesaving information and medication is mass murder. This crime against humanity has been willfully perpetrated by a group of sociopathic despots that possess a delusional "G-d complex" and perceive themselves as superhumans with the right to enslave others.

It is my strong hope and prayer that they will be brought to justice in both the earthly and heavenly courts.

COVID-19 and the Global Predators is an absolutely authoritative book describing the root causes of the genocide and crimes against humanity that we are still living through. It is a bright light of clarity in a world of dark confusion. God should bless Dr. Breggin and his wife with good health and a long life. Thank you for this gift.

—*Vladimir "Zev" Zelenko MD, February 2021*

Dr. Zelenko is the creator of Zelenko Protocol for the early treatment of COVID-19, which has saved countless lives worldwide. In March 2020, his team was one of the first in the country to successfully treat thousands of COVID-19 patients in the prehospital setting. He graduated *summa cum laude* with a B.A. degree in chemistry from Hofstra University and received his medical degree from the School of Medicine, State University of New York at Buffalo. He showed the world how to successfully treat COVID-19.

INTRODUCTION

by Elizabeth Lee Vliet MD

The Breggins' book on COVID-19 is a powerful wakeup call that our core freedoms are under attack from all directions, including from medical institutions we previously trusted to guide us saving lives during illness emergencies. The travesty of orchestrated suppression of lifesaving medicines to treat COVID-19 has resulted in the societal murder of over two million people globally.

This massive death toll has been horrific in the U.S., United Kingdom, Western Europe, Canada, and Australia where physicians are still largely *prevented* from access to lifesaving medicines and threatened with loss of license or prosecution if they go against bureaucratic and political suppression of early home-based treatment.

Through it all, who is standing up for sick patients? During the pandemic, not one single *national* leader in the pandemic medical response is saying: "COVID-19 is terrible, and I am going to work with the best experts from many fields of medicine to find ways to treat the virus early to reduce these hospitalizations and deaths, as we have *always* done with *any* viral illness."

No national public health leader on the COVID response task force has ever mentioned that physicians and patients could work together and treat people *at home* to prevent the high mortality rate that came from waiting until people were critically ill and sending them to the hospital to die alone. Their tunnel vision focus was and is solely on contagion control through mask mandates and lockdowns, while waiting in fear for a promised experimental vaccine. Early treatment at home remains a complete blank in every leadership conference and media discussion.

Anthony Fauci, Deborah Birx, "experts" at CDC, NIH, and FDA, have never treated a single COVID patient and show absolutely no empathy for patients sick at home or families who lost loved ones. These individual human beings—with lives, loves, and a soul—have simply been, and still are, dehumanized "cases" or "positive tests" or "deaths." These national public health officials have lost all connection with the *human* dimension of the disease crisis they have been paid to lead as they pontificate on TV with their constantly shifting messages designed to instill and maintain a fearful public to control people. Their lack of compassion and failure to recommend early treatment was and is unconscionable.

Many of us on the front lines taking care of patients simply refused to sit back and watch our patients die. We knew it was possible to prevent most of the deaths if we treated people at home before they were critically ill. Cardiologist Dr. Peter McCullough and family physician Dr. Zev Zelenko were two of the earliest pioneers and strong advocates for early home treatment and both continue to lead the early home-based treatment approaches. We have worked closely together sharing information and treatment approaches and speaking out in the media since March 2020. I agree with the points in their cogent and powerful introductions for the Breggins' book.

COVID-19 and the Global Predators is literally a must-read. I want to prepare and warn you, however, it is not "easy" to stomach the enormity of true evil that unfolds as you read the details of the Breggins' thoroughly documented research, showing step by step how the global predators knowingly weaponized the SARS-CoV-2 virus and suppressed access to simple treatments to push their mass vaccination agenda aimed at fulfilling their malevolent "Great Reset" agenda.

Who are the puppet masters behind the curtain causing loss of life, loss of economic and social stability, and loss of freedom? The Breggins' book reads like a Machiavellian detective story as they explain who these global predators are and exactly what their intentions are for their end game.

At times, you may feel overwhelmed by the degree of deadly decisions being made in the shadows with full knowledge of disastrous consequences to people around the world. It is tough emotionally to face such malevolence that took place prior to and during 2020, and continues to occur daily.

Yet, you *must* face this reality: ONLY if you know the truth about what was done with COVID-19 and how it is being used against us will you be able to take steps to defend yourself and fight for your life and liberty. You can no longer sit back and depend on a doctor to give you straight truth. You are going to have to be proactive and research your options.

In my search to find meaning in all this, I turned to a book I first read as a college student and again as a medical student: *Man's Search for Meaning* by Viktor Frankl, MD. Dr. Frankl was an Austrian Jewish physician and psychiatrist who wrote about living through and surviving the pure hell of four Nazi concentration camps during World War II. As a survivor of the Holocaust, he continued his career in psychiatry teaching *Logotherapy*—the therapy he developed focused on helping people find their individual meaning and purpose in life. He touched the lives of millions over his career.

At each point in my life that I read and reread his gut-wrenching, terrifying experiences, I gained critical insights for my own life challenges. But it was during my own major adversities as a physician-turned-patient—facing major spinal surgeries four times to avoid paralysis—that his extraordinarily profound message struck my core like a lightning bolt: When all else is lost, and all around us is devastation and death, the *last human freedom is the freedom to choose one's mental attitude in the face of adversity.*

Dr. Frankl's words have even more power as we face death and devastation on all fronts orchestrated in the COVID-19 pandemic. I faced personal loss of loved ones, and I faced the professional choice of letting my patients die if I gave in to the political and bureaucratic control of my medical decisions. Or I could stand up to medical tyranny and risk my career while doing my best to save lives.

I had additional choices to make. How would I use this adversity? What *good* am I able to personally bring from it that will help others? My belief in God, my Christian belief in the sanctity of life, and Viktor Frankl's words all guide me daily. I continue to choose to stand with my courageous colleagues to fight against medical and social tyranny. We continue to heed the words of Dietrich Bonhoeffer, the Lutheran minister ultimately put to death for fighting against the Nazi evil:

> **"Silence in the face of evil, is itself evil. God will not hold us guiltless. Not to speak is to speak. Not to act is to act."**

Together, we choose to speak. We choose to act. We choose to fight against medical and social tyranny. Together, our group of frontline physicians has saved thousands of lives, using early, home-based treatment with existing medicines we have safely used for decades. Some of us lost jobs; some of us faced threats to our medical licenses; some of us lost hospital privileges for treating patients outside the hospital. We have never stopped trying to help save our patients.

You, dear reader, now face that same hard choice:

Will you give up reading the Breggins' book because it is difficult to face?

Will you simply read this book, set it aside, and submit to the "new normal" of the radical globalists' new restrictions on your life?

Will you choose to let your mind be controlled by fear, becoming one of the prey for the totalitarian Global Predators seeking to control us?

Or will you use the Breggins' critical information to help you overcome the traumas of the COVID pandemic?

Will you use this book to understand the battle ahead, plan your individual action to restore your health, and regain your freedom to choose how you live your life?

Will you seek to find ways to live fully and find your new meaning and purpose in your own life from the devastation and destruction that COVID has wrought in our lives?

Will you choose to get involved, face your fears, use your mind and energy to speak out and help preserve all freedoms, and to change the world around you for the better?

The world is watching and waiting for us Americans to do exactly that.

—*Elizabeth Lee Vliet, MD*
Tucson Arizona, February 14, 2021

Dr. Vliet is a board-certified independent physician, patient advocate, and proponent of medical freedom, free of third-party and government interference in private medical practice focusing on Preventive and Climacteric Medicine (www.ViveLifeCenter. com). Dr. Vliet is the coauthor of *A Guide to Home-Based COVID Treatment.* (www. CovidPatientGuide.com). She is also the author of seven consumer health books, as well as many editorials on medical freedom and restoring the sanctity of the physician-patient relationship. She is a recipient of the 2014 Ellis Island Medal of Honor and the 2007 Voice of Women Award from the Arizona Foundation for Women.

PART ONE

Anthony Fauci and the Making of COVID-19

CHAPTER 1

Fauci, Globalism, and China: Discoveries that Drove Us into Action

T here is a persistent sense of urgency, anxiety, and confusion in America that is unprecedented in our lives and perhaps unknown since the War of Independence or the Civil War. In addition to the overall personal, societal, economic, and political stress caused by COVID-19 policies and restrictions, the pandemic continues to be used to justify the imposition of vast fear and oppression on the American people. All this is being done in the name of public health, but it represents authoritarian politics with worldwide ramifications.

As we tried to understand what continues to be happening to America during COVID-19, we pulled on one medical, political, and economic string after another—and found ourselves unraveling a coalition of global predators. They are individual billionaires, political leaders, government agencies, giant corporations, industries, universities, the major media, the social media, big tech, and even the international medical and scientific establishment. All of them were and are actively benefitting from COVID-19 at the expense of the American people and most of the world.

Nearly all of these globalist individuals and institutions are continuing to prevent America and other Western nations from benefitting from the already available early treatment regimens for COVID-19 that are safe, effective, and very inexpensive. The globalists want to profit from COVID-19, in part by developing government-sanctioned, very expensive drugs and vaccines that are potentially ineffective, dangerous, and highly experimental—and worth billions of dollars.

The predatory globalists also want to weaken America. When we are no longer in the way, they can further impose their own agendas on humanity.

Meanwhile, something more threatening has been coming out into the open. Every time we investigated one of these globalist individuals and institutions—and we will clearly document this—they turned out to have strong ties to China and its ruling Chinese Communist Party (CCP). China is a major and growing source of wealth, self-aggrandizement, and power for those globalist parasites who feed on our ailing great nation. Much more to our surprise, the wealth of China's 400 billionaires increased by 60% from March to December 2020, greatly outdistancing the rise in wealth among America's 600 or more billionaires.

In addition, China has made stunning leaps in venture capital or unicorn technology companies.[13] ByteDance, the owner of TikTok, is number one in the world with a valuation of $140 billion. A ride-sharing and ride-hailing company, Didi Chuxing, is in second place, and its rival ByteDance in seventh.

We began coming to conclusions that others had reached before us[14]—the Chinese Communist Party is winning an undeclared war against the United States as part of its intention to create a Communist Empire. Our window into all this began with investigating Anthony Fauci and his scientific collaborations with Bill Gates, the World Health Organization (WHO), the giant pharmaceutical industry, and ultimately China's Wuhan Institute of Virology and the Chinese Communist Party. Overall, the Chinese Communists are winning victories in multiple sectors from drawing American billionaires into their orbit and transforming our universities with their propaganda[15] to financing infrastructure and technology in the poorest nations of the world.[16]

Dr. Anthony Fauci Takes the World Stage

By now, nearly everyone has seen and heard Anthony Fauci on television. He is the slender, smallish, older man with grey hair and a gentle smile who represents himself as the authority on science in all matters related to COVID-19. For almost 40 years, he has been the director of the National Institute of Allergy and Infectious Diseases (NIAID), and he has become the expert most relied upon for COVID-19 guidance. On the organizational chart, NIAID is among the 27 institutes[17] under the National Institutes of Health (NIH). Another institute under NIH is the National Institute of Mental Health (NIMH), where early in my career I was a full-time consultant for two years.

For many people, Fauci exemplifies the best in a scientist and public servant. We became immediately suspicious of him because he carries himself as if he were the scientific "authority" on epidemics and public health measures when, in fact, "science" and "authority" are incompatible.

Science is an evolving human enterprise with many shifting viewpoints. Instead of authorities, there are often conflicting, self-interested viewpoints. In the area of public health, which is so chaotic and constantly changing with untold human factors involved, the idea of a person who is an authority—or even worse, *the* authority—is untenable. Any person who presents as a scientific authority is trying to compel you with arguments that are probably not true and almost certainly not in your self-interest.

We were also concerned because Fauci often contradicted himself and was clearly pushing the agendas of wealthy and powerful drug companies. But we never imagined that tracking his activities would lead us to Bill Gates and then to what we now call the global predators. We never imagined we would find Fauci directly funding projects making deadly viruses with American researchers openly and even proudly collaborating with the greatest global predator of all, the Chinese Communist Party, ultimately fueling its chilling biological weapons strategy called "unrestricted warfare." None of this made any sense until we began to understand the psychology and the goals that hold together this loose coalition of global predators, each member pursuing its own agenda to the detriment of humanity.

FAUCI GOES FULLY POLITICAL

We skip ahead to October 31, 2020, only 72 hours before the presidential election day.[18] Some background on federal law helps us appreciate what occurred. The federal Hatch Act does not allow government employees to "use their official title or authority when engaging in political activity."[19] In addition, Fauci is a

Career Senior Executive Service (SES) employee,[20] which further limits his political activities:[21]

> Career Senior Executive Service (SES) employees are subject to further restrictive rules and cannot engage in partisan political activities even during off-duty hours or while away from work. These employees are largely limited to exercising the most basic rights of civic participation, such as voting, making political contributions, and expressing individual opinions.

In defiance of the Hatch Act, at that critical moment in American history, two days before a highly contested presidential election, Anthony Fauci interfered with the election by issuing dire warnings about COVID-19 and President Donald Trump's handling of it. *The Washington Post*, in obvious collaboration with Fauci and the global predators, offered this threatening headline:[22]

> "A whole lot of hurt": Fauci warns of COVID-19 surge, offers blunt assessment of President Trump's response

The *Washington Post* went on to repeat and explain the ominous headline:

> "We're in for a whole lot of hurt. It's not a good situation," Anthony S. Fauci, the country's leading infectious-disease expert, said in a wide-ranging interview late Friday. "All the stars are aligned in the wrong place as you go into the fall and winter season, with people congregating at home indoors. You could not possibly be positioned more poorly."

Fauci, a leading member of the government's coronavirus response, said the United States needed to make an "abrupt change" in public health practices and behaviors.

The Washington Post clearly identifies Fauci as the leading government official, making Fauci's remarks anything but "personal," and Fauci is explicitly calling for an "abrupt change" in the leadership of the country, which is profoundly political.

Fauci's extraordinary and terrifying, and patently illegal, warnings echoed and reinforced those of candidate Joe Biden, who predicted in the second presidential debate that under a Trump administration, we were facing "a dark winter." This anti-Trump warning was also widely carried in mainstream media.[23]

Ironically, "dark winter" is a reference to a May 2001 wargame scenario conducted at Andrews Air Force Base involving a limited smallpox biowarfare attack against the United States.[24] It was sponsored by the Johns Hopkins Center for Civilian Biodefense Studies.

The last-minute surge of dire threats about a second term for President Trump might have been explained away as the final desperate efforts of Democrats trying to unseat the Republican incumbent. To us, it had become clear that something much greater and more ominous was going on with no end in sight except the destruction of the United States of America.

Anthony Fauci of NIAID and Jeff Bezos, the owner of *The Washington Post*, are like-minded participants in an international movement we are calling predatory globalism. The alliance remains at present an informal governance; however, it has sufficiently shared goals and more than enough power to withhold lifesaving COVID-19 treatments from America and much of the world, while forcing us to accept destructive, counterproductive restrictions on our personal, educational, business, and religious lives. In the process, these predators are vastly increasing their wealth, glory, and power.

Our Hesitation to Take on the Horror

Early in 2020, our attention was riveted on the gripping descriptions by overwhelmed physicians struggling to help patients in Northern Italy.[25] The anguished testimonials from grief-stricken doctors made them seem like witnesses to Dante's *Inferno*. Hospitals flooded by a torrent of very sick and dying patients with doctors exhausted by a disease unlike anything they had ever seen. Patients with high fevers and a pneumonia that seemed to drown them from the inside out. Strange skin lesions that puzzled specialists. Blood clotting so thick it stopped up needles as it was being withdrawn. The dying were unresponsive to any known treatment and made worse by traditional ventilators.

As we began our research into COVID-19 policies and practices, we became dismayed by the absence of any rational science to justify the shutdowns and to prevent the use of the very safe and only useful medication available, the inexpensive hydroxychloroquine—and now including ivermectin[26]—along with other therapeutics. But we remained reluctant to become involved in controversies that might distract us from our lifetime psychiatric reform work. Knowing how much we have been attacked and censored during our decades of shedding light on the aberrant behavior of the drug companies and my psychiatric colleagues who serve their interests, we were also realistically afraid of taking on more. The personal risks involved in throwing light on an even more widespread and ultimately more catastrophic aspect of human corruption was daunting.

The Discovery that Ramped Up Our COVID-19 Work

Ginger Breggin unearthed a 2015 scientific publication disclosing something so bizarre she could hardly believe it was real.[27] The shock from what we learned from this single, seemingly innocuous research study in a respected journal made us determined to develop the Coronavirus Resource Center[28] and to participate in bringing sound science, ethics, and politics to bear on COVID-19 policies and practices.

The disturbing paper, published in December 2015 by Vineet D. Menachery and his colleagues at the University of North Carolina, was titled "A SARS-like cluster of circulating bat coronaviruses shows potential for human emergence."[29] *The article showed that leading up to COVID-19, the U.S. was collaborating with and funding two of China's top gain-of-function scientists making SARS-CoV pathogens.*

Gain-of-function is a misleading euphemism for taking viruses out of nature and making them more infectious or virulent, commonly turning harmless viruses into deadly human pathogens capable of causing pandemics. The term should be *gain-of-lethal-function.*

A follow-up paper in 2016 in the respected *Proceedings of the American Academy of Sciences*[30] listed no Chinese authors. Nonetheless, in the Acknowledgments it detailed how Chinese and American scientists were continuing to work closely together to create what was clearly a *precursor to SARS-CoV-2.* It was also a potential biological weapon.

The Acknowledgments to the 2016 paper begin by describing the Chinese contribution: "ACKNOWLEDGMENTS. We thank Dr. Shi Zhengli of the

Wuhan Institute of Virology for access to bat CoV sequences and plasmid of WIV1-CoV spike protein." The Chinese were working with us to insert the deadly spike protein into the otherwise harmless coronavirus to enable it to penetrate human cells to cause a SARS-CoV pandemic. Like the 2015 research, this 2016 report was also funded by Fauci's NIAID as well as other NIH institutes.

The word "coronavirus" is commonly abbreviated as CoV. SARS stands for Severe Acute Respiratory Syndrome. The 2015 research paper was describing the laboratory making of deadly SARS-CoV pathogens in collaboration with the Chinese Communist regime.

Coronaviruses are found in many animals and humans in numerous varieties, almost always benign. The "corona" refers to the appearance of a halo under an electron microscope. A few strains of coronavirus have been identified as a cause of mild and occasionally moderate upper respiratory infections in humans, including many cases of the common cold.[31] One strain, retrospectively called SARS-CoV-1—far deadlier than SARS-CoV-2—had already caused a highly lethal but quickly suppressed epidemic in 2002-2004. SARS-CoV-1 was a fact of recent history that was largely ignored by the authorities and the media, and its origin—whether in nature or in a lab—was never determined (Chapter 3).

The Funding of the U.S./Chinese Collaboration

Along with National Natural Science Foundation of China, several American agencies were funding the collaborative research. U.S. funders listed in the Acknowledgments included Fauci's "National Institute of Allergy and Infectious Disease," the "National Institute of Aging of the US National Institutes of Health (NIH)," USAID funding through EcoHealth Alliance, and "the National Institute of Diabetes and Digestive and Kidney Disease of the NIH." We will untangle these and other grants from the U.S. to Chinese gain-of-function researchers in Chapter 3, when we examine whether or not Anthony Fauci lied to Congress about them.

In the 2015 Menachery paper, in addition to the University of North Carolina, the American researchers included representatives from Harvard and the FDA's National Center for Toxicological Research. This powerful group of American institutions and scientists were collaborating with Chinese researchers from the Wuhan Institute on gain-of-function research that was providing a scientific basis for China's biological warfare program. Remember, gain-of-function refers to gaining more dangerous functions, such as lethality.

One of the two Chinese coauthors of the 2015 collaboration was Wuhan researcher Shi Zhengli,[32] who has been given the affectionate and respectful title of "Bat Woman." Her name is often spelled Zhengli-Li Shi in the scientific literature. She identified herself as from the "Wuhan Institute of Virology, Chinese Academy of Sciences, Wuhan, China."[33] The second Chinese author, Xing-Yi Ge, is among the top Chinese gain-of-function researchers. *By involving these two Chinese researchers in the making of SARS-CoV pathogens, we were directly involving the Chinese Communist Party and the People's Liberation Army in their development of the basic science of viral biological warfare.*

Anthony Fauci was the inspiration and persistent supporter of the U.S./Chinese collaboration. Put simply, Anthony Fauci has been a huge benefactor to China's Wuhan Institute of Virology, its projects, and its individual researchers. Anyone with knowledge of China would realize that Fauci was ultimately funding China's Communist Party, its military, and its biological warfare program—with money from U.S. taxpayers.

The Wuhan Institute of Virology has long been known as "a center of China's declared biowarfare/biodefence capacity."[34] Its current director is China's "top military bio-warfare expert."[35] No one can hold positions as high as the two Wuhan scientists working with potential weapons without being closely involved with the military and under the absolute control of the Chinese Communist Party (CCP).

The Chinese researchers worked actively with the virus in their own labs. The 2015 article described how "X.-Y.G." and "Z.-L.S." were assisting in the central processes of enabling viruses to infect and damage the cells of living humans. Thus, the Americans and the Chinese collaborated in learning how to make harmless coronaviruses into SARS-CoV viruses, one of which the Chinese would eventually engineer into SARS-CoV-2, the cause of COVID-19.

WE REACH OUT TO PRESIDENT TRUMP

On April 15, 2020, we published on our website breggin.com an analysis titled "2015 Scientific Paper Proves U.S. & Chinese Scientists Collaborated to Create Coronavirus that Can Infect Humans."[36] We criticized the research as extremely risky in respect to potential leaks of deadly viruses and in enabling the Chinese Communist Party to have and to develop biological weapons. The next day, April 16, 2020, we followed up with a video on our YouTube Channel. The title of the video declared "U.S. and China Collaborated to Make a Deadly Coronavirus."[37] Almost overnight it had 40,000 viewers, and numbers continue to grow.

Ginger and I sent links for the blog report and video to every scientific, media, and political contact available to us, including some with close contacts in President Trump's inner circle.

Two days after we began our public disclosures, on April 17, 2020, a reporter at a press conference asked President Trump about the funding of the collaborative project with China.[38] The President indicated he was already aware of it and said, "We will end that grant very quickly. It was granted quite a while ago." Then, on April 20, 2020, the administration demanded "a list of all Chinese participants."

Shortly after this, President Trump stopped the funding for research that was making virulent coronaviruses in collaboration with China with its risks of accidental leaks and the certainty of China making bioweapons from it. However, President Trump did not stop Fauci and American researchers from conducting this research in the U.S. or with nations other than China.

When asked in June 2020 by a Congressional committee why the grant had been canceled, Fauci said that he did not know why.[39]

This early success in helping to bring some sanity into the insane world of COVID-19 encouraged us to continue with our research and educational campaigns. We felt proud of our mid-April blog/report and YouTube video exposing this ongoing disastrous collaboration between the U.S. and China,[40] especially because it had a significant impact when President Trump went over Fauci's head and cancelled its funding.[41]

We discovered that, unfortunately, Fauci doesn't go away easily.

Meanwhile, no major media we could locate, such as ABC News,[42] supported the President's decision to stop developing and sharing lethal biotechnology with the Wuhan Institute and the government of China. All of this began to make us wonder even more what was going on. Why were we the first to blow the whistle on the obvious treachery involved with our government funding collaborations with China on building deadly viruses? Why was there no national display of outrage in support of President Trump's cancellation of the collaboration? We were beginning to see the larger implications of what was going on in America and the world—and found it an almost overwhelming task to confront the depth of evil and the horrifying threat to America. Nonetheless, having contributed to President Trump's decision to stop Fauci's China collaboration inspired us to give our full attention to COVID-19 corruption.

BACKGROUND FOR THE 2015 STUDY

The research can be traced to an announcement by the University of North Carolina on September 9, 2013, about a $10-million award from NIH to a program led by Ralph Baric.[43] The purpose was to study and manipulate "highly pathogenic human respiratory and systemic viruses which cause acute and chronic life-threatening disease outcomes." The justification for this ominous project was, among other things, to develop vaccines. The description said nothing about how they were going to create these monstrous pathogens in the laboratory. *As eerily unreal as it seemed to us, our research for this book moved us inexorably toward the conclusion that COVID-19 has always been about making fortunes from rushed-through vaccines while weakening America and strengthening globalism.*

Ralph Baric's Fauci-funded research goes back years earlier to making infectious clones of infectious human coronaviruses in September 2008. It was supported by grants from NIH and from Vaccines for Global Health from the UNC School of Public Health. The study was titled "Systematic Assembly of a Full-Length Infection Clone of Human Coronavirus NL63."[44] Once again, the justification for the research is the development of vaccines and other therapeutics.

Shortly after, in December 2008, another article was published from Vanderbilt and the University of North Carolina, again with Ralph Baric.[45] They synthesized "a bat SARS-like coronavirus," a "SARS-CoV," that infected live mice and "human ciliated airway epithelial cells" in the lab. Human antibodies attacked the virus in the human cell experiments. There was one Swiss researcher but no Chinese. Funding included Fauci's NIAID. It makes us wonder how many deadly SARS-CoV viruses have been engineered since 2008 and now lurk in laboratories throughout the world.

The two 2008 publications indicate that interested globalists like Bill Gates had much earlier signals that American researchers were able to create the makings of both a SARS-CoV pandemic and vaccines for it. Two years later, in 2010, Moderna would be founded for the expressed purpose of making RNA vaccines for coronaviruses. In 2015, Bill Gates would mention in a video that he was already working with them (Chapter 15).

Early Recognition of the Origin of SARS-CoV-2

The actual origin of SARS-CoV-2 was quickly recognized by scientists soon after the epidemic struck. In February 2020, a Chinese researcher published a study confirming the novel coronavirus was man-made in a Chinese laboratory; soon afterward, the paper disappeared from the usual search engines.[46] Other Chinese experts confirmed the high probability that the virus came from one of their own labs.[47] Like so much else, it was "made in China." Also, predictably, China used our technology, but they did not have to steal it. In this case, it was given to them by Anthony Fauci along with U.S. taxpayer money to further develop it.

China's Well-Known Commitment to Biological Warfare and Stealing U.S. Secrets

Western observers have known for some time that China has an aggressive biological warfare program. A CNBC report[48] citing John Demers on May 13, 2020, declared:

> "Biomedical research has long been at the heart of something the Chinese have wanted and something they have engaged in economic espionage to get," according to John Demers, assistant attorney general for national security. He told CNBC on Monday, "It would be crazy to think that right now the Chinese were not behind some of the cyberactivity we're seeing targeting U.S. pharmaceutical companies and targeting research institutes around

the country that are doing coronavirus research, treatments, and vaccines."

The CNBC report was based on a May 13, 2020, FBI press release titled, *People's Republic of China (PRC) Targeting of COVID-19 Research Organizations.*[49] The FBI gave instructions to all facilities related to COVID-19 to increase cybersecurity. FBI director Christopher Wray said in August that "no country poses a broader, more severe intelligence collection threat than China," adding, "China has pioneered a societal approach to stealing innovation in any way it can, from a wide array of businesses, universities, and organizations."[50]

The tragic reality is the Chinese did not need to plant a secret spy inside our programs that were making deadly coronaviruses. Fauci invited them in and helped to pay for their personal participation and for their flagship research institute as well.

Chinese "Military-Civil Fusion" Guaranteed Fauci Was Funding China's Military

We have all heard about the Chinese government stealing technology from Western companies that do business with China, and Fauci must have been aware of this when he purposively enmeshed China in our research attempts to create pathogenic viruses. But, unlike most of us, Fauci had to be further aware that the China's Military-Civil Fusion policy means that American technological research in Chinese hands is *automatically* and *inevitably* shared with their military.

Turning harmless viruses into lethal pathogens is a militarily important technology any Chinese researcher or institute would be required to report to authorities in the Chinese Communist Party as part of China's Military-Civil Fusion. Faced with military superiority of the United States—especially in combination with U.S. allies like Japan, Taiwan, and Australia—pandemic bioweaponry is a relatively cheap and yet devastating weapon for China to possess. It can be used as a deterrent or a threat. *Anthony Fauci, his NIAID, and the umbrella organization, NIH, have been colluding with China, providing that violent dictatorship with a significant upgrade in its military threat to the United States and the world.*

China's overarching policy of Military-Civil Fusion means there is a *constant* and *required* sharing of all civilian technology with the military.[51] The U.S. Department of State has specifically warned about the danger in an official release titled *What is Military-Civil Fusion (MCF)?*[52]

> Military-Civil Fusion is a national strategy of the Chinese Communist Party (CCP) to develop the People's Liberation Army (PLA) into a "world class military" by 2049. Under MCF, the CCP is acquiring the intellectual property, key research, and technological advances of the world's citizens, researchers, scholars, and private industry in order to advance the CCP's military aims. The CCP is systematically reorganizing the Chinese science and technology enterprise to ensure that new innovations simultaneously advance economic and military development.

In the same publication, Secretary of State Michael Pompeo warned, "If the Communist Party gives assurances about your technology being confined to peaceful uses, you should know there is enormous risk to America's national security."

Would Anthony Fauci have understood that the Chinese Communist Party and its military were deeply involved in his funded research? Or is Military-Civil Fusion a new policy that escaped his attention? Actually, the military's seizure and use of civilian technology is a policy that has been part of the Chinese government since Mao Zedong. A report from the Air University of the U.S. Air Force stated:[53]

> China's Military-Civil Fusion program is not new. Every leader since Mao Zedong has had a program to compel the "commercial" and "civil" parts of Chinese society to support the PLA. It has gone by different terms, Military-Civil Integration, Military-Civil Fused Development, etc. General Secretary Xi Jinping has elevated the concept to Military-Civil Fusion. But in all cases, it is the "Military" that comes first. ...

> Since Xi Jinping's assumption of power, the role of the military, and the importance of MCF have markedly increased.

The basic concept behind Military-Civil Fusion has been around at least since the early 1980s.[54] Xi Jinping, who has placed increasing emphasis upon it, became General Secretary of the Communist Party of China in 2012 and President of the People's Republic of China in 2013. Individuals like NIAID director Anthony Fauci MD and NIH director Francis Collins MD must be aware that they are funding the biowarfare program of the Chinese Communist Party when they fund Chinese researchers, research projects, and the Wuhan Institute in their efforts at engineering deadly coronaviruses.

WAS FAUCI INNOCENT OR UNINFORMED?

Was Fauci naive, innocent, or uninformed about the dangerous implications of his work? That is not possible. Beginning at least as far back as 2012, there was growing concern in the scientific community about the risks associated with making deadly viruses (gain-of-function research), including the "hostile misuse of the life sciences" and the "wholesale militarization of the life sciences."[55]

The ongoing controversy is described by the Center for Infectious Disease Research and Policy of the University of Minnesota:[56]

> Amid the controversy, a group of 40 prominent flu researchers announced a moratorium in January 2012 on further H5N1 gain-of-function research. The group called off the moratorium on Jan. 23 of this year [2013], saying the pause had allowed time for scientists to explain the potential benefits of the research and for governments and others to review relevant policies.

The 40 prominent flu experts[57] may have withdrawn their request for a moratorium because of Fauci's efforts to protect his NIAID-funded projects. However, in 2013, another group, the Foundation for Vaccine Research (FVR), wrote a letter to President Obama asking him to stop gain-of-function research:

> The FVR letter was signed by the group's chair, Simon Wain-Hobson PhD; its executive director, Peter Hale, and 15 other scientists. The list includes such prominent names as Paul A. Offit MD, of the Children's Hospital of Philadelphia; Marc Lipsitch PhD, of the Harvard School of Public Health; Sir

Richard Roberts PhD, of New England Biolabs, a 1993 Nobel laureate; and Michael J. Imperiale PhD, of the University of Michigan, a member of the National Science Advisory Board for Biosecurity (NSABB).

In September/October 2012, a month before Barack Obama's election to a second term, Fauci, as the single author, published *Research on Highly Pathogenic H5N1 Viruses: The Way Forward.*[58] He documented in the abstract that scientists at the time had a voluntary moratorium on his research:

> The voluntary moratorium on gain-of-function research related to the transmissibility of highly pathogenic H5N1 influenza virus should continue, pending the resolution of critical policy questions concerning the rationale for performing such experiments and how best to report their results. The potential benefits and risks of these experiments must be discussed and understood **by multiple stakeholders, including the general public**, and all decisions regarding such research must be made in a transparent manner. (Bold added.)

Notice the reference to "multiple stakeholders, including the general public." The concept of stakeholders, as confirmed in Dr. McCullough's pointed use of the term in the introduction to this book, refers to the array of global predators working together on any given project. Even in Fauci's remarks, we see that "the general public" is just one among many. And who represents the "general public"? Undoubtedly, "experts" will represent them. Wholly absent is the concept of a democratic republic in which the people, through voting and other lawful means, control the major directions taken by society.

Despite what Fauci says, we will find that he has assiduously avoided transparency, not only on this issue of gain-of-function research, but also on any of his activities surrounding COVID-19.

Fauci, in fact, was deceptive in the article itself when he stated:

> Scientists working in this field might say—as indeed I have said—that the benefits of such experiments and the resulting knowledge outweigh the risks. It is more likely that a pandemic would occur in nature, and the need to stay ahead of such a

threat is a primary reason for performing an experiment that might appear to be risky.

The degree of controversy surrounding Fauci's gain-of-function studies was considerable in 2012 to 2014.[59] The mounting data would undermine his contention that the greatest danger was from a pandemic arising out of nature rather than out of a human lab. We shall find that lab leaks of pathogens far outweigh any spontaneous emergences from nature (Chapters 1-4).

WHO OWNS MODERNA'S VACCINE AND OTHER GLOBALIST ENTANGLEMENTS WITH VACCINES

Fauci had to know years ago that any potential military technology shared with China would go straight to the Chinese Communist Party and its military. If Fauci had thought about it for a split second, he would have been compelled to realize he was acting treacherously toward the United States.

What could motivate a person of such high stature in America to behave in this manner? It cannot be poor judgment. He is much too shrewd, intelligent, and worldly for that. We believe the simplest and most probable explanation is that Fauci participates in a widespread group of individuals and entities who accept this behavior as normal as they pursue their global enterprises to increase their wealth and power. Fauci is an exemplar of the implementers of global predation. All probably share his willingness to ignore or to sacrifice the interests of the United States in the pursuit of wealth, self-aggrandizement, and power. Now with Joe Biden, Fauci has a deep admirer in the Oval Office of the United States of America.

Are we suggesting Fauci became personally wealthy from his support of China? According to Justia Patents, Fauci has a total of 12 patents pending or in effect.[60] Many are related to HIV, but there is an overlap with coronavirus research. We have found no information demonstrating Fauci *is or is not* making money from them, but according to federal regulations, he is limited annually to an additional $150,000 outside income.

Even if Fauci does not collect the patent profits, they would go to his NIAID which he controls and whose power he wields. This is somewhat analogous to how Gates benefits from the Bill & Melinda Gates Foundation. His investments through the Foundation continue to bring more wealth and power to him as he grows the Foundation that he controls.

In a telling moment on TV on April 25, 2021, the interviewer on *Meet the Press*, when saying goodbye to Francis Collins, teases him about not really being Fauci's boss. He jests, "And for those that are wondering, technically, [he's] Dr. Fauci's boss."[61]

NIH and NIAID also own many patents surrounding the coronavirus and have made no disclosures concerning how much money they make from drug companies or from governments who use the Institute's patented processes and viruses. It appears that NIH may be a huge stakeholder with great financial interests in making money from vaccines. Indeed, NIH appears to own a patent on Moderna's vaccine![62]

Under the Emergency Use Authorization, Fauci and his colleagues make "investments" into pharmaceutical companies with our *U.S. taxpayer money*, but the investments are actually gifts from us to them, since the money is an outright award. For example, Moderna, whose vaccine was highly touted in advance by Fauci, received a grant of nearly half-a-billion dollars ($438 million) from the U.S. Government Agency BARDA to accelerate development of its vaccine.[63]

In addition, the Defense Advanced Research Projects Agency (DARPA) has pitched in up to $56 million of taxpayer money for Moderna's vaccine "to fund the development of a mobile manufacturing prototype," which is designed to speed up vaccine and therapeutic production.[64] The grant is part of DARPA's "Nucleic Acids on Demand Worldwide (NOW)" initiative. DARPA is the premier government agency for secret cloak-and-dagger research and may tie Fauci to the biological warfare sector of the military to which his research obviously contributes.

Pfizer's vaccine is the main competitor to Moderna's, but they seem to be happily sharing the predators' largesse rather than competing. The Bill & Melinda Gates Foundation is also heavily invested in mRNA vaccines, up to $1.75 billion as of December 2020.[65] Pfizer and Moderna's mRNA vaccines were the first and second ones authorized by the FDA through the Emergency Use Authorization.

As we will document in Chapter 15 and as reviewed in the *Chronology and Overview with Pandemic Predictions and Planning Events* at the end of the book, early in 2017, Bill Gates boasted that he was already involved in researching the manufacture of "RNA vaccines" for the coronavirus.

The Chinese communists are invested in Pfizer's vaccine, making them a most powerful stakeholder. In a tangled web that we have been unraveling, Bill Gates is partnered with Germany's BioNTech,[66] which is partnered with China's Shanghai Fosun Pharmaceutical Group.[67] BioNTech is also partnered with Pfizer[68]—all of them (the U.S., Germany, and the Chinese Communists) working together on mRNA vaccines. Once again, the U.S. global corporations are bilking America while fueling the Chinese Communists with money and dangerous biomedical know-how.

Meanwhile, Bill Gates is deeply and directly invested in the Pfizer vaccine through stock purchases in both Pfizer and BioNTech—leading one well-known investment advisor to "bet on" Bill Gates and Pfizer.[69]

The U.S. Government is also very much invested in Pfizer. In July 2020, the U.S. Department of Health and Human Services and Department of Defense announced a $1.95 billion agreement with Pfizer to produce 100 million doses of a Covid-19 vaccine.[70] On December 10, 2020, the FDA announced it was approving Pfizer's vaccine under the Emergency Use Authorization. Notice something odd: the U.S. was hungrily buying up Pfizer's vaccine more than four months before the FDA approved it?[71]

Fauci is close to Bill Gates and serves on his important vaccine committee. We do not know what, if anything, he is paid for these services. A search of the usual internet sources about the wealth of famous people reveals that Fauci is exceptional in his privacy. Since there is a ceiling on how much federal employees can make from patents and other government-related activities, we will find that NIH and NIAID are the major beneficiaries of revenues from patents held in relation to vaccine and coronavirus research by Fauci, Collins, and many other science-oriented government employees. The Chinese Communists also hold patents in the vaccine work on which they collaborate with U.S.

Setting aside corrupt financial gain, Fauci already has a government salary of $417, 608 with extremely good benefits and an amazing retirement. That's a package probably worth in excess of half-a-million dollars each year. Many of us could relax and enjoy the power of doling out billions of dollars to favorite corporations while hobnobbing with the likes of corporate presidents, U.S. presidents, and Bill and Melinda Gates. You can even become the "favorite hero" of Julia Roberts and get an "Award of Courage" personally from her![72] Maybe his power and fame, along with a huge guaranteed-to-grow salary, is all the man needs.

Maybe we will never figure out all of Fauci's twisted motives. Does it really matter? Among numerous American traitors to Communist regimes, many have lived modestly and sacrificed their lifestyles and lives for the cause.[73]

We do know this: Fauci has greatly increased the wealth and hence the power of his National Institute of Allergy and Infectious Disease, as well as the umbrella organization, the National Institutes of Health. He has wrapped himself in considerable glory and "partnered" with billionaires and CEOs of great companies. That might be enough motivation to corrupt many a man. There can also be hidden ideological motives. He has grandiose plans for authoritarian or totalitarian governments to make humankind safe from viruses by limiting the activities of people in nature, despite the grave cost to civilization. However, searching for motivation can be a misdirection in complex situations like this where an individual is obviously sacrificing his own country—and the wellbeing of humanity—for the benefit of another country, which is a monstrous threat to humanity.

How deep is Anthony Fauci's betrayal of America? As we further examine his conduct, consider that Fauci is not nearly the most powerful of the global predators; instead, he is acting on their behalf. Through Fauci's words and actions, we will come to a fuller understanding of how much globalist leaders in America and throughout the world are exploiting and undermining our nation.

CHAPTER 2

Fauci Builds a Reckless Research Empire and Lies to the U.S. Senate

B at Woman, Shi Zhengli, whom we met as a coauthor of the Menachery et al. 2015 collaborative project between the U.S. and China,[74] continued to play a key role in the 2016 Menachery et al. publication as indicated in their Acknowledgments. She was thanked for working from the Wuhan Institute of Virology on the key "spike protein" that enables a harmless coronavirus to become an invasive pathogen. Shi Zhengli was directly funded by Fauci through the EcoHealth Alliance, confirming the ongoing close working relationship between the American and Chinese researchers under Fauci's watchful eye (See ahead in this chapter.).

Dr. Shi Zhengli is director of the Wuhan Institute lab that extracts these viruses from bats and makes them pathogenic for humans. She has been fawned upon in the American scientific media, accompanied by a chorus of denials by her, by China, and by the scientific community that SARS-CoV-2 escaped from her facility.[75] She is, in fact, an ambassador of lies for the Chinese Communist Party.

Western scientists and scientific publications feel so deeply embedded in the global research community, including and especially China, that they do everything they can to quash criticism of the brutal Communist dictatorship. They go along with China's refusal to acknowledge the origin of SARS-CoV-2 in Chinese labs. We will find Bill Gates, who surely knows better, publicly pandering to the supposed trustworthiness of the Chinese.

Xing-Yi Ge, the other Chinese coauthor of the 2015 research collaboration, is also a scientific powerhouse in China. He was the lead author of an important article in 2013 about the laboratory process that led to SARS-CoV-2.[76] The 2013 paper is titled "Isolation and characterization of a bat SARS-like **coronavirus** that uses the ACE2 receptor." (Bold added.)

It is worth slowly pondering the above title: The Chinese were openly writing about removing and studying "SARS-like coronaviruses" from bats that can potentially invade human cells through the "ACE2 receptor" in human beings. This study by itself removes the Chinese origin of SARS-CoV-2 from the realm of "conspiracy theory" to the realm of scientific likelihood. With U.S. cooperation, China was developing the technology to build innumerable deadly SARS-CoV bioweapons, including ones more deadly than SARS-CoV-2.

Extensive Fauci Funding of the U.S./Chinese Collaboration

The three research studies we described as leading up to China's creation of SARS-CoV-2—one in 2013, one in 2015, and one in 2016—were among several funded by Fauci's institute. Peter Daszak, who heads an organization that was funneling money from Fauci's institute to the Wuhan Institute, was listed as an author on the 2013 paper. When Daszak's EcoHealth Alliance is listed as a funder, the money is being channeled from Fauci's institute.

As a part of the research chain of events leading up to SARS-CoV-2, Fauci also funded a 2016 study from China involving about one dozen Chinese scientists,[77] nine of whom listed their affiliation as Wuhan Institute of Virology. The one obvious Westerner among the many authors was Peter Daszak.

Who Are Daszak and His EcoHealth Alliance?

As we researched the growing list of players involved in promotion and funding of the collaborative research with China making deadly viruses, we found that the EcoHealth Alliance is deeply involved worldwide with globalist organizations, many with huge resources and influence, including *The Lancet* and the World Health Organization. The EcoHealth Alliance website offers the following information about the nonprofit's influence and interconnection with many of the biggest movers and shakers in the globalist community:[78]

- Johnson & Johnson, the huge pharmaceutical company that has developed a vaccine against COVID-19, and a globalist supporter of the UN 2030 goal.[79]

- Boehringer Ingelheim, which is the second largest animal health company in the world.

- At least 20 academic university "partners" from the U.S. and worldwide, including China. Among the American universities are the John Hopkins Bloomberg School of Public Health, Princeton, Columbia, UC Davis, and Pittsburgh. Others include Saudi Arabia's King Saud University, Thailand's leading Chulalongkorn University, and many more.

- Governmental partners, including the CDC, NIH, United States Geological Survey (Wildlife Health Center), as well as departments in Malaysia and New York State.

- Conservancy organizations, such as Future Earth, the Global Health Security Agenda Consortium, and the Planetary Health Alliance.

Peter Daszak and his EcoHealth Alliance are primary worldwide drivers for gain-of-function research. Daszak recently declared, according to his ally Bloomberg.com, that "viral outbreaks are 'happening more frequently, they're spreading quicker, they're killing more people and they're crushing our economies'.... He estimated there are 1.7 million unknown viruses in the world. It would cost over a $1 billion to identify 70% of them, but it would be money well spent, Daszak said."[80] Such spending would also put money in his pocket.

Daszak and his organization are operating on the hypotheses that it is possible to track down and identify most existing viruses—whether the host is an animal, a vegetable, or a person—and humans can then prepare defenses against the viruses. Meanwhile, he ignores how rapidly these and other unidentified viruses are mutating, much as we predicted and now witness in the transformations of COVID-19. This is one reason why an effective coronavirus vaccine has been elusive and may defy optimistic claims now being made.

Another of Daszak's hypotheses or self-justifications is his claim that the intrusions of humankind are disrupting natural habitats around the world and causing humans and wildlife to encounter each other. This increasing intimacy allegedly leads animals and humans to exchange pathogens more often. Unfortunately, in this burgeoning war against the viruses, Daszak, Fauci, and the Chinese are making many, many more viruses with the potential for leaking from labs and/or being released in secret or open warfare.

We shall document that the long-term "solution" for Fauci and other global predators is to bring humankind "into harmony" with nature, a feat that involves a drastic shrinking of the human population and a stripping of many of civilization's most advanced and comforting achievements. Meanwhile, as this chapter will demonstrate, the risk of a coronavirus in nature seriously infecting a human being is barely worth consideration compared to the multiple infections and escapes from research facilities, especially labs in China.

How difficult will it be to provide vaccines for the potentially toxic viruses in the world? In a six-year period (2006-2013), a survey of bat virus research disclosed over 248 unique viruses from 24 "viral families."[81] And that survey only covers viruses in bats.

It is time to call a definitive halt to gain-of-function research. Many well-funded scientists will want to argue the need for this work. But their livelihoods, their reputations, and their self-worth all depend upon the continued financing of a field of science that is immeasurably destructive.

Daszak and the Globalists Defend China

On February 19, 2020, *The Lancet*, one the world's leading medical journals, published a letter by a large number of professionals defending the Chinese from each and every accusation against them in respect to COVID-19.[82] The writers specifically denounce the "conspiracy" theories that SARS-CoV-2 came from the Wuhan Institute rather than from nature or that the Chinese failed to do everything possible to cooperate with the world.

As if offering a Communist manifesto for global scientists, they speak of "solidarity" with "the scientists, public health professionals, and medical professionals of China, in particular..." They describe China as a paragon of cooperative and honorable behavior.

The letter was secretly organized by Daszak, a man who actually funded the Chinese in their making dangerous SARS-CoVs in their labs and who knew the Communists had a history of accidental releases of these very same viruses (Chapter 3). It took Daszak only a few weeks to organize the letter with nearly 30 signatures.

Overall, the fawning support of China demonstrates a spectacular allegiance to the Chinese Communist dictatorship, even to the disregard of scientific truth and the safety of humanity. Their loyalty is a tribute to the degree the Communists have bamboozled and bedazzled the elite of the West, including a major journal like *The Lancet*. Indeed, we shall find *The Lancet* attempting to insert itself as a leader among the global predators, not merely in respect to COVID-19, but in the whole arena of global progress (Chapter 19).

Fauci Justifies Dangerous Gain-of-Function Research

The main justification for engineering deadly viruses seems to be the goal of developing vaccines for potentially deadly coronaviruses that might emerge from nature. Fauci is a strong advocate for this, and he works closely with Bill Gates who is betting a chunk of his wealth on vaccines for COVID-19. Fauci is on Gates' high-profile international vaccine board called the Leadership Council.[83] We will find he is also extremely close with the pharmaceutical industry, in which Gates is also involved, and that Fauci is willing and able to do anything to get their COVID-19 drugs onto the market, however ineffective and dangerous they may be.

The following concept, held dearly by Fauci and other defenders of gain-of-function research and promoters of unlimited vaccine research, appears through the literature: "In considering the threat of bioterrorism or accidental release of genetically engineered viruses, it is worth remembering that nature is the ultimate bioterrorist."[84] To which Ginger Breggin replies, "Nature is random— evil with purpose and intention is the ultimate bioterrorist."

The creation of SARS-CoV-2 was a scientific achievement and not an accident of nature. The series of events leading up to the laboratory creation of SARS-CoV-2 required enormous manpower and funding, and the overcoming of political obstacles, including a moratorium placed on gain-of-function research by President Barack Obama in 2014, which we will further describe. Fauci remained undaunted and managed to keep funding research in which Americans collaborated with Chinese and hence the Chinese Communist Party, culminating in China's ability to engineer SARS-CoV-2 from bat viruses as part of their building a stockpile of potential biological weapons.

Fauci Denies Gain-of-Function Funding

On May 11, 2021, Senator Rand Paul (R-KY), while questioning Anthony Fauci at a hearing, pushed him about funding collaborative gain-of-function research with the Wuhan Institute in China. Here is a report from the *Washington Times*,[85] which we have verified from the hearing video: [86]

> During a Senate hearing on the coronavirus, Paul first pointed to the work between Baric and Shi as evidence of U.S. support for gain-of-function research in China.

> "This gain-of-function research has been funded by the NIH," Paul contended. "The collaboration between the U.S. and the Wuhan institute continues." He asked: "Dr. Fauci, do you still support the NIH funding of the lab in Wuhan?"

> Fauci, the director of the U.S. National Institute of Allergy and Infectious Diseases and the chief medical adviser to President Joe Biden, replied: "Sen. Paul, with all due respect, you are entirely and completely incorrect — that the NIH has not ever and does not now fund gain-of-function research in the Wuhan Institute of Virology."

> He added: "Dr. Baric is not doing gain-of-function research, and if it is, it is according to the guidelines, and it is being conducted in North Carolina, not in China … If you look at the grant and you look at the progress reports, it's not gain-of-function."

Notice that Fauci emphatically denies that NIH, which includes his NIAID, has ever funded Chinese gain-of-function research "in the Wuhan Institute."

More amazing to us, Fauci goes on to say that Baric has not been performing gain-of-function research. In short, he is denying any NIH funding of gain-of-function research.

In the following sections, we will look at NIAID and NIH funding of Wuhan scientists working with Americans and Wuhan scientists working by themselves in labs in the Wuhan Institute. We will also document the collaboration that was taking place between the Americans and Chinese in research under the direction of NIAID and NIH.

Evidence that NIAID and NIH Have Been Collaborating with and Funding Chinese Gain-of-Function Research

In Chapter 1, we discussed the 2015 Menachery publication that includes Baric and two Chinese Wuhan Institute researchers among the authors and participating researchers.[87] They are the "Bat Woman" Shi Zhengli and the highly respected Chinese gain-of-function researcher, Xing-Yi Ge. These two Chinese researchers are the backbone of basic research for biological warfare by the Chinese Communist regime.

The Acknowledgments to the 2015 Menachery article state that Fauci's institute was funding the gain-of-function research directly through NIAID and also from NIAID through EcoHealth Alliance. Three specific recipients were named: Baric, Menachery, and Wayne A. Marasco from Harvard.

Whether or not any U.S. agency gave money directly to the scientists from the Wuhan Institute, the Fauci-funded Americans have invited the two top Chinese researchers for the Communist regime into an intimate relationship with U.S. research. However, there is funding from another American agency, USAID, which financed Shi Zhengli through the intermediary EcoHealth Alliance.

It is worth excerpting the Acknowledgments from this paper, which confirm that it is considered gain-of-function:

> Research in this manuscript was supported by grants from the **National Institute of Allergy and Infectious Disease** and the **National Institute of Aging of the US National Institutes**

of Health (NIH) under awards U19AI109761 (R.S.B.), U19AI107810 (R.S.B.), AI085524 (W.A.M.), F32AI102561 (V.D.M.) and K99AG049092 (V.D.M.), and by the National Natural Science Foundation of China awards 81290341 (Z.-L.S.) and 31470260 (X.-Y.G.), and by **USAID-EPT-PREDICT funding from EcoHealth Alliance** (Z.-L.S.). Human airway epithelial cultures were supported by the **National Institute of Diabetes and Digestive and Kidney Disease of the NIH** under award NIH DK065988 (S.H.R.). ... *Experiments with the full-length and chimeric SHC014 recombinant viruses were initiated and performed before the* **GOF** *[gain-of-function] research funding pause.* (Bold and italics added.)

The "experiments" described are classic gain-of-function studies. We have included the identification numbers of the grants because they can often be found through search engines on the internet.

Summarizing, NIAID and NIH awards go to R.S.B., who is Ralph Baric at the University of North Carolina. In the Senate hearing, Fauci denied that Baric was conducting or being funded for gain-of-function research. NIH awards also go to VDM, referring to Menachery, who works under Baric. Additional NIH funding goes to S.H.R., who is Scott H. Randell from a separate department at the University of North Carolina. "Human airway epithelial cultures were" also supported by NIH.

A 2016 Follow-Up Menachery Report Confirms Direct Involvement with the Wuhan Lab

A 2016 follow-up paper[88] describes in detail how Chinese and American scientists were working together to create what was clearly a *precursor to SARS-CoV-2, and also a potential biological weapon.* Much like the 2015 paper, this one lists funding by NIAID and other NIH agencies. Although Shi Zhengli is not listed as an author, she is thanked for performing very important tasks in providing the "spike protein" from her lab that makes the coronavirus capable of attacking humans:

ACKNOWLEDGMENTS. We thank Dr. Zhengli-Li Shi of the Wuhan Institute of Virology for access to bat CoV sequences and plasmid of WIV1-CoV spike protein. Research was supported by the National Institute of Allergy and Infectious Disease and the National Institute of Aging of the NIH under Awards U19AI109761 and U19AI107810 (to R.S.B.), AI1085524 (to W.A.M.), and F32AI102561 and K99AG049092 (to V.D.M.). Human airway epithelial cell cultures were supported by the National Institute of Diabetes and Digestive and Kidney Disease under Award NIH DK065988 (to S.H.R.). Support for the generation of the mice expressing human ACE2 was provided by NIH Grants AI076159 and AI079521 (to A.C.S.)

Fauci Funds Chinese Scientists in Their Own Labs

A 2018 paper by Wang et al.[89] includes almost a dozen Chinese researchers, most notably the "Bat Woman" Shi Zhengli and five others also from the Wuhan Institute. Joint authors include Daszak from EcoHealth Alliance and several colleagues. Numerous American agencies funded the research: "This study was jointly funded by the National Natural Science Foundation of China Grant (81290341) to ZLS; the National Institute of Allergy and Infectious Diseases of the National Institutes of Health (Award Number R01AI110964) to PD and ZLS; United States Agency for International Development (USAID) Emerging Pandemic Threats PREDICT project Grant (Cooperative Agreement No. AID-OAA-A-14-00102) to PD."

PD is Peter Daszak and ZLS is Zhengli-Li Shi, also known as Shi Zhengli. Not only was Fauci funding Chinese researchers inside the Wuhan Institute, he specifically funded Shi Zhengli, who is the face of the Chinese Communist propaganda machine in the Western popular and scientific media as well as a committed specialist in gain-of-function research. *The factual evidence utterly demolishes Fauci's denial of funding scientists working in the Wuhan Institute of Virology.*

NIAID and NIH also funded a 2017 paper by a largely Chinese group led by Lei-Ping Zeng, mostly from the Wuhan Institute, where the research was conducted. The NIAID funding again went to Shi Zhengli (R01AI110964), and NIH funding went to two other scientists, Shibo Jiang and Lanying Du (R01AI098775).[90] In addition, NIH provided antibodies to test on the infected human cells.[91] Four of seven authors are from the Wuhan Institute, including the first two authors, Zeng and Xing-Yi Ge, and the corresponding author Zheng-Li Shi. The paper describes very elaborate and highly technical lab experiments involving SARS-CoV-1 and variants with which they infected human cell preparations and studied the effectiveness of antibodies against them. This is gain-of-function research with the aim of testing antibodies against pathogenic variations of coronaviruses.

A 2016 paper with many Chinese authors also confirms NIAID funding for Chinese gain-of-function research. It is called "Isolation and Characterization of a Novel Bat Coronavirus Closely Related to the Direct Progenitor of Severe Acute Respiratory Syndrome Coronavirus." It was published in the American-based *Journal of Virology*.[92] As stated, "HHS | National Institutes of Health (NIH) provided funding to Peter Daszak and Zheng-Li Shi under grant number NIAID R01AI110964." This is one of multiple Wuhan Institute gain-of-function projects that will lead to SARS-CoV-2.

In 2015, a paper called "Bat origin of human coronaviruses" by Ben Hu, Xingyi Ge, Lin-Fa Wang, and Shi Zhengli was funded by the National Institutes of Health (NIAID R01AI110964). Three of the authors were from Wuhan Institute and one from a medical school in Singapore. The paper is a review rather than a lab study but very relevant to gain-of-function research in which the researchers are involved.[93]

In addition, we found an earlier 2013 Chinese research publication, "Isolation and characterization of a bat SARS-like coronavirus that uses the ACE2 receptor," which was funded by Fauci's NIAID.[94] The first author is Xing-Yi Ge and the last is Shi Zhengli, the same two elite Wuhan researchers who were funded by NIAID in the 2015 Menachery gain-of-function research. They capture bats and experiment on them in their Wuhan labs. Most of the authors are Chinese, but a few are not, including Peter Daszak. This 2013 article again illustrates that Fauci's NIAID and other NIH institutes were funding the Chinese Communist Wuhan Institute gain-of-function research that was carried out entirely at the Wuhan Institute.

We were able to trace Fauci's funding of EcoHealth Alliance further back to 2008 under a group of projects called "Risk of Viral Emergence from Bats." [95] The official NIAID summary of methodology for each of these studies includes "using a range of *in vitro* techniques (including infection in bat cell culture), examine the pathogenesis of these new viruses, and a pool of available bat viruses which have not yet emerged in humans." This, too, is gain-of-function research.

This review of Fauci-funded gain-of-function research is not comprehensive. We believe we are only scratching the surface.

Fauci Is Accused of "Outsourcing" Gain-of-Function Research to the Wuhan Institute

In mid-2020, Fauci was accused of "outsourcing" gain-of-function research directly to the Wuhan Institute to avoid Obama's moratorium. We recently concluded that this accusation has two flaws. First, the funding of the Wuhan Institute research was formulated before President Obama's moratorium in late 2014. Second, Fauci did not need to outsource; he simply continued funding gain-of-function research in the U.S. in complete defiance of Obama's moratorium as exemplified by the Menachery et al. publications in 2015 and 2016 (Chapter 1).

Nonetheless, it is probably true that Fauci's NIAID was directly funding China's Shi Zhengli in 2014. During the controversy about outsourcing, on April 28, 2020, *Newsweek* investigated the charge and made the following observation:[96]

> In 2019, with the backing of NIAID, the National Institutes of Health committed $3.7 million over six years for research that included some gain-of-function work. The program followed another $3.7 million, 5-year project for collecting and studying bat coronaviruses, which ended in 2019, bringing the total to $7.4 million. ...

> The NIH research consisted of two parts. The first part [linked] began in 2014 and involved surveillance of bat coronaviruses and had a budget of $3.7 million [from Fauci's NIAID]. The program

funded Shi Zhengli, a virologist at the Wuhan lab, and other researchers to investigate and catalogue bat coronaviruses in the wild.[97] This part of the project was completed in 2019.

A second phase of the project, beginning that year, included additional surveillance work but also gain-of-function research for the purpose of understanding how bat coronaviruses could mutate to attack humans. The project was run by EcoHealth Alliance, a non-profit research group, under the direction of President Peter Daszak, an expert on disease ecology. NIH canceled the project just this past Friday, April 24th, Politico reported. Daszak did not immediately respond to *Newsweek* requests for comment.

The words "first part" are linked to the 2014 NIAID project summary.[98] Shi Zhengli does not merely collect bat viruses; she mainly manipulates and engineers them into pathogens in the laboratory that she directs at the Wuhan Institute.

The summary for the 2014 NIAID project confirms that one of its purposes was funding for gain-of-function research:

Test predictions of CoV inter-species transmission. Predictive models of host range (i.e. emergence potential) will be tested experimentally using reverse genetics, pseudovirus and receptor binding assays, and virus infection experiments across a range of cell cultures from different species and humanized mice.

This gain-of-function, Fauci-funded research was aimed at one specific scientist, Shi Zhengli, working in the Wuhan Institute, and the "Abstract Text," which is "provided by the applicant," is about her work in China.

Other NIAID Funding for Chinese Gain-of-Function Research

Since Fauci seems to be distancing himself from all gain-of-function research, it is worth reviewing NIAID's history in this respect. Since 2014, NIAID has been giving money directly to Peter Daszak and EcoHealth Alliance for "conducting laboratory experiments to analyze and predict which newly-discovered viruses pose the greatest threat to human health."[99] This research project was shared with Shi Zhengli as a collaborator, as reported in Menachery's 2015 and 2016 publications. Sharing research with her, and therefore with the Chinese Communist Party, was at least as dangerous as funding her directly—which was also done by Fauci and NIH.

We found summaries for the seven grants from NIAID to Peter Daszak and EcoHealth Alliance between 2014 and 2020 that involve obvious gain-of-function research, including descriptions of creating "franken-mice" called "humanized mice" that are modified with human receptors to enable them to be infected with human CoV: "Test predictions of CoV inter-species transmission. Predictive models of host range (i.e. emergence potential) will be tested experimentally using reverse genetics, pseudovirus and receptor binding assays, and virus infection experiments across a range of cell cultures from different species and humanized mice."[100] This is very dangerous gain-of-function research.

Here are some excerpts from the official NIAID summaries for this gain-of-function research spanning 2018-2020: "These will be characterized to assess risk of spillover to people, and a series of *in vitro* (receptor binding, cell culture) and *in vivo* (humanized mouse and collaborative cross models) assays used to assess their potential to infect people and cause disease" (2020); "We will use S protein sequence data, infectious clone technology, *in vitro* and *in vivo* infection experiments and analysis of receptor binding," (2019); and "Predictive models of host range (i.e. emergence potential) will be tested experimentally using reverse genetics, pseudovirus and receptor binding assays, and virus infection experiments across a range of cell cultures from different species and humanized mice" (2018).

More About Ralph Baric and Fauci's Lies

When being grilled by Senator Paul, Fauci dismissed the idea that Baric's North Carolina research was gain-of-function. Fauci has a long history of supporting gain-of-function research involving Baric, going back at least to a publication in December 2008 that states, "Synthetic recombinant bat SARS-like coronavirus is infectious in cultured cells and in mice."[101] In this study, a SARS-CoV-like coronavirus was cultured in *human* lung epithelium cells and in live mice—a definitive gain-of-function study. The Acknowledgments give details that the National Institute of Allergy and Infectious Diseases supported four of the authors, including Michelle M. Becker (the first author) and Ralph S. Baric.

CONCLUSION THAT FAUCI LIED TO THE U.S. SENATE

Baric also has his name on the Menachery et al. 2015 and 2016 gain-of-function studies and, indeed, he is the force behind that research at the University of North Carolina.

There is absolutely no doubt whatsoever that Fauci was lying when he declared neither his NIAID nor NIH had funded gain-of-function research at the Wuhan Institute. Not only NIAID, but also NIH has been involved. NIAID in particular has been funding research on gain-of-function with coronaviruses at the Wuhan Institute since at least 2014. The funding often goes directly to Shi Zhengli and sometimes it is awarded to other Chinese researchers. Some of the research activities are conducted mainly or entirely at the Wuhan Institute, and at other times they are centered in the University of North Carolina with involvement from the Wuhan Institute. Working together, Dr. Baric and a large team of Chinese scientists have developed a SARS-CoV virus that is probably the deadliest we have documented. It causes an inflammation of the brain (encephalitis) that kills franken-mice who have implanted human receptors for the virus.[102]

FAUCI AND THE U.S. MILITARY

In a different context from "public health," Fauci has expressed another justification for this high-risk research in making pathogens. In *Biological Weapons*, Jeanne Guillemin, senior fellow at the Security Studies Program at MIT, described Fauci's two justifications for this work—public health and military defense:

> Anthony Fauci, NIAID's director since 1984, claimed that important discoveries about infectious diseases in general could be made while investigating pathogens that posed little or no current public health threat. For example, the study of smallpox

or Ebola might yield an antiviral drug that worked against other viruses, or research on the anthrax toxins might yield an antitoxin with alternate uses. Fauci compared the transition to counter-bioterrorism research to the development of AIDS research in the 1980s, when the public and young scientists had to be educated about its importance. He concluded that "there needs to be an accelerated and sustained effort to draw more competent scientists/ researchers into the field of research on counter bioterrorism."

Guillemin then criticized Fauci's expressed motive of defense against bioterrorism:

NIAID director Anthony Fauci has advised that a whole new generation of biologists should be engaged in the field of biodefense research. But increased numbers of scientists trained to work specifically with biological agents may in itself pose a threat to civilian safety. As discoveries move from the laboratory to final standardized products, more scientists and technicians will gain experience in animal and human testing of new vaccines and drugs and other products. This experience would presumably include tests and trials that simulate aerosol exposure during a biological weapons attack. Yet biodefense expansion proceeded with little consideration by scientists of its risks or why, after fifteen years of government warnings of bioterrorist attacks, none had occurred. Apparently, federal funding for technological solutions to bioterrorism largely precluded political analysis and action.

In other words, Fauci wants and gets to have *his own way* in conducting very dangerous research free of restraint or oversight. Remember, biodefense is inseparable from biowarfare because both depend upon making deadly viruses in labs. Fauci's justification of his research for "defense" purposes leads us to suspect that he has deep connections to the military-industrial complex and intelligence agencies.

There is an even more puzzling aspect to Fauci's avidly expressed interest in defense against biological warfare. If defending America and humankind from epidemics is Fauci's goal, how can he justify going around Presidents Obama and Trump to collaborate with and to fund Chinese virus researchers, the Wuhan Institute of Virology, and ultimately the Chinese Communist Party and its military? Enhancing China's ability to attack us is hardly a defense against them!

None of Fauci's behavior makes sense unless we view Fauci for what he is—an archetype of the global predator who feels no genuine identification or emotional ties with any nation or any specific religious or ideological viewpoint and instead remains forever devoted to increasing his own self-aggrandizement and his power, while adding a veneer of progressive idealism to his nefarious activities. Without this lens of predatory globalism, so much of what is going on in relation to COVID-19 remains clouded, but with understanding predatory globalism, a great deal more becomes painfully clear.

Have Fauci and Daszak Created Their Own Military-Civil Fusion?

Peter Daszak's EcoHealth Alliance appears to provide a nexus for bringing together America's own Military-Civil Fusion. On December 16, 2020, *Independent Science News* displayed a telling headline: "Peter Daszak's EcoHealth Alliance Has Hidden Almost $40 Million in Pentagon Funding and Militarized Pandemic Science."[103]

> Four significant insights emerge from all this. First, although it is called the EcoHealth Alliance, Peter Daszak and his nonprofit work closely with the military. Second, the EcoHealth Alliance attempts to conceal these military connections. Third, through militaristic language and analogies, Daszak and his colleagues promote what is often referred to as, and even then somewhat euphemistically, an ongoing agenda known as "securitization." In this case, it is the securitization of infectious diseases and of global public health. That is, they argue that pandemics constitute a vast and existential threat. They minimize the very real risks associated with their work and sell it as a billion-dollar solution. The fourth insight is that Daszak himself, as the Godfather of the Global Virome Project,[104] stands to benefit from the likely outlay of public funds.

The Department of Defense, at nearly $40 million, is the largest funder of EcoHealth Alliance, followed by Health and Human Services (HHS), including the CDC, NIAID, and NIH. *We suspect that one of Fauci's most powerful defenders in his long tenure has been the military-industrial complex, in particular, the Department of Defense.*

How does any of this ultimately make sense? Why would so many powerful American agencies and individuals—including politicians, bureaucrats, and scientists—tolerate and even support Fauci's collaboration with the Chinese Communist Party's biowarfare program? This widespread treachery only becomes understandable as we continue to delve into the dominant, often-covert influence in the world today of predatory globalism backed by the Chinese Communists. The final goal of the Communists is the weakening and then destruction of America as a patriotic, free, and democratic republic that will thwart globalism and ultimately the new Chinese Empire.

We did not imagine that this would be our conclusion—that the global predators, many of them Americans, have no allegiance other than their belief that the future is with Communist China. It is a conclusion that will be documented throughout this book.

How the 2015 U.S./China Virus Is Clinically Similar to Its Derivative, SARS-CoV-2

As already described, the collaborative 2015 virus and the Chinese version that created COVID-19 are from the same SARS-CoV line made in labs from bat viruses. They both have the characteristic human-engineered spike protein, making them capable of infecting human cells.

What struck me as a physician on first reading the 2015 research paper was the eerie similarity between the clinical effects of the 2015 lab virus and those of SARS-CoV-2.[105] The 2015 lab virus left younger mice remarkably unharmed but proved deadly to older mice or mice with compromised immune systems. The virus showed a strong tendency to attack lung tissue in mice and, in the lab, it was also shown to invade *human* bronchial epithelium cells. It produced weight loss in mice, one of the common effects of SARS-CoV-2 infection in humans.

Like SARS-CoV-2, the 2015 lab-made deadly coronavirus was resistant to all standardized treatments, although it was not tried on the successful early COVID-19 treatments described in this book and used by frontline doctors.

Then, in a warning for the future that still goes unheeded, the researchers were unable to find an effective vaccine: "Evaluation of available SARS-based immune-therapeutic and prophylactic modalities revealed poor efficacy; both monoclonal antibody and vaccine approaches failed to neutralize and protect from infection with CoVs using the novel spike protein." In short, none of the available vaccines worked.

Much more threatening, when the available vaccine made of "inactivated whole SARS-CoV" was given to older animals, they became sicker when exposed to their newly-minted SARS-CoV. In the older animals vaccinated and then exposed to the virus, "augmented immune pathology was also observed, indicating the possibility of the animals being harmed because of the vaccination."

The exaggerated immune response that erupts after vaccination and exposure to the virus is one of the mechanisms that is now killing tens of thousands of patients around the world. *Instead of protection, the vaccine increased their vulnerability to become seriously ill.* In Chapter 11, which is about vaccines, we discuss how we warned about this vaccine risk in April 2020 after reading the Menachery et al. article. As we go to press, the tragedy is now unfolding.

Celebrities or Dangerous Enemies?

Shi Zhengli, the Wuhan Institute scientist who plays such an important role in the collaboration with American researchers, was made into a celebrity by *Scientific American* online in April and in print in June 2020. The magazine's headline touts her heroism: "How China's 'Bat Woman' Hunted Down Viruses from SARS to the New Coronavirus."[106] It is subtitled, "Wuhan-based virologist Shi Zhengli has identified dozens of deadly SARS-like viruses in bat caves, and she warns there are more out there." *One might never have guessed that the "more out there" were lurking not in bat caves, but in the even darker recesses of the Wuhan Institute of Virology and other facilities in China, America, and other countries.*

Scientific American's purpose was "to address rumors that SARS-CoV-2 emerged from Shi Zhengli's lab in China." This exemplifies the collusion between the global scientific community and other predators, especially Communist China, with whom all or nearly all have deep ties. Shockingly, they continue to do this even at the human cost of flooding the world with additional pathogenic viruses from China's unsafe research institutes.

A Scientist Speaks Out

In November 2020, Kamran Abbasi— executive editor of the respected British journal, *The BMJ*—wrote a remarkable critique of the deterioration of science during COVID-19.[107] Abbasi titled his editorial, "COVID-19: politicization, 'corruption,' and suppression of science. When good science is suppressed by the medical-political complex, people die."

Abbasi concluded:

> Politicization of science was enthusiastically deployed by some of history's worst autocrats and dictators, and it is now regrettably commonplace in democracies. The medical-political complex tends towards suppression of science to aggrandize and enrich those in power. And, as the powerful become more successful, richer, and further intoxicated with power, the inconvenient truths of science are suppressed. When good science is suppressed, people die.

CHAPTER 3

Was SARS-CoV-2 "Made in China"?

Since early in the pandemic, the globalists have vigorously sung a loud chorus of "SARS-CoV-2 emerged from nature and not from a Chinese lab." The major media, social media, scientific journals, Bill Gates, Fauci, WHO, Communist China, and other global predators want us to believe that COVID-19 must have been contracted from bats in nature, perhaps through an intermediate animal host, and passed on to humans without any human tampering in the process. This enforced conformity of worldwide opinion and communications has demonstrated the colossal control that the global predators and Communist China exert over the most important institutions of the Western world.

Very recently, Fauci and some of the press have begun to waffle about the source of the pandemic. Facebook has ended its policy of censoring all attempts to discuss challenges to the establishment theory.[108] However, we suspect the globalists will never fully and unequivocally declare the truth that—with considerable financial and scientific help from the United States government—SARS-CoV-2 was made in China and released from the Wuhan Institute.

There Are No Known SARS-CoVs in Nature and None Are Known to Have Emerged from Nature

During and ever since the epidemic of SARS-CoV-1 in 2002-2004 in southern China, an enormous effort has been made to find a SARS-CoV in nature. *Yet no SARS-CoV has ever been found in nature. Furthermore, no SARS-CoV epidemic or other emergence of SARS-CoV among humans has ever been traced to nature but at least seven have been traced to labs, including SARS-CoV-2.*

On May 20, 2021, after coming to these conclusions, we found a book-length scientific monograph that confirmed our conclusions. The treatise by Steven Carl Quay MD, PhD[109] is dated January 29, 2021 and concludes, "SARS-CoV-2 is not a natural zoonosis but instead is laboratory derived."[110] Quay concludes, "the probability a laboratory origin for CoV-2 is 99.8%" while "the probability of a zoonotic [emergence from nature] is 0.2%."

Other scientists have also documented in statistical detail that the risk of an epidemic coming from a lab that is producing dangerous viruses is very high compared to a rare emergence from nature.[111]

The problem is not deadly coronaviruses lurking in "nature"—the problem is Anthony Fauci and the global predators whose interests and ambitions he serves by making SARS-CoV pathogens in American and Chinese labs while feigning that the risk comes from nature.

SARS-like Coronaviruses

Efforts to find SARS-CoVs in nature have found instead what Chinese scientists call SARS-*like* coronaviruses,[112,113] but these harmless viruses would require intricate laboratory engineering to make them capable of causing SARS or any other dangerous disease in an animal or human. As one report summed up:[114]

> In the new research, that held true again—none of the viruses from the cave *by themselves* displayed the genetic traits of the SARS coronavirus that spread to humans, infecting more than 8,000 people during the 2002-2003 emergency.

Even if the original SARS-CoV-1 did emerge from nature, it would then be a rare, one-in-a-million occurrence from evolution. A second emergence from nature would also be a very rare event. By comparison, there is a vastly greater likelihood of additional future releases of an engineered SARS-CoV from a lab, either by accident or intentionally.

Lab Origins of Seven SARS-CoV Leaks and Outbreaks

In 2004, WHO showed considerable concern about the security of laboratories around the world, and especially in China. Very ominously, it noted that they had no idea how many labs worldwide were storing SARS-CoVs. Beyond that, the implication is that SARS-CoVs are being experimented with in labs, possibly creating new versions that have been resulting in lethal infections in one or more of the accidental releases:[115]

> The recent outbreak of Severe Acute Respiratory Syndrome (SARS) in China which infected nine people and killed one of them is now over said WHO on 19 May. The outbreak, which began in April at a laboratory in Beijing, has raised questions about storing and handling the killer virus in laboratories.

> The recent outbreak was the third of four SARS outbreaks associated with a laboratory since 5 July 2003...

> "The big question is: is it safe to handle SARS?" said Dr Cathy Roth from the Dangerous and New Pathogens team at WHO's Global Alert and Response unit, adding: "Countries need to identify all the laboratories holding the virus and to ensure that all correct biosecurity measures in those labs are in place." Roth said it was impossible to know how many laboratories across the world were storing SARS virus stocks.

Of course, WHO would never say anything like that now. That's practically a conspiracy theory aimed at the abused Chinese Communists.

Dangerous lab mistakes with pathogens are very common. A 2011 government review found that between 2003 and 2009, a total of 395 incidents in *American* labs involved the potential release of pathogens, with seven of the incidents causing the infection of humans.[116]

The Wuhan Institute, which in 2015 became China's first laboratory to achieve the highest level of international bioresearch containment (known as BSL-4), had a well-known record of poor security, [117, 118] making a dangerous leak highly probable. Numerous leaks of other pathogens were also reported in December 2019 in China, around the time SARS-CoV-2 was leaked from the Wuhan Institute.[119] Indeed, leaks and other mishaps involving dangerous infectious agents had been occurring at U.S. CDC facilities,[120,121,122, 123, 124, 125, 126, 127, 128] as well as the United States Army Medical Research Institute of Infectious Diseases, in Fort Detrick, MD, which was temporarily shut down by the CDC.[129]

How Many SARS-CoV Escapes Have There Been?

More relevant than lab accidents in general, how many accidental releases of *SARS-CoV*-engineered pathogens involving infected humans have there been from labs worldwide? The best summary is provided by Martin Furmanski MD, PhD in "Threatened pandemics and laboratory escapes: Self-fulfilling prophecies" published in 2014 in the *Bulletin of Atomic Scientists*. Here is his summary,[130] which we have verified from numerous sources:

> SARS has not reemerged naturally, but there have been six escapes from virology labs: one each in Singapore and Taiwan, and four separate escapes at the same laboratory in Beijing.

A 2004 report from the *China Daily* described two of the viral leaks from the Beijing lab:[131]

> The small outbreak began in March and the World Health Organization declared it contained in May. ... Official investigation shows that it is an accident due to negligence.

The cases had been linked to **experiments using live and inactive SARS** coronavirus in the CDC's virology and diarrhea institutes … (Bold added.)

In at least two of the episode of escapes from the National Institute of Virology in Beijing a few people were infected by a human carrier after leaving the facility.[132]

Six plus one makes seven. SARS-CoV-2 is the seventh laboratory release. Since there have already been so many accidental releases of deadly variants of SARS-CoV from labs, and the odds of an emergence from nature are extremely small, it is reasonable to conclude that the threat from a SARS-CoV made in a lab are monstrously larger than the threat from a natural emergence from nature. It makes absurd Fauci's claims to be protecting humankind with gain-of-function research, when clearly, in itself, his research is the massive threat to humankind.

Even *Bloomberg*, on May 27, 2021, published an independent *Opinion* that described the six SARS-CoV leaks or escapes, without mentioning that SARS-CoV-2, whether intentional or not, is number seven, and SARS-CoV-1 may well be number eight. *Bloomberg* omits mentioning SARS in the headline and instead buries this most important data toward the end of the report.[133] The globalists still do not want the world to know the origin of SARS-CoV-2.

A Former Israeli Intelligence Officer Describes the LAB Origin of SARS-CoV-1

A very informed expert, Dr. Eyal Pinko,[134] has stated:[135]

At the Biological Weapons Conference held in October 2002, China declared that it had never developed, manufactured, or stockpiled biological weapons of any kind…

About a month after her declarations of the non-existence of biological weapons research and development, the SARS epidemic outbreak starts **in the city of Wuhan** in November 2002, which continued until February 2003. **China reported to the World Health Organization for the first time in the outbreak that the**

virus had burst due to a virus treatment malfunction during a study at its research institute.

It was impressive to see China's ability to contain, treat, and overcome the outbreak of the SARS epidemic in about three months. [Bold added.]

This is worth repeating: According to Dr. Pinko, China originally told WHO that SARS-CoV-1 "burst due to a virus treatment malfunction during a study at its research institute." The Chinese were already doing gain-of-function research in 2002 with coronaviruses, and a deadly variety of SARS-CoV escaped due to a malfunction in the lab. We have not been able to confirm that statement about the report to WHO about a lab leak or the statement that SARS-CoV-1 began in a Wuhan research facility, of which there are several.

Eyal Pinko is a retired Navy Commander[136] who served in the Israeli Navy for twenty-three years, including two years in managing missile development programs and six years in the intelligence branch. He has a PhD in Modern Naval Asymmetric Warfare and has received Israel's Security Award and other honors. He has written a detailed analysis of China's longstanding, systematic, and seemingly thorough involvement in developing biological warfare:[137]

> The Chinese strategic concept, called "Non-Contact Warfare," holds that China must fight its rivals as far away as possible from Chinese territory, with minimal risk of Chinese forces, in order for China not to fight on its land.
>
> …China operates additional biological weapons institutes besides the Wuhan Institute, *all under the command of the Chinese military.* …
>
> In order to develop biological weapons knowledge, China operates on several levels. The first level is cooperation and knowledge transfer with other countries such as Russia, Iran, and North Korea.
>
> *The second level is bringing experts from the Western countries to China and their assistance in establishing research infrastructures and knowledge development in China.* (Italics added.)

We have tried to contact Dr. Pinko by email and by phone but have been unable to do so. More recently, his email and phone number are no longer valid, and the report from which we were quoting was subsequently taken down and we had to recapture it on the Wayback Machine.

Chinese Researchers and Workers Show No Fear of Bat Viruses

We now have strong evidence that Shi Zhengli and her colleagues who work in the bat caves and in the Wuhan labs routinely handle bat excrement and endure bat bites without bothering to wear much, if any, protective clothing and without being worried. An article published by the *Taiwan News* on January 15, 2021, has the headline, "Video shows Wuhan Lab scientists admit to being bitten by bats—Chinese scientists shown using little or no PPE while handling bats in wild, samples in lab."[138]

The *Taiwan News* article begins:

> A video released two years before the start of the Wuhan coronavirus pandemic shows Wuhan Institute of Virology (WIV) scientists being cavalier toward protective equipment and being bitten by bats that carry deadly viruses such as SARS, demonstrating a lax safety culture in the lab.
>
> On Dec. 29, 2017, Chinese state-run TV released a video designed to showcase Shi Zhengli, also known as "Bat Woman," and her team of scientists at the WIV in their quest to find the origin of SARS. Despite the fact that the scientists work in a biosafety level 4 lab, they show a shocking disregard for safety when handling potentially infectious bats both in the wild and in the lab.
>
> From 4:45 to 4:56, a scientist can be seen holding a bat with his bare hands. Team members from 7:44 to 7:50 can be seen collecting potentially highly infectious bat feces while wearing short sleeves and shorts and with no noticeable personal protective equipment (PPE) other than gloves.

The report also describes workers being bitten by bats and showing little concern:

> He said that the bat's fangs went right through his glove, which was likely nitrile. He described the feeling as "like being jabbed with a needle." The video then cuts to a person's limb showing swelling after a bat bite.

According to the newspaper, the researchers and technicians do routinely get preventive rabies shots before going into the caves, so they are aware of taking real threats seriously.

The Chinese would not act so cavalierly with bats if they did not know there is virtually no risk of catching SARS-CoV from the cave dwellers. Zoonotic transmission of SARS-CoV is a farce maintained to justify all the expensive and truly dangerous risks associated with engineering harmless bat viruses into pandemic pathogens.

When Fauci and others call for a stop to widespread human interference in nature, the single most important measure would be to stop the Fauci-funded scientists from mucking about in places like bat caves and then purposely creating pathogens in their labs.

SARS-CoV-1: More Deadly—
But It Was Treated and Contained

The overall death rate for SARS-CoV-1 was in the range of 9%-10%, and the death rate in people 65 and older was up to 40%-50% or more. Both rates were extremely high and probably more than ten times that of SARS-CoV-2.[139, 140]

Perhaps, the high death rate of the original SARS-CoV-1 accounts in part for the mistaken dire predictions initially made about SARS-CoV-2 in early 2020. Those exaggerated predictions were also intended as part of the campaign to terrify the world population.

Fortunately, this earlier coronavirus epidemic like COVID-19 spared children and youth to a remarkable degree. In Hong Kong, where 298 people died from SARS-CoV-1, with a high percentage of death in the elderly, the mortality rate was 0% for children aged 0–14 years. For COVID-19, death rates for children and youth are also at or near zero in the U.S. and around the world.

During SARS-CoV-1 in 2003, only eight people in the United States had "laboratory evidence" of the virus, and they had traveled from other infected areas, according to the CDC.[141] Because it made people so ill, and because it was less contagious, it was easier to contain than SARS-CoV-2.

Laboratories quickly became able to study and to work with the virus, and on April 17, 2003, a team of scientists submitted the complete genome for a strain of SARS-CoV-1 to the CDC.[142] Whether they made the virus or it came from nature, given the high fatality rate, the Chinese military must have thought they had the makings of a very potent biological weapon.

Doctors Did Their Best to Treat Without Government Interference

SARS-CoV-1 was a much deadlier virus than SARS-CoV-2, yet there was no attempt by governments to tell doctors or patients what medications they could use. Like every other novel disease in modern times, doctors were free to explore various treatments. Some patients were helped in China, for example, with azithromycin and prednisone,[143] as they are similarly helped today with SARS-CoV-2.

In this initial SARS-CoV epidemic, no public health official or government agency prevented doctors from doing their best in the United States and seemingly not anywhere else in the world. It was desirable for doctors to experiment with new treatments within the general guidelines of medical experience with respiratory viruses. This is confirmed in the U.S. by an Institute of Medicine report in 2004.[144]

Secretary of State Pompeo: COVID-19 Originated in Wuhan Institute in Autumn 2019

On January 15, 2021, Secretary of State Pompeo released an official warning about the Wuhan Institute's history of deception in light of the upcoming WHO investigation.[145] In the document titled "Ensuring a Thorough, Transparent Investigation of COVID-19's Origin," he declared:

Illnesses at the Wuhan Institute of Virology (WIV): The United States government has reason to believe that several researchers inside the WIV became sick in autumn 2019, before the first identified case of the outbreak, with symptoms consistent with both COVID-19 and common seasonal illnesses. This raises questions about the credibility of WIV senior researcher Shi Zhengli's public claim that there was "zero infection" among the WIV's staff and students of SARS-CoV-2 or SARS-related viruses.

Unfortunately, Pompeo provided no information to back this important allegation. If it is true, it hammers home the origins of the virus and China's deceitful coverup, which includes Shi Zhengli. However, the only published evidence we could find is a Harvard study of increased traffic patterns surrounding the Wuhan Institute in the fall of 2019, along with increased reports of patients suffering from flu-like illnesses along with diarrhea, which is not a symptom unique to or characteristic of COVID-19.[146] The study by itself is not convincing.

Pompeo's hope for an investigation of the origin of SARS-CoV-2 was in vain. Not only is the Chinese Communist Party impermeable to investigations from the West, WHO inevitably exonerates China. This was confirmed by WHO's selection of Peter Daszak as a member of its team of ten investigators.[147] We shall repeatedly meet Daszak throughout this book as the world's cover-up specialist for denying the origin of SARS-CoV-2, including his chairing a similar commission by *The Lancet* aimed specifically at exonerating China and the Wuhan Institute (see Chapter 19 for more about the *Lancet* and WHO commissions).

When the announcement of Daszak's appointment to the WHO commission was made, scientists "condemned" his participation, one rightfully claiming that Daszak's conflicts of interest "unequivocally" disqualified him.[148]

Chinese Were Actively Engineering Dangerous Coronaviruses Right Up to the Pandemic

We received additional confirmation that the Wuhan labs and the University of North Carolina labs—the sites of American/Chinese collaborations that we helped to stop—were continuing to experiment with making coronaviruses up to the time of the SARS-CoV-2 outbreak in China.[149] The collaborators were genetically modifying mice with human-like receptors that made the mice infectable with SARS-CoV-2.

Ralph S. Baric, perhaps the leading American in gain-of-function researcher, was one of the numerous authors. He was lending his name to this research published in the midst of COVID-19—long after the Chinese had begun acting like enemies toward the U.S.[150]

The *Taiwan News* reported on and released a video interview with Peter Daszak on December 9, 2019,[151] describing the ongoing research. In the video, Daszak stressed the importance of his "collaborations" with China, claiming science had to be conducted out in the open. He said that over 80 % of his support came from federal funding. Contrary to naysayers who say that making deadly viruses in labs is "too difficult" to be concerned about, Daszak explained, "You can manipulate them pretty easily in the lab."[152]

The newspaper refers back to an earlier report by Daszak about making genetically modified "humanized" mice that can be infected with SARS-CoV. These viruses can also infect humans, and, probably for that reason, Daszak reportedly observed that humanized mice experiments were more "dangerous" than other kinds of research. Yet these are precisely the experiments that Fauci has been funding and led to SARS-CoV-2. Mice are humanized by creating receptors for the virus that are similar to those in humans.

From an even more recent source, on July 9, 2020, Chinese researchers, along with one American, published a collaborative paper describing how they made mice susceptible to SARS-CoV-2 infection and how these mice developed full-blown COVID-19.[153] They accomplished this feat by creating genetically engineered mice with human-like receptors (ACE2) in their cells. They reported, "In this study, we successfully established a SARS-CoV-2 infection model using these transgenic mice. In our results, the infected mice had two different outcomes: recovery and death."

> The infected mice generated typical interstitial pneumonia and pathology similar to those of COVID-19 patients. Viral quantification revealed the lungs as the major site of infection, although viral RNA could also be found in the eye, heart, and brain in some mice.

Their variant of SARS-CoV was unusually lethal to the mice, causing a generalized inflammation of the brain (encephalitis) and death: "The infected mice that lost >20 percent body weight maintained robust replication viral RNA copies in the lung and brain, although some mice succumbed to lethal encephalitis" (p. 51). Encephalitis in human COVID-19 is rare.[154]

Seemingly blinded by their ambitions, the authors voice no concerns about the potentially catastrophic results of an accidental release of infected mice that could then infect humans and cause deadly encephalitis. Instead, they conclude in their Abstract, "Our results show that the hACE2 mouse would be a valuable tool for testing potential vaccines and therapeutics." The "h" in hACE2 refers to its human origin.

The article has more than 20 authors, including one American, Ralph Baric, the powerhouse behind the North Carolina studies that led to our informing President Trump and his cancellation of the funding by Fauci. The "Bat Woman," Shi Zheng-Li (Shi Zhengli), is acknowledged as a key person in the "conceptualization" of the study.

This ongoing research, published in 2020, was primarily conducted in the Wuhan Institute and involved 18 of its staff. This once again makes clear that an accidental or purposeful release of the original SARS-CoV-2 from the Wuhan Institute should have been seen as the greatest likelihood at the onset of COVID-19.

Suppressed Chinese Research Identified the COVID-19 Leak from a Laboratory

Botao Xiao and Lei Xiao are Chinese scientists with numerous scientific publications who early in COVID-19 linked it to the Wuhan Institute. Botao Xiao[155] received his PhD from Northwestern University in 2011. He then became a postdoctoral research fellow at Harvard Medical School from 2011 to 2013. From 2017 to the present, he has been professor at the highly ranked South China University of Technology. Botao Xiao's research was partly supported by a grant from the National Science Foundation of China. The second author, Lei Xiao, is a published researcher at the Hubei University of Technology in Wuhan.

On February 6, 2020, in ResearchGate publication,[156] Xiao and Xiao review confirmed the connections that we found between China's capacity to create

SARS-CoV-2 and a series of research efforts funded by Anthony Fauci and his institute. The Chinese authors, one of whom was described as living in Wuhan, began by rejecting the idea that the virus came from a bat at the city's food market: "The probability was very low for the bats to fly [more than 900 kilometers] to the market. According to municipal reports and the testimonies of 31 residents and 28 visitors, the bat was never a food source in the city, and no bat was traded in the market."

Xiao and Xiao documented that the Wuhan Institute was working with Chinese horseshoe bats as a source of coronaviruses for gain-of-function research. They observed the Wuhan "principal investigator," Xing-Yi Ge,[157] had already succeeded in making a SARS-CoV with "the potential for human emergence" and observed, "A direct speculation was that SARS-CoV or its derivative might leak from the laboratory." Xing-Yi Ge was one of the two Chinese coauthors in the Menachery et al. study of 2015. Thus, without intending to, Xiao and Xiao linked the new pandemic to the main gain-of-function project funded by Fauci as an American/Chinese collaboration. Xiao and Xiao punctuated their conclusions, stating, "the killer coronavirus probably originated from a laboratory in Wuhan," and they urged greater safety measures.

After learning from the press that Botao Xiao had been "disappeared" by the Chinese Communist Party, we searched the news and found nothing to indicate he was alive. On February 26, 2020, he reportedly sent a brief email to the *Wall Street Journal*,[158] stating he had withdrawn his paper because it was "not supported by direct proofs." No one should believe his remarks. They easily could have been involuntary or forged.[159] We hope he and his coauthor, brave and honorable scientists, are alive and well.

After our initial investigation, we found that Botao Xiao's scientific paper had disappeared from ResearchGate, an uncommon phenomenon that a Fox News analyst commented on.[160]

BLOCKBUSTER SCIENTIFIC PAPER
BARELY SURVIVES GLOBALIST CENSORSHIP

Poorly covered in the U.S. press, the former head of MI6 reported in early June 2020 he had seen an unidentified scientific paper that proved that SARS-

CoV-2 was produced in a lab.[161] He also warned about the imbalance in what China was learning from our science establishment.

The scientific paper to which he referred was rejected from some journals and had to be rewritten to remove "explicit claims against China." It finally did get published by Sørensen and Dalgleish with a mouthful of a title that obscured the controversy, "Biovacc-19: A Candidate Vaccine for COVID-19 (SARS-CoV-2) Developed from Analysis of its General Method of Action for Infectivity."[162]

Remember, science has gone global, and predatory globalists reject all criticism of their ally, Communist China. This means it is practically impossible, or perhaps entirely impossible, to get any scientific paper published if it tells the truth about the coronavirus story in a way that implies that China is at fault.

When the Sørensen and Dalgleish paper did come out, one newspaper happily concluded, apparently relying on their own untrained eyes, that the report did not say it was man-made. They were mistaken. The authors managed to sneak into their published paper their certainty that it was engineered in a laboratory when they called it a "chimeric virus" in both the abstract and the text. A chimeric virus is a man-made one.

Sørensen and Dalgleish then go on to describe their own laboratory procedures similar to those carried out in the 2015 collaborative paper involving both U.S. and Chinese labs. They describe "insertions placed on the SARS-CoV-2 *spike* surface" to enable it to enter into human cells. (Italics added.)

Innumerable scientists should have been able to pounce on the Sørensen and Dalgleish article as an amazing confirmation; first, that the virus was created in a Chinese lab in the same or similar fashion as the *American-Chinese* collaboration in 2015, and, second, that this demonstrated SARS-CoV-2 was the result of a long series of gain-of-function lab studies by Americans and Chinese scientists, many or most of them financed in part by Fauci's NIAID. Yet this manuscript, which we are widely circulating, and our book may be the first clear and full discussion of this paper's important conclusions and implications.

We followed another newspaper lead in which Dr. Dalgleish, the second author in the above paper, stated he had one more coming out that was more direct in stating SARS-CoV-2 was made in a lab. We then located the paper titled, *The Evidence which Suggests that This Is No Naturally Evolved Virus: A Reconstructed Historical Aetiology of the SARS-CoV-2 Spike* by Sørensen, Dalgleish and Susrud. The paper was typeset for publication but with no indication of the publisher or likelihood of publication by a journal. We cite it here[163] in the hope researchers will add it to overwhelming evidence that SARS-CoV-2 is a laboratory creation.

It now appears that the authors Sørensen and Dalgleish may have an update to their work coming out soon, describing how the Chinese retro-engineered SARS-CoV-2 in a fraudulent effort to make it look as if it had an evolutionary history.[164]

Meanwhile, the nonprofit environmental group U.S. Right to Know (USRTK) in December 2020 began publishing a series of disclosures about behind-the-scenes shenanigans among researchers trying to dismiss the lab origin of SARS-CoV-2, including efforts to revise their published articles to maintain the coverup in collusion with journals.[165] The lock-step effort of the scientific community to protect the Chinese Communist Party remained a frightening source of bewilderment, and in earlier public versions of the manuscript for this book, we said such actions would require further investigation by others. Instead, we delayed the publication of the book to do a more in-depth investigation, tracking the lines of influence and financial entanglements, which, in general, go to China. Many are described in the text and others are summarized in the back of the book in the *Chronology and Overview*.

Why a Deadly Bat Coronavirus Jumping to a Human is Especially Unlikely

In defiance of common sense, a 2017 paper ominously titled, "Jumping species—a mechanism for coronavirus persistence and survival," Menachery gave his rationalization for doing the dangerous research we would highlight and President Trump would stop:[166]

> *Zoonotic transmission [jumping from an animal to a human]of novel viruses represents a significant threat to global public health and is fueled by globalization, the loss of natural habitats, and exposure to new hosts. For coronaviruses (CoVs), broad diversity exists within bat populations and uniquely positions them to seed future emergence events. In this review, we explore the host and viral dynamics that shape these CoV populations for survival, amplification, and possible emergence in novel hosts.*

It is astonishing that Menachery and apparently all those associated with the research claim to be heading off the rare event of a novel coronavirus jumping to and seriously harming humans, while they themselves intentionally make it happen—creating a "jumper" virus in the lab—while giving it wide distribution to labs around the world, including China.

The ability of the project to make a pathogen out of the coronavirus in no way indicates there is even the slightest chance of the same thing happening in nature. After all, it took a multi-million-dollar, several-year collaborative research effort involving many extraordinary technologies and a large number of scientists from two nations, and additional help from others, to purposely turn this harmless bat virus into a virulent one. Along the way, the process required many intermediate steps, each step necessitating careful reasoning and considerable trial and error, all with a very specific purpose in mind.

Unlike humans, viruses do not have conscious purposes. A virus living in a bat in a remote cave does not have a purpose to evolve into a human pathogen. In an extremely rare event, it may accidentally mutate into one that is contagious to humans, but the mutation would have no adaptive value in the cave and die out. This is especially true in a cave where the only human contact is largely confined to Chinese researchers.

If the chance of a virus spontaneously mutating into a human pathogen is infinitesimal, then imagine how unusual it would be for that extremely rare mutation to find a human in a cave to infect, thereby reproducing. As for infecting a vector animal who then infects a human, the probability of that additional step might even be more slight. Of course, bats do fly outside their caves, but what's the likelihood of flying into a human and biting it? Not much, unless the humans are researchers repeatedly tampering with them, and they have on occasion bitten researchers; but the victims of the bites do not seem much bothered about, as we have seen in this chapter.

As already noted with multiple citations from 2004 to the present, there have been many accidental releases and other dangerous mishaps involving potentially deadly viruses from laboratories in China and the United States. The escape of a man-made virus, intentionally or not, is infinitely more likely than a lone bat virus evolving into a pathogen in nature and then finding a human host.

One Source Early Recognized the Origin of SARS-CoV-2

The Epoch Times was founded by Chinese dissidents and expatriates who fled Communism. An investigative report filmed by the conservative newspaper includes American experts and scientific citations from Chinese publications.[167] It came out June 28, 2020. In the film, the Bat Woman, Dr. Zhengli-Li Shi, describes how she was appointed by the Chinese government in 2003 to find the source of the 2002-2004 SARS epidemic virus. She found the famous bat cave, 1,100 miles from Wuhan, and located ten families of SARS-CoV viruses among the creatures.

She explains how she brought back "thousands" of samples to her lab in the Wuhan Institute of Virology. After SARS-CoV-2 appeared, she feared it might have escaped from her lab but personally compared it to all her samples—remember, she has thousands—and found it was not from one of the SARS-CoV in her lab.

The entire story sounds preposterous. The task of ruling out that the thousands of viruses brought back to her lab from caves were not the source of SARS-CoV-2 sounds insurmountable. More importantly, she completely avoids discussing the unknown but significant numbers of SARS-CoV that she was engineering into pathogens in her lab, which are far more likely the source of pandemic than the harmless viruses typically found in bats in nature.

In addition, the story strongly confirms that the origin of SARS-CoV-1 has never been located. That increases the likelihood that SARS-CoV-1 was also a creation of a Chinese laboratory under the constant supervision of the Communist Party and its military.

Overview of the High Risk of a SARS-CoV Virus Escaping the Wuhan Institute

(1) Estimates of accidental releases or mishandlings of pathogenic coronaviruses by China are high.[168,169,170,171,172,173]

(2) Fauci had to know the State Department, following two visits to the Wuhan Institute, had warned in 2018 that the safety measures, specifically for coronaviruses, were inadequate.[174]

(3) The first SARS-CoV appeared in China[175] in late 2002-2003 with no known origin and could probably have originated in a lab.

(4) *In 2004 there were four documented accidental releases of SARS-CoV from the Beijing Institute of Virology.[176, 177] There was also one in Taiwan and one in Singapore.[178] These six releases of SARS-CoV from labs, plus SARS-CoV-2, make seven documented lab releases compared to zero documented leaps from nature. SARS-CoV-1 would make it eight releases from labs with none from nature.*

(5) The United States itself suffers from many episodes of dangerous mishaps with pathogenic viruses, including from CDC labs.[179] The gain-of-function research funded by Fauci at the University of North Carolina (UNC) at Chapel Hill has had six security lapses in handling SARS-CoV.[180] *ProPublica* received the information through Freedom of Information (FOIA) requests from NIH. It provides some grim details:

> From Jan. 1, 2015, through June 1, 2020, the University of North Carolina at Chapel Hill reported 28 lab incidents involving genetically engineered organisms to safety officials at the National Institutes of Health, according to documents UNC released to ProPublica under a public records request. The NIH oversees research involving genetically modified organisms.
>
> Six of the incidents involved various types of lab-created coronaviruses. **Many were engineered to allow the study of the virus in mice.** (Bold added.)

If there were any doubt about it, the "many" experiments that "were engineered to allow the study of the virus in mice" indicates gain-of-function studies—the creation of pathogens from harmless viruses. According to a 2019 report in the *Taiwan News* which interviewed Peter Daszak, early in 2015 he gave a report "Assessing Coronavirus Threats" in which he said that "experiments involving humanized mice have the highest degree of risk."[181]

The following typical statement made by NIH to a newspaper about the much greater likelihood of an "emergence from nature" than a leak from a Chinese lab is simply false.[182]

"Most emerging human viruses come from wildlife, and these represent a significant threat to public health and biosecurity in the US and globally," the statement read.

"Scientific research indicates that there is no evidence that suggests the virus was created in a laboratory."

(6) In 2015, an article in *Nature Medicine* by Menachery et al.[183] reported the creation of pathogenic SARS-CoVs from harmless bat viruses in a Fauci-funded collaboration between the U.S. and China, adding to China's capacity to make them on its own. The 2015 man-made virus behaved similarly to SARS-CoV-2 in that it killed elderly and impaired mice but spared younger ones. The lab virus infected samples of human respiratory epithelium and was impervious to treatment and vaccines.[184] Therefore, we know the Chinese had the capability of engineering SARS-CoVs in their laboratories.

(7) Soon after the start of the epidemic in China, Chinese researchers proved the virus was man-made by comparing it to natural, unmodified coronaviruses and those made to penetrate human cells, which requires special laboratory modifications and were being worked on by the Chinese military.[185] Another group of Chinese researchers proved the virus was man-made by comparing it to natural, unmodified coronaviruses and those made to penetrate human cells, which requires special laboratory modifications and were being worked on by the Chinese military.[186]

(8) More recently, Dr. Yan, who escaped from China, and her colleagues published a paper linking SARS-CoV-2 to a chain of Chinese and American research going back to 2013 and to the Wuhan Institute (See Chapter 4 for more about Dr. Yan).[187] The research studies they cited were funded by Fauci directly through his NIAID and indirectly through Peter Daszak's EcoHealth Alliance. Yan and her colleagues confirmed that Chinese scientists now have no serious impediments to making these viruses.[188,189] They estimate China now has the largest storehouse of coronavirus bioweapons in the world.[190]

(9) More than 20 Chinese scientists, nearly all from the Wuhan Institute, published a paper in July 2020 in *Cell* documenting their ability to genetically engineer mice who were then infected with SARS-CoV-2 and displayed unusually severe and lethal cases of COVID-19. This confirmed China's continuing determination and capability to conduct dangerous virus research with grave

military implications.[191] The Bat Woman, Shi Zhengli, was listed as the final author, reflecting her role as lab director. Ralph Baric, the scientist behind America's gain-of-function program, collaborated with the Chinese on this high-risk project that was directly tied to the Chinese military through the policy of Military-Civil Fusion.

(10) After the original 2002-2004 SARS-CoV epidemic, if there was a leap from nature to humankind causing COVID-19, the odds would have been infinitesimal for evolution causing it compared to the high risk of an accidental or intentional release from Chinese labs.[192]

(11) The Wuhan market China officially described as the source of SARS-CoV-2 does not sell bats, and there are no such colonies within hundreds of miles. Scientific papers have shown the virus infected humans much earlier than originally thought with no connection to the market.[193]

(12) The cavalier manner in which Chinese researchers go unprotected into bat caves and in their labs, handling feces and even sustaining bites, confirms they have no fear of being infected with SARS-CoV from direct contact with bats. The fear of transmission from nature is concocted to justify highly dangerous research on creating SARS-CoV pathogens in the lab.

The almost universal efforts of the scientific community to protect China and its Communist Party in respect to the origins of the pandemic from within their laboratories has included the collaboration of key figures like Fauci and Gates, worldwide scientists, the most highly reputed scientific journals such as *The Lancet*, the mainstream media, and larger organizational structures such as WHO, the CDC, and NIH. This blight indicates a degree of collaborating corruption that boggles the mind.

We have only begun to grasp the size of this successful corruption of the scientific community and its transformation into an ally of the greatest global predator of all, Communist China, with its cunning, violent dictatorship. It speaks to the invasiveness of the Chinese Communist Party into Western culture and science, and the willingness of our society to sacrifice itself and its highest values on behalf of forces that remain evil beyond our understanding and even our imagination.

Now we turn to the largely untouchable question, "Did China *purposely* release SARS-CoV-2?"

CHAPTER 4

Did China *Intentionally* Unleash COVID-19 on the World?

The Wuhan Institute has been known as "a center of China's declared biowarfare/biodefense capacity."[194] Its director is Chen Wei, a Major General in the People's Liberation Army, and China's top bioweapons expert.[195] In January 2020, she was appointed head of the Wuhan Institute of Virology. The media almost never mentions that the Wuhan Institute is fundamentally a biowarfare military research facility.

China's policy of Military-Civil Fusion, described in Chapter 1, guarantees a hand-in-glove relationship between the Chinese military and everything important taking place at the Wuhan Institute of Virology. The Chinese Communist Party and its military could easily have arranged a purposeful "accidental" release of SARS-CoV-2 from the Wuhan facility. But what advantage would China gain from such a seemingly self-destructive act as unleashing COVID-19 on its own population?

What We Know with Certainty

We have overwhelming evidence that SARS-CoV-2 was manufactured and then released by the Wuhan Institute. We also know the Chinese Communist Party intentionally went out of its way to spread it around the world. The government halted domestic flights to and from Wuhan, a city of 11 million people and the surrounding province of Hubei, while *intentionally* promoting flights from the region to the rest of the world. The same also happened in other cities, including Shanghai and Beijing—shutdowns of domestic flights while pushing international flights.[196, 197] *The Economic Times* summed up the Chinese lockdown:

> While China continued to protest against international travel bans, it successfully quarantined Wuhan and other affected cities. The total domestic lockdown of Hubei province and the flight ban imposed inside China had immediate effect. As per data from TomTom Traffic Index, Wuhan had a traffic density of 60% in January while Shanghai and Beijing had nearly 80% density. After the total lockdown, the average traffic density fell to below 10% in Wuhan and Shanghai during February and below 5% in Beijing. While implementing a total domestic lockdown in February, China kept assuring the world that the situation was not serious and fully under control.

The policy persisted for more than six months. At the end of August, the *BBC News* reported, "Air travel has been gradually picking up since the coronavirus grounded the majority of planes in February."[198]

China continued in early February to demand other countries stop banning flights from China or Wuhan, even though they had already implemented a ban on domestic flights and other forms of travel to and from Wuhan.[199] In the face of increasing bans on flights to China by other nations, China continued for months afterward to operate and to press for increased flights from China to the world.[200]

The Chinese locked down Wuhan on January 16, 2020, but through all of January 2020, an estimated 4,000 people flew directly from Wuhan to the United States.[201] Nineteen largely filled flights went to San Francisco and New York with no enhanced screening.

Additionally, in January, according to a *New York Times* report, there were over 1,300 direct passenger flights from all of China to 17 airports in the United States, for a total of 381,000 travelers.[202] About one-quarter were Americans returning home. In addition, a large, uncounted number of people flew from China to the U.S. through intermediate stops.

According to the *Times* report, a significant number of people also flew in through indirect routes:

> In addition, untold others arrived from China on itineraries first stopping in another country. While actual passenger counts for indirect fliers were not available, Sofia Boza-Holman, a spokeswoman for the Department of Homeland Security, said they represented about a quarter of travelers from China.

In a remarkable insert, *The Epoch Times* on January 20, 2021, produced a diagram of the initial reported COVID-19 cases in about two dozen different countries that traced back to flights from Wuhan.[203] The newspaper commented:

> For at least three weeks, Beijing knew about the severity of the CCP virus outbreak in the central Chinese city of Wuhan, but downplayed the crisis and suppressed any information at odds with the official narrative that the virus was containable. This coverup allowed the virus to spread beyond China's borders, sparking a global pandemic. ... Authorities were aware of the outbreak by at least late December and publicly confirmed its existence on Dec. 31, 2019.

> For the next 19 days, Chinese authorities insisted the disease was "preventable and controllable" and that there was little to no risk that it could spread among humans. The World Health Organization (WHO) echoed those statements.

> Doctors who tried to warn others about the outbreak were punished for "rumor mongering." ...

> Meanwhile, Chinese social media platforms scrubbed references to the virus and comments critical of the regime's handling of the outbreak from December. Propaganda organs issued at least 18 directives to domestic news outlets in January ordering them to limit coverage of the outbreak and stick to official statements.

It was not until Jan. 20 that the Chinese regime confirmed human-to-human transmission.

In summary, starting early in January 2020, China knowingly flooded the U.S. with potentially infected people until President Trump—against the advice of Fauci and other globalists—stopped all flights from China at the end of the month. China nonetheless continued flooding the world, which also put America at risk of infected travelers from other countries. It was an act of "unrestricted warfare."

China Corners Market on PPEs and President Trump Takes Action

In another large-scale deadly economic manipulation, China stopped *exporting* necessities for the control of the COVID-19 pandemic, and as the major manufacturer, began depriving the world of them. Simultaneously, it began *importing* vast amounts of protective gear, including two billion masks and more than 25 billion items of protective clothing. Executives from companies like Honeywell and 3M disclosed that in January, China not only stopped all exports of N95 respirators and Personal Protective Equipment (PPE) such as gloves, masks, and booties, but increased buying from U.S. manufacturers like themselves. From late January 2020 through the end of February 2020, China's cornering of the market created extreme shortages in an unprepared world, including the United States.[204]

Jenna Ellis, a senior legal adviser to President Trump's re-election campaign, declared, "People are dying. When you have intentional, cold-blooded, premeditated action like you have with China, this would be considered first-degree murder."[205] President Trump expressed outrage at both China and the conspiring U.S. companies, and on April 2, 2020, the President invoked the Defense Production Act to force 3M to prioritize the production of respirators for the Federal Emergency Management Agency.[206]

The New York Times covered a portion of the story in July 2020 with an article titled "China Dominates Medical Supplies, in This Outbreak and the Next."[207] It read like an advertisement for China's superior strategic business planning. In a typical fashion, the globalist newspaper made no mention of how China systematically cornered the market to the detriment of humanity nor about the President invoking the Defense Production Act to protect and substantially increase American supplies. This is one of many examples of how an individual or group receiving information from *The New York Times*, *The Washington Post*, CNN, or other globalist predator news media would have a very distorted view of the activities surrounding COVID-19, skewed toward favoring or admiring China and denigrating the United States.

All the global predators are siding with the Chinese Communist Party in its unrestricted warfare against us.

Fauci, Tedros, and China

In its nefarious activities at the start of the pandemic in Wuhan, China was backed by the World Health Organization (WHO) and its Director-General Tedros Adhanom Ghebreyesus, a patently corrupt totalitarian politician from Ethiopia.[208,209,210] Tedros has been accused of covering up devastating cholera outbreaks in his own country by *The New York Times*[211] and other sources.[212] Tedros, who has had a conflicted relationship with the U.S., is closely allied to China, which is WHO's second biggest donor and became its largest donor after President Trump pulled the U.S. funding.

Consistent with the importance of WHO within the circles of global predators, Bill Gates upped his long-time contribution to WHO after President Trump withdrew U.S. support. According to *US News and World Report*:[213]

> The Gates Foundation has been a key donor to the WHO over the past decade, accounting for as much as 13% of the group's budget for the 2016-17 period. In February, the foundation pledged $100 million to fight the coronavirus pandemic, and it upped that to $250 million in April after an order from President Trump brought U.S. government funding to a halt.

Meanwhile, Anthony Fauci, like the Director-General of WHO, was against President Trump's ban on air travel. Fauci called the ban "irrelevant" because it could not prevent the virus from eventually spreading worldwide.[214] In the same interview during which Fauci resisted any travel bans to and from China, he suggested the virus might diminish (like the flu) when the weather changed:

"The wild card here is that this is a brand new virus, this novel coronavirus, and we do not know if it's going to diminish as the weather gets warm. We can't count on that."

If Fauci is indeed such a good scientist, how do we account for his dramatic downplaying of the original COVID-19 reports? The best predictor of the potential risk was SARS-CoV-1, which, as we have seen, had a very high 10% death rate. The answer is that Fauci, first and foremost, is a predatory globalist. Making light of the virus while China spread it around the world was in the deepest interests of predatory globalists. We will document that they were already planning for a coronavirus outbreak in a public forum and predicting widespread deaths and economic shutdowns.

Fauci Defies President Trump in Front of America

Anthony Fauci has been outspoken in his support of WHO's Tedros. On March 25, 2020, at a critical moment early in the crisis, while standing beside President Trump at a nationally televised Coronavirus Task Force presentation, Fauci openly and publicly undermined President Trump's concerns about Tedros.

The following pithy, revealing excerpt from the official White House transcript[215] of the televised discussion demonstrates Fauci's willingness to undermine the President. Fauci refuses to comment on the lack of transparency from China, a problem that led to China's covert infliction of the pandemic on the world. Fauci describes how he has known Tedros since Tedros was in Ethiopia—a time during which Tedros was accused of extraordinary corruption and even indifference toward epidemics in his homeland. Fauci gives us a big hint about just how close he is with Tedros, saying he had just gotten off the phone with him a few hours earlier in the day when he was leading a WHO phone call. "Tedros is really an outstanding person," Fauci announced. President Trump replies:

> PRESIDENT TRUMP: But the fact is that I have heard for years that [WHO] is very much biased toward China, so I don't know. Doctor, do you want to you — do you want me to get you into this political mess?

DR. FAUCI: No, I don't want you to do that. But I will. (Laughs.) So, Tedros is really an outstanding person. I've known him from the time that he was the Minister of Health of Ethiopia. I mean, obviously, over the years, anyone who says that the WHO has not had problems has not been watching the WHO. But I think, under his leadership, they've done very well. He has been all over this. I was on the phone with him a few hours ago leading a WHO call.

QUESTION FROM THE PRESIDENT; Praising China's transparency, sir?

DR. FAUCI: No. No, I'm not — I'm not talking about China. You asked me about Tedros.

QUESTION FROM THE PRESIDENT: The World Health Organization was praising China for its transparency and leadership on their response to the pandemic.

DR. FAUCI: You know, I can't comment on that because — I mean, I don't have any viewpoint into it. I mean, I don't—I don't even know what your question is.

It is telling—even chilling—that a few hours before the task force meeting, Fauci was on the phone with Tedros and now praises him to the President's face on national television. The ominous connections among the triad of Tedros' WHO, China, and Anthony Fauci's NIAID help to explain how China initially was praised rather than condemned for its handling of the coronavirus.

Although very recently he has begun to waffle,[216] Fauci has done everything he can to wholly exonerate the Chinese Communist Party and the Wuhan Institute of Virology of any possible wrongdoing. He continues to promote the discredited claim[217] the coronavirus infection originated in a Wuhan wet market, where, in fact, no bats are sold and none can be found for hundreds of miles, except at the Wuhan Institute itself, which collects and experiments on bats, extracting coronaviruses from them.[218, 219, 220] Meanwhile, simultaneously and coincidentally, the Wuhan Institute of Virology was making deadly coronaviruses similar to this supposed "leap from nature."

While unleashing the virus on the world, China's government, the Communist Party, also intentionally withheld the existence of the internal epidemic, then claimed the virus came from the wet food market, while initially denying it could be transmitted by humans.[221] However, on February 20, 2020, as already discussed, two Chinese researchers published a study proposing the novel coronavirus was man-made, probably in a Chinese laboratory:[222]

> We noted two laboratories conducting research on bat coronavirus in Wuhan, one of which was only 280 meters from the seafood market. We briefly examined the histories of the laboratories and proposed that the coronavirus probably originated from a laboratory.

Other experts confirmed the probability that SARS-CoV-2 originated from the Wuhan Institute,[223] where the Chinese were known to be engineering SARS-like bat coronaviruses into virulent human pathogens similar to SARS-CoV-2.

Recently Disclosed Emails Show Fauci and NIH Supporting Praise for China's Handling of Pandemic

In an October 2020 Press Release[224] and in its January 2021 issue of *The Verdict*, Judicial Watch announced, "Fauci Emails Show WHO Entity Pushing for a Press Release 'Especially' Supporting China's Response to the Coronavirus." After going to court to force NIH to respond to its Freedom of Information Act (FOIA), Judicial Watch started receiving a stream of Fauci emails, and these were some of the more important. They showed how WHO coordinated with Fauci, NIH, and the Global Preparedness Monitoring Board to approve WHO's unqualified support of China's devious and misleading behavior regarding the initial outbreak of the dangerous coronavirus.

The Global Preparedness Monitoring Board (GPMB) is a combined arm of WHO and the World Bank. More than a dozen board members include Anthony Fauci as well as a former head of the Chinese CDC and the director of the giant Wellcome Foundation. The mandate of the GPMB is to work with "key policymakers" around the world towards "increased preparedness and response capacity for disease outbreaks and other emergencies with health consequences."[225]

When Fauci, NIH, and the Global Preparedness Monitor Board all approved WHO's support of China, it was a triumph of organized global predators.

Tom Fitton, the head of Judicial Watch, commented, "These Fauci emails show how praising China was the odd priority of the WHO in the face of a novel and dangerous coronavirus." *China First* is the motto at WHO and seemingly at NIH and with Fauci, as well.

President Trump was critical of WHO's collaboration and coverups with China and pulled U.S. funding. Subsequently, on January 21, 2021, within hours of Joe Biden taking over as President, Fauci declared by video to WHO, "I am honored to announce the United States will remain a member of the World Health Organization." Fauci was now identifying himself as President Joe Biden's "Top Medical Advisor." Tedros, the unscrupulous, corrupt politician who was made Director-General of WHO with China's support, praised Fauci as "my brother Tony." It was a good day for China.

I warned of this day coming in my October 29, 2020, video, "Fauci and Biden Together: A Scientific Dictatorship."[226]

Would China Have Been Motivated to Purposely Release the Pathogenic Coronavirus?

We know the Chinese Communist regime intentionally spread the virus around the world, but did the Communists purposely unleash it to start the pandemic?

For most of us, it seems impossible that leaders of a nation would unleash a potential pandemic virus on its own people. Nevertheless, China has a modern history of lashing out against its own people when it meets the needs of the Communist Party, most dramatically under Mao and the "Gang of Four" Communist leaders for a full decade, 1966-1976. It was like a plague in *Britannica*'s brief description of the cost:[227]

> The revolution left many people dead (estimates range from 500,000 to 2,000,000), displaced millions of people, and completely disrupted the country's economy.

The basic attitude often expressed by Communist leaders is the community comes before the individual. My experience with young Chinese students in our community of Ithaca, New York, confirms they are brought up and schooled to suppress their own needs to those of authority. The Chinese people have been taught compliance and conformity by government suppression, such as limiting the size of the family and controlling the universities and careers that young people can pursue.

I was surprised to learn from Chinese citizens in America that China has no safety net for the elderly—no equivalent of social security. One told me, "We have a saying, 'You're born alone and you die alone.'"

It seems highly probable that under the stress of President Trump's America-First policies, China wanted to impose its own version of a Great Reset on the world. Since China produced so many of the world's medical necessities and medical supplies, the CCP probably thought it was in a better position to survive a pandemic than the United States. Also, it could use whatever repressive measures were required to stem the tide of the epidemic in their own country, and they did so in Wuhan and elsewhere. There were widespread accusations that people were being locked wholesale in their apartment compounds and even individual apartments.[228]

Other than winning its conflict with America, there was another strong motivation to unleash this strain of virus on its own people. As we have seen from SARS-CoV-1 and from animal lab experiments, pathogenic coronaviruses tend to spare children and target older people. This made these viruses less than an ideal biological weapon, but perhaps a good one given China's circumstances. China's one-child policy, and its more recent two-child policy, left the society with too few young people to support the growing older population. It is axiomatic that a modern state requires constant growth of its working-age population, and this growth was insufficient in China. Furthermore, the limitation on children led families to abort females in preference for males who could take care of them, leading to an insufficient number of women, again impeding the birth rate.

Because of the negative effects of their limits on births, in 2018, China announced it was considering lifting its limitation on children born to families. The change in regulations were slated to come into effect in March 2020,[229] but as of November 24, 2020, they were still being considered amid growing worries about the nation's shrinking workforce.[230]

There is yet another reason China might have wanted to intentionally release SARS-CoV-2 on its own people. The Chinese Communist Party has been expanding its control through increased surveillance through monitoring behavior and giving out social credits for good behavior. Despite their heavy controls over the internet and social media, reinforced with help from depraved American globalist firms, there is increased unrest, culminating in the serious protests in Hong Kong against the loss of its autonomy from China. COVID-19 provided an opportunity to shut down any protests in the name of public health and, more importantly, allowed China to flex its muscles, displaying its ability to confine people to their own homes against their will. Totalitarian public health joined totalitarian political policies to make clear to the Chinese who's in charge—and it's not "the people."

The Chinese economy was also flagging, and the government could have hoped they would come out economically stronger than the United States. If this was part of their planning calculus—to harm America's economy much more than their own, they seem to have already succeeded. Although China's economy declined, it remained the only country with an actual positive growth rate (2.1 percent) in 2020 with more bullish predictions for 2021.[231] If the Chinese did release SARS-CoV-2 as an attack on the U.S., they probably believe they have done well.

Finally, China might have seen spreading the pandemic to the U.S. as a way of ruining President Trump's presidency and preventing a second term. To accomplish that, China would need the cooperation of many American politicians, business leaders, and media willing to blame it all on President Trump while exonerating China. With candidate Joe Biden and the Democrat Party in the lead at the time, the Chinese Communist Party anticipated an avalanche of support inside the U.S. They probably got more than they could have hoped for with the second impeachment. China may not be far away from "recoupling" with the United States and re-establishing its self-serving, lopsided trade relations of the past, enriching the global predators along the way.

Experts on China Share Our Perspective

We are not alone in drawing attention to the harmful effects of China's initial handling of the epidemic. An article in the *Washington Examiner* cites one prominent critic:[232]

Former British spy chief says China and WHO bear responsibility for flawed coronavirus response

According to the *Washington Examiner*:

> The former head of the United Kingdom's foreign intelligence service said the Chinese government and the World Health Organization bear responsibility for a flawed response to the coronavirus pandemic.

> Sir Robert John Sawers, who was chief of the Secret Intelligence Service, or MI6, from 2009 through 2014, told BBC Radio on Wednesday the Chinese Communist Party is "evading" its responsibility for the current global pandemic. He also said the WHO has "serious questions to answer."

> Sawers, who was also the former British permanent representative to the United Nations from 2007 through 2009, warned that China has been moving toward becoming a surveillance state for years, especially under the half decade of leadership of Xi Jinping, and that it has increased its influence over the United Nations. ...

> The former MI6 leader argued, "It would be better to hold China responsible for those issues rather than the World Health Organization" because the WHO is "only as good as its member states." He also lamented that China's role at the U.N. has "steadily grown as China's power has grown," and noted that "heads of U.N. agencies are wary of offending one of the major powers."

> Still, Sawers said, "That doesn't excuse the head of the WHO for failing to stand up for the facts and the data and making the right demands of the Chinese."

Escaped Chinese Scientist Confirms the Worst

In late April 2020, Li-Meng Yan, PhD, an experienced Chinese virologist at the Hong Kong School of Public Health, escaped to the United States.[233,234] Dr. Yan explained to the media that she left China to tell the world about China's coverup of the real source of the deadly pandemic, the Wuhan Institute. She has remained in hiding since, while talking to newspapers and appearing on Tucker Carlson on the Fox News Channel on September 15, 2020,[235] and at other times. She told Tucker, "This virus, COVID-19 SARS-CoV-2 virus, actually is not from nature. It is a man-made virus created in the lab."

I had a personal conversation with Dr. Yan on April 27, 2021, on a secure video connection before we both joined a video conference. In discussing her firm belief that the Chinese Communists purposely released SARS-CoV-2, she reminded me that in the past when there were several SARS-CoV and additional other leaks of deadly pathogens from their labs, the Communists always responded quickly, traced and isolated the carriers, and kept the virus in check. I had already written about these multiple leaks and immediately recognized the truth of what she was observing. In this conversation and with her three publications on SARS-CoV-2, she helped convince me that the Chinese purposely released SARS-CoV-2 as a bioweapon. I believe the Communist dictatorship chose this moment in time specifically to interfere with the presidency of Donald Trump.

Dr. Yan's First of Three Scientific Reports

For the first time, on September 14, 2020, Dr. Yan and three colleagues put a prepublication version of their new paper online, confirming SARS-CoV-2 is a man-made product of Chinese laboratories.[236, 237]

To describe China's background in developing the spike protein, Yan et al. explained this method has been "repeatedly" used in laboratories to create "human-infecting" coronaviruses of non-human origin. To document their observation, the authors cite four research publications. Two of the four citations are to the 2015 and 2016 Menachery papers that involved collaborations with Chinese researchers. A third paper is also American in origin and does not involve Chinese researchers; nevertheless, like the first two, the referenced study is supported by Fauci's institute.[238] Three papers show the direct connection between Fauci's funding and China's ability to build SARS-CoV-2. The fourth paper involved neither U.S. researchers nor U.S. funding.

Yan et al. also cites the 2015 Menachery paper to show the Wuhan Institute of Virology has been working on studies to make these "human-infecting viruses." This links the Chinese success in gain-of-function research directly to their collaboration with the U.S. project, funded by Fauci.

Yan et al. then go on to make the chilling observation that the Wuhan Institute now possesses "the world's largest collection of coronaviruses." The authors follow this with another critical observation that there is no longer any "technical barrier" to the "engineering" of viruses to enable them to infect humans. They are speaking from their professional experience working with Chinese colleagues in the field.

This means the Chinese now have the unlimited ability to keep manufacturing pandemic viruses. This should not be a surprise given their collaboration with the U.S. and the millions of dollars the collaboration brought to them, plus their own independent publications, and the inevitable desire of the Communist Party of China to create and stockpile biological weapons.

Additionally, Yan and her colleagues link the engineering history of SARS-CoV-2 directly to China's military. Here is one of their seven references to the military's involvement in developments leading to SARS-CoV-2:

> The genomic sequence of SARS-CoV-2 is suspiciously similar to that of a bat coronavirus discovered by military laboratories in the Third Military Medical University (Chongqing, China) and the Research Institute for Medicine of Nanjing Command (Nanjing, China).

Yan and her colleagues took great risks putting their scientific paper online and linking SARS-CoV-2 to the Chinese military. In her media interviews, many in August 2020, she has been very direct in blaming the Chinese Communist Party and its military. The headline of one interview makes Yan's view unmistakably clear: "Li-Meng Yan: Coronavirus was developed in Chinese military labs. The Chinese virologist, who claims she fled to the U.S. after receiving threats due to her research, has accused the Chinese military of creating COVID-19."[239]

A number of media sources have dismissed Yan and her articles. Some have challenged her credentials, although it appears she has an MD and a PhD.[240] Snopes.com, which considers itself the "internet's definitive fact-checking site,"[241] gathered experts to refute her. One claimed that to make a SARS-CoV pathogenic virus in a lab was "a feat that would be nearly impossible"—when it had been going on for years. Another said it would be less probable than shredding a Shakespeare sonnet, then shredding a dictionary, mixing them together, and getting back Shakespeare's sonnet. Apparently, none of them had any idea that the supposedly impossible had already been accomplished by the collaborative Chinese-American effort, funded by Fauci and documented in numerous published papers, and, indeed, described by Yan in her first paper. It is a wonder that China has sufficient resources or connections to buy the souls of so many scientists around the world.

Li-Meng Yan's research background is also overlooked by critics of her two articles about SARS-CoV-2. She is the second of nine authors on a scientific paper published by *The Lancet Infectious Diseases* in March 2020 on "Viral dynamics in mild and severe cases of COVID-19."[242] She is also second of more than a dozen authors for an article in *Nature* pertaining to the transmission of SARS-CoV-2 in animals.[243] Both of these are in top Western journals. In addition, she is a coauthor on another article concerning flu vaccines.[244] All three of these are highly technical articles related to viruses, demonstrating a significant body of expertise on the part of Dr. Yan.

DR. YAN'S SECOND REPORT DRAWS ATTENTION TO UNRESTRICTED WARFARE AGAINST THE UNITED STATES

In October 2020, Yan and her three colleagues put a second paper onto the internet, titled "SARS-CoV-2 Is an Unrestricted Bioweapon: A Truth Revealed through Uncovering a Large-Scale, Organized Scientific Fraud." [245] They concluded, "The scientific evidence and records indicate that the current pandemic is not a result of accidental release of a gain-of-function product but a planned attack using an Unrestricted Bioweapon."

The implication of the terms "unrestricted bioweapon" and "unrestricted biowarfare" at the time escaped us as it did most other analysts.

The term "unrestricted warfare" has enormous implications within the Chinese military, and Yan was specifically referring to this in her title and several times in her article. *Unrestricted Warfare* is the title and theme of a book written by two Communist Chinese military officers, Qiao Liang[246] and Wang Xiangsui. Published in 1999 by the Chinese government, it was reportedly used as a handbook by the People's Liberation Army (PLA).[247] We believe that for Yan it is a given fact that bioweapons are allowed or encouraged under the principle of unrestricted warfare and that the coverup they are describing in their two articles convinced them that the Chinese Communist Party used this as an unrestricted weapon and is trying to keep it secret.

In explaining the need for the new term, unrestricted warfare, the Chinese military authors stated:

> When we suddenly realize that all these non-war actions may be the new factors constituting future warfare, we have to come up with a new name for this new form of war: Warfare which transcends all boundaries and limits, in short: unrestricted warfare (p. 12).

The authors further explained:

> In terms of beyond-limits warfare, there is no longer any distinction between what is or is not the battlefield. Spaces in nature including the ground, the seas, the air, and outer space are battlefields, but social spaces such as the military, politics, economics, culture, and the psyche are also battlefields. And the technological space linking these two great spaces is even more so the battlefield over which all antagonists spare no effort in contending (p. 206).

On the final page of *Unrestricted Warfare*, the military authors discuss "globalization" as a new and immensely complicating factor. Every aspect of life has become a battlefield and the "chasm" between warfare and non-warfare is "nearly filled up." The answer to this complexity, they conclude, is "unrestricted warfare," much of which may be carried out in secret without a declaration of war and without ethical restrictions.

In the context of unrestricted warfare as a Chinese policy, it makes sense for Yan and her colleagues to have grasped the release of pathogens from the

Wuhan Institute, the cover-up by the Communist government, and its intentional spread of the pandemic (Chapters 1-4). It seems reasonable for them to have concluded that it was an act of China's ongoing undeclared war upon America and the Western world.

Dr. Yan's Third Report Re-Emphasizes that China's Release of SARS-CoV-2 Was an Act of War

Dr. Yan's third paper was again written with the same three coauthors.[248] Here is an excerpt from its "Opening Statement:"

> The defeat of humans by COVID-19 is for two fundamental reasons. First, SARS-CoV-2, the causative agent of COVID-19, is not a naturally occurring pathogen but an Unrestricted Bioweapon. It has designed and significantly enhanced functions and therefore could not be controlled easily using strategies that would normally work for naturally occurring pathogens. It is a product of the bioweapons program of the Chinese Communist Party (CCP) government, the network of which includes not only the CCP scientists but also certain overseas scientists and organizations. SARS-CoV-2 was created based on template viruses ZC45 and ZXC21, which were originally discovered in bats by scientists of the People's Liberation Army (PLA). The subsequent laboratory modifications had enabled its ability to infect humans as well as had enhanced the virus in its pathogenicity, transmissibility, and lethality.

> The second fundamental reason of our defeat was that the world was made to look away from the true nature of SARS-CoV-2 and therefore responded inadequately on multiple aspects and occasions. A massive misinformation campaign has been undertaken by the CCP government to cover up the true origin of SARS-CoV-2, which involved destroying data and samples, publishing fabricated viruses on top scientific journals, controlling the narrative of the origin debate through bribed top scientists and organizations, amplifying the falsified natural origin theory through media control,

labeling all other origin theories as "conspiracy theories", and defaming individuals who reveal the truth of SARS-CoV-2. As a result of the CCP's efforts here, the true, weaponized nature of SARS-CoV-2 has been obscured and was not known by most of the public.

This is the concluding statement by Dr. Yan and her colleagues:

> Ultimately, the realization of SARS-CoV-2 being an Unrestricted Bioweapon and the current pandemic an Unrestricted Biowarfare makes it clear that the CCP regime is the one responsible for this brutal attack on humanity. The CCP government must be held accountable. If not, it can only be encouraged to commit more crimes and continue to hurt the global community.

Dr. Yan's paper reviews the manipulations of information carried out against her and also against an understanding of the origin and purpose of the COVID-19 pandemic. The paper adds fascinating and valuable details to our overall description of the relationships among the global predators. She also cites confirmatory Chinese-language military literature on the use of stealth biological warfare.

Concluding Thoughts on Whether or Not the Release from Wuhan Was Intentional

We know that SARS-CoV-2 was made in China at the Wuhan Institute of Virology with technology developed in Fauci-funded collaborative efforts with the United States. We also know SARS-CoV-2 was part of the line of SARS-CoVs developed in labs at the University of North Carolina and at the Wuhan Institute with some involvement of other facilities in the U.S. and China.

We further know that soon after the release of the virus, the Chinese dictatorship began to deny the dangers associated with SARS-CoV-2 while it purposely spread the virus around the world with thousands of civilian airplane flights. At the same time, the Chinese Communists denied and continue to deny that the virus escaped from any of their research facilities, a fraudulent hoax still defended by Fauci and other global predators. Therefore, whether or not the Chinese Communists intended the release, they took advantage of it as a part of their undeclared, unrestricted warfare against the United States of America.

It seems highly probable that the release of the virus was not only purposeful, but it was in fact premeditated and planned over several years. The planning goes at least as far back as 2017, when Gates announced the equivalent of what would in 2020 be called Operation Warp Speed and the Great Reset involving corporate-government fusion.

If they did not have foreknowledge of a coming SARS-CoV pandemic, why would billionaires like Gates and global corporations like Moderna and Pfizer make such enormous high-risk investments beginning at least as long ago as late 2016 and early January 2017? Why would the Johns Hopkins Bloomberg School of Public Health in 2017 describe in book form the details of an anticipated SARS-CoV pandemic? Why would Gates boast about fast-tracking "novel" vaccines and "RNA vaccines" in January 2017, showing that he and Moderna were already working on a SARS-CoV vaccine, if he did not have strong assurances that a coronavirus pandemic was a near certainty? At the start of COVID-19, why would Anthony Fauci have gone to such great lengths to promote and rush the approval specifically of Gilead's medication Remdesivir and Moderna's mRNA vaccine—both dangerous, experimental, and wholly unproven at the time—if he and Bill Gates were not sure of the need for them? All this and more will be discussed in Chapter 15 and 16, and in the *Chronology and Overview with Pandemic Predictions and Planning Events*.

It seems certain that the Chinese Communists purposely delayed reporting accurately about COVID-19 and then spread it around the world as stealth, unrestricted warfare against the Western democratic republics and to strengthen themselves.

We cannot prove with the same certainty that the Chinese intentionally released SARS-CoV-2. Even so, as Dr. Li-Meng Yan has cogently argued, if the release were not intentional, why didn't the Communists take immediate actions to stop the virus dead in its tracks as they had done with several other SARS-CoV releases in earlier years? There is a strong probability—it is more likely than not—that the Chinese Communist regime released SARS-CoV intentionally as a covert act of war against the United States and the democratic republics of the world.

CHAPTER 5

Fauci Deceives Presidents Obama and Trump About His Deadly Research

We have seen how, for many years, Anthony Fauci funded collaborative research with China on engineering benign bat CoVs into highly infectious viruses, like SARS-CoV-2. Fauci's institute gave direct funding to these gain-of-function research projects in America and in China. He funded individual researchers and directly funded the Wuhan Institute—all toward the aim of turning benign bat coronaviruses into deadly pathogens.

The stated goal was always to find potentially dangerous viruses in nature, to upgrade them into deadly pathogens, and then to practice making vaccines for them—all to prepare the world for an inevitable outbreak from nature. The unstated goal was to contribute to American bioweapons development. We are also discovering that the unspeakable goal was to further increase the wealth and power of global predators with China as the leader among them.

President Obama Stops Fauci's Gain-of-Function Research

Fauci and his institute NIAID, as well as the umbrella organization NIH, have overcome serious obstacles to keep alive their experiments in creating deadly viruses. In 2014, President Obama declared a moratorium on research exactly like those being conducted by Menachery et al. in collaboration with China on gain-of-function research. At the time, the Menachery studies, which were actively moving along, should have come to a halt. Here is the opening of the October 17, 2014, declaration from the "White House: President Barack Obama:"[249]

Doing Diligence to Assess the Risks and Benefits of Life Sciences Gain-of-Function Research

Summary:

The White House Office of Science and Technology Policy and Department of Health and Human Services today announced that the U.S. Government is launching a deliberative process to assess the potential risks and benefits associated with a subset of life sciences research known as "gain-of-function" studies.

Following recent biosafety incidents at federal research facilities, the U.S. Government has taken a number of steps to promote and enhance the Nation's biosafety and biosecurity, including immediate and longer-term measures to review activities specifically related to the storage and handling of infectious agents. …

Because the deliberative process launching today will aim to address key questions about the risks and benefits of gain-of-function studies, during the period of deliberation, the U.S. Government will institute a pause on funding for any new studies that include certain gain-of-function experiments involving influenza, SARS, and MERS viruses. Specifically, the funding pause will apply to gain-of-function research projects that may be reasonably anticipated to confer attributes to influenza, MERS, or SARS viruses such that the virus would have enhanced pathogenicity and/or transmissibility in mammals via the respiratory route.

This description exactly fits the collaborative studies between U.S. researchers and Chinese scientists from the Wuhan Institute, who were only months away from publishing a scientific paper on the creation of a gain-of-function virus by engineering a bat virus to make it virulent and able to infect humans. *No other research has surfaced which so perfectly fits what President Obama was trying to stop—but Fauci would avoid stopping it!*

The Obama government's description of the moratorium continues:

> During this pause, the U.S. Government will not fund any new projects involving these experiments and encourages those currently conducting this type of work—whether federally funded or not— to voluntarily pause their research while risks and benefits are being reassessed.

This White House order could not be clearer. During the time when the government was investigating the risks of gain-of-function research, it would not start funding any new projects, and it asked all ongoing projects to "voluntarily pause their research while the risks and benefits are being assessed."

FAUCI SIMPLY IGNORES
PRESIDENT OBAMA'S MORATORIUM

As brazen, unconscionable, and inexplicable as it seems, Fauci never stopped funding the single most important and most dangerous gain-of-function research addressed by Obama's moratorium.

From 2014 through to the present, Fauci's NIAID has continued funding the work of Menachery under Baric at the University of North Carolina, Chapel Hill—the research that led directly to SARS-CoV-2. The available figures span 2014 to an ongoing grant in 2019 with annual awards varying between $581,646 and $666,442.

Perhaps to give a lower profile to the grant, the money was passed through Peter Daszak's EcoHealth Alliance.[250] Under "Other Information," it states that the project start date was June 1, 2014, shortly before the moratorium. The current budget is slated to end June 30, 2021, but the project ends June 30, 2025.

We are unaware of any major media making it known that Fauci flat out defied President Obama's moratorium.

Fauci Also Avoids Obama Moratorium by Outsourcing to China

A number of small, critical media described Fauci as outsourcing his gain-of-function ambitions to China during the controversy over banning it in the U.S. However, as described in Chapter 2, he was *continuing* to directly fund a project with China he had begun before the moratorium (Chapter 2). In addition, he more brazenly continued to fund the Chinese collaboration through American grants to the University of North Carolina (Chapters 1 and 2). Nonetheless, it is important to see that some media outlets—but not the mainstream media—recognized the controversy and even the threat contained in funding research with China that provided the basic science for biological warfare.

The *Asia Times*[251] published an analysis with this disturbing headline, "Why U.S. outsourced bat virus research to Wuhan," followed by the subhead, "US-funded $3.7 million project approved by President Trump's COVID-19 guru Dr. Anthony Fauci in 2015 after U.S. ban imposed on 'monster-germ' research."

In April 2020, as COVID-19 spread, the British newspaper *Daily Mail Online* quoted a U.S. lawmaker's outrage over directly funding the Wuhan Institute:[252]

> US Congressman Matt Gaetz said: "I'm disgusted to learn that for years the U.S. government has been funding dangerous and cruel animal experiments at the Wuhan Institute, which may have contributed to the global spread of coronavirus, and research at other labs in China that have virtually no oversight from U.S. authorities."

An analysis[253] published in May 2020 by M. Dowling in the *Independent Sentinel* again described Fauci as outsourcing gain-of-function research to China. It helped break ground as we had done a month earlier, by pointing to the danger of giving biological weapons to China:

TRUSTING MAOISTS

Dependency on Communists, trusting Communists, what could possibly go wrong?

President Trump's administration is investigating the $3.7 million in tax dollars that went to the Wuhan lab, and Matt Gaetz called for an immediate end to NIH funding of Chinese research. Whether anything will come of it is questionable.

The ban on GOF [gain-of-function] research in the USA has been lifted. Maybe the USA shouldn't do it either. When mankind plays with nature, it usually doesn't go well.

Unfortunately, our press doesn't investigate or even ask pertinent questions. Some reporters are just too stupid or biased to bother.

Imagine if President Trump said our CDC is incompetent so I will pay Russia to do our GOF research?

When the original research paper was published in 2015, it did not go unnoticed. The inherent dangers in creating new human coronavirus pathogens in the Menachery research were discussed in a commentary by Jef Akst in *The Scientist* on November 16, 2015, along with NIH's decision to allow the research to continue.[254] Unfortunately, as it so often happens, the danger of the Chinese collaboration went unmentioned! Instead, there is an addendum added to the original report trying to dismiss any such association or concern:

> Update (March 11, 2020): On social media and news outlets, a theory has circulated that the coronavirus at the root of the COVID-19 outbreak originated in a research lab. Scientists say there is no evidence that the SARS-CoV-2 virus escaped from a lab.

Notice that in the update, the Wuhan Institute or China is not even mentioned. Instead, the idea of an escape from a lab is dismissed without a hint about the potential offender. Reports from *Nature*, *The Scientist*, and *Scientific American*, among many others, confirm the progressive media and the scientific community were desperately trying to avoid throwing suspicion on the Chinese Communists for any role in COVID-19, including avoiding any mention that the U.S. government was collaborating with the Wuhan Institute in turning routine bat viruses into pathogens, deadly to humans, thus directly helping the Chinese Communist Party develop bioweapons.

Here we see the influence of globalism. People knew each other, people made money from each other, science trumped national security, and any kind of funded collaborative research with China became untouchable and beyond criticism.

If asked, some of the individual scientists might have said "science" is pure and should be shared among competing and even hostile nations for the sake of science and peace, but that innocence is not what rules globalism. The prevailing attitude among globalists seems to be: Never put America First. Put our global friends and interests ahead of everything.

Of all the technologies we have given to China, how to make highly infectious and lethal viruses from bat viruses may be the most dangerous. Incredibly, however, a nearly total blackout on U.S. funding for China building biological weapons was displayed by the media, science commentators, and politicians. This confirmed the pervasiveness of the globalist viewpoint that has no special interest in protecting American interests or America's survival, and perhaps not even in the world's survival, while fortunes are being made and power is being accrued.

Globalists, when using science to justify totalitarian control, talk about "science" as if it were a universal spirit or god. Since science is a creation of human beings, it is neither perfect nor pure, but always depends on the human source with all the biases and corruptions and, yes, idealism that humans live by. [255]

FAUCI'S CONTINUED FUNDING THE MOST DANGEROUS GAIN-OF-FUNCTION RESEARCH

In their 2015 publication, Menachery et al. acknowledged the existence of Obama's moratorium on gain-of-function studies but expressed the belief that it did not necessarily apply to them because they did not initially anticipate they could have succeeded in creating a virulent virus! However, a November 9,

2015, interview[256] given while Menachery's article was on the way to publication, indicates he had been stopped, temporarily at least:

> He [Menachery] and his co-authors noted they had to stop some of their work because of U.S. government policies. The U.S. has a moratorium on so-called gain-of-function research, which includes some research that enhances the ability of a pathogen such as a virus to infect people or spread among them.

We found no other evidence that they slowed down their research. They published their results shortly thereafter in December 2015 and again in 2016 without indicating any further delays.

By the end of 2015, NIH had already granted an exception to the dangerous Menachery study, allowing it to continue. According to *Nature*: [257]

> The latest study was already underway before the U.S. moratorium began, and the U.S. National Institutes of Health (NIH) allowed it to proceed while it was under review by the agency, says Ralph Baric, an infectious-disease researcher at the University of North Carolina at Chapel Hill, and a co-author of the study. **The NIH eventually concluded that the work was not so risky as to fall under the moratorium, he says.** (Bold added.)

It is most remarkable that NIH and Fauci decided that Fauci's highly-prized project—the epitome of a dangerous gain-of-function study—was not dangerous enough to be stopped under President Obama's edict. It was, in fact, the exact classification of gain-of-function research that Obama intended to stop and the scientists worldwide had protested—and it resulted in China creating a group of deadly coronaviruses of which SARS-CoV-2 was a member. The result of Fauci and NIH's arrogant self-importance was a continuation of the research leading to SARS-CoV-2 and the COVID-19 pandemic.

Fauci Subverted New Controls at NIH in 2017

More recently, a retrospective report by Andrew Kerr in the *Daily Caller*[258] shed new light on Fauci's defiance of President Obama's intentions and of federal regulations. In 2017, the Department of Health and Human Services (DHS) formed a review board called the Potential Pandemic Control and Oversight (P3PC) specifically for the purpose of monitoring gain-of-function research. In effect, this meant monitoring Anthony Fauci, so Fauci simply failed to send his gain-of-function grants to the Wuhan Institute through the committee. Fauci treated his funding of the Wuhan Institute as if it were his own fiefdom, untouchable by presidents of the United States or any special oversight committee.[259] NIH and Fauci denied that the research at the Wuhan Institute pertained to gain-of-function, but the fiscal year 2019 abstract for the project included the following gain-of-function activities: "We will use S protein sequence data, infectious clone technology, *in vitro* and *in vivo* infection experiments and analysis of receptor binding …" That is a detailed description of gain-of-function research.

Fauci Awards Huge Grants for Dangerous International and Collaborative Viral Research

We move ahead to 2017 and the end of President Trump's first year as president. Up to this point in time, Fauci had simply been ignoring President Obama's moratorium, leading to the U.S. aiding China in its development of SARS-CoV-2.

On December 19, 2017, NIH announced it was lifting the ban set by President Obama on gain-of-function research.[260,261] Unlike Obama's ban, which came from the White House on the President's official stationery, we have thus far found no indication President Trump was involved in Fauci's decision making.

The New York Times commented, "There has been a long, fierce debate about projects — known as gain-of-function research—intended to make pathogens more deadly or more transmissible."[262] *Fauci simply did an end run on the dissent, ignoring it, and quietly lifting the ban—a ban he had already been ignoring since its onset.*

Ignoring the China Collaboration

In the various establishment analyses, we have read about the controversy surrounding gain-of-function research, including one in *The Lancet* in early 2018,[263] that there is no hint or complaint about the studies involving collaboration with the Chinese Communists and their military. Even more striking, no mention is made of the most important gain-of-function studies of all, the American-Chinese collaboration by Menachery, ongoing since 2014 and published in 2015 and 2016. A *New York Times* article on the same subject in December 2017[264] mistakenly claimed that under Obama's ban, all research was halted, including research on SARS, when, in fact, the SARS research by Menachery et al. was being funded, and the key scientific articles had already been published. It then mentions exceptions that were allowed to proceed, but none are SARS related. Fauci and the globalists had placed an invisible curtain over this extremely dangerous research that eventually helped the Chinese to engineer SARS-CoV-2. The invisible curtain is controlled by the global predators who were looking forward to a disaster to make even greater fortunes and to enhance their power through the Great Reset of corporate-government fusion.

The *Times* article does warn against publishing dangerous information, comparing it to the risk of publishing atomic secrets. It fails to mention we were actually *collaborating with China*, creating a much greater risk than merely publishing information about research. It is no exaggeration to say collaborating with China on building virulent, epidemic viruses was at least as dangerous as collaborating with them or the Russians on building atomic weapons. Indeed, some experts have opined that spreading pestilences is a far greater threat now, and in the future, than nuclear warfare.[265]

Fauci Increases His Spending on Making Pathogens

In August 2020, Fauci began a funding spree for gain-of function research. Maria Godoy from NPR reported Fauci's *fait accompli* with the headline, "Group Whose NIH Grant for Virus Research Was Revoked Just Got a New Grant." Her report summarized:

The National Institutes of Health has awarded a grant worth
$7.5 million over five years to EcoHealth Alliance, a U.S.-based
nonprofit that hunts emerging viruses. The award comes months
after NIH revoked an earlier grant to EcoHealth, a move scientists
widely decried as the politically motivated quashing of research
vital to preventing the next coronavirus pandemic.

EcoHealth Alliance is one of 11 institutions and research teams receiving
grants from NIH, announced August 27, 2020, to establish the Centers for
Research in Emerging Infectious Diseases. The global network will monitor
pathogens emerging in wildlife and study how and where they go on to infect
humans.

Peter Daszak told NPR none of the new money was going to China. Daszak
also claimed China's ability to do gain-of-function research had been stopped by
President Trump's withdrawal of funding from it, an extremely unlikely outcome.
While we believe Fauci and the U.S. enabled China to further accelerate its
efforts, we have no illusion that the Chinese Communists cannot now proceed
without U.S. support. Instead, we believe Dr. Yan's insider assessment that the
Chinese no longer have any impediments to making pathogenic coronaviruses.

Daszak further confirmed how his Fauci funding will be used for the
controversial gain-of-function research originally stopped by President Obama.
He said:

The next step in that research is to sequence the whole genome
of those viruses and say, could they bind to human cells? Does
this look like a virus that could potentially emerge?

How do researchers like Daszak determine if a virus found in nature can
become a pathogen, i.e., "bind to human cells?" The laboratory scientists engineer
it into a pathogen and use their success to claim it could also emerge naturally
from nature—a conclusion which makes no sense. Engineering a benign virus
into a lethal one is a complex, time-consuming, highly technical process, thereby
making an accidental change of that sort in nature extremely unlikely.

Here is how it is done. First, the genome of the virus is mapped. Then an
attempt is made to see if the virus can be engineered to invade human cells.
This is not determined by simply eyeballing the virus. In numerous research
studies carried out in the U.S. and China—many funded by NIAID and other

U.S. agencies—researchers physically modify the harmless coronavirus in the lab to make a SARS-CoV.

The process involves creating a new lab-created spike protein and attaching it to the surface of the coronavirus. If that works, the new spike protein enables the virus to gain entry into mouse and human lung cells in lab preparations. At the same time, living mice are genetically and biologically modified with a human receptor that will enable the potential SARS-CoV to enter their cells. When the researchers succeed in infecting human lung epithelial cell preparations *and* making the living mice sick with a SARS-CoV illness, they believe they have made coronaviruses capable of producing a Severe Acute Respiratory Syndrome (SARS) in humans. Then they absurdly claim that natural evolution is likely to do the same thing and use that claim to scare us into giving them further amounts of our taxes to support their dangerous Frankenstein-like activities.

Fauci Currently Funds Americans Who Are Closely Affiliated with China

Days after the *PRN* article announcing Fauci's massive new funding of viral research, on September 6, 2020, Col. Lawrence Sellin (Ret.) of the Citizens Commission on National Security warned about the overall NIAID grant. It totaled $82 million, including many institutions in addition to EcoHealth Alliance. Sellin cited virology research at the University of Texas as specifically involving multiple Chinese-connected researchers.[266] As a retired military officer concerned with security, Col. Sellin did not deem this a good idea.

The grant, which was officially announced by NIAID on August 27, 2020,[267] did indeed fund Peter Daszak of the EcoHealth Alliance, Inc. for its Emerging Infectious Diseases-South East Asia Research *Collaboration* Hub. This new grant strongly suggests that Fauci is continuing to fund collaboration with China (or some of her neighbors) on gain-of-function research. If not directly funded, China could gain information through other Southeastern Asia nations that are neighbors and depend on the goodwill of the Communist giant. Or, as we will now see, the Chinese Communist Party could benefit from this research through Chinese scientists already embedded in American universities and laboratories, but beholden to the Party.

Sellin drew upon his military intelligence background to make observations on

Fauci's funding of the Chinese military. With his permission, here is the entire text of his blog, with the links included as endnotes:[268]

DID FAUCI'S NIH INSTITUTE FINANCIALLY
ASSIST CHINA'S MILITARY?
by Col. Lawrence Sellin (Ret.) September 6, 2020

A disturbing pattern of cooperation between Dr. Anthony Fauci's NIAID and the Chinese military raises questions about technology transfer and the origins of the current COVID-19 pandemic.

U.S. patent number 8933106[269] entitled "2-(4-substituted phenylamino) polysubstituted pyridine compounds as inhibitors of non-nucleoside HIV reverse transcriptase, preparation methods and uses thereof" is assigned to the Institute of Pharmacology and Toxicology, Academy of Military Medical Sciences of China's People's Liberation Army.

One of the inventors of that patent, Shibo Jiang, is a graduate of the First and Fourth Medical University of the People's Liberation Army, Xi'an, China. He is a long-time collaborator with institutions associated with the Chinese military and, since 1997, a recipient of U.S. government research grants from the National Institute of Allergy and Infectious Diseases (NIAID), headed by Dr. Anthony Fauci.

In one of the two scientific references used to support the above-mentioned patent "Discovery of diarylpyridine derivatives as novel non-nucleoside HIV-1 reverse transcriptase inhibitors," Shibo Jiang is listed as a co-author, along with the four other inventors on the patent.

In the Acknowledgments section of that scientific publication, which supports the patent application, three separate NIAID grants are cited, two of which, AI46221 and AI33066, were awarded to co-inventors on the patent, Shibo Jiang and Kuo-Hsiung Lee, respectively.

Shibo Jiang and Kuo-Hsiung Lee are co-inventors on another U.S. patent, 8309602,[270] also assigned to the Institute of Pharmacology and Toxicology, Academy of Military Medical Sciences of China's People's Liberation Army.

Although no scientific publications are listed in the 8309602 patent, you can compare the chemical compounds with those in "Diarylaniline Derivatives as a Distinct Class of HIV-1 Nonnucleoside Reverse Transcriptase Inhibitors," which has as co-authors all the co-inventors of the patent.

That research was also supported by three separate NIAID grants, two of which, AI46221 and AI33066, were awarded to co-inventors on the patent, Shibo Jiang and Kuo-Hsiung Lee, respectively.

NIAID funding of China's military research programs does not appear to be restricted to those two patents.

Since 2004, Shibo Jiang has had scientific collaboration with Yusen Zhou, who was a professor at the State Key Laboratory of Pathogen and Biosecurity, Institute of Microbiology and Epidemiology, Academy of Military Medical Sciences in Beijing.

It is unclear whether Yusen Zhou also received his education at one of China's military medical universities, but his early scientific work was associated with the Department of Infectious Disease, 81st Hospital[271] of the People's Liberation Army, Nanjing Military Command and the Fourth Medical University of People's Liberation Army, Xi'an, Shibo Jiang's alma mater.

Shibo Jiang and Yusen Zhou are listed as co-inventors on at least eight U.S. patents, the references[272] supporting those patents, for example, 9889194,[273] was research funded[274] by NIAID.

Until his recent death, Yusen Zhou's collaboration with Shibo Jiang continued into the COVID-19 pandemic, publishing a July

30, 2020 *Science* article[275] together with institutions associated with China's military.

In a 2014 article,[276] Shibo Jiang was working with the Institute of Biotechnology, Academy of Military Medical Sciences, Beijing.

In 2017, he conducted[277] research with the Translational Medicine Center, People's Liberation Army Hospital No. 454 and the Department of Epidemiology, Medicinal Research Institute, Nanjing Military Command.

Between 2012 and 2020, Shibo Jiang has published twelve scientific articles with the Wuhan Institute of Virology and eleven articles between 2013 and 2020 with the University of Texas Medical Branch (UTMB), Galveston, Texas.

The UTMB has been designated one of the ten Centers for Research in Emerging Infectious Diseases[278] newly funded by a NIAID grant totaling $82 million. UTMB has at least two permanent faculty members trained at China's Military Medical Universities, has had connections to or former employees from the Wuhan Institute of Virology and Yusen Zhou's State Key Laboratory of Pathogen and Biosecurity, Institute of Microbiology and Epidemiology, Academy of Military Medical Sciences in Beijing, as well as other Chinese institutions.

Another new center is the EcoHealth Alliance, a long-time collaborator with the Wuhan Institute of Virology, which has been awarded $7.5 million.

Given the history described above and before any new funding is allocated, an investigation and auditing of previous NIAID grants should be undertaken to determine exactly how much U.S. taxpayer money has benefitted China's military.

Chinese Scientists Are Required to Steal U.S. Biomedical Research

Even before we began further investigation of the University of Texas, we found widespread concern about scientists with foreign connections, mostly Chinese, stealing biomedical research secrets and materials during their work and training in the U.S. Writing for *The New York Times* in November 2019 in an article headlined "Vast Dragnet Targets Theft of Biomedical Secrets for China," Gina Kolata reported:[279]

> The N.I.H. and the F.B.I. have begun a vast effort to root out scientists who they say are stealing biomedical research for other countries from institutions across the United States. Almost all the incidents they uncovered and that are under investigation involve scientists of Chinese descent, including naturalized American citizens, allegedly stealing for China.

> Seventy-one institutions, including many of the most prestigious medical schools in the United States, are now investigating 180 individual cases involving potential theft of intellectual property. The cases began after the N.I.H., prompted by information provided by the F.B.I., sent 18,000 letters last year urging administrators who oversee government grants to be vigilant.

There is at least one reported incident of a Chinese researcher caught smuggling SARS in a vial into the country.[280] This and numerous other dangerous activities raise serious questions about the security of any of Fauci's grants to U.S. institutions. With that in mind, we looked into the University of Texas and its Galveston National Lab to see if it might be especially vulnerable to leaks or to theft.

The University of Texas Medical Branch and China's Wuhan Institute

We located an April 15, 2020, University of Texas Medical Branch press release with the arresting title, "The Galveston National Lab and Wuhan Institute of Virology."[281] It was disturbing to see such a proud headline pairing the American lab with its Chinese equivalent. It was consistent with Fauci's disregard for American security interests to continue gain-of-function research at the Galveston National Lab (GNL).

Here are excerpts from the Galveston National Lab's press release boasting about its relationship with the Wuhan Institute:

> **April 16, 2020**—The Galveston National Laboratory, located on the campus of the University of Texas Medical Branch, is one of two university-based maximum containment (BSL-4) laboratories in the U.S. focused on the study of highly infectious diseases and the development of medical countermeasures. … The lab is part of National Institute of Allergy and Infectious Diseases (NIAID) Biodefense Laboratory Network … Through our Biosafety Training Center, **UTMB has provided laboratory safety and security training for scientists and operations personnel in more than 45 countries, including China.** The relationship with the Wuhan Institute of Virology and the GNL dates back to 2013 and has been facilitated through an ongoing dialogue co-sponsored by the Chinese Academies of Science and U.S. National Academies of Science, Engineering and Medicine, with cooperation from the Chinese CDC and others.
>
> In recent years, we have provided training to scientists, biosafety and engineering professionals, including many from China. [Bold appears in the original.]

On further research, it turns out the Galveston National Lab at UTMB, the recipient of new funding by Fauci, has been an object of special federal concern because of its many ties to China. For example, a detailed Fox News report headlines: "Prominent university bio lab urged to reveal extent of relationship with Wuhan lab at center of coronavirus outbreak."[282]

As Fauci right now pumps money into research on creating deadly viruses in America, at the Galveston National Lab and lesser programs, how hard will it be to determine where the money ends up? How much information related to our national security will continue to flow to China? So far, it looks as if it will be nearly impossible to know where the research money will be going—and that is something to worry about.

Here is an excerpt from an April 24, 2020, letter from the General Counsel of the U.S. Department of Education insisting on more information from UTMB and its Galveston National Lab about its complex ties to China and specifically to its Wuhan Institute of Virology:[283]

> Between June 6, 2014, and June 3, 2019, UT reported approximately twenty-four contracts with various Chinese state-owned universities and ten contracts with Huawei Technologies, all purportedly worth a reported total of $12,987,896. It is not clear, however, whether UT has in fact reported all gifts from or contracts with or relating to the Wuhan MCL, *the Wuhan Institute of Virology*, and/or all other foreign sources, including agents and instrumentalities of the government of the *People's Republic of China.* Therefore, to verify UT's compliance with Section 117, the Department requests that your Institution produce the following records ... (Italics added.)

The letter from the General Counsel of the U.S. Department of Education to the University of Texas also demands information about almost two dozen suspected Chinese businesses, universities, and other entities. The list concludes with this ominous demand for information about the University, its Galveston National Lab, *and its relationship to the Communist Party of China:*

> The Communist Party of China, its agents, employees, representatives, and instrumentalities (including but not limited to the agents, employees, representatives, and instrumentalities of entities such as the Communist Party of China's Central Committee, Central Office, and Politburo Standing Committee; the General Office of the Central Military Commission; the Chinese Ministry of Education; the Chinese Ministry of Science and Technology; the People's Liberation Army; the Chinese Ministry of State Security; the Chinese Ministry of Industry and Information Technology; the Chinese Ministry of Foreign Affairs; the Chinese Ministry of National Defense; the Central Bank of the People's Republic of China; and any People's Republic of China province, autonomous region, or municipality).

Sellin was right to focus on new Fauci funding for the Galveston National Lab with its multiple, probably inextricable, and, at times, obscured ties to China. If Fauci wants to keep his much valued and carefully cultivated relationship with the Chinese Communist Party, then he may have picked the right place in America to award funding for this viral research that doubles as preparation for biological warfare.

At this time, we suspect it is impossible to fund virus-related research at American universities and facilities while guaranteeing the Chinese government will not be gaining information relevant to our national security and even to humanity's survival.

The U.S. Is Not Alone in Empowering China's Bioweapons Program

The seemingly cavalier empowering of China's biological warfare program is not limited to the United States. Globalist influences may be equally or more powerful in Canada. In the summer of 2019, a scandal erupted when it was discovered Canadian scientists had shipped samples of Ebola[284] to China, justifying it as part of Canada's "efforts to support public-health research worldwide. Sharing of such samples internationally is relatively standard practice." The report went on to cite warnings from experts around the world, including one from the U.S.:

> James Giordano, a neurology professor at Georgetown University and senior fellow in biowarfare at the U.S. Special Operations Command, said it's worrisome on a few fronts.

> China's growing investment in bioscience, looser ethics around gene-editing and other cutting-edge technology and integration between government and academia raise the spectre of such pathogens being weaponized, he said.

> That could mean an offensive agent, or a modified germ let loose by proxies, for which only China has the treatment or vaccine, said Giordano, co-head of Georgetown's Brain Science and Global Law and Policy Program.

The Chinese Communist Party must think we Westerners are a pack of naive fools ... or traitors.

FAUCI AND NIAID WERE CENTRAL TO THE CHAIN OF ACTIVITIES THAT LED TO SARS-CoV-2

As we have noted, Gates and other globalists in the pharmaceutical industry are hugely invested in vaccines, and Fauci has opened opportunities for fast tracking. [285] In the case of remdesivir, which Fauci promised would become the "standard" for treatment of hospitalized COVID-19 patients, we shall find he funded and then manipulated a dangerous, ineffective drug into being accepted—for a time, at least—as the first-line treatment.

It is telling that Fauci was so key to enabling the Chinese to make SARS-CoV-2, and then became the international management czar for the worldwide affliction he helped to create. Now, he wants to lead us down the road of worldwide public health totalitarianism to save us from "human-made" environmental destruction that allegedly encourages the emergence of pathogens from nature.[286] Fauci is making the most of COVID-19 to enhance the wealth of his institute, his personal glory, and his already immense influence.

WHAT WERE THE AMERICAN COLLABORATORS THINKING?

We have already suggested that the Chinese Communists must think we Westerners are a pack of naive fools. Or maybe they realize many of our leaders are bedazzled by ambitions for wealth, self-aggrandizement and power, rendering them unable to identify China as a dangerous foe. Or perhaps progressives in the West have been so indoctrinated with Marxist principles they are blind to the danger of Communism. Whatever the reasons of those who participate— they are scientists, billionaires, and the CEOs of giant corporations, globalists want to get into the action with China, while China prepares to take over and to eat them alive at its earliest opportunity. No Communist regime has ever tolerated independent opportunists in their midst.

CHAPTER 6

Fauci's Grandiose
Utopian Ambitions

I n a recent "scientific" article in *Cell* that Fauci authored with one of his assistants, he lied, claiming without reservation or qualification that COVID-19 emerged from nature on its own and not from laboratory tinkering.[287] Then he did more than ignore his own role in funding the engineering of coronaviruses with China: he blamed us—you, me, and humanity—for causing the virus by disrupting nature:

SUMMARY AND CONCLUSIONS

SARS-CoV-2 is a deadly addition to the long list of microbial threats to the human species. It forces us to adapt, react, and reconsider the nature of our relationship to the natural world. Emerging and re-emerging infectious diseases are epiphenomena of human existence and our interactions with each other, and with nature. As human societies grow in size and complexity, we create an endless variety of opportunities for genetically unstable infectious agents to emerge into the unfilled ecologic niches we continue to create. There is nothing new about this situation, except that we now live in a human-dominated world in which our increasingly extreme alterations of the environment induce increasingly extreme backlashes from nature.

Science will surely bring us many lifesaving drugs, vaccines, and diagnostics; however, there is no reason to think that these alone can overcome the threat of ever more frequent and deadly emergences of infectious diseases. Evidence suggests that SARS, MERS, and COVID-19 are only the latest examples of a deadly barrage of coming coronavirus and other emergences. The COVID-19 pandemic is yet another reminder, added to the rapidly growing archive of historical reminders, that in a hu-man-dominated world, in which our human activities represent aggressive, damaging, and unbalanced interactions with nature, we will increasingly provoke new disease emergences. We remain at risk for the foreseeable future. COVID-19 is among the most vivid wake-up calls in over a century. It should force us to begin to think in earnest and collectively about living in more thoughtful and creative harmony with nature, even as we plan for nature's inevitable, and always unexpected, surprises.

Fauci declares that COVID-19 is the result of the "human-dominated" world in which we live, and he promotes an extreme progressive ideology that massive changes must be made in how we relate to nature, including how we build—or do not build—our economies. He wants a vast, progressive political program to evaluate and change human activity on a global basis:

Disease emergence reflects dynamic balances and imbalances, within complex globally distributed ecosystems comprising humans, animals, pathogens, and the environment. Understanding these variables is a necessary step in controlling future devastating disease emergences.

Fauci blames humanity, this "human-dominated" environment, for causing COVID-19, when he is the single American who contributed most directly to the development of potentially epidemic coronaviruses that he put into the hands of the Chinese Communist Party and its military. He, above all politicians masquerading as scientists, knew there had been one questionable emergence from nature of pathogenic SAR-CoVs, compared to seven known accidental or purposeful releases from Chinese labs. He is also among the men most to benefit from the catastrophe through the growth of his institutional domain and his close relationships to Bill Gates and the pharmaceutical industry.

With his own nefarious activities in funding Chinese and American laboratories, Fauci's stated position may be the most colossal misdirection in history: the American most responsible for enabling the Chinese Communists to engineer SARS-CoV-2 in their Wuhan Institute is blaming COVID-19 on humanity's indiscretions in nature. Working with China, Fauci himself has funded and promoted taking viruses out of nature and engineering them to become pandemic viruses, but he wants us to take his advice on transforming widespread human activity to make us less disruptive! Fauci is the Great Disruptor of Nature, whose work enabled China to unleash COVID-19 on the world.

Fauci is announcing himself as radical totalitarian with his new political vision:

> Living in greater harmony with nature will require changes in human behavior as well as other radical changes that may take decades to achieve: rebuilding the infrastructures of human existence, from cities to homes to workplaces, to water and sewer systems, to recreational and gatherings venues. In such a transformation we will need to prioritize changes in those human behaviors that constitute risks for the emergence of infectious diseases. Chief among them are reducing crowding at home, work, and in public places as well as minimizing environmental perturbations such as deforestation, intense urbanization, and intensive animal farming. Equally important are ending global poverty, improving sanitation and hygiene, and reducing unsafe exposure to animals, so that humans and potential human pathogens have limited opportunities for contact. … Since we cannot return to ancient times, can we at least use lessons from those times to bend modernity in a safer direction? These are questions to be answered by all societies and their leaders, philosophers, builders, and thinkers and those involved in appreciating and influencing the environmental determinants of human health.

In Fauci's world, concerns such as democracy, the U.S. Constitution and Bill of Rights, or liberty as a primary personal and political principle simply do not exist. Indeed, something even more basic, the importance of love and human relationship, seems beyond his concern or understanding.

Jeffrey A. Tucker of the American Institute for Economic Research (AIER) finds Fauci's vision as dangerous and appalling as we do.[288] In a report titled "Lockdown: The New Totalitarianism," he characterizes Fauci's philosophy and the lockdowns very eloquently:[289]

> This is sheer fanaticism, a kind of insanity wrought by a wild vision of a one-dimensional world in which the whole of life is organized around disease avoidance. And there is an additional presumption here that our bodies (via the immune system) have not evolved alongside viruses for a million years. No recognition of that reality. Instead the sole goal is to make "social distancing" the national credo. Let us speak more plainly: what this really means is forced human separation. It means the dismantlement of markets, cities, in-person sports events, and the end of your right to move around freely. ...
>
> The lockdowns are looking less like a gigantic error and more like the unfolding of a fanatical political ideology and policy experiment that attacks core postulates of civilization at their very root. It's time we take it seriously and combat it with the same fervor with which a free people resisted all the other evil ideologies that sought to strip humanity of dignity and replace freedom with the terrifying dreams of intellectuals and their government sock puppets.

FAUCI ADMITS TO SYSTEMATICALLY MISREPRESENTING DATA TO MANIPULATE PUBLIC

We have discussed how public health officials openly discuss among themselves the use of fear to have people do what they want them to do; however, until Fauci, we had not heard anyone advocate lying. *The New York Times*[290] recently obtained an admission from Anthony Fauci that for many months he has been systematically raising the estimates of the number of vaccinations needed in America to produce herd immunity because he knew people were feeling negatively toward vaccinations and did not want to frighten them off:

> Recently, a figure to whom millions of Americans look for guidance—Dr. Anthony S. Fauci, an adviser to both the Trump administration and the incoming Biden administration—has begun incrementally raising his herd-immunity estimate.

> In the pandemic's early days, Dr. Fauci tended to cite the same 60-to-70 percent estimate that most experts did. About a month ago, he began saying "70, 75 percent" in television interviews. And last week, in an interview with CNBC News, he said "75, 80, 85 percent" and "75-to-80-plus percent."

> In a telephone interview the next day, Dr. Fauci acknowledged that he had slowly but deliberately been moving the goal posts. He is doing so, he said, partly based on new science, and partly on his gut feeling that the country is finally ready to hear what he really thinks.

> Hard as it may be to hear, he said, he believes that it may take close to 90-percent immunity to bring the virus to a halt …

> Dr. Fauci said that weeks ago, he had hesitated to publicly raise his estimate because many Americans seemed hesitant about vaccines, which they would need to accept almost universally in order for the country to achieve herd immunity.

The New York Times did not call this lying, but several conservative news groups did. Fauci has probably been manipulating everything he says to have one or another political effect. Consider this January 24, 2020, CNN report:[291]

> Fauci … said that he feels China is being more transparent with the world this time than it was during previous outbreaks, such as the 2003 SARS outbreak. "This time around from my perception they look like they're being quite transparent," he said. Fauci said that China's decision to shut down travel may not have a huge

impact on containing the spread, but "it's their judgment that this is something that in fact is going to help. That's something that I don't think we could possibly do in the United States, I can't imagine shutting down New York or Los Angeles. Whether or not it does or does not is really open to question because historically when you shut things down it doesn't have a major effect."

This one short commentary is filled with distortions or lies. First, by January 24, 2020, it was very clear China was not being transparent about many things, such as the initial infectiousness of the virus, its origin in its Wuhan Institute and not a marketplace, and flooding America and the world with potentially infectious passengers from Wuhan, Hong Kong, and elsewhere while it was shutting down domestic air travel. Second, he doubts shutting down air travel will stop the epidemic from spreading, a position he later shifted. This is the man who would eventually try to shut down everything, including traveling during Thanksgiving, Christmas, and New Year's Day. Third, he says he cannot imagine shutting down New York or Los Angeles because "historically when you shut things down it doesn't have a major effect."

TIME FOR US TO ACT

America's leadership and the American people need to know that Anthony Fauci—working in the service of global interests other than the United States—funded research that eventually unleashed COVID-19 upon the world. In addition, this same Fauci funding has enabled China to possess the largest store of coronaviruses in the world, along with the technology to continue turning them into human-infecting agents. Meanwhile, despite its obvious dangers, Fauci continues to fund gain-of-function research that creates deadly viruses which can leak from labs or be released as biological weapons. Under a Biden administration, Fauci will thrive, and research collaboration with China on creating deadly pandemic viruses is likely to increase, and the world … the world will face disastrous scenarios beyond the imagination.

Unfortunately, our research into Fauci's role in the pandemic disclosed he is not nearly the only one collaborating with China. There is a network we are calling the global predators. We can go directly from Fauci to Bill Gates to begin tracking them down.

CHAPTER 7

Fauci, Bill Gates, *Forbes,* the Koch Brothers— and Their Victims

Many of us are beginning to realize COVID-19 is not the catastrophe we must resist. The worldwide globalist leadership wants us to believe we must unite against the virus, but it is far more important to unite against them. We must oppose the ongoing totalitarian measures and their plans for a "New Normal" and a "Global Reset" that come with an acute escalation on America's liberty as the combine of progressive and capitalistic predators exert their power to crush freedom of speech.

THEIR VICTIMS

We must not forget the suffering caused by global predators during COVID-19:

- People who have died from the government interfering with their access to early, effective treatments and being isolated from their bereaved families.

- Health professionals, especially frontline doctors, who have been persecuted and prohibited from providing their patients lifesaving treatments and whose careers and professional status have been harmed for speaking out against these gross interferences with doctor-patient relationships.

- Individuals who became sicker or died because of delayed treatments for cancer and other disorders when hospitals and clinics were closed to them.

- Residents of nursing homes and long-term care facilities whose lives have been made miserable and lonely, and the vast numbers of them who have been killed by confinement without care and treatment.

- Individuals injured by their experimental vaccines and medications.

- Populations on whom they are imposing coercive public health policies with a demoralizing loss of autonomy and personal freedom.

- The hundreds of millions around the world who have been driven into loneliness and terrorized by unnecessary public health threats, warnings, and abuses.

- Small business owners whose life's work they are destroying.

- The many who have lost their jobs.

- The poor, marginalized, and chronically ill who have fewer community resources and less to fall back on.

- Churchgoers deprived of their spiritual sustenance and community.

- Extended families and loved ones who are separated.

- Children whose school and social lives they have taken away.

- Young athletes and scholarship students whose careers may have been derailed.

- Families and partners impaired and broken up by the stress of living under confined, abnormal conditions, and their children who have suffered from abuse or isolation.

- People who live alone and now find themselves utterly without companionship.

- Parents who lost special services for handicapped children, as well as all parents who must cope with no babysitters, no daycare centers, and no school to take care of and help their children.

- The many ordinary people who are suffering more anxiety and other emotional trauma than they can easily handle.

- Untold millions of people robbed of the experiences of once-in-a-lifetime, special occasions such as weddings, bar mitzvahs, proms, graduations and funerals.

- The societies and nations disabled in infinite ways.

The global predators and perpetrators will feel little or nothing except indifference and disdain toward the people on whom they are inflicting so much suffering to advance their own agendas.

WHO ARE THE GLOBAL PREDATORS?

The global predators are an international movement of rich and powerful entities that are pushing us toward more top-down government with the accumulation of increasing power and wealth among powerful individuals and groups commonly identified as globalists or the elite.

They claim to represent humanity when instead they represent their own voracious appetites for wealth, self-aggrandizement, and power. Some of them are well-known individuals like Anthony Fauci and Bill Gates; some of them are powerful agencies, like the Food and Drug Administration (FDA), the various National Institutes of Health (NIH), the CDC (Centers for Disease Control and Prevention), and the World Health Organization (WHO). The non-government groups of banks, corporations, industries, and philanthropic organizations often closely align themselves with big governments, and to a surprising extent, with China and its Chinese Communist Party.

The description of them as elite is misleading. They may think of themselves that way, but we do not. We prefer to call them global predators. For them, COVID-19 has been a boon beyond all expectations. Their predations account for much of the confusion and anguish surrounding the pandemic, its prolongation, and its deaths.

We shall document how all these individuals and their institutions commit egregious acts against America and humankind which make no sense except that their allegiances and goals are different from ours. Having this predatory globalist outlook, they openly express enthusiasm for China and side with the violent dictatorship over America and the Western world. Like fish accustomed to swimming in foul water or people stuck living and breathing in extreme air pollution, Fauci and his fellow global predators no longer know what it means to be a normal human being with a normal sense of concern for the lives of other human beings.

We shall document how they have been openly planning for this moment for several years under the guise of an emergency public health response to a pandemic. It does not matter if they are Bill Gates and Michael Bloomberg, who parade themselves as capitalists, or if it is Xi Jinping and other Chinese leaders, who are devoted Communists. Their ultimate aims are the same—to increase their wealth, self-aggrandizement, and power, while dominating the rest of us. That the Communists may be the only ones genuinely dedicated to their ideology makes them all the more powerful and dangerous.

In Chapter 22, we further elaborate on the specific characteristics of global perpetrators and identify the top American billionaires and corporations that act like global predators. Predators dominate the lists of billionaires and wealthy corporations.

Fauci and Gates

Bill Gates is among the leaders of the predatory globalists, and he has known and worked with Anthony Fauci for years. Fauci is one of only five other powerhouses, including the director general of WHO and the executive director of UNICEF, who sit on the Leadership Council of the Bill & Melinda Gates Foundation to head "a collaboration to increase coordination across the international vaccine community and create a Global Vaccine Action Plan."[292] Fauci has been on this Leadership Council since its inception in 2010, indicating that he was involved with vaccine planning for at least ten years before COVID-19, a point we emphasize in the *Chronology and Overview*. This is globalism near or at its summit, and both Gates and Fauci are growing in power, influence, and the control of wealth as a result of COVID-19. On a number of occasions, Gates has made clear he talked less and less to President Trump and more and more to Fauci.[293] It is no coincidence the Bill & Melinda Gates Foundation has one of its foreign offices in China, and both Gates and the Chinese Communists fund and control the World Health Organization (WHO). Anthony Fauci relates to both of them and carries out their mutual ambitions, about which we will continue to learn.

Bill Gates and the Medical Industry

We can only touch upon the myriad ways the Bill & Melinda Gates Foundation uses organized medicine worldwide to increase Gates' wealth, aggrandizement, and power. We have documented that Gates contributes money to Fauci's NIAID and to Tedros's WHO, two of the most influential agencies for directing American and international policies surrounding COVID-19. Gates' Foundation also gives to the CDC, another influential member of the group, including a $13.5 million award to the CDC Foundation for global vaccine programs.[294]

Now we learn Gates has been pouring money into yet another member of the worldwide medical-military complex: the *Chinese CDC*, including its vaccine programs. We found it confirmed on the website of the Bill & Melinda Gates Foundation.[295] One Gates grant for $500,580, dated August 2018, was purposed "to accelerate introduction of safe, effective affordable vaccines into China's Expanded Program on Immunization."

There is no essential difference between China's vaccine program and China's military bioweapons program. The ruling Communist Party's practice of Military-Civil Fusion makes certain that any corporate, medical, or other civil activities with even remote potential for military application must be given to and controlled by the People's Liberation Army (Chapter 1).

Experiments for the development of coronavirus vaccines have been going on for many years in the U.S./China collaborative gain-of-function research that Fauci funded through NIAID. These are the dangerous projects we publicized and President Trump then cancelled. The engineered deadly viruses become potential offensive military weapons and the vaccines become potential defensive weapons.

Now we add a new wrinkle—the connections between Pfizer and the Chinese pharmaceutical industry and therefore to the Chinese Communist Party. Since the 1980s, Pfizer has been investing huge amounts through China Pfizer,[296] including "actively collaborating with health service providers, academia and governments to support the rapid growth of healthcare in China."[297] Investors applauded and stocks went up when Pfizer and their German partner joined with the Shanghai Fosun Pharmaceutical Group in late 2020, gaining an infusion of $85 million from the Chinese firm for the development of Pfizer's mRNA vaccine.[298] Involvement of a Chinese company means that China's military will have direct access to any useful information obtained by the company, whether openly or by stealth.

We doubt if one in a million recipients of Pfizer's vaccine knows it is a joint development venture with the Chinese Communist Party. This huge financial arrangement was undoubtedly long in the making. Probably to avoid the Trump administration criticizing or canceling it, the news was released on November 10, 2020, days after Biden's seeming election victory for President of the United States.

To finish this mini analysis of global predators at work in the medical field that began with Bill Gates and ended with Pfizer, guess who Bill Gates has invested in and promoted as "the clear coronavirus vaccine leader?" Yes, it's Pfizer and their vaccine program.[299]

On a June 5, 2020, program,[300] Laura Ingraham described how Gates "has his fingers" on everything relating to COVID-19. She commented on how he manipulates WHO and became WHO's biggest donor when President Trump withdrew U.S. support. She described how WHO resisted Gates' wish for the agency to declare a pandemic and how Gates donated $50 million to it, resulting the next day in their declaration. She also noted that Gates has spent several billion dollars on activities related to the UN agency.

Ingraham wisely warned about the dangerous consequences of one man having so much power over an organization that is supposed to be medically objective and scientific. Notably, Director-General Tedros of WHO is not a physician or medically trained. He is a corrupt politician ultimately under the control of the Chinese Communist Party.

BILL GATES: CHINA PERFECT, AMERICA FLAWED— LET'S GET ON WITH THE QUARANTINES

On March 25, 2020, Bill Gates did a TED interview with Chris Anderson and Whitney Pennington Rodgers called "How We Must Respond to the Coronavirus Pandemic."[301] The first half of the interview dealt in some detail with what went wrong in January and February that delayed the world's response to COVID-19.

Gates talks extremely vaguely about delays by governments around the world and his efforts to warn them, never implicating China, Fauci, or any other global partner in initially withholding the seriousness of the China outbreak. Gates mentions setting up a working group with Wellcome Trust and MasterCard and contributing $100 million of his own money. Later in the interview, he talks about "backing" the development of the new mRNA vaccines with no mention of any potential risks or the wealth and power it will accrue to him.

In the TED interview, Gates is generally critical of America's response to the epidemic. In marked contrast, without a hint of any criticism, he is very positive toward China's responses to COVID-19. Gates praises their draconian measures. Without a hint of doubt, he expresses trust for China's statistics on their supposedly successful "extreme" shutdown.

At the time of the Gates' interview, China was claiming it had "no new cases" of COVID-19 among its 1.5 billion citizens. That is, the rate of infection had dropped to "zero" because of their alleged amazing public-health prowess.

Chris Anderson gently casts doubt on China's claim to having no new cases. Without hesitation, Gates replies that he has "absolutely"[302] no doubt China's claim is valid and is due to their rigorous quarantining. In the face of further gentle skepticism by Anderson, Gates rapidly mutters some logarithmic calculations that end in zero, supposedly proving how China could have reached zero cases, a clever ploy which shuts down further questions.

Gates' manipulation of the truth could not change reality, including further pandemic outbreaks in China in 2021. They have been so large that they have interfered with their economy[303] and food supplies,[304] put 22 million into lockdown,[305] and then another 32 million[306] and even more millions[307] into lockdowns, resulting in a renewal of drastic suppression as China has gone into "Wartime Mode" trying to deal with the resurging virus.[308]

Overall, China is not doing nearly as well with COVID-19 as their global advocates like Bill Gates had predicted.

What Bill Gates Already Knew While Exonerating China of Any Wrongdoing

While praising China at the expense of the United States, Bill Gates must have been aware of China's purposeful stealth war against America, its own Chinese people, and humanity.

Beginning with the earliest manifestations of the deadly coronavirus in its country, the Chinese government betrayed the rest of the world, especially the United States:

> China's early denials to WHO and the world about the virus's communicability, which delayed adequate worldwide responses.[309]

> China's self-described destruction of the viral materials at the Wuhan Institute, suggesting foul play, which slowed down the world's knowledge about the virus and its true origin.[310]

> China's shutdown of its own internal airline traffic and internal travel, while flooding the U.S. with more than 400,000 airline passengers, spreading the virus upon an unsuspecting America, as well as the rest of the world.

> China's hoarding of materials for masks and other protective supplies while importing massive numbers, depleting the world's supply.[311]

Then there are China's assaults on its own people:

> China's warlike military assault on its own people—only possible under a violent totalitarian state—may have been used with some success in slowing down the spread of the disease.[312]

> The cost of the oppression has been tragic. There is further suppression of political dissidents.[313] Ordinary Chinese citizens are sealed en mass into large apartment complexes with gates welded shut and doors of homes locked closed, resulting in death from disease, starvation, and thirst.[314] A few people diagnosed with COVID can lead to a neighborhood or community being locked down with everyone trapped inside their houses.

Globalists like Bill Gates, Anthony Fauci, Klaus Schwab and Michael Bloomberg have been and continue to be uniformly indifferent toward these atrocities. Meanwhile, the Chinese Communists have openly described how they have used the opportunity of COVID-19 to impose even more strangulating party rule on the people.[315]

So, what is going on with Bill Gates? We will find he shares all the characteristics of today's predatory globalists from American billionaires and the giant tech corporations to WHO and the Chinese Communist Party. They uniformly present an unerring defense of China's violent and threatening dictatorship while casting aspersion on the U.S., especially America First policies which thwart their globalist ambitions.

China, we will continue to document, works closely with American globalists. An America First policy in the U.S. is the major, if not only, impediment to the success of these international predators. America alone, they believe, can prevent them from growing even wealthier, more self-aggrandizing, and more powerful because of COVID-19 public health policies and practices.

Bill Gates' March 25, 2020, TED interview is further noteworthy for his promotion of totalitarian public health measures while showing no concern for their threat to democracy or freedom. He repeatedly emphasizes that America needs a radical shutdown despite its acknowledged devastating effects on the economy. He believes the most important measures are to test as many people as possible and to isolate them. This, he claims, is the only way to slow down the virus, and he cites China's alleged success as evidence. Gates is adamant America needs "an extreme shutdown," declaring "there is no middle course."

Gates also talks about the ID cards so many people rightfully see as a big step toward worldwide social control, much as China is imposing growing control over its people. He explains, "Eventually what we will have to have is certificates of who's a recovered person, who's a vaccinated person."[316] He briefly explains the aim is to control the movement of people, and then, before he says more, Chris Anderson quickly changes the topic.

Gates imagines the wealthier nations getting out of difficulty in a matter of weeks—a very inaccurate prediction. In contrast, he sees worldwide recovery taking up to three years. Throughout this discussion of the "costs" and the difficulty of recovery, he never touches upon or is asked about the suffering caused by vast limitations on going to school or work, participating in or watching sports and entertainment, and visiting with family and friends, as well as generally being confined at home.

Again, in total consistency with other global predators, Gates pays no heed to the principles of personal and political freedom that are the bedrock of America's founding and which make it the world's great fortification, however shaky, against a takeover by global totalitarianism. Gates represents one of the great tragedies of America: we have raised several generations of children in progressive schools without their gaining any understanding of the heroism and wisdom of our founders, the economics and politics of freedom, and the stunning accomplishments embodied in the Declaration of Independence, the Constitution, and especially the Bill of Rights.

Globalists like Bill Gates have little humility and lack judgment about their own limited capacity to plan for the entire world. They do not grasp or see any benefit to them in the principle of individual liberty as the best driver of human creativity. They have no respect for personal responsibility and freedom as the generators of progress.

FORBES, THE FINANCIAL MAGAZINE, AS A GLOBAL PREDATOR

A great deal of debate has occurred concerning nationalism versus globalism, mostly advocating and defending globalist principles. As a *Forbes* article in 2017 stated, "Nationalism is a dirty word for most globalists."[317] Among "globalist" political leaders, *Forbes* lists President Barack Obama, French President Emmanuel Macron, former Italian Prime Minister Matteo Renzi, Canadian Prime Minister Justin Trudeau, and some Arab "princelings."

Not surprising, *Forbes* displays its own typical globalist mentality in assigning the conflict to competing interests—in essence, the selfish dunderheaded nationalists and the more enlightened globalists. American globalists seem to have no idea what drives so many Americans to fear and resent them. It's not a mere conflict of "selfish" economic interests; it's more about principles like loving the flag, patriotism, a belief in the American Dream, and personal and political freedom.

The words "freedom" and "liberty" cannot be found in the *Forbes* report. The word "democracy" appears once, but not in a meaningful context. *Forbes* does not grasp that for most opponents of globalism, the real issues are preserving traditions and their democratic republic, as well as political and personal freedom. Not any of the globalists seem to have the idea that they are threatening something good, wonderful, and unique embodied in this Land of the Free and Home of the Brave.

On January 7, 2021, as the political assault on President Trump and the political right heated up, *Forbes* took a step so brazen, bold, and hostile to America it illustrates how far our nation is moving toward totalitarianism. In an article titled "A Truth Reckoning: Why We're Holding Those Responsible Who Lied for Trump Accountable," *Forbes* announced its intention *to ruin* any company that so much as hired one of President Trump's closer allies or supporters.[318]

> Let it be known to the business world: Hire any of Trump's fellow fabulists above, and Forbes will assume that everything your company or firm talks about is a lie. We're going to scrutinize, double-check, investigate with the same skepticism we'd approach a Trump tweet. Want to ensure the world's biggest business media brand approaches you as a potential funnel of disinformation? Then hire away.

What *Forbes* says matters to many people, and within a short time the article had over one million viewers. By this time in January 2021, we had become informed enough to make a quick check for the China connections that all global predators seem to have, and we discovered that in 2014, *Forbes* had been sold to an Asian investor group based in Hong Kong, China.[319] For us, as our book will continue to show, this was no surprise given the Chinese Communist Party influence on *The New York Times*, *The Washington Post*, CNN, and other major globalist media.[320]

The Koch Brothers and the Cato Institute.
Globalist Libertarians?

The Koch brothers are well-known international investors, but they are libertarians, so committed to liberty and supportive of conservatives they are hated by the left. They have to be true-blue Americans, don't they?

Charles and David Koch, known as the Koch brothers, are among the wealthiest and most powerful people in the world. The brothers were tied for 11th place among billionaires, before the younger brother David died in 2019 at age 79.[321]

Koch Industries increased its revenue by 4.5% during the first year of COVID-19, newly placing it number one among America's privately held companies with annual revenues of $115 billion.[322] They are invested in 50 countries,[323] and their multiple corporations enable them to tailor their investments in areas with growth opportunities created by COVID-19.[324]

The Koch brothers, as self-identified libertarians, are a rarity among well-known billionaires. They regularly supported selected Republican and Conservative politicians. But starting with his campaign for presidency, they undermined Donald Trump, who responded by accurately labelling them globalists.[325] Key policy differences include what the critics have called President Trump's "trade war with China" and his insistence on a controlled southern border.[326] The CATO Institute, originally called the Charles Koch Foundation in honor of its cofounder and main donor at the time, follows the principles laid down by their patrons.[327]

Conflicts over open borders and coddling China led President Trump to openly stand up to the Koch brothers, risking the support of many Republican and Conservative politicians and alienating a major potential donor. On July 31, 2020, the President tweeted about the Koch brothers:[328]

> They want to protect their companies outside the U.S. from being taxed, I'm for America First & the American Worker – a puppet for no one. Two nice guys with bad ideas. Make America Great Again!

On the same day, the President also tweeted:[329]

The **globalist** Koch Brothers, who have become a total joke in real Republican circles, are against Strong Borders and Powerful Trade. I never sought their support because I don't need their money or bad ideas. They love my Tax & Regulation Cuts, Judicial picks & more. (Bold added.)

Conservative patriotic organizations have also been critical of the Koch brothers as globalists rather than as patriots and supporters of America.

The Koch Family, Libertarianism, and Inhumanity

Along with Charles Koch, another of the three cofounders of the Cato Institute was Murray Rothbard. A libertarian economist, Rothbard believed parents should be legally allowed to sell their children on a free market and, furthermore, parents were under no legal obligation to feed or otherwise keep their children alive.[330] Humans, Rothbard maintained, had no positive legal obligations to one another and instead were only obligated to respect the rights of others to be equally free. As another extreme example of callousness, I personally heard Rothbard give a speech at a libertarian conference in the 1970s where he displayed a shocking lack of concern for older Americans thrust into extreme poverty by catastrophic swings in the bond or stock markets. When asked specifically, "What about an elderly woman whose corporate or municipal bonds might crash, leaving her destitute," Rothbard made a quirky smile and said she, too, would have to take the consequences of freedom.

It also became obvious to me that the large corporations, such as found in the pharmaceutical industry or global banking, did not operate based on voluntary exchange—without force or fraud. They routinely commit fraud, starting with unconscionable manipulations of their controlled clinical trials and their marketing practices, and they buy favors within the government to support their fraudulent behaviors.[331] They also threaten and intimidate their critics. It was further obvious that large governments were not going to shrink in size, government-induced poverty and gross misfortunes were not going away, and a large modern society requires more than voluntary charity to provide safety nets.

As noted earlier, the Cato Institute and the Koch brothers support open borders with unlimited immigration into the United States.[332] Here is the Cato Institute's explanation in their 2019 annual report:[333]

> The fight for free and open trade doesn't stop at goods. Free trade in international labor is also an important component, and one where the gains to human well-being are most appreciable. Immigration has been part of the American Dream since the Founding. Thomas Paine famously called for making the United States "an asylum for mankind." Free movement across national borders is one of the policies fundamental to human flourishing and produces some of the greatest benefit.

Actually, "free trade in international labor" would turn America into one more cheap labor market to fatten global predators. Inevitably, wage scales would begin to approximate those of the average poor nation. *As Ginger's mother expressed in one of our War Room discussions, free trade in human labor ignores the sacred value of the individual.* Ginger added, *free trade values profit over the people enslaved to make that profit.* In the concluding chapter, we shall return to this overarching importance of treasuring each other while joining the fight for freedom.

Furthermore, libertarian principles do not require opening borders; only the greed of some libertarians demands them. Closed or controlled borders are easily and rationally justified on the libertarian principle of a nation's right to self-defense, exactly as the military or police are justified by libertarians to protect the freedom of our citizens. We should use the minimal necessary force, like a big fence, to defend our borders and our lives.

The biggest gift we can give to the world is not flooding ourselves beyond capacity with desperate people who often do not share our values. The biggest gift we can give the world is by once again becoming the torch of liberty throughout the world while resisting all that would suppress humanity.

In respect to COVID-19, the Koch brothers and their institutions do promote liberty in calling for an end to forcing people to stay at home and to forcing companies to shut down. They have also been critical of the CDC;[334] but that does not make up for their larger harms. In respect to corporations, they do not expect them to fight for freedom; rather, they should learn to adapt to the situation imposed upon them by the pandemic politicians.

Instead of rejecting big government and demanding ethical behavior from politicians, the Koch brothers became masters of manipulating the system by buying politicians. Most importantly, the Koch brothers meet what might be considered a litmus test for being a global predator: the willingness to sacrifice and even destroy the United States with unlimited immigration to increase their global wealth and power. The Koch brothers have undermined America and President Trump's populist presidency by pushing the Republican Party into favoring open borders, international predatory industry, and globalist aspirations.

Is the Koch family tied to China? We have noted throughout our book that predatory globalists always end up having ties to China and to favor that violent dictatorship over our own nation. The Koch brothers vigorously opposed President Trump's tariffs against Chinese goods and simultaneously fought against punishing China for their tariffs on our goods or other offenses.[335] This is typical predatory globalist behavior, selfishly and dangerously siding with China against the United States.

Yes, the Koch Industries are deeply invested in China, including building a $1-billion chemical factory in China and selling one of their companies for $2 billion to Chinese investors.[336] According to *The New York Times*, "Koch Industries also has big investments in China, where it employs more than 23,000 people."[337]

There are serious moral and political questions about American citizens, especially libertarians, who should know better yet use their influence and their investments to protect and strengthen China—the world's largest totalitarian predator and the greatest existential threat to our nation.

We repeat: globalism respects no boundaries. It masquerades as progressive, conservative, libertarian, or Marxist. Globalists place the accumulation of wealth, glory and/or power at the top of their ambitious agendas and do so without regard for national identities or human freedom. Only a few, mainly the Chinese Communist Party or Jihadists, are strongly motivated by political or religious ideologies, and that gives them special advantages over the increasingly relativist and spiritless leaders of the West.

Living in a Moral Vacuum

From Joe Biden and John Kerry to Bill Gates, Klaus Schwab, and corporate CEOs worldwide, globalist leaders are now operating in a moral vacuum that has deprived them of wisdom and the knowledge of their own limitations. They have no humility or maturity about their own fallibility when making long-term predictions or when promoting worldwide totalitarian measures. Their personal and political failures are ignored or blamed on others. A future Genghis Khan is now a geeky kid inventing something new on his computer, and his future fields of battle will be fought digitally—with as much or more ruthlessness than the emperors of the Mongol Empire.

It is stunning to listen, watch, or read endless discussions about the need for massive global change as if America and its history as a beacon of liberty never existed. Our heritage and accomplishments as the freest people in history have been erased from the minds of the new global leadership, including Bill Gates, Michael Bloomberg, and Klaus Schwab. They pay almost no attention to: (1) How will world governance be implemented for these vast programs that "must" happen within a decade or two? (2) What political processes are required for its achievement? and (3) What are the consequences of the loss of individual and political freedom? Like all totalitarians, they plan on handling all that when they get there and doing so with utter ruthlessness.

China's Spreading the Virus Was Preceded by China's Covert Longstanding Undermining of America

The voracious globalists are many in number. They find sympathetic accomplices, including the FDA, DEA, CDC, and the so-called "Deep State" in America, which we call the embedded globalist bureaucracy. *Only through understanding predatory globalists can we finally make sense out of the seeming craziness or irrationality surrounding the destructive, international response to COVID-19. What is being inflicted on us is not "crazy;" it is simply not in our self-interest but in the interest of others who wield enormous power over us.*

Behind the Scenes: China

Information about globalism and COVID-19 has been furiously bubbling to the surface for some time. Initially, we thought we were beyond being surprised. However, something of overarching significance has been coming up. Besides casting a shadow over the handling of COVID-19, it has descended over all of America's international economic life. It is the looming presence of the People's Republic of China and its mutually convenient alliance with the global predators against the United States.

The *Epoch Times* on November 23, 2020, published an article, "U.S. Invested Billions into Companies with Ties to China."[338]

> President Trump's latest executive order bars U.S. investors from holding ownership stakes in a list of 31 Chinese companies designated to have ties with various Chinese Communist Party (CCP) military apparatuses. [339]

> This order, while somewhat limited in scope, ensnares several well-known Chinese companies including non-public companies such as Huawei and publicly traded companies such as China Mobile and Hangzhou Hikvision. The 31 companies were previously designated by the Pentagon as being "owned or controlled" by the People's Liberation Army (PLA), the official name of the Chinese military. …

> In addition, examining the foreign ownership base of these companies reveals that a number of them are partially owned by prominent U.S. investment firms, insurance companies, and pension funds.

> President Trump's latest measure at restricting Chinese companies would prevent U.S. capital from funding China's military and intelligence agencies that could hurt U.S. interests.

Although we might not be openly at war with them, the Chinese Communist Party has been at war with us since it took power during and after World War II. China has insidiously invaded our institutions, including our academic and scientific establishments. Meanwhile, American globalist predators have opened their arms to the Chinese Communist Party. We shall see this tragedy woven throughout this book in a way we never anticipated, but can no longer ignore.

> We are not the only ones to express serious concerns about China's predatory attitude toward us and their participation in a global existential threat to America. Brian T. Kennedy, president of the American Strategic Group, wrote "Facing Up to the China Threat" in *Imprimis* for its September 2020 issue:[340]

> We are at risk of losing a war today because too few of us know that we are engaged with an enemy, the Chinese Communist Party (CCP), that means to destroy us. The forces of globalism that have dominated our government (until recently) and our media for the better part of half a century have blinded too many Americans to the threat we face. If we do not wake up to the danger soon, we will find ourselves helpless.

> That is a worst-case scenario. I do not think we Americans will let that happen. But the forces arrayed against us are many. We need to understand what we are up against and what steps must be taken to ensure our victory.

We came to similar conclusions through our analysis of the seeming madness surrounding the nation's policies and programs for dealing with COVID-19. These policies are not in the interest of the American people; instead, they are in the interests of what Brian Kennedy calls "the forces of globalism," including China. President Donald Trump met extraordinary resistance and pressure, and outright fraud, perpetrated by predatory globalism and its adherents embodied in Fauci and in nearly every federal institution, including Congress, the embedded globalist bureaucracy, and the NIH, CDC, and FDA.

But Is the Worst Yet to Come?

The predatory globalists have not yet come near to accomplishing their aims. They do not seem to have a central organization. They are still cooperating and have not yet begun to war among themselves for dominance over humanity. Much like progressives in America, they have not faced that their empowering the Chinese Communist Party will ultimately turn them from predators into prey.

Perhaps even more ominous than predatory globalism itself, we will find, while they are enthusiastically embracing COVID-19 to advance their wealth, self-aggrandizement, and power, they are anticipating even more globally disastrous scenarios than COVID-19 to further their aims.

PART TWO

COVID-19
Policies
and the End
of Science

CHAPTER 8

What Is *Your* Risk of Death if You Catch COVID-19?

Most people have very unrealistic fears about the risk of dying from COVID-19. This is due in part to the CDC and to Dr. Anthony Fauci who continue to inflate the risk of COVID-19 deaths. We therefore begin by examining the most fundamental issue of all: If you or a loved one are infected with SARS-CoV-2 and become ill, what is your risk of death? It is probably much lower than you think or fear.

The CDC bases its estimated death rates from COVID-19 on death certificates, and this method is accepted as authoritative by Dr. Anthony Fauci and many others. However, the CDC has recently revealed only 6% of COVID-19 death certificates list the disease as the *sole* cause of death, while 94% have two or three additional listed causes.[341] Furthermore, there is no way to ascertain what the primary cause of death was among the 96% with multiple listed causes of death.

Most people who die while being positive for SARS-CoV-2 are near to or past the average longevity. In addition to being old, the great majority are already ill with heart disease, cancer, or some other chronic illnesses that may have caused them to die. Even using the CDC's biased data, the risk of death from COVID-19 for most people is too small to require them to sacrifice the quality of their lives as the government demands under the threat of catching COVID-19.

Using their exaggerated data, the CDC made a "best estimate" for the risk of dying *after infection* and *becoming ill* with COVID-19. The CDC reported the following estimates at the height of the pandemic on September 10, 2020:[342]

Current CDC Best Estimates for Infection Fatality Ratio

0-19 years:	0.00003 (0.003%)
20-49 years:	0.0002 (0.02%)
50-69 years:	0.005 (0.5%)
70+ years:	0.05 (5%)

By far the greatest risk is 70+ years old. If you are in that age range, and you become infected and ill with COVID-19, you will have a 5 in 100 risk of dying according to the CDC. Given the age and potential infirmity of these people, this is not an extraordinary death rate. Furthermore, the death rate of the next most vulnerable is one-tenth of that or 0.5 in 100. *This data provide no reason or justification to impose totalitarianism on America, to crush our society and ravage our economy, to retard the education of our children, and to force deadly wealth-draining vaccines on our entire nation.*

Furthermore—and this is extremely important—these CDC-inflated death rates are based on patients who never received adequate treatment at home because it was prohibited by Anthony Fauci, the FDA, the CDC, WHO, the drug companies and other global predators who have been feasting on COVID-19. When patients receive the safe and effective treatments used by frontline doctors in the U.S. and around the world, there are very few deaths, even within the older population. The widespread use of effective early treatments would have done away with the need for Operation Warp Speed and all the draconian measures.

In summary, the overall risk of anyone at any age dying from COVID-19 is infinitely less than these already low CDC figures indicate because (1) The statistics are highly inflated, (2) the numbers reflect the risk of dying *after you become infected and symptomatic* with SARS-CoV-2, and (3) the government has deprived most of these people of readily available, inexpensive, safe, and effective treatments for COVID-19 as developed and refined by front line medical doctors like those who wrote the introductions to this book, Zev Zelenko, Peter McCullough, and Elizabeth Lee Vliet.

If you receive appropriate early treatment for COVID-19, how much will your risk of hospitalization and death be reduced? The results of a recent study confirm what frontline physicians have been reporting since early 2020:[343] "As a result, our early ambulatory treatment regimen was associated with estimated 87.6% and 74.9% reductions in hospitalization and death respectively, p<0.0001."

In short, with a very high degree of statistical certainty, if you receive proper early treatment, you will reduce your chance of hospitalization by 87.6% and your risk of death by 74.9%.

We repeat, early in the pandemic, withholding these treatments from patients became a crime against humanity that resulted in many millions of deaths throughout the world.

Yet Another Cause for Over-Counting Cases and Death Rates

Federal programs and insurance companies give bonuses to doctors and hospitals for working with COVID-19 patients. The result: physicians "say they are being pressured to add COVID-19 to death certificates and diagnostic lists—even when the novel coronavirus appears to have no relation to the victim's cause of death."[344] Merely adding COVID-19 as a "possible" contributor to the hospitalization or the death brings in more money, literally at the stroke of a pen or the tap of a keyboard.

It appears that Medicare COVID-19 incentives may range from around $13,000 for a hospitalized patient to $39,000 for one who is put on a ventilator.[345] Given how harmful ventilators can be for COVID-19 sufferers, this huge $39,000 incentive is almost like a bounty for killing patients.

Defenders of the COVID-19 establishment call it a conspiracy theory to imagine that financial incentives can have a negative impact, but even the CDC agrees it can be a problem.[346]

> Republican Rep. Blaine Luetkemeyer of Missouri questioned [CDC Director] Redfield during a House Oversight and Reform subcommittee hearing on coronavirus containment. He asked about the "perverse incentive" for hospitals to classify deaths as being coronavirus immune-compromised related when the virus didn't cause the death.

> Luetkemeyer voiced concern about how an automobile accident-related death could be recorded as a COVID-related death. In this scenario, the death was recorded as COVID-related because

the virus was in the victim's system, even if the car accident was the major factor.

"As long as you have COVID in your system you get to claim it as a COVID death, which means you get more money as attending physician, hospital, whatever," Luetkemeyer said during the hearing. "Would you like to comment on that, about the perverse incentive? Is there an effort to try and do something different?"

Redfield responded by telling the congressman, "I think you're correct in that and we've seen this in other disease processes too."

"In the HIV epidemic, somebody may have a heart attack but also have HIV," Redfield explained. "The hospital would prefer the DRG [diagnostic code related to reimbursement] for HIV because there's greater reimbursement. So I do think there is some reality to that."

There are other pressures as well coming down from the CDC and a federal bureaucracy that obviously believe their institutional power is tied to exaggerating the risks associated with COVID-19.

THE RISK OF DEATH IN CHILDREN WITH COVID-19

The above CDC data states the risk of infected children up to age 19 dying from COVID-19 is 0.00003 (0.003%). But how many children are *actually* dying from COVID-19? The CDC makes it very difficult to figure this out.

Fortunately, on October 10, 2020, the American Academy of Pediatrics and the Children's Hospital Association published data submitted from the individual states.[347] Based on 42 states reporting, they found that "0%-0.16% of all child COVID-19 cases resulted in death." Sixteen of the 42 states reported *no deaths among children.*

The age of the "children" went up to 17, 18, or 19, depending on the state's criteria, making many of them young adults. The risk of death in children and young adults with COVID-19 is truly small.

These risks do not justify drastic lockdown measures imposed on children and young adults. Most tragically, they do not justify keeping children and youngsters out of school. *Still, Dr. Fauci and other public health officials continue to act as if there is a grave risk of exposing children and young adults to SARS-CoV-2, when there is not.*

The Risk of COVID-19 to the Elderly Is Serious

The CDC data listed in the previous table indicates that at 50-69 years of age, the risk of dying when infected with SARS-CoV-2 is 0.005 (0.5%) and for 70-plus years old it is 0.05 (5%). People 65 and older account for nearly all the deaths—70% to 94% of them, depending on the reporting of the state.[348]

The higher death rate among the elderly is tragic, but it is considerably lower than most people imagine. Many elders seem to think getting COVID-19 is a death sentence when it certainly is not. A 5% death rate for people 70 and older, many of whom are very ill and near the end of life, does not demand the imposition of extraordinary, disabling lockdowns and other drastic transformations on the entire population, including the children.

There is also agreement that any serious risk of lethality from the coronavirus is limited to older people and those with multiple comorbid disorders. The death rate increases at age 50, where it remains small and begins escalating around age 65 to 75, in part because older people have a much higher proportion of multiple physical infirmities, including weaker immune systems.

People 65 and older account for 70 to 94% of all deaths, depending on the state.[349] This means the overall death-rate estimates are mostly accounted for by deaths of people 65 and older with the mean somewhere around age 80. This indicates that most preventive efforts can and should be focused on the relatively small population of older people, especially those with preexisting illnesses.

It is difficult to obtain reliable statistics on the actual annual rate of death from the coronavirus, probably because it is much lower than the experts and advocates of top-down government control predicted, and they do not want the public knowing how low it is. Former *New York Times* science reporter, Alex Berenson, whose work seems both independent and reliable, made an annual rate estimate of 0.26% in June, which is far below estimates from public health pundits who were predicting up to 5.0% or more.[350] Berenson's relatively low estimate of a 0.26% annual death rate was identical to the CDC's own best estimate made in June 2020; a figure which disappointed and frustrated the mainstream pundits.[351]

On June 27, 2020, Berenson summarized:

> The actual figure could be as low as 0.1 percent or as high as 0.4 to 0.5 percent, though treatment advances should mean it will trend lower over time. Even at 0.26 percent, the rate is still significantly higher than influenza most years; more comparable to a bad flu strain like the 1968 Hong Kong flu.

Additionally, the virus hopefully will lose its lethality as time goes on, which is consistent with the virus's inability to maintain a stable form and the tendency for less lethal forms to spread more easily to push out more lethal ones that kill off their hosts.

Meanwhile, all these figures, even those from the reliable Berenson, are based largely or entirely on CDC reporting, and these CDC estimates are turning out to be wildly inflated. In the last few days, as of this writing, the CDC reported that the coronavirus was the sole or only cause of death in a mere 6% of coronavirus deaths reported to the CDC and that 94% had an average of 2.5 other listed causes and most of them were elderly.[352]

It seems probable that a significant proportion of deaths were actually caused by other diseases, and much of the threat from COVID-19 has been manufactured by experts, public health scientists, and government agencies. These special interest groups have been working hard to frighten Americans into conforming to drastic suppressive measures while the government subsidizes the pharmaceutical industry in its massive efforts to find a more expensive and remunerative drug than hydroxychloroquine and an elusive and perhaps unobtainable vaccine.

An Example of Confusion Surrounding How to Count Deaths

Collin County in Texas is a good example of how these changing standards affect the numbers of COVID-19 cases counted. During the May 18, 2020,

Collin County Commissioners Court, Administrative Services gave a PowerPoint presentation on the new case definition for COVID-19.[353,354] Their presenter referred to the Texas Department of State Health Services guidelines distributed on May 11, 2020 as her source.[355] Prior to the changes, the case definition for a confirmed case of COVID-19 was a positive PCR (swab test for present infection) lab result for COVID-19.

Now positive lab results are no longer required for the counting of a COVID-19 death. In each category, the counted cases will be many times higher. According to the *CDC FAQ: COVID-19 Data and Surveillance*, "A COVID-19 case includes confirmed and probable cases and deaths."[356]

The Great Barrington Declaration and Our Petition for Older People

Three epidemiologists from Harvard, Oxford, and Stanford wrote the Great Barrington Declaration, and by February 2, 2021, it was signed by 741,952 concerned citizens. In addition, there were signatures from 13,540 medical and public health scientists and 40,962 medical practitioners.[357] Those of us who signed it agree with their statement: "We know that vulnerability to death from COVID-19 is more than a thousand-fold higher in the old and infirm than the young. Indeed, for children, COVID-19 is less dangerous than many other harms, including influenza." We agree that the near zero or zero death rate among children and youth is too low to justify measures that deprive them of a normal social life and their schooling.[358]

The risks associated with COVID-19 do not justify the measures being imposed on America. The motivation for the lockdowns comes from predatory globalists and not from genuine scientists or legitimate science.

Our household includes a husband and wife who are 84 and 69 years old, and the wife's mother who is 94. None of us want to lockdown the nation or the world on account of us. For this reason, we sponsored a separate petition by Stand for Health Freedom specifically for older people to sign stating we don't want the schools, society, and the economy shutdown because of us. You can still sign up for this remarkable petition to have your support for opening our nation again and your declaration will *automatically* be sent to your state and federal representatives.[359]

There is a place for older people taking extra precautions and for the government offering special services, but this can be done without vastly impairing the lives of everyone else. We do not need to inflict such enormous harm on the economy and on society or to spend such huge sums of money to protect our vulnerable older population.

CHAPTER 9

Killing Patients to Make Inexpensive Good Drugs Look Bad

We begin this chapter with a quote from one of the most outstanding scientists in the world in the arena of COVID-19:

> COVID-19 is the only medical problem that comes into public view every day with no mention of early treatment. Patients are handed their positive SARS-CoV-2 result and given no advice on treatment, no hotline for available research protocols, no prescriptions, and no medical follow up to adjust regimens.

It is a colossal blunder by Dr. Fauci and all who have been involved with the NIH/CDC/FDA/White House Task Force to ignore the high-risk patient and his/her journey with COVID-19 and to understand that such a serious medical problem always required the highest priority and constant agency attention and engagement with the medical community to reduce the risk of hospitalization and death.

Because of obvious shortcomings, the U.S. has had record morbidity and mortality, the vast majority of which was completely avoidable. This American tragedy will undoubtedly become the focus of investigative historians who will reel through days and weeks and months of press briefings, media interviews, and public statements without a single mention of how doctors should have and could have treated desperately ill patients in their progressively terrifying days at home before succumbing to hospitalization, painful procedures, hypoxia, and finally dying alone in hospital isolation.

Peter A. McCullough MD, MPH[360]

Dr. McCullough has testified under oath that he has effectively treated all of his high-risk COVID-19 patients at home with compassion to achieve the goals of reducing the intensity and duration of symptoms, hospitalization, and death. Through the crisis, he has directly treated and advised on many hundreds of cases worldwide. Dr. McCullough's full impact on the world will never be known; however, speaking to thousands of physicians and scientists and standing as a beacon for compassionate clinical care, he, along with Zev Zelenko, may have saved millions of human lives.

TRYING TO KILL A LIFESAVING MEDICATION

The research and scientific community, in the service of the pharmaceutical industry and the global predators, conducted one study after another in which they gave COVID-19 patients toxic and even lethal doses of either chloroquine or hydroxychloroquine. Often, they used the older drug, chloroquine, when hydroxychloroquine is "a less toxic metabolite of chloroquine."[361]

To further discredit these good medicines, they gave them to patients on death's door, when their proven effective use is as a prophylaxis or early in the treatment of viral diseases, including COVID-19. In short, doctors have murdered patients to discredit hydroxychloroquine, and the medical/scientific establishment has celebrated them and published their articles and opinions. This is such a dreadful accusation, we will examine and cite examples.

Sadly, the older, cheaper, and more effective drugs are suppressed precisely because they result in rapid, home-based cures, doing away with the need or the justification for the Emergency Use Authorization (EUA) and rushing vaccines through the pipeline. The EUA authorizes government agencies to push through new drugs and vaccines at breakneck speed without regular FDA approval and with huge financial help to the drug companies—but only if there are no other available safe and effective treatments.

MEGADOSE OR MASS MURDER

We became so incensed by one of the deadly overdose studies that we titled our report "Research Study—Or Megadose Mass Murder."[362] We were among the first to publicize this horrendous medical betrayal but never expected the response that the overdose study ended up receiving from the establishment.

The Brazilian authors of the study[363] must have known they were treading on dangerous territory by purposely causing many deaths. Coming from a poor area of the country, they may have felt they could get away with sacrificing their patients without local reprisals. They simply gave lethal doses of chloroquine to patients to prove that the drug and its derivative hydroxychloroquine were too dangerous to treat COVID-19.

Respected sources, such as all recent editions of the classic textbook *Goodman & Gilman's The Pharmacological Basis of Therapeutics* (2011, p. 1405), make the same basic observation about the extraordinary safety of both chloroquine and the even safer derivative hydroxychloroquine:

> **Toxicity and Side Effects.** Taken in proper doses, chloroquine is an extraordinarily safe drug; however, its safety margin is narrow, and a single dose of 30 mg/kg may be fatal.

The Brazilian study used enormous, repeated doses of chloroquine: *1200 mg daily for ten days.* This dose is so large the authors could not cite a single other clinical study that approximated this megadose range, except in a single study in which hydroxychloroquine was given in the hope of suppressing cancer.

The potentially lethal single dose of chloroquine begins at 30 mg/kg for a 40 kilogram[364] or 89-pound patient. Since the patients in this study were extremely sick, many had comorbid illnesses, and a number were elderly, it is likely some were probably under 89 pounds. *An 89-pound patient given 1200 mg of hydroxychloroquine is already within the fatal dose range of 30 mg/kg. Indeed, a patient with a normal weight but with serious comorbid issues, would also be within the fatal dose range.*

Furthermore, that potentially lethal dose, as described in Goodman & Gilman's textbook, is calculated on the basis of a one-time ingestion of 1200 mg. The Brazilian patients were given this potentially lethal dose for ten straight days!

In addition, the potentially fatal dose range is calculated for people who are in reasonably good health, while all the patients in the Brazil study were extremely ill with COVID-19, some were elderly, and many had comorbid diseases, including heart disease. The lethal dose for them would be considerably below 30 mg/kg.

Given their physical condition and frailties, all the patients in this study were at risk of death from the megadoses of chloroquine administered to them for ten days.

Beyond all those considerations, these doses over a period of ten days are higher than they even seem. Chloroquine has an extremely long half-life, measured in days and weeks rather than hours.[365] There is also evidence the half-life increases with the dose.

Altogether, this means the high doses over ten days would accumulate in increasingly greater concentrations that would persist well beyond the termination of drug treatment—leading to increased blood concentrations and even greater lethality. This medical study was murderous.

Many of the patients were probably too ill to properly metabolize or break down the drug, increasing the effect of the dose through its increased concentration in the blood and hence, again, intensifying its lethality. Worse yet, the patients were already at death's door in "intensive care units" for the "treatment of severe COVID-19 patients." They were described as "critically ill" (p. 1). Some were "unconscious," according to the prepublication version.[366]

No wonder 39% of the patients died. Sixteen of the 41 patients were dead. It would have been worse if an oversight committee had not intervened to stop the madness. It took the intervention of an independent monitoring group to prevent the researchers from continuing their slaughter. In their final version of the paper, the authors put the independent monitoring group on the author's list, probably to falsely claim that they themselves, the authors, had stopped the study, when they had not.

What happened when *The New York Times* and the *Journal of the American Medical Association* got prepublication versions of this negligent and murderous study? Did they call for a civil or criminal investigation?

Here's how the study was greeted. It was released prepublication online on April 11, 2020.[367] The globalist *New York Times* was so happy to thump President Trump's drug that it published a big story to publicize the Brazilian study on April 12, 2020, one day after the prepublication report appeared on the internet.[368] The article was then rushed to formal publication online on April 24, 2020, by the *Journal of the American Medical Association (JAMA)* on its *JAMA Network Open*.[369] The journal of the AMA even gave online Continuing Medication Education (CME) credits to doctors who read it.[370]

Neither *JAMA* nor *The New York Times* made any suggestion the doctors had done anything wrong, such as kill their patients with huge overdoses. Instead, *The New York Times* and *JAMA*, on behalf of their avaricious global masters, praised the study for showing how dangerous hydroxychloroquine really was.

To this day, this one study in the AMA's respected journal has probably contributed the most to many doctors grimacing when a patient asks for hydroxychloroquine. They have been erroneously taught by two of their most respected sources of information, *JAMA* and the *NYT*, that one of the safest drugs in the world actually kills people.

Simultaneously, on April 24, 2020, the FDA ramped up its attack on hydroxychloroquine, limiting its use to hospitals. Eventually, the agency would tell doctors to never use it at all. A coordinated effort like this involving *JAMA*, *NYT*, and FDA—instantly raising an obscure Brazilian study to global importance—speaks to the predatory power that can be brought to bear by the most threatening forces in the world today.

The Brazilian doctors concocted this study not only to discredit hydroxychloroquine, as well as President Trump who had correctly endorsed it, but they also probably wanted to discredit their own President, Jair Bolsonaro, a supporter of both President Trump and hydroxychloroquine.[371]

When the FDA joined forces against hydroxychloroquine, it used fraudulent studies like the Brazilian example to declare the drug too dangerous for clinical use—although chloroquine and hydroxychloroquine are among the safest drugs in the world with experience spanning many decades in treating tens of millions of patients for malaria, rheumatoid arthritis, lupus, and other afflictions.

The Brazilian study was not the only one to use megadoses of chloroquine or hydroxychloroquine, literally sacrificing human lives to prove the "people's drug" was too dangerous. Other studies in Great Britain have also been accused by reputable scientists of using excessively high doses of hydroxychloroquine in order to achieve negative results, including excessive death rates.[372]

We All Need Hope and Faith
to Face So Many Betrayals

After reading the beginning of this chapter, you probably need a ray of hope about doctors. For us, hope and inspiration began with Vladimir "Zev" Zelenko MD. We heard about this Orthodox Jewish doctor, as he was often described, saving lives in Brooklyn. At about the same time, we learned about his persecution. Having been through this process ourselves, especially in the early years of our work, we thought something important must be taking place. It turned out to be incredibly important!

Dr. Zelenko had developed a protocol for treating the early stage of COVID-19 that was so successful the establishment had to suppress it and him. Nevertheless, so brave and determined is this man, he did not let them succeed. Dr. Zelenko's treatment protocol, starting with hydroxychloroquine and azithromycin and now with new variations, remains the basis for the successful treatment of early COVID-19.

For one of our family members, who was vulnerable as a result of pre-existing respiratory problems, it may have been lifesaving, turning around a rapidly worsening case of COVID-19 within the first day of treatment with hydroxychloroquine. Here is Dr. Zelenko's original work[373] and two recent scientific papers he coauthored.[374] We thank you, Dr. Zelenko; you are a genuine "American Hero."[375]

On January 23 and March 3, 2021, I interviewed Dr. Zelenko on *The Dr. Peter Breggin Hour*, my weekly radio/TV interview and talk show.[376] They are two of the most profound hours I have ever spent with anyone. Many viewers have been deeply moved by Dr. Zelenko's medical and scientific integrity. His courage and strength are unparalleled in facing his own life-threatening illness while speaking truth to save lives in the face of extreme retaliation, not only from the global predators, but his own community. His concept that those who "serve God" have nothing to fear from "humans" was one among many spirit-shaking insights.

Informed Physicians Testify
before the U.S. Senate

On November 19, 2020, members of the Association of American Physicians and Surgeons (AAPS), to which I belong, organized presentations to a senate committee on the subject of "Early Outpatient Treatment: An Essential Part of a COVID-19 Solution."[377] Physicians and an epidemiologist with top credentials and affiliations from Harvard, Yale, and elsewhere discussed the effective early treatment of COVID-19. This was followed by "Part II" of the hearing on December 8, 2020, which emphasized studies supporting ivermectin as a key antiviral drug.[378, 379] The major media coverage of both hearings was pitifully sparse and unfavorable, and the Democratic Party excoriated Senator Johnson.

For a recent, powerful scientific review confirming the value of home-based treatments, see the extraordinary scientific publication by Peter McCullough, Paul Alexander et al. (2020), written by 57 authors with overwhelming support for early treatment of COVID-19 in *high-risk* adults.[380] Another, more recent 2021 review, shows that early treatment with multidrug regimens reduces hospitalization by 87% and death by 75%.[381] *These results are extraordinary and remind us that withholding these treatments is indeed a crime against humanity.*

All those actively opposing these early treatments have participated, knowingly or unknowingly, in an organized withholding of lifesaving treatments from many millions of patients worldwide. An unknown but large portion of them have died.

You can watch my hour-long separate interviews with physicians Peter McCullough, Elizabeth Lee Vliet, and Vladimir "Zev" Zelenko on *The Dr. Peter Breggin Hour,* as well as shorter videos I have done with them and other physicians and experts.[382]

All our videos were originally posted on Dr. Breggin's YouTube Channel, but that is close to being permanently shut down. You will probably have to go to www.Brighteon.com, www.BitChute.com, and other platforms. Check our website, www.breggin.com, for further information and sign up there for our Free Frequent Alerts.

Once again, the free early treatment guides are available from The Truth for Health Foundation and AAPS without cost and with an accompanying list of resource doctors around the U.S.[383] For readers with a scientific bent, the protocol has also been published in medical journals.[384]

Vladimir Zelenko MD, who wrote one of the three introductions to our book, is one of three outstanding physicians who authored another report in support of hydroxychloroquine-based treatments. The other two are Harvey Risch MD and George Fareed MD. They have published a 248-page monograph which assembles and comments on dozens of studies demonstrating the effectiveness of hydroxychloroquine.[385]

The review is titled *Medical Studies Support MDs Prescribing Hydroxychloroquine for Early Stage COVID-19 and for Prophylaxis*. It can be downloaded without cost from our Coronavirus Resource Center at www.breggin.com.[386]

One of the three authors, Harvey A. Risch MD, PhD, professor of epidemiology at Yale School of Public Health, wrote a definitive public defense of the drug in *Newsweek*.[387] On May 27, 2020, Dr. Risch separately published a review, "Early Outpatient Treatment of Symptomatic, High-Risk COVID-19 Patients that Should be Ramped-Up Immediately as Key to the Pandemic Crisis."[388] It appeared in the world's leading journal in the field, the *American Journal of Epidemiology* (AJE).

A retrospective detailed study from China has found that low-dose oral hydroxychloroquine (HCQ) "was associated significantly with the reduced fatality of critically ill patients with COVID-19."[389] Laboratory data also showed the medication "greatly lowered the levels of IL-6, one of the most inflammatory cytokines." The dose of 200 mg twice per day for seven to ten days was indeed low and well within the range at which it has proven extremely safe for more than half a century. The scientists concluded, "HCQ treatment significantly reduced the fatality of critically ill COVID-19 patients" compared to patients who were not treated with the medication. They concluded:

> In summary, this retrospective study demonstrates that HCQ significantly reduces death risk of critically ill COVID-19 patients without apparent toxicity, and its mechanism of action is likely mediated through its inhibition of inflammatory cytokines on top of its ability in inhibiting viral replication.

The authors believe their study is strong enough to recommend its use "to save lives" in seriously ill patients. They also recommended its use as an option for patients at an early stage of treatment because of its safety in a long history of treating malaria infections.

Another positive retrospective study has come from South Korea.[390] The authors found that "HCQ [hydroxychloroquine] with antibiotics was associated with better clinical outcomes in terms of viral clearance, hospital stay, and cough symptom resolution" compared to another antiviral drug or to "conservative treatment." Nothing like this can be said for remdesivir, an expensive, dangerous, and worthless drug Fauci made sure received FDA approval for treating COVID-19.

Like the Chinese study, hydroxychloroquine did not result in any deaths or other serious adverse events, further adding to its remarkable safety profile. Despite this, unpublished studies are still touted by the media and the scientific press to discourage the use of hydroxychloroquine.[391]

Treating physicians from all over the U.S. are speaking out about their successes in prescribing hydroxychloroquine for patients with COVID-19 and the need to be able to continue doing so. On July 28, 2020, physicians identifying themselves as America's Frontline Doctors gathered in Washington, D.C. to speak out on behalf of their patients.[392] The video of the doctors' press conference has been taken down and banned by Twitter, YouTube, and Facebook. You can see the video and hear from these frontline doctors directly.[393, 394, 395]

Why the Government and Medical Globalists Attack Lifesaving Hydroxychloroquine and Ivermectin

The worldwide suppression of hydroxychloroquine and now ivermectin, even at the direct cost of lives, is so inexplicable that before describing the situation further, it may help to answer the question, "Why is it so necessary for the establishment to suppress inexpensive, easily available treatments for COVID-19?" The answer is this: *If effective treatments are already available, the FDA cannot issue Emergency Use Authorizations (EUAs), allowing the FDA to skip its usual safety and effectiveness studies to push expensive, highly remunerative drugs and vaccines down the pipeline as experiments to be inflicted on the population.*[396]

If they cannot issue EUAs, which exempt drugs and vaccines from going through the usual FDA multi-year approval process, they cannot create Warp Speed or any other process aimed at giving the drug companies government help in slamming through their new, expensive, and dangerously experimental drugs and vaccines.

The EUA statute permitting a Warp Speed approach explicitly states, *"For FDA to issue an EUA, there must be no adequate, approved, and available alternative to the candidate product for diagnosing, preventing, or treating the disease or condition."*[397]

In short, any existing lifesaving drugs must be shot down, and people must be allowed to die so drug companies and their big investors like Bill Gates and their supporters like Anthony Fauci can gain more wealth, self-aggrandizement, and power.

Under the Emergency Use Statute, All Drugs and Vaccines Are Experimental When Given to the People

As noted above, under the EUA system, when the vaccine or drug reaches the stage of approval, it only needs to meet this weak criterion of "may be effective." The statute explains:[398]

> The "may be effective" standard for EUAs provides for a lower level of evidence than the "effectiveness" standard that FDA uses for product approvals.

It is important to know that "may be effective" means the treatment has not been proven to be effective by the FDA. That a drug or vaccine "may be effective" is a very low standard—much too low to gain approval under normal FDA conditions or to have doctors or patient advocates take enthusiastically to the treatment. Despite what Fauci, NIH, and the FDA say, the drugs and vaccines coming out under Operation Warp Speed have not been proven to be safe or effective. They have a high likelihood of being unsafe and useless given the short-term, rushed, and politically corrupt circumstances of their rapid development and distribution under the EUA.

Remember, without denying the effectiveness of hydroxychloroquine and ivermectin, Fauci and his associates could not have created or continued Operation Warp Speed in May 2020 to use taxpayer money to help them rush medications and vaccines through the usually more thorough and markedly slower FDA approval process.

DEEP STATE CORRUPTION AND OPPOSITION

In 2006, federal legislation created BARDA—Biomedical Advanced Research and Development Authority—an agency devoted to organizing and funding partnerships with drug companies to create medications and vaccines during national emergencies.[399] Much too late, President Trump removed and reassigned the head of BARDA, Rick Bright.[400] Bright, who was appointed by President Obama shortly before President Trump took over from him, publicly boasted about disrupting President Trump's policies for the treatment of COVID-19:[401]

> Specifically, and contrary to misguided directives, I limited the broad use of chloroquine and hydroxychloroquine, promoted by the administration as a panacea, but which clearly lack scientific merit.

Bright then bolstered his argument with a false statement based on the corrupt research from the global predatory medical and scientific establishment that we are examining in this chapter:

> These drugs have potentially serious risks associated with them, including increased mortality observed in some recent studies in patients with COVID-19.

Bright's brazen public admission of interfering with Trump administration policies exemplifies how immune and how proud Deep State bureaucrats felt in undermining the presidency of Donald Trump. His decision to block hydroxychloroquine continues to cause or contribute to untold numbers of unnecessary deaths in America and elsewhere.

> *In evaluating the inadequacy of President Trump's efforts to promote hydroxychloroquine for COVID-19, consider the malicious resistance the former president was up against from the global predators, Fauci, and the Deep State within his own executive branch.*

One More Hint that the Globalists Knew a Pandemic Was Coming

Here is one more eerie suggestion of an international leadership prescient about the coming pandemic, a topic we explore in depth in Chapters 15 and 16, and present as a timeline in our *Chronology and Overview with Pandemic Predictions and Plans*. In January 2017, Homeland Security published a 135-page grand plan for an interagency response to "biological incidents."[402] It is possible the government was tooling up in anticipation of something it knew was inevitable—a leak from one of the many laboratories experimenting with coronaviruses as pathogens, probably originating in China with its poorly secured facilities. It's also possible there was some sort of advance notice, a possibility we will explore further throughout this book.

Think of it this way. On the one side, we have a virtual army of multibillionaire globalists allied with the giant, worldwide drug industry, all poised to become incredibly wealthier and more powerful with the advent of a pandemic by having the government support and carry out their research for them while reducing the standards of approval, so their products do not have to be proven effective.

On the other side stands a limited number of honorable, ethical, devoted physicians, most of whom are risking their careers to save lives by defying NIH, the FDA, and the CDC. Among the dedicated physicians we know, none have ever seen such an unethical and deadly restraint imposed on the rights of patients and physicians to seek the most effective treatments available. It is a ghastly horror show that most people barely see because the media and the scientific community have thrown in their lot, not with lowly hydroxychloroquine, but with the global predators.

The Safety of HCQ

In my many decades of experience reviewing drug side effects, hydroxychloroquine is one of the safest drugs I have evaluated. The drug has been FDA approved for 65 years, so its safety profile is well known. The FDA-Approved Full Prescribing Information has no black-box warning about lethal risks as many other drugs do, including many psychiatric drugs.[403]

Hydroxychloroquine is on the World Health Organization's *List of Essential Medicines*.[404] It has been known for decades as being among the safest and most effective medicines needed in any health care system. Almost all problems are with larger or longer-term amounts than doses used to treat the current epidemic. Deaths are extremely rare, and the WHO states the following:

> "Despite hundreds of millions of doses administered in the treatment of malaria, there have been no reports of sudden unexplained deaths associated with quinine, chloroquine, or amodiaquine, although each drug causes QT/QTc interval prolongation."[405]

The cardiac issue, QT interval prolongation everyone warns about, is extraordinarily common—*found in 247 other drugs including many commonly used psychiatric drugs.*[406] Many U.S. doctors who use it for various FDA-approved purposes—malaria, lupus, rheumatoid arthritis—have announced publicly they have never seen a death from it over many years.

ORGANIZED GOVERNMENT ATTACKS ON LIFESAVING TREATMENTS

FDA Breaks Its Own Rules to Stop Hydroxychloroquine

Believe it or not, we have barely begun to unravel for you the globalists who have come out in the open to stop hydroxychloroquine. Now we look at the FDA and then NIH. The FDA itself has agreed until now that U.S. doctors are allowed to prescribe medicines for off-label use. The agency has stated, "From the FDA perspective, once the FDA approves a drug, healthcare providers generally may prescribe the drug for an unapproved use when they judge that it is medically appropriate for their patient."[407] The right of doctors and their patients to agree to use off-label drugs is established by tradition and law, something I can attest to as a practicing physician and as a forensic psychiatrist.[408, 409] Once a medication is approved, physicians commonly use it for a wide variety of other purposes. There is probably more off-label use of medications in the world today than on-label use.

In my specialty of psychiatry, it is routine for doctors to use a drug, psychiatric or not, if in their judgment they believe it will be helpful. I have not had a case in which I testified that a doctor committed malpractice because he gave a drug off-label.

Despite traditional prohibitions on interfering with treatment decisions between doctors and their patients, the FDA joined the Pharmaceutical Empire's attack on hydroxychloroquine and has tried to terminate its off-label prescription. On April 24, 2020, the FDA issued a "Drug Safety Communication" titled *FDA Cautions Against Use Outside of the Hospital Setting or a Clinical Trial Due to Risk of Heart Rhythm Problems.*[410] The agency gave a dire warning to any doctors' clinic or private practice doctors about any use of hydroxychloroquine, stating it "should be used for COVID-19 only when patients can be appropriately monitored in the hospital as required by the EUA (Emergency Use Authorization) or are enrolled in a clinical trial with appropriate screening and monitoring."[411]

The FDA also warned doctors not to use the drug except in a preestablished controlled clinical trial: "If a healthcare professional is considering use of hydroxychloroquine or chloroquine to treat or prevent COVID-19, FDA recommends checking www.clinicaltrials.gov[412] for a suitable clinical trial and considering enrolling the patient."

To make sure physicians would not even have emergency access to the drug, on June 15, 2020, the FDA cancelled the Emergency Use Authorization of hydroxychloroquine which it had previously approved.[413] This nailed the coffin shut on tens of thousands of people whose lives might have been saved by taking safe and effective doses of hydroxychloroquine early in their illnesses.

Why would the FDA take such an illegal, drastic, and deadly stand against a safe and effective drug already in use worldwide against a pandemic? The global medical/pharmaceutical complex lives in dread of a cheap drug like hydroxychloroquine that will squash its mad goldrush toward making expensive drugs and vaccines for COVID-19. This has happened before, when a vaccine developed for other coronaviruses remained unfunded for human trials in 2016 because of a lack of immediate need or interest.[414] The FDA has put the pharmaceutical empire above all other human and scientific concerns. The NIH joined the fray on behalf of the global predators.

NIH Goes Further to Shut Down *All* Preventive and Early Treatment of COVID-19

Half a year into the pandemic, amid rising public and professional concerns, on October 9, 2020, the National Institutes of Health (NIH) took another step toward medical totalitarianism with an extraordinary announcement. It further took treatment out of the hands of American doctors and their patients by declaring there were *no good treatments for prevention or early treatment of COVID-19*. Patients must wait for treatment from a doctor until they are so sick that they need hospitalization *and* oxygen. They probably took this dramatic step in part as a response to all the doctors who were organizing to promote the treatment they were using despite establishment pressure.

Since then, patients all over the United States have called or seen their doctors with early symptoms of COVID-19—such as a dry cough, tiredness, cold-like symptoms, and a loss of smell and taste—and their doctors have turned them away. Literally turned them away, oftentimes mouthing NIH by saying, "There are no outpatient treatments for COVID-19," or "We're not allowed to treat COVID-19 outpatients. Go to the hospital if you get worse."

To confirm what we are reporting, the following is taken from a section of the NIH treatment guidelines for COVID-19 titled "For Patients with COVID-19 Who Are Not Hospitalized or Who Are Hospitalized with Moderate Disease but Do Not Require Supplemental Oxygen":[415]

Recommendations

- The Panel does not recommend any specific antiviral or immunomodulatory therapy for the treatment of COVID-19 in these patients. [This is an attack on hydroxychloroquine.]

- The Panel **recommends against** the use of **dexamethasone** or **other corticosteroids** for the treatment of COVID-19 unless a patient has another clinical indication for corticosteroid therapy.

Never in the history of medicine has such deadly rubbish been accepted by the medical establishment. Doctors cannot even try to treat their patients until they are on death's door.

By the new standards, even President Donald Trump would not have been given any antiviral or anti-inflammatory agents to save his life because he did not require oxygen in the hospital. By these standards, lives that could be saved by excellent preventive or early treatment will now die. That is happening right now!

People are dying because NIH, the FDA, and the CDC are each, in their own ways, preventing Americans from receiving good treatment. The federal agencies do this in collaboration with drug companies and multibillionaires who are gaining wealth, self-aggrandizement, and power by developing incredibly expensive drugs and vaccines with the financial support of the government.

It is not easy to suppress and censor such vital information as the availability of lifesaving medications during a pandemic. To do that the global exploiters have used tech media such as Facebook, Twitter, and YouTube to censor comments by President Trump[416, 417] and by physicians who support the use of hydroxychloroquine.[418] Only the power of predatory globalism can account for this, and we will continue to describe who they are and what they are doing.

The leaders of globalism, many directly involved with the Pharmaceutical Empire, are doing everything in their power to stifle hydroxychloroquine, and now ivermectin.[419] Draconian measures by state governors have made it increasingly difficult for physicians to use these medications.[420] Fortunately, there are legal ways to step around these restrictions, as indicated in the front of our book in "How to Find Lifesaving COVID-19 Treatment Guides and Doctors."

Ivermectin Partially Overcomes Its Initial Rejection by NIH

On August 27, 2020, NIH used its stacked COVID-19 Treatment Guidelines Panel to recommend "against the use of ivermectin for the treatment of COVID-19, except in a clinical trial."[421] As it did with NIH's assault on hydroxychloroquine, the media and the medical/scientific establishment reacted as if NIH had the legal power to stop doctors from prescribing a medication for a disorder, when it does not. Nevertheless, doctors in America and many other countries stopped using it. Once again, the pharmaceutical industry was protected from an inexpensive, safe, and effective drug.[422]

For reasons we cannot ascertain, on January 14, 2021, NIH somewhat softened its stance on ivermectin, issuing an opinion that it did not recommend *for or against* using the drug. Instead, it stated, "There are insufficient data to recommend either for or against the use of ivermectin for the treatment of COVID-19."[423] Again, this had no legal power, but some informed, courageous, and honorable doctors began to use effective medications. The media and the medical/scientific community continue to reject them, cowering before the broadsides directed at them by their establishment with its links to global predators who will benefit from many more deaths.

Pierre Kory MD testified in front of U.S. Senator Ron Johnson's Homeland Security Committee Meeting on "Focus on Early Treatment of COVID-19" on December 8, 2020. In his written testimony, he began with a brief review and summary of "Data Supporting Ivermectin as a Potential Global Solution to the COVID-19 Pandemic." He wrote, "Ivermectin is already eradicating coronavirus infections in multiple regions of the world. Dozens of studies demonstrate its efficacy …" Here is his summary with all its multiple citations:

1) Since 2012, multiple in-vitro studies have demonstrated that Ivermectin inhibits the replication of many viruses, including influenza, Zika, Dengue and others.[424]

2) Ivermectin inhibits SARS-CoV-2 replication, leading to the absence of nearly all viral material by 48h in infected cell cultures.[425]

3) Ivermectin has potent anti-inflammatory properties with in-vitro data demonstrating profound inhibition of both cytokine production and transcription of nuclear factor-κB (NF-κB), the most potent mediator of inflammation.[426]

4) Ivermectin significantly diminishes viral load and protects against organ damage in multiple animal models when infected with SARS-CoV-2 or similar coronaviruses.[427]

5) Ivermectin prevents transmission and development of COVID-19 disease in those exposed to infected patients.[428]

6) Ivermectin hastens recovery and prevents deterioration in patients with mild to moderate disease treated early after symptoms.[429]

7) Ivermectin hastens recovery and avoidance of ICU admission and death in hospitalized patients.[430]

8) Ivermectin reduces mortality in critically ill patients with COVID-19.[431]

9) Ivermectin leads to striking reductions in case-fatality rates in regions with widespread use.[432]

10) The safety of Ivermectin is nearly unparalleled given its near nil drug interactions along with only mild and rare side effects observed in almost 40 years of use and billions of doses administered.[433]

11) The World Health Organization has long included Ivermectin on its "List of Essential Medicines."[434]

Dr. Kory then gives a more detailed presentation of ivermectin's "existing clinical studies in the prevention, early, and late treatment phases of COVID-19 … All studies are positive, with considerable magnitude benefits, with the vast majority reaching strong statistical significance."

It appears that the safety and effectiveness of ivermectin is similar to or better than hydroxychloroquine. There is also some experience with combining the medications. Dr. Kory explained in his report that the widespread dissemination of his finds did not result in a single media or governmental contact, except from Uganda's government which was planning to use the medication. There was scant news coverage of his testimony. Occasional mainstream media "news" coverage with negative headlines such as "No Evidence Ivermectin is a Miracle Drug Against COVID-19."[435]

The Short Tenure of Dr. Atlas

After President Trump chose physician Scott Atlas as a new advisor and member of the White House Coronavirus Task Force, Dr. Atlas declared, "The drug hydroxychloroquine has gotten a bad rap thanks to politics, media hype, and some 'garbage' medical research."[436] He elaborated:

> I sort of make the analogy that we all know objective journalism is basically dead in this country, I'm very cynical about that, and now what we're seeing is that objective science is nearly dead. ... Hydroxychloroquine is super safe. It's a complete myth, it's a total distortion, to say that, oh, my God, this drug is very dangerous for people. It's been used for 65 or 70 years, not just prophylactically for malaria, which I used it myself for that many years ago, but also used for people with things like rheumatic arthritis, autoimmune-type diseases. Very safe drug.

Dr. Atlas's statements are consistent with the best available science. What happened next? After three tempestuous months, Dr. Atlas left the White House Coronavirus Task Force.[437]

Open Letter to Fauci about His Opposition to Hydroxychloroquine for COVID-19

The cutting edge of public health totalitarianism in the U.S. and the world today is Anthony Fauci MD, a man who has successfully taken the reins to control what happens during COVID-19. Fauci exemplifies how totalitarianism has erupted through positions of power during the current pandemic.

The following open letter to Dr. Fauci characterizes and challenges the behavior of Fauci in a striking fashion but has been utterly ignored by the major media. It can only be found in smaller newspaper outlets. Indeed, most people are unaware of how many physicians are outraged by what Fauci is doing because it is not covered in the major media. The entire letter should be read by anyone interested in public health policy and its implementation by Anthony Fauci. It was written by three physicians: George C. Fareed MD,

Brawley, California; Michael M. Jacobs MD, MPH, Pensacola, Florida; and Donald C. Pompan MD, Salinas, California. It was dated August 13, 2020, and updated August 22, 2020. Here is the opening statement of the Letter to Dr. Fauci,[438]

> Dear Dr. Fauci:
>
> You were placed into the most high-profile role regarding America's response to the coronavirus pandemic. Americans have relied on your medical expertise concerning the wearing of masks, resuming employment, returning to school, and of course medical treatment.
>
> You are largely unchallenged in terms of your medical opinions. You are the de facto "COVID-19 Czar." This is unusual in the medical profession in which doctors' opinions are challenged by other physicians in the form of exchanges between doctors at hospitals, medical conferences, as well as debate in medical journals. You render your opinions unchallenged, without formal public opposition from physicians who passionately disagree with you. It is incontestable that the public is best served when opinions and policy are based on the prevailing evidence and science, and able to withstand the scrutiny of medical professionals.

The doctors' letter goes on to describe the well-known outstanding physicians who support COVID-19 treatment based on hydroxychloroquine. It then challenges Fauci:

> Yet, you continue to reject the use of hydroxychloroquine, except in a hospital setting in the form of clinical trials, repeatedly emphasizing the lack of evidence supporting its use. Hydroxychloroquine, despite 65 years of use for malaria, and over 40 years for lupus and rheumatoid arthritis, with a well-established safety profile, has been deemed by you and the FDA as unsafe for use in the treatment of symptomatic COVID-19 infections. Your opinions have influenced the thinking of physicians and their patients, medical boards, state and federal agencies, pharmacists, hospitals, and just about everyone involved in medical decision making.

Indeed, your opinions impacted the health of Americans, and many aspects of our day-to-day lives including employment and school. Those of us who prescribe hydroxychloroquine, zinc, and azithromycin/doxycycline believe fervently that early outpatient use would save tens of thousands of lives and enable our country to dramatically alter the response to COVID-19. We advocate for an approach that will reduce fear and allow Americans to get their lives back.

We hope that our questions compel you to reconsider your current approach to COVID-19 infection.

That this trenchant letter has been so ignored by the media and major medical organizations indicates the hold globalism and the pharmaceutical industry, in league with government agencies, have upon the world.

We Urge President Biden to Release Hydroxychloroquine and Ivermectin

In a time full of political fraud and deception, one of the most unfair charges thrown at President Trump was that under his direction the United States had a relatively high fatality rate. This was not President Trump's fault. Blame belongs to those who opposed using "Trump's Drug," hydroxychloroquine and, more recently, ivermectin. President Biden is showing no inclination to resume President Trump's lead in advocating for the release of hydroxychloroquine from its vast government stores.

President Biden, you are not being so thoroughly undermined by the world globalist exploiters of humanity, so you are in a better position to cancel the FDA and NIH directives that kidnapped hydroxychloroquine and ivermectin. You could then make available the tens of millions of hydroxychloroquine doses stockpiled by the federal government.[439]

CHAPTER 10

Fauci Cheats to Make Expensive Bad Drugs Look Good

F auci has called *randomized controlled trials* (*RCTs*) the Gold Standard for testing drugs. The problem is that science is only as reliable as the institutions and people who plan, monitor, and analyze the data and later promote the results of the clinical trials. Fauci displayed outrageous bias from the beginning in his handling of his NIAID-funded, remdesivir clinical trial. Fauci conducted the trial on behalf of and for the benefit of the manufacturer, Gilead. He criticized President Trump for encouraging the use of a very safe and helpful hydroxychloroquine for COVID-19 and then proceeded to push remdesivir, which had failed in two earlier trials: one because too many patients were dying and the other because it did not help and worsened the respiratory condition of a significant number of patients.

By Fauci's Gold Standard of clinical trials, remdesivir should have been trashed before its clinical trial began, and other more hopeful drugs should have replaced it at the head of the line. Fauci had to know from the beginning that remdesivir was a failed antiviral drug and would probably do more harm than good. An earlier, famous remdesivir trial for Ebola was stopped because remdesivir was causing a significantly higher mortality rate than other antiviral drugs in the same trial.[440] As one recent medical source noted, the remdesivir arm of the Ebola trial had to be aborted "because of an increase in death among patients taking it, meaning it did not help those patients."[441]

Then, shortly before Fauci began his remdesivir clinical trial, a strong peer-reviewed article with a double-blind study was published indicating that remdesivir is both useless and dangerous in treating COVID-19. *The Lancet* study found remdesivir devoid of any statistically significant clinical improvements:[442]

> **Interpretation:** In this study of adult patients admitted to hospital for severe COVID-19, remdesivir was not associated with statistically significant clinical benefits (p. 1, bold in original).

In addition, the study found *no antiviral effect for the drug in human patients*:

> … remdesivir did not result in significant reductions in SARS-CoV-2 RNA loads or detectability in upper respiratory tract or sputum specimens in this study (p. 9).

More frightening, remdesivir produced very severe adverse reactions in the form of "respiratory failure" or "acute respiratory distress syndrome" in 5% of patients. In other words, five of every 100 patients taking remdesivir developed a life-threatening decline in their condition. This echoes the findings of the disastrous remdesivir Ebola trial.[443]

Long before his promotion of remdesivir and his corruption of its clinical trial, Fauci was already professionally invested in Gilead, the manufacturer of remdesivir. Fauci's institute already had a panel for the purpose of developing treatment guidelines for the pandemic on an ongoing basis. It is called "NIH COVID-19 Treatment Guidelines." Hundreds of drugs were coming up the pipeline seeking federal approval for treating the novel coronavirus.[444] On a government treatment advisory panel, would you expect to find a dominant membership block that was financially indebted to one drug company with a single drug, while almost every other drug company and their drugs were left out?

Of the 50 members on Fauci's committee to set treatment guidelines, only 11 admitted to financial ties to drug companies.[445] Nine of the 11 revealed financial ties to Gilead, remdesivir's manufacturer. The ties consist of "Research Support," "Consultant," and "Advisory Board." Put simply, from the beginning, Fauci was stacking his treatment guidelines committee with those who had ties to Gilead, remdesivir's manufacturer. The committee's actions confirmed that all of them were dancing to Fauci's tune of making remdesivir the world's first and most remunerative officially approved drug for COVID-19.

The Fauci-controlled panel—before any clinical trials were completed and analyzed—nonetheless came out strongly in favor of the highly experimental remdesivir and wholly against the commonly used hydroxychloroquine. The panel declared, "On the basis of preliminary clinical trial data, the Panel recommends the investigational antiviral agent remdesivir for the treatment of COVID-19 in hospitalized patients with severe disease."[446] From the start, it was willing to recommend a highly expensive, high-risk "investigational" drug—but not older, safer drugs that clinicians had been using for many decades to treat viruses and that had more positive clinical trials than remdesivir.

An Illustration of Evil Intent

To begin with, Fauci's drug remdesivir had a largely unrevealed encumbrance. Fauci and his globalist media rarely mention that remdesivir *must be given intravenously over a five or ten-day period.* This procedure itself has risks. It also limits its use to patients sick enough to be hospitalized. As a result, it cannot be used around the world like hydroxychloroquine and ivermectin as a prophylaxis and as a treatment to be given at the earliest signs of the disease, long before the patient becomes sick enough to need hospitalization or to die.

Fauci's evil intent to enrich global predators is nowhere better illustrated than this scenario: He prematurely pushes remdesivir—a proven-dangerous, probably ineffective, highly expensive intravenous drug, while he suppresses hydroxychloroquine and ivermectin which are proven-safe, highly effective, abundant and inexpensive drugs that can be obtained over-the-counter for pennies in many countries.

The Boondoggle

What did Gilead get from Fauci? As previously described, Gilead received the approval of Fauci's NIH committee for use of remdesivir in severely ill patients, even though there was no evidence for its effectiveness and, more remarkably, the NIH had no authority to recommend and, in effect, approve drugs for specific treatments.

Then, Fauci used Emergency Use Authorization for making remdesivir available before any successful studies had been completed. The combination of rejecting the inexpensive, safe, and effective hydroxychloroquine for emergency use while accepting the extremely expensive, highly unsafe, and proven ineffective remdesivir was an egregious display of the power of Fauci and the many global predators feeding on drug company profits.

Next, the remdesivir clinical trial aimed at obtaining FDA approval was "sponsored by the National Institute of Allergy and Infectious Diseases (NIAID)" and was "the first clinical trial launched in the United States to evaluate an experimental treatment for COVID-19."[447] That's a very powerful but wholly misleading endorsement and in itself probably guaranteed the FDA would approve the drug in deference to its colleagues at NIH and NIAID.

Beyond all that, there was the financial windfall to Gilead: NIAID organized and funded their trial, saving Gilead in the range of $70-100 million dollars. More importantly, NIAID saved Gilead years of time, guaranteed its place first in line ahead of innumerable competitors, and positioned Fauci to clean up any messy results.[448, 449]

How Fauci Cheated in Handling the Trial

Once Fauci got the remdesivir study going, the drug was doing so poorly that he started to manipulate it. One example was highlighted by a May 1, 2020, headline in *The Washington Post*: "Government researchers changed metric to measure coronavirus drug remdesivir during clinical trial."[450] The "metric" is the "endpoint" or standard used to determine the effectiveness of a drug, such as reducing the death rate, severity of the illness, time to complete recovery, or number of days in the hospital. During the trial, Fauci vastly lowered the standard for measuring success because the drug was not proving effective.

Before a scientifically based clinical trial begins, researchers are required to establish and to state in advance the endpoints or standards for determining effectiveness. They are not supposed to manipulate those endpoints before the trial is over. The endpoints for remdesivir were listed in the original plan for the trial. These initial standards included two truly meaningful criteria for calling remdesivir a success—lowering the fatality rate and increasing the rate of complete recovery.[451]

Lowering the endpoints twice is a sign of a high degree of bias and manipulation in a clinical trial.[452] It also indicates that the researchers knew the drug was failing, which indicates that they were not conducting a truly double-blind study where the researchers do not know who is getting the drug and who is getting the placebo. Fauci was cheating by figuring out in advance that the study was going poorly for remdesivir and Gilead (and the patients!), and then tried to cover it up by discarding the original and more reasonable standards for success.

The new Fauci standards of care were bizarrely concocted. The only standard for success became "time to recovery," with a unique standard for recovery that did not mean actual recovery. Fauci's criteria for "recovery" included patients who remained hospitalized or who were at home requiring limitations on their activities and/or requiring oxygen.[453]

Here are the three categories for "success" for remdesivir:

1) Hospitalized, not requiring supplemental oxygen—no longer requires ongoing medical care; 2) Not hospitalized, limitation on activities and/or requiring home oxygen; 3) Not hospitalized, no limitations on activities.[454]

Put simply, a patient was judged to have been successfully treated and "recovered" even when the patient remained in the hospital or returned home partially disabled with limited activities and with compromised respiratory function requiring supplemental oxygen.

Fauci's NIH press release estimated remdesivir shortened the recovery time for the COVID-19 illness by a mere four days (from 15 to 11).[455] As inadequate as that sounds for a wonder drug's only achievement, it is utterly fallacious when realizing Fauci still had to reinvent the concept of recovery in order to squeeze these meager results out of the massively complex data. That Fauci would intervene to manipulate and distort the trial was both unconscionable and arrogant, as well as unethical. Remarkably, it was not the most radical or extreme of Fauci's actions.

Fauci Massively Intervenes to Prematurely End the Trial

The remdesivir trials must have been a complete bust because Fauci terminated them ahead of time by breaking the double-blind. In an interview, Fauci explained that the results of the remdesivir trial were so promising he had "an ethical obligation to immediately let the placebo group know so they can have access" to the drug.[456] This was a lie because the results were so bad that the endpoints or standards for success had to be ridiculously contorted and lowered.

Furthermore, as described earlier in this chapter, two earlier clinical trials showed remdesivir was potentially lethal. By encouraging placebo patients to switch to remdesivir in the NIH clinical trials, Fauci exposed them to potentially lethal adverse effects.

The remdesivir trial, like its Brazilian chloroquine trial (Chapter 9), demonstrated how globalists are willing to sacrifice patients to their greater good of increasing the wealth, self-aggrandizement, and power of the global elite. In this case, they were sacrificing humanity by fraudulently pushing onto the world market a drug that is unsafe and ineffective.

Fauci's remdesivir has never been allowed to meet Fauci's own Gold Standard—a completed, randomized, double-blind, placebo-controlled clinical trial. This turned Fauci's "Gold Standard" into one of the weakest kinds of study, an open observational study that was not allowed to finish as planned. What a bizarre ending to the miracle drug Fauci so highly touted two weeks earlier, declaring, "This will be the standard of care."[457]

The globalist betrayals of humankind did not end here with Fauci's fraudulent remdesivir clinical trial.

The FDA Approves Remdesivir

After Fauci prematurely terminated the NIAID remdesivir study, I imagined the drug would have difficulty getting FDA approval. With the changing of its stated goals for success, its failure to demonstrate any significant positive results, its lack of attention to adverse effects, and finally its premature breaking of the double-blind, thus ending the clinical trial, it was both a failed study and a non-study.

Nonetheless, on October 22, 2020, the FDA approved remdesivir, now with the trade name Veklury, for the safe and effective treatment of COVID-19.[458] The FDA also based its conclusion on two admittedly even weaker studies, but the flagship NIAID study by Fauci was present as "substantial evidence" with no mention of any of the malfeasance described in this chapter.

The FDA acceptance of these fraudulent conditions shows that the leaders of the supposedly "watchdog" agency would sink as low as Fauci to pander to drug companies, billionaire investors, and their own embedded globalist bureaucracy.

New England Journal of Medicine (NEJM) Promotes Fauci's Failed Study of Remdesivir

The *New England Journal of Medicine* is rated by authorities as the most respected and impactful journal in the world, and it published Fauci's study. Like the FDA approval, it, too, made no mention of any of the cheating and manipulation, including the changing of the endpoint standards for success and the premature termination.[459] Ultimately, all three of the world's most prestigious medical journals,[460] *The New England Journal of Medicine*, *The Journal of the American Medical Association*, and *The Lancet* would display their globalist, predatory nature in treacherous ways.

Fauci's Inevitable Career on Behalf of the Pharmaceutical Empire

Thirty-six years ago, Dr. Anthony Fauci was appointed director of NIH's National Institute of Allergy and Infectious Diseases (NIAID). According to his Institute profile,[461] "He oversees an extensive portfolio of basic and applied research to prevent, diagnose, and treat established infectious diseases," and he "has advised six presidents on HIV/AIDS and many other domestic and global health issues." He is the "author, coauthor, or editor of more than 1,300 scientific publications, including several textbooks." Dr. Fauci is probably the most influential American in respect to anything having to do with epidemic diseases, including our current struggle with COVID-19.

In evaluating Dr. Fauci's motives, keep in mind it is impossible to remain director of the NIH Institute of Allergy and Infectious Diseases for 36 years and a consultant to six presidents without being deeply beholden and obedient to pharmaceutical companies who manufacture and sell antibiotics, antiviral agents, and vaccines. I had a brief two-year career as a physician and a full-time consultant to the National Institute of Mental Health (NIMH) with the rank of lieutenant-commander in the U.S. Public Health Service. It was clear from the beginning that government health officials work as if industry, powerful lobbying groups, and politicians are sitting on their shoulders, pulling strings to make their heads nod, and their influence and their budgets grow.

Fauci's longevity is a testimonial to his responsiveness to the enormous power wielded by the pharmacological industry. Their influence and control over him explains Fauci's suppression of a tried-and-true treatments while promoting remdesivir, a very expensive and experimental alternative with known, severe, adverse effects and no evidence of effectiveness. It accounts for the egregious support he gave to research in collaboration with China that created deadly viruses, which gave the pharmaceutical industry more information about developing vaccines.

Most sinister of all, Fauci's funding and support eventually gave China the SARS-CoV-2 virus that turned into a pandemic. That pandemic has already become a multi-billion-dollar boondoggle as Fauci pumps huge bundles of money into the coffers of his friends and partners in industry.

WHO FINDS REMDESIVIR INEFFECTIVE AND WITHDRAWS SUPPORT

Surprisingly, WHO recently completed a worldwide study confirming remdesivir did not improve the lifespan of patients and was not worth its high cost. ABC News reported:

> A large study led by the World Health Organization suggests that the antiviral drug remdesivir did not help hospitalized COVID-19 patients, in contrast to an earlier study that made the medicine a standard of care in the United States and many other countries.

> The results announced Friday do not negate the previous ones, and the WHO study was not as rigorous as the earlier one led by the U.S. National Institutes of Health. But they add to concerns about how much value the pricey drug gives because none of the studies have found it can improve survival.

Notice how ABC News leans over backward not to discredit Fauci's NIH study, even suggesting it was more rigorous than WHO's study, when, in fact, it was a disgraceful scandal.

Why did WHO admit remdesivir does not work? Probably only an insider could know. I suspect they were finding the drug was losing its support in the medical community, given its lack of effectiveness, its costliness, and the difficulty of giving it intravenously. More importantly, by discrediting remdesivir, they ramp up the need for more drugs to be approved on a rush basis by the FDA and other drug agencies around the world. Remember, the key to understanding global predators is their obsessive devotion to increasing their wealth, self-aggrandizement, and power.

Fortunately, there are other physicians and scientists in the news and on Twitter, who have spotted Fauci was untrustworthy in his handling of the remdesivir trial.[462] However, unless the political tide turns against top-down, coercive, public health policies and measures, the nation will continue to move toward increasingly totalitarian rule under the guise of treating COVID-19. Under the Biden administration, it is getting much worse.

CHAPTER 11

COVID-19 "Vaccines" – A Giant and Fatal Experimentation on Humanity

It is horrifying to consider, but in the short time between editing this chapter and finishing the book, the number of vaccine-related deaths reported to the CDC has leaped from 4,434 on May 12, 2021, to 10,991 in early July 2021, a mere two months. We further evaluate this growing crisis in the endnotes of the *Chronology and Overview*.

If it were humanly possible to put numbers on the alarm we already expressed in this chapter, the number of deaths is now 2 ½ times greater and our feeling of alarm grows. Rather than delay the book by updating the numbers in this chapter, we ask our readers to realize that the COVID-19 vaccine death rate has vastly increased and will be yet more swollen by the time you read this. The COVID-19 vaccines are causing a worldwide catastrophe of unimaginable, unpredictable proportions.

The warnings we wrote in earlier drafts of this chapter have been preempted by three growing tragic realities in May and June 2021:

First, as of May 12, 2021, 4,434 deaths have been reported to the CDC among people shortly after receiving COVID-19 vaccines.[463] *Toward the end of May, it rose to over 5,000.*

These deaths occurred in the U.S. within a few days of being jabbed with Moderna or Pfizer's mRNA vaccines, many on the first day and most within the first several days. No one in authority is paying any attention to it, even though a tiny fraction of so many deaths would have led the CDC and FDA to pull any other vaccine off the market.

Second, it is clear that the SARS-CoV-2 spike protein that the vaccines force the human body to make are highly toxic to humans, often causing bleeding and clotting in the endothelium or lining of blood vessels. Contrary to claims by advocates, the vaccine-induced antigenic material does not remain localized at the vaccine site but instead is produced throughout the body.

On June 1, 2021, a review of the dangers of the vaccines was published with senior author and former White House COVID-19 advisor to President Trump, Dr. Paul Elias Alexander. The publication put together a new and distressing body of information and research about the toxicity of the SARS-CoV-2 spike proteins that the vaccines force our RNA to produce throughout our bodies.[464] It is now clear that mRNA and DNA vaccines spread their effects throughout the body and brain, causing cells to make the deadly SARS-CoV spike protein. That spike then damages the endothelium of cells and in the blood vessels, causing inflammation, bleeding, and clotting. In addition to these harms, these areas of destruction and the spike protein induce the immune system to attack not only the spike protein but also the afflicted areas of the body. The result can be cytokine storms—hyperactive overreactions of the immune system that can be lethal.

Third—and this may be the most disturbing conclusion we have drawn in this book—the lethality of the COVID-19 vaccines was planned. The danger of vaccinated human patients dying after exposure to SARS-CoV-2 has been repeatedly warned since 2006 based on animal studies. These ominous reports continued to be published after the start of Operation Warp Speed.

This is the first time we are discussing this mind-numbing conclusion and so we will now go into more detail.

The Background of Intentionally Making
and Distributing Lethal Vaccines

The search for DNA vaccines began as early as 2003 in the midst of the SARS-CoV-1 epidemic.[465] In 2006, two years after the original SARS-CoV worldwide epidemic subsided, Damon Deming and a group of American researchers, including Ralph Baric from North Carolina, were already able to characterize many of the insurmountable problems with SARS-CoV vaccines.[466] Using a spike protein similar to the ones in current vaccines, a "SARS-CoV strain spike (S) glycoprotein protein spike," they found the "vaccines not only failed to protect" from a wide range of SARS-CoV, they "resulted in enhanced immunopathology with eosinophilic infiltrates within the lungs of SARS-CoV–challenged mice." The "pathology presented at day 4, peaked around day 7, and persisted through day 14, and was likely mediated by cellular immune responses." During this research, they made "chimeric" viruses, confirming that it was gain-of-function research. It was funded by Fauci's NIAID.

At no point in the future were animal researchers able to overcome this fatal flaw—that animals vaccinated for SARS-CoV infections with mRNA or DNA vaccines experienced serious illness and death when challenged by SARS-CoV pathogens. The illness often resembled what we now call COVID-19.

In 2008, a large Japanese research team discovered a similar reaction, which they summarized in the title of their paper, "Prior Immunization with Severe Acute Respiratory Syndrome (SARS)-Associated Coronavirus (SARS-CoV) Nucleocapsid Protein Causes Severe Pneumonia in Mice Infected with SARS-CoV1."[467] The Acknowledgments show no U.S. funding, but they thank a University of Pittsburgh colleague for help.

A 2011 research article funded by Fauci's NIAID and the University of North Carolina, with Baric's name on it, reconfirmed and elaborated on the conclusions of earlier efforts to make vaccines for coronaviruses. Vaccinated older mice exposed to SARS-CoV are not protected and instead are vulnerable to human-like severe immune reactions. *Furthermore, the culprit is the spike protein that the animal or human body makes in response to vaccine.*

The report declared, "Importantly, aged animals displayed increased eosinophilic immune pathology in the lungs and were not protected against significant virus replication" and "When challenged with zoonotic and human chimeric SARS-CoV incorporating variant spike glycoproteins, the aged BALB/c mouse model reproduces severe lung damage associated with human disease, including diffuse alveolar damage, hyaline membrane formation, and death."

Notice the mention of "zoonotic and human chimeric SARS-CoV," indicating that they are doing gain-of-function research and problems making more than just one chimeric or lab-made virus.

More research continued to confirm that mRNA or DNA vaccines for SARS-CoV posed this threat of illness or death when the animals were exposed to a SARS-CoV and the threat was especially strong to impaired or older animals.

Then, in September 2020, in the midst of the Operation Warp Speed, a major review warned that the vaccines "being expedited through preclinical and clinical development" could "exacerbate COVID-19 through antibody-dependent enhancement (ADE)."[468] Antibody-dependent enhancement is a euphemism for when vaccines make the disease worse.

In case you doubt that those in authority know about this risk of antibody-dependent enhancement, here is a statement from NIH that was on the internet in response to a search of the term on June 8, 2021:[469]

Antibody-dependent enhancement
and SARS-CoV-2 vaccines

> **Antibody**-based drugs and vaccines against severe acute respiratory syndrome coronavirus 2 (SARS-CoV-2) are being expedited through preclinical and clinical development. Data from the study of SARS-CoV and other respiratory viruses suggest that anti-SARS-CoV-2 antibodies could exacerbate COVID-19 through **antibody-dependent enhancement** (ADE). [Emphases in original.]

There is an unbroken pattern of mRNA and DNA viruses sensitizing animals to have extreme immunological over-reactions, causing illness and death. The problem was never overcome and as the research continues, it is apparent that no one is trying to overcome it. Thus, the lethal dangers of these vaccines were documented in the scientific literature at least as far back as 2006 and summarized in a review article in 2020 that overlapped in time with Operation Warp Speed. Nonetheless, the global predators never flinched and pushed the vaccines into production and worldwide emergency use.

People like Fauci and Gates, as well as drug company scientists, also would have known that these lethal threats would not be limited to old people. They would likely afflict anyone with comorbidities or compromised immune systems and probably some healthy people, too. They knew in particular that any exposure to SARS-CoV after vaccination would produce potentially lethal hyperactivity of the immune system all the way up to a cytokine storm. They knew this was going to happen and went ahead despite it—*and therefore almost certainly because of it.*

To add to the purposiveness of pushing forward with the lethal vaccines, I found no evidence that the globalists put any funding, let alone the hundreds of millions of dollars required, into trying to solve the problem. It is no longer tenable to blame the vaccine deaths on sloppiness or rushing to make money. All those participating in the planning for COVID-19 were also planning for the vaccine to further the aim of the ridding humanity of its infirm or old people.

There seems to be no other explanation for continuing full speed ahead despite this ominous research which spanned 14 years right on through the EUA in late 2020. There can be no other explanation for allowing the vaccine deaths to pile up in the U.S. and around the world, far exceeding that of any other vaccines. There can be no reason why the over 5,000 deaths reported in association with vaccination are being ignored by the CDC, the FDA, NIH, and NIAID.

What is going on here? Those who fully understand what they are doing are probably certain they are doing the right thing—culling humanity. It is a principle with a long and honored history among eugenicists and euthanasia advocates, including many scientists, philanthropic organizations, and political leaders. It is twenty-first century eugenics. In the case of COVID-19 and the vaccines, there is nothing voluntary about it—and that makes it mass murder.

How the New Vaccines Work

The quotes around "vaccine" in the title of this chapter indicate that these injections bear no resemblance to anything else in human history that has ever been called a vaccine. They most nearly resemble a Trojan horse.

DNA (deoxyribonucleic acid) contains our genetic code—the blueprint for our biological existence. It is found in the nucleus of every cell in our bodies. An enzyme enables the DNA to morph into RNA (ribonucleic acid), which can become mRNA, the messenger with instructions for the cell to make proteins.

Both types of vaccines use genetic instructions copied from SARS-CoV-2 which enable the deadly virus to make the spike protein that gains entry into human cells. Instead of taking the genetic material directly from SARS-CoV-2—an expensive and painstaking task—scientists reproduce the genetic material artificially in the lab. In both DNA and mRNA vaccines, the genetic material induces our body to make the same spike protein that is on SARS-CoV-2.

In effect, as the recipient of the vaccine, our own body receives an injection with a "message" to create an alien SARS-CoV spike protein known to be very toxic to mammals, including humans. Hence, our comparison to the infamous Trojan horse. The theoretical concept is that our immune system will then attack the spike protein that its own body has just produced, which, hopefully, will induce some degree of immunity to SARS-CoV-2 if it challenges us. But that is not what happened when the hidden Greeks crept out of the gift horse in the middle of the night, slaughtered the Trojan guards, and opened the gates to the Greek army.

We will find a number of problems associated with all the cells in the body being directed to produce the alien spike protein. The spike protein itself is part of the disease process of COVID-19. Among other harmful effects, it damages endothelial cells in many organs of the body and in the blood vessels this can cause bleeding and clotting, and even death. In addition, the spike protein can elicit severe immunological overreactions, leading to the kind of potentially deadly cytokine storm seen in COVID-19.

There are two important differences between mRNA and DNA vaccines, the last of which are called viral vector vaccines. The first difference is the method of delivering the piece of genetic code into the human body. The mRNA, which is sealed in a nanoparticle package, enters the blood stream and finds its way to the RNA. The DNA vaccines, as the term viral vector suggests, instead use a non-replicating animal adenovirus to carry the strand of DNA genetic code to the cells in the body.

The second difference is that the mRNA vaccines send their genetic strands into the blood stream directly to the RNA in the cells with no intermediate steps. The mRNA provides genetic code with the orders and the plan to make the SARS-CoV-2 spike protein. In contrast, the DNA vaccine involves two steps. First it finds its way to a messenger RNA in the body which then tells the RNA to produce the same SARS-CoV-2 spike protein.

It is untrue that the injected materials and their effects are limited to the injection site. The messenger coding goes throughout the body, potentially affecting all cells in the body.[470]

So far there is no evidence that the mRNA or DNA vaccines directly impinge upon DNA in the cell nucleus, but thoughtful journalists[471] and scientists[472] do worry that the process could end up modifying human DNA and the genome. The process may be similar to the "reverse-transcription" by which COVID-19 may be changing the genome of some patients.[473] The cavalier dismissal of any such concerns is incompatible with ethical medicine, science, or public health policy.

Since the outcome of both DNA and mRNA vaccines is the same—forcing the body to make SARS-CoV-2 spike proteins—the two types of vaccines have many similar adverse effects. *Because the spike protein causes some of the more severe harms in COVID-19, we would expect at times to see those same harmful effects as a result of vaccination.*

Moderna and Pfizer-BioNTech make mRNA vaccines, the main ones marketed in the U.S. Through its subsidiary Janssen, Johnson & Johnson makes a DNA vaccine that has been more commonly distributed in Europe. AstraZeneca is marketing a similar DNA vaccine.

Numerous other vaccines are coming down the pipeline using these and other technologies.[474] All are being rushed through without normal regulatory approval and must be considered dangerous and experimental. All will have their own hazards which, because of Operation Warp Speed, will only begin to become obvious after they have been given to millions of people over many months and years.

Another problem has been arising with the mRNA vaccines—the nanoparticle polyethylene glycol (PEG) sealer that is put around the messenger RNA to package and protect it in the human body before it enters the cells. It is being implicated in severe allergic reactions (anaphylaxis) and other problems.[475]

If all this sounds more like science fiction than science, in some ways it is. There is an expression about buying stock or other investments "on spec," meaning purchasing them as a high-risk investment that might miserably fail. We, the people of world, are being forced by our governments to buy these vaccines "on spec." Indeed, they are not even vaccines in the usual sense, but intricate experiments with RNA and

DNA and the production of potentially deadly alien protein. Tragically, we can lose more than money on these radical speculations.

Because of the gravity of exposure to these injections that tamper with our RNA, and potentially with our DNA—one which also shoots nanoparticles into our bodies to circulate throughout and the other which injects alien DNA attached to an animal virus—I asked physician and scientist Peter A. McCullough MD, one of the most dedicated and knowledgeable professionals in the world on COVID-19, if he would write a comment for our readers. On June 2, 2021, Dr. McCullough wrote:

> Never before have we used technology where we trick our own bodies to make a dangerous protein which is a laboratory product of bioterrorism research. Where does the Wuhan spike protein go in the body? What stops our cells from making too much or producing for too long a period of time? Most importantly, if the SARS-CoV-2 kills humans with the spear of the unbreakable GOF Wuhan spike protein, why would the vaccinologists not consider the same weapon causing similar types of death in some of those coaxed forward for their injections?

Recent May and June 2021 scientific reports related to increased detection of severe reactions to mRNA and DNA vaccines can be found in the *Review and Chronology* which was updated shortly after completion of the main text of this book.

We will return to the built-in hazards of the vaccines but first the increasing numbers of deaths must be examined.

Deaths From COVID-19 Vaccines

The Growing Number of Reported Deaths in the U.S. and Worldwide

Spain,[476] Sweden,[477] and the whole of Europe[478] have been reporting their own concerns about excessive deaths following vaccination. In Israel, based on Ministry of Health data, an independent scientist found more deaths from the mRNA vaccines than from COVID-19 and described it as a "new Holocaust."[479]

In the U.S., vaccine-related deaths are reported to the CDC's adverse reporting system called VAERS. It is well known and generally agreed upon that reported adverse vaccine events to VAERS typically represent a very small fraction of the actual events, as low as 1% or even less of the actual events taking place. This is true as well for the FDA system called FAERS (and earlier called AERS), although estimates are better for the FDA's system, suggesting that 1-10% of serious adverse drug events are actually reported.

In respect to the FDA FAERS reporting system for medication adverse effects, in 2017 an authoritative review declared:[480]

> There is no existing solid methodology for estimating the reporting rate in FAERS. When forced, most drug safety professionals will cite a single line in a GAO report from 2000 that estimates that between 1 and 10% of adverse events are reported into FAERS.

The GAO or U.S. General Accounting Office report was submitted to Congress as part of testimony before a Congressional Committee in 2000 by the internal watchdog agency.[481] The GAO's actual conclusions are even more revealing:

> *It is well known that all spontaneous reporting systems experience a high level of underreporting.* For example, FDA believes that its system for gathering information about ADEs, the Adverse Event Reporting System (AERS), receives reports for only about 1 to 10 percent of all ADEs. Indeed, FDA relies on AERS primarily to generate "signals" of new adverse drug events that the agency can then investigate through other data sources. (Italics added.)

In a report by Harvard Pilgrim Healthcare, Inc., the CDC's vaccine reporting system, VAERS, is found to do even more poorly in reflecting the true frequency of the problems reported:[482]

> Adverse events from drugs and vaccines are common, but underreported. ... Likewise, *fewer than 1% of vaccine adverse events are reported.* Low reporting rates preclude or slow the identification of "problem" drugs and vaccines that endanger public

health. New surveillance methods for drug and vaccine adverse effects are needed. Barriers to reporting include a lack of clinician awareness, uncertainty about when and what to report, as well as the burdens of reporting: reporting is not part of clinicians' usual workflow, takes time, and is duplicative. (Italics added.)

What Is the Actual Number of Deaths Being Caused by the COVID-19 Vaccines?

As bizarre and outrageous as it seems, the number of actual deaths from the vaccines in the U.S. could be 100 times greater than 4,434! That number is stratospheric; I will not even write it down. Instead, let us take an extremely conservative estimate that not one in 100 vaccine deaths, but one in 10 vaccine deaths are reported. *This would mean that in excess of 40,000 people have been killed by the vaccines in America as of mid-May 2021.* This must be integrated into our thinking and actions: A conservative, minimalistic interpretation of current VAERS data indicates that more than 40,000 people have died shortly after vaccination in the U.S. for COVID-19.

How bad is that in terms of historical vaccine disasters? It is unprecedented because, until now, vaccines have never been rushed through under EUAs. In addition, as the next section demonstrates, vaccines have been withdrawn for far, far lesser threats.

The Certainty that COVID-19 Vaccines Should Be Stopped

If the CDC or FDA were operating under normal conditions, reports of over 4,000 deaths would have completely halted the distribution of these lethal agents. This has happened with many lesser adverse effects than death. Tragically, in today's climate of "do any harm necessary" to sell vaccines and to crush the American spirit, this slaughter goes unabated while the CDC refuses even to examine the catastrophe.

Exactly what did it take for the federal government to withdraw the swine flu vaccine from use? It was not deaths, as some people believe, but a rare neurological adverse effect. According to the CDC itself: [483]

> In 1976 there was a small increased risk of a serious neurological disorder called Guillain-Barré Syndrome (GBS) following vaccination with a swine flu vaccine. The increased risk was approximately 1 additional case of GBS for every 100,000 people who got the swine flu vaccine. When over 40 million people were vaccinated against swine flu, federal health officials decided that the possibility of an association of GBS with the vaccine, however small, necessitated stopping immunization until the issue could be explored.

When the risk was confirmed, the vaccine was permanently withdrawn.

In fact, vaccines can be recalled when there is a safety concern *in the absence of reported events*. Again, according to the CDC: In 2007, Merck & Company, Inc. voluntarily recalled 1.2 million doses of Haemophilus influenzae type b (Hib) vaccines due to concerns about potential contamination with bacteria called *B. cereus*. The recall was a precaution, and after careful review, no evidence of *B. cereus* infection was found in recipients of recalled Hib vaccines.

Clearly, the rules have been drastically changed, pushing us into a new era of letting Americans die from COVID-19 vaccines by the thousands without the CDC so much as calling for an investigation. Instead, beyond all reason and ethics, the CDC has declared that none of the deaths, even ones within hours of the injection, were vaccine related.[484] This is severe dereliction of duty by the CDC and by other government agencies and politicians who do not intervene to stop this murderous policy of ignoring such an overwhelming pattern of reports of death.

In earlier years, a vaccine was withdrawn for causing a rare but severe neurological disorder and another was withdrawn for an unverified concern about bacterial contamination. Today, the COVID-19 vaccines will not be investigated, let alone withdrawn, by the U.S. government when we have over 4,000 reported deaths that are but the tip of the iceberg of total deaths.

These events, along with many others, confirm a basic conclusion of our book that COVID-19 is a cover story for global predators to extract untold billions or trillions of dollars from our economies and peoples worldwide while imposing top-down government and docility upon all of us.

The Chronology and Overview at the end of the book makes it easy to track how the inexorable plans for vaccinating the world for COVID-19 were put into action as early as 2015-2017.

Why Do So Few Adverse Drug and Vaccine Events Get Reported

In my research and forensic work, I have some experience with VAERS and a career of experience with FDA Adverse Event Reporting System (FAERS). Healthcare providers are required to report vaccine injuries but not drug injuries; but, in my considerable experience, healthcare providers almost never report deaths related to treatment. In more than 50 years as a medical expert, I have reviewed hundreds of injury cases related to medical treatment, many involving death. *Not one of them was reported to the FDA by healthcare providers.*

There are many reasons why healthcare providers do not report injuries to their patients. Besides the omnipresent fear of lawsuits, there are also fears of repercussions from oversight committees or from outraged families. In addition, there is the natural inclination to deny their potential role in someone's death. Beyond that, healthcare providers are so overworked that they avoid taking on something new and unfamiliar, such as making an official report to a federal agency. Many of them have never made a report to FAERS or VAERS and do not know how.

There are even more reasons why *vaccine* injuries will not be reported. First, the environment is now extremely hostile to any recognition of vaccine-related injuries. Websites are being taken down or censored for simply discussing vaccine injuries.[485] Most healthcare providers live in fear of discussing vaccine injuries among themselves or with the patients. They do not dare mention the problem on social media, let alone make an official report about them. This is very similar to their fear of discussing alternatives to vaccines, such as well-established protocols for the early treatment of COVID-19.

Second, most people being vaccinated probably have no relationship with the individual vaccinating them and have no follow up afterward. They have been jabbed at a drugstore or in a tent after lining up in a parking lot. If the vaccinated individual dies, the family would not even know whom to tell about it, and none of the peripheral healthcare providers would take it on themselves to report it.

No single report to the CDC or FDA's reporting system by itself means anything until it is investigated more thoroughly. However, as I describe in my scientific articles and medical textbooks, such as *Brain-Disabling Treatments in Psychiatry, Second Edition* (2008), patterns of death reports are exceedingly important and may in themselves lead the FDA to withdraw a drug or add to its warning. *Usually, it takes years for these patterns to emerge, so the rapid accumulation of these vaccination death reports should lead to a moratorium on all vaccinations in order to save lives.*

Early Disclosures about Vaccine Hazards

These vaccines were and remain too hazardous for human experimentation. Because of the great interest the globalists were showing in the mRNA vaccines, many researchers began animal research on them several decades ago. The conclusive results are straightforward: The vaccines are too deadly in animals to be given to humans, even experimentally.

An important review article was published in 2012, a mere eight years before Moderna and Pfizer's vaccines, warning not to proceed with human experimentation, let alone with worldwide vaccination.[486] The review concluded:

> These SARS-CoV vaccines all induced antibody and protection against infection with SARS-CoV. However, challenge of mice given any of the vaccines led to occurrence of Th2-type immunopathology suggesting hypersensitivity to SARS-CoV components was induced. Caution in proceeding to application of a SARS-CoV vaccine in humans is indicated.

In other words, the vaccines made many animals too hyperactive in their immune responses, making many of them ill or killing them. In their discussion in the text of the article, they also emphasized that their review "supports caution for clinical vaccine trials with SARS-CoV vaccines in humans." This is a very thorough review with many citations to the scientific literature spanning several years. The conclusions confirm the fears that many scientists have expressed from early on about vaccines for COVID-19.

It now appears that one reason for rushing these vaccines through the FDA on an "emergency basis" under the EUA was to avoid a real examination of their known dangers. Under normal circumstances, the existing animal literature would have prohibited continuing with experiments on humans.

THE DANGERS OF THE ALIEN SPIKE PROTEIN CREATED IN THE BODY AFTER VACCINATION

This is the second new threat, along with vaccine deaths, summarized at the beginning of the chapter. In December 2020, two scientific articles were published indicating the spike protein (S protein) alone can cause inflammation and damage to the endothelial cells in blood vessels in the body and brain of animals and humans.[487, 488] The spikes can cause both hemorrhages and clotting. Therefore, the mRNA and DNA vaccines, which force the human body to make the same spike protein attached in the lab to make SARS-CoV-2 lethal, were very likely to be harmful to people.[489] In the rush to market vaccines, this threat was ignored.

The damage from isolated spike proteins found in animals is similar to the damage found in humans from COVID-19. Endothelial damage leads to a clotting disorder in the capillaries of the lungs which can be seen on autopsy.[490] In a more widespread manner, "Prior evidence has established the crucial role of endothelial cells in maintaining and regulating vascular homeostasis and blood coagulation."[491] In animals, damage from the spike protein has also caused injury to the brain.[492] The evidence increasingly points to the role of spike protein both in COVID-19 and in COVID-19 *vaccine* injuries.

The vaccines are forcing the RNA in cells throughout the human body to make spike proteins like those found in COVID-19 and so injuries from the vaccines can mimic COVID-19. This is consistent with reports that some post-vaccination individuals become positive on tests for COVID-19. It also confirms that vaccination before or after exposure to or illness from COVID-19 may create or worsen a COVID-19-like condition. Much of the evidence for this is presently anecdotal, but when anecdotal evidence is confirmed by the known actions of a vaccine, in this case, it must be taken very seriously.

On May 31, 2021, I interviewed Uwe Alschner on the "Dr. Peter Breggin Hour" on radio and by video about a recent discloser of new information that Pfizer had withheld about its vaccine. Dr. Alschner is a German historian who has been investigating COVID-19. Despite earlier pharmaceutical company and vaccine advocate claims that the vaccinations only affected an area of the arm near the injection site,[493] the internal Pfizer study demonstrated that the mRNA vaccination forced cells *throughout the human body* to produce the highly toxic artificial spike proteins.

Unexpectedly intense concentrations were found in the ovaries,[494] posing a threat to fertility that was already being raised anecdotally based on apparent hormonal disruptions in women.[495] Because we were already under a "First strike" threat from YouTube for posting an interview with attorney Tom Renz about the threat of COVID-19 vaccines to children, we posted Dr. Alschner's interview on Brighteon.com and elsewhere.

A review by Paul Elias Alexander, Parvez Dara, and Howard Tenenbaum was published on June 1, 2020, with the title," The COVID-19 spike protein may be a potentially unsafe toxic endothelial pathogen."[496] They too found the vaccine to be a special threat to pregnant women, starting, "We never ever administer an untested biological substance to a pregnant woman." They showed grave concerns about future adverse effects that have not even been tested. They noted that the vaccines effects are especially active in the ovaries. For many people that brings echoes of Bill Gates and other global predators to mind—the ones who so urgently want to reduce the population of the world.[497]

The Vaccines Are Not FDA-Approved for Safety and Effectiveness

Newspaper analyses abound with statements like this: "A Food and Drug Administration analysis finds Moderna's COVID-19 vaccine to be safe and effective …"[498] or "FDA experts say Moderna's COVID-19 is safe and effective."[499] In fact, no such official determination has ever been made by the FDA for the vaccines arriving through Operation Warp Speed, and none is feasible without corruption. Exactly as we have documented in Chapter 10 in respect to Warp Speed drugs like remdesivir, the Emergency Use Authorization (EUA) requires only that the drug or vaccine "may be effective."

FDA approval for a vaccine takes 8-12 years, and sometimes 20 years, and not eight months. The mRNA and DNA vaccines are approved for limited "emergency use" based on the weak unscientific standard that they "might be useful," and it seems unlikely that they could ever pass a legitimate FDA safety standard.

FORESEEN RESULTS OF RUSHING AHEAD EXPERIMENTALLY

What could possibly go wrong with an experimental vaccine for a coronavirus? We began our COVID-19 reform work in earnest when our widely viewed video led President Trump to stop Fauci's funding of collaborative studies with China making deadly SARS-coronaviruses in the lab. In that video,[500] we described how vaccinations did not work and then warned about a specific problem with vaccinations for the deadly SARS-CoV that had been created in the lab. *Prior standard vaccination with the killed virus made the viral disease even worse in older or ill mice! Instead of being protected by vaccination, the older and immune-compromised animals became more ill than expected when they were exposed to the live virus after vaccination.* Notably, "vaccination resulted in robust immune pathology" or harmful hyper-responses of the immune system also seen in SARS-CoV-2 infection.

Data in one state reported that half the deaths were occurring in a senior living facility,[501] and most likely that is a consistent pattern. Arguments that "these patients are just old and already sick" do not mitigate the reports. Also, if almost half the deaths occur outside these facilities, it means that younger and less incapacitated people are dying from the vaccines.

The mRNA and DNA vaccines do not involve killed or weakened viruses, and many experts feel that they should not be called vaccines. These vaccines induce our bodies to make a protein very similar to the spike protein of SARS-CoV-2, posing a risk of a more serious immune response than traditional vaccines. They also cause a risk from the spikes which are being produced in cells throughout the entire human body. These spikes are now proving to be serious risks for harming the lining of small blood vessels, causing inflammation and clotting.[502] A research team from several American medical centers found, "It is concluded that ACE2+ endothelial damage is a central part of SARS-CoV2 pathology and may be induced by the spike protein alone." It was published online December 24, 2020, and is dated April 2021.[503]

A Gates Donation Ousts a Courageous Vaccine Scientist

Dr. Gøtzsche is an extraordinarily conscientious doctor and exemplified in his role as cofounder and board member of the Cochrane, a consortium aimed at monitoring the quality of published medical science. I know Dr. Gøtzsche, and he is scrupulously scientific. Over a long and active career, he has published hundreds of scientific articles, many concerning vaccines. He is a strong proponent of vaccines, in general.[504] He also has background as a laboratory scientist working on vaccines.

In 2018, Dr. Gøtzsche was fired from Cochrane, the organization he founded, and then from his hospital in Denmark whose research department he had directed for many years. Successfully removing him from his career as professor and department director threw a chill of fear in any establishment scientist who wants to or is doing research in the vaccine field.

Fortunately, Dr. Gøtzsche is a pillar of strength and is continuing to play a significant role from more outside the establishment. Nevertheless, the vaccine establishment, which lurks at the top of the vaccine food chain, continues to win the larger battle, in part by excluding any criticism of vaccines from Cochrane and intimidating the rest of the medical/scientific establishment. Meanwhile, Cochrane, the organization whose illustrious founding goal was to monitor the quality of scientific publications, has now adopted the heinous goal of serving the Master of Vaccines, Bill Gates. This tale of greed and betrayal exemplifies what is happening throughout the entire medical/scientific establishment and everywhere else where the predatory globalists have become influential and dominant.

What Was Behind the Firing of Dr. Gøtzsche?

In September 2016, the Bill & Melinda Gates Foundation gave $1,156,829 to the Cochrane,[505] the organization that Dr. Gøtzsche helped found and where he continued to serve on the board of directors. The Bill & Melinda Gates Foundation, as already described, is a strong multibillion-dollar supporter of vaccinations, so much so that Gates recently announced, "We must make this the decade of vaccines."[506]

Two years later, in 2018, the Cochrane Foundation came under severe criticism from *The BMJ* and independent experts for its publication of an article minimizing the risks of the HPV vaccination. Dr. Peter Gøtzsche was one of the scientists who raised these questions before he was summarily thrown off Cochrane's Governing Board of Trustees on September 25, 2018. The *BMJ* linked Dr. Gøtzsche's dismissal to his criticism of the Cochrane HPV vaccination paper and to its support from the Gates Foundation with its deep commitment to the vaccination.[507]

What happened to Cochrane after the firing of Dr. Gøtzsche? It quicky deteriorated from a serious critic of scientific abuses to a publicist for the vaccine producers. In 2020, it published an annual review of vaccines that was supposed to be scientific, but it used a multi-color format similar to a very expensive drug company brochure with content to make a drug marketer's heart sing. But unlike drug ads, which are monitored by the FDA, this masquerades as scientific information that transcends any regulations pertaining to advertisements.

My Response to the Attack on Dr. Gøtzsche and Scientific Freedom

I was in close communication with Dr. Gøtzsche during and after the attack on him, worked with him on the establishment of his independent Institute for Scientific Freedom, and spoke at its founding conference in Denmark.[508]

These attacks on scientific inquiry inspired me to write my first scientific vaccine article.[509] I began writing it before COVID-19 and submitted my work for publication in mid-2020 to the *International Journal of Risk and Safety in Medicine*, which published it with great alacrity one month later. The title is "Moving past the vaccine/autism controversy—to examine potential vaccine neurological harms." My article makes no indictment of vaccines in general and is critical of some who make unsubstantiated claims about vaccine injuries.

My focus in the scientific publication is on the systemic cover-up of vaccine harms by the medical/scientific establishment and the drug companies. This is not a conspiracy theory. I document how it takes place in plain sight. My article calls for better and more open scientific research at a time when so many infants and children are being given multiple vaccines that have never been tested in combinations and never been tested using clinical trials with placebo control groups. Researching and writing the article turned out to be a fortuitous tune up for me for many issues surrounding COVID-19.

Bill Gates and Vaccines

Did Gates have a special interest in 2018 in suppressing criticism of vaccines? The next year, 2019, Bill Gates began to promote and to spend millions for a "universal flu vaccine."[510] Then, as already noted, he tried to make this the Decade of Vaccines. Attempting to crush Dr. Gøtzsche, one of the world's braver and most respected scientists, made less prestigious and less courageous scientists terrified of whispering even a hint of criticism of Gates' plans.

Does Bill Gates have patents and other industrial know-how that bring profits to the foundation named after him and his wife and which he personally directs? In short, do vaccines and their manufacture enrich his foundation, which is now his life's work? According to Intellectual Property Watch,[511] the answer is a definite *yes*. Bill Gates appears to be a vaccine tycoon. He is a special kind of tycoon, one who is so powerful he wants to declare a "Decade of the Vaccine" to gain worldwide support for his business.

Bill Gates is no longer a businessman; he's a philanthropist now, isn't he? No, not exactly. Here are excerpts from an interview with Gates by *Intellectual Property Watch*:

> "In terms of IP [intellectual property] what we do is actually quite simple," Gates said in response to a question from Intellectual Property Watch. "We fund research and we actually, ourselves or our partners, create intellectual property so that anything that is invented with our foundation money that goes to richer countries, we're actually getting a return on that money."

"By doing that we have more money to devote for research into neglected diseases and the diseases of the poor," he said. "Now when our medicines go into the poor countries, they are always going in without any intellectual property fee, at very lowest cost pricing."

"In fact," he added, "we're a pioneer of going to vaccine manufacturing to making volume commitments to allow them to build high volume facilities that are very, very low cost – so working with people on getting the prices down."

"But," he said, "the intellectual property system has worked very well to protect our investments so that when they are used in rich countries we get a payback and then we have the control to make sure that it is not creating any financial burden on the countries that are the poorest."

Bill Gates' proindustry, profit-making approach to healthcare is common knowledge and often appreciated in the financial press, such as *The Wall Street Journal.* [512]

Provided it is legal, is there anything unethical about a person making a great deal of money on healthcare which he funnels back into his own personal foundation? This might be a difficult or complicated question if the person in question were not one of the wealthiest and most powerful people in the world. [513] When one individual, who is already worth billions of dollars, becomes such a moving force within a single industry and specialized healthcare sector like vaccines and is able to enrich his foundation in the process, gaining yet more wealth and power, that presents an enormous public health risk to the world.

This section is but a small aspect of the overall involvement of Bill Gates in COVID-19 which this book investigates.

How Far Will Scientists Bend the Truth to Promote Vaccines?

When President Trump stopped supporting the World Health Organization, Bill Gates became its largest donor. [514] WHO then deleted and redefined scientific concepts to promote Bill Gates and the vaccine industry in which he invests so heavily and in which Communist China has such a great interest.

Jeffrey Tucker of AIER has written another insightful essay that begins by describing everyone's familiar experience of getting vertigo trying to follow conflicting public health announcements:[515]

> Coronavirus lived on surfaces until it didn't. Masks didn't work until they did, then they did not. There is asymptomatic transmission, except there isn't. Lockdowns work to control the virus except they do not. All these people are sick without symptoms until, whoops, PCR tests are wildly inaccurate because they were never intended to be diagnostic tools. Everyone is in danger of the virus except they aren't. It spreads in schools except it doesn't. ... Well, now I have another piece of evidence to add to the mile-high pile of fishy mess.

Tucker's remarkable report is titled "WHO Deletes Naturally Acquired Immunity from Its Website." Get rid of a basic scientific concept—acquired community or herd immunity—one with great explanatory power that has helped us save lives? Why would WHO do that? Because Bill Gates and his vaccine industry partners do not want people waiting around for their communities to develop herd immunity or community immunity. They want to make us so hopeless about ever achieving herd immunity that we will buy Bill's vaccines.

WHO deleted the original and accurate definition of herd immunity as something that can be achieved through vaccination or through "immunity developed through previous infection." Now WHO has removed from its website the definition of herd immunity as deriving from exposure to the disease itself. Instead, herd immunity loses its scientifically accurate and historically perfected definition and becomes something achieved only through vaccination.

This is no joke and shows how far the global predators will go. In their hands, science has become nothing more than a marketing tool. This puts our civilization one step further down the road to totalitarian rule.

The Special Standing of the Vaccine Manufacturers

Many professionals have told me the vaccine manufacturers are the nastiest of all those in the medical field. This is not a coincidence. They are making among the biggest bucks while being among the most unconscionable, least regulated, and most protected by federal legislation and a special vaccine court.

The vaccine manufacturers are so powerful that the FDA allows them to test vaccines without using *placebo*-controlled clinical trials. Remember, Fauci touted these formal clinical studies in which drugs are randomly compared to a harmless "sugar pill" as the "gold standard" and the only acceptable standard for drug approval. Nevertheless, for decades Fauci has gone along with the FDA not requiring placebo controls for the approval of any vaccines or for any post-marketing follow-up studies.

Add to that the extreme rush to develop vaccines lickety-split, plus the general lack of ethical restraints within the drug companies themselves, and there's more reason to be cautious.

Finally, and this is a shocker to most people, if you're injured by a vaccine, you cannot sue the manufacturer. Instead, you must sue the federal government in a special vaccine court which functions more like a kangaroo court than a legitimate American court.

The Risk of the New Vaccines Defies the Imagination

Even the establishment American Association of Medical Colleges (AAMC) has voiced concerns about the rush to make COVID-19 mRNA vaccines, while citing previous tragedies with other new vaccines.[516] Not only are mRNA and DNA vaccines dangerously complex and experimental and being rushed through an emergency approval process, but the FDA has never before approved one.

Because the mRNA and DNA interfere with the cellular regulatory mechanism called RNA, myriad unexpected adverse reactions are possible. These vaccines are essentially priming the immune system, while providing it a SARS-CoV-like "spike protein" as an internally created alien substance for the immune system to attack.

On December 5, 2020, I interviewed Peter A. McCullough MD, MPH, who is a leader in the field of treating COVID-19, especially in the critical early period of the illness. The one-hour discussion was videoed for *The Dr. Peter Breggin Hour* on the radio and as "TV" on my YouTube Channel on December 9, 2020.[517, 518] Dr. McCullough described the unique and largely untested mechanism of action of these new vaccines which bond with the individual's cellular RNA, affecting the function of the major guidance system for cell function and reproduction in us humans. Because this is such a new and radical intervention, and because the usual many years of testing have been reduced to two months, its potential dangers are unpredictable and potentially enormous on the health of the individual and the whole human population.

Frontline Doctors Raise Scientific Concerns

An excellent overall critique and warning about the SARS-CoV-2 vaccines is found in *America's Frontline Doctors Position Paper on COVID-19 Experimental Vaccines.* The organization, America's Frontline Doctors, like many good sources of information on COVID-19, has been subject to a mixture of media assaults and media blackouts but, in fact, represents the responsible practice of medicine and good science. Its Mission Statement includes:[519]

> The doctor-patient relationship is being threatened. That means quality patient care is under fire like never before. Powerful interests are undermining the effective practice of medicine with politicized science and biased information. Now more than ever, patients need access to independent, evidence-based information to make the best decisions for their healthcare. Doctors must have the independence to care for their patients without interference from government, media, and the medical establishment.

A fundamental conclusion of the doctor group's *Position Paper* is worth every American's attention. Under the heading of "Call to Action," it makes two simple potentially lifesaving points:

1. Always use the correct language. COVID-19 EXPERIMENTAL Vaccines.

2. Immediately make it known that you will refuse to consent with any attempt to mandate an experimental vaccine by an employer, school, or business.

The critical analysis section, "IV. COVID-19 Experimental Vaccines Controversy," of the Position Paper is divided into "Safety Concerns" and "Questions Regarding the Effectiveness." Below is a list of the categories discussed which provides a template for evaluating these dangerous agents at the present time and as more information becomes available:

Safety Concerns Regarding the Experimental COVID-19 Vaccines

1. Brand New Technology.

2. Failure of Previous Coronavirus Vaccines.

3. No Independently Published Animal Studies.

4. Known Complications.

5. Unknown Complications.

6. Pharmaceuticals are Immune from All Liability.

7. An Experimental Vaccine Is Not Safer Than a Very Low IFR. [This means that dangerous, experimental vaccines are not a safer alternative to improving treatments to bring about an increasingly low IFR or Infection Fatality Rate, which America's Frontline Doctors and other groups are achieving with medication combinations that include hydroxychloroquine or ivermectin.]

Questions Regarding the Effectiveness of the COVID-19 Experimental Vaccines

1. No Proof the Vaccine Stops Transmission of the Virus.

2. Unknown Mortality or Hospital Admission Benefit.

3. The Vaccine Lasts Unknown Duration.

The *Position Paper* produced by America's Frontline Doctors includes a serious warning that other sources and experts in virology have personally confirmed to us. Some fear this potential adverse reaction might make these experimental vaccines worsen the effects of the epidemic:

> Antibody Dependent Enhancement (ADE), is when anti-COVID antibodies, created by a vaccine, instead of protecting the person, cause a more severe or lethal disease when the person is later exposed to SARS-CoV-2 in the wild. The vaccine amplifies the infection rather than preventing damage. It may only be seen after months or years of use in populations around the world. This paradoxical reaction has been seen in other vaccines and animal trials.

On November 27, 2020, Britain's *The BMJ* published an opinion piece titled "COVID-19 vaccines: where are the data?" signed by four professors in the fields of medicine, preventive medicine, public health, and epidemiology.[520] They decried the widespread use of experimental vaccines before there was sufficient published data to allow the scientific community to evaluate the safety and effectiveness.

WebMD, an industry-oriented website,[521] nonetheless published an analysis called "Doctors Wary of Rushed COVID Vaccine." They reported, "Only 17% of doctors say they will get a COVID-19 vaccine if it is authorized before all clinical trials have been completed, according to results of a Medscape poll."

Conclusions about Vaccines

Some medical experts who are generally in favor of rushing the vaccine onto the market nonetheless have warned about the lack of information about long-term safety data and have suggested it might be best for people to wait a year before taking the vaccine.[522]

However, these recommendations were made before the recent disclosure of more than 4,000 vaccine-related deaths reported to the official VAERS reporting system. We are now calling for a moratorium on the vaccinations until the data can be studied. Older people, who are most vulnerable to vaccine injuries, must be included in the moratorium. Children must also be protected from the vaccines because COVID-19 poses no significant threat to them and because they are not a serious source of transmission of the virus. For all people, the risk of taking mRNA vaccines remains greater than the risk of doing without them.

In the midst of this debacle, the global predators are already making untold billions of dollars for the drug companies and for private investors including Bill Gates.[523,524] This is a global predatory feast for private investors and the pharmaceutical companies, in addition to those in banking and related industries. It is also a power party, a celebration, for government officials like Fauci who have been supporting them.

At a national conference days before President Trump's inauguration, Fauci bragged that he would reduce the "risk" for drug companies during the anticipated coming pandemic (Chapter 16). He more than kept his word by guaranteeing that his favorite companies would start earning billions right out of the starting gate. For many companies, Fauci and the government eliminated their risk by awarding billions of dollars to cover much of their initial costs, by guaranteeing huge purchases, and by protecting them from any lawsuits for the harms they would inevitably cause.

It was a sure thing for the global predators, but nothing was guaranteed for the public, including the safety or effectiveness of the vaccines. It has turned into a tragic disaster whose proportions are still unfolding, and the vaccinations must be stopped now.

We originally recommended that people wait a year for the giant vaccine experiment to unfold before daring to participate. Now we are calling upon the government to permanently withdraw all mRNA vaccines and all DNA vaccines (viral vector vaccines) from distribution and use. Given the already vast numbers of reported deaths and the mounting information on the dangers of the human body creating a SARS-CoV-2 spike protein within itself—these vaccines cannot be rationally or legitimately approved for continued injection into human beings even on an experimental, research, or emergency basis.

CHAPTER 12

Masks Are Not
About Your Health

Yesterday, a group of 3- and 4-year-olds were walking in a line outside my office window in Omaha. They were all spaced three feet apart, all in masks, heads bowed, not talking, hands uniformly behind their back. It looked like a dystopian procession of tiny prisoners—which, in essence, is what it was. No more laughing smiling faces. No holding your classmate's hand. Just masked children learning absolute obedience.

Lee Merritt, MD *Journal of American Physicians and Surgeons,* Winter 2021

The wearing of face masks became a political litmus test that was used to challenge President Trump and continues to be used to subdue the citizenry. The internet is filled with photos of strong advocates of face masks taking them off in restaurants or dropping them to their chins to speak clearly and to relieve discomfort. Meanwhile, we see intimidated people driving alone in their cars wearing masks and walking or jogging alone in parks with their faces covered and their breathing hampered.

In most situations, face masks are probably more harmful than helpful. Overall, there is a shocking lack of evidence for their usefulness in preventing viral disease transmission.

At the very start of the campaign for wearing face masks, there was a scientific pushback that was largely ignored. On April 20, 2020, a Rapid Response was published in *The BMJ* by Antonio Lazzarino, a physician and epidemiologist, University College London, UCL Institute of Epidemiology and Health Care, and two others. Their summary makes the following points which I have elaborated on in brackets:

> The two potential side effects that have already been acknowledged are:
>
> (1) Wearing a face mask may give a false sense of security and make people adopt a reduction in compliance with other infection control measures, including social distancing and handwashing.
>
> (2) Inappropriate use of face mask: people must not touch their masks, must change their single-use masks frequently or wash them regularly, dispose of them correctly and adopt other management measures, otherwise their risks and those of others may increase.
>
> Other potential side effects that we must consider are:
>
> (3) The quality and the volume of speech between two people wearing masks is considerably compromised and they may unconsciously come closer. While one may be trained to counteract side effects, this side effect may be more difficult to tackle.
>
> (4) Wearing a face mask makes the exhaled air go into the eyes. This generates an uncomfortable feeling and an impulse to touch your eyes. If your hands are contaminated, you are infecting yourself.
>
> (5) Face masks make breathing more difficult. [For people with COPD, asthma, a simple cold, or other respiratory problems, face masks worsen their breathlessness and increase their vulnerability to illness. Any viruses already in the body can be activated. Decreased oxygen concentration and increased carbon dioxide in the blood has many adverse metabolic effects and can trigger

a stress response with increased cortisol release that can impair the immune system. These breathing problems can also lead to lightheadedness, anxiety, and panic.][525]

(5B) The effects described at point 5 are amplified if face masks are heavily contaminated (see point 2).

(6) While impeding person-to-person transmission is key to limiting the outbreak, so far little importance has been given to the events taking place after a transmission has happened, when innate immunity plays a crucial role. The main purpose of the innate immune response is to immediately prevent the spread and movement of foreign pathogens throughout the body. The innate immunity's efficacy is highly dependent on the viral load. [Water vapor collecting in the face mask can support the activity of SARS-CoV-2, increasing the viral load and the risk of initiating or intensifying infection.]

In conclusion, … It is necessary to quantify the complex interactions that may well be operating between positive and negative effects of wearing surgical masks at population level. It is not time to act without evidence. [Citations omitted.]

D. J. Rancourt conducted a thorough review of randomized clinical studies of the value of face masks and concluded there is no compelling evidence to support them in the treatment of viral respiratory illnesses:[526]

Conclusion: By making mask-wearing recommendations and policies for the general public or by expressly condoning the practice, governments have both ignored the scientific evidence and done the opposite of following the precautionary principle. In an absence of knowledge, governments should not make policies that have a hypothetical potential to cause harm. The government has an onus barrier before it instigates a broad social engineering intervention or allows corporations to exploit fear-based sentiments. Furthermore, individuals should know that there is no known benefit arising from wearing a mask in a viral respiratory illness epidemic, and that scientific studies have shown that any benefit must be residually small, compared to other and determinative factors.

Masking Ourselves from Each Other

Wearing face masks cuts people off from each other's added facial expressions and cues. It reduces verbal communications. Today, when people go outdoors or into buildings wearing a mask, they often make no attempt to make a friendly gesture and indeed they commonly avert eyes. More than anything, masks create discomfort and alienation between people, which is a hallmark of the totalitarian society.

Decades ago, *In Crowds and Power*, Nobel Prize Winner in Literature, Elias Canetti, wrote astonishing passages about the negative effects of wearing masks in general:

> People's attitude to this play of the features varies. In some civilizations the freedom of the face is largely restricted; it is thought improper to show pain and pleasure openly; a man shuts them away inside himself and his face remains calm. The real reason for this attitude is the desire for personal autonomy: no intrusion on oneself is permitted, nor does one intrude on anyone else. A man is supposed to have the strength to stand alone and the strength to remain himself. The two things go hand in hand, for it is the influence of one man upon another which stimulates the unending succession of transformations. They are expressed in gestures and the movements of the face and, where these are suppressed, all transformation becomes difficult and, in the end, impossible.

> A little experience of the inflexibility of such unnatural "stoics" soon leads one to understand the general significance of the mask: it is a conclusion; into it flows all the ferment of the yet unclear and uncompleted metamorphoses which the natural human face so miraculously expresses, and there it ends. Once the mask is in

position there can be no more beginnings, no groping towards something new. The mask is clear-cut; it expresses something which is quite definite, and neither more nor less than this. It is fixed; the thing it expresses cannot change. ...

A mask expresses much but hides even more. Above all, it separates. (p. 374)

In our culture in ordinary times, the wearing of a mask or kerchief in public presents several threats, most notably, the person is sick and infectious, or the person is a criminal about to threaten us. In a time of great turmoil, with riots occurring nightly in cities around the country, and acts of vandalism in the name of political retribution or righteousness, perpetrators happily wear masks, giving them anonymity during their perpetrations.

ALEX BERENSON ON MASKS

Alex Berenson, the former science writer for *The New York Times*, has been doing excellent analyses of the false science surrounding COVID-19 and has published part 3 of his booklets *Unreported Truths about COVID-19*.[527] He covers the science and the waffling back and forth by people like Fauci, and concludes:

As Anthony Fauci said at a press conference in May, "It's not 100 percent effective. I mean, it's sort of respect for another person, and have that person respect you." He added that he wears masks "because I want to make it be a symbol for people to see that that's the kind of thing you should be doing."[528]

Of course, encouraging people to take actions that are (supposedly) symbolically valuable is different than forcing them. I may want to wear a pink pin to show I care about beating breast cancer, but Governor Cuomo can't make me.

At least I don't think he can, though I'm not so sure anymore.

The not-so-good reason is that making people wear masks frightens them. Frightens us. Masks are warnings none of us can escape. This virus is different. This virus is dangerous. This virus is not the flu. We had better hunker down until a vaccine is ready to save us all.

But the worst reason of all is that mask mandates appear to be an effort by governments to find out what restrictions on their civil liberties people will accept on the thinnest possible evidence. They are the not-so-thin edge of the wedge. Today, we must wear masks. Tomorrow we'll need negative COVID tests to travel between countries. Or vaccines to go to work.

I wish masks worked. I wish we didn't have to fight about them.

But they don't.

On December 16, 2020, health writer Conan Milner published a review in *The Epoch Times*, "Science of Masks: Masks have become mandatory in many places, despite conflicting evidence." Milner's report provides a balanced perspective which, I believe, confirms there is insufficient scientific evidence to force people to wear masks and there is a genuine risk of masks making some people much sicker. He points out the CDC has waffled on the subject, at first citing ten studies indicating they were not useful and then changing its mind. Fauci, too, has waffled, at first saying we did not need them and most recently saying masks would be required well into 2022.

At the very least, we can say the argument for the effectiveness of masks is weak, and the cost of wearing them is high in terms of causing and aggravating cardiovascular respiratory illnesses, diminishing social interactions, causing alienation, fostering docility and conformity, and making dictatorial politicians feel they can demand of the citizens ever more submission and forsaking of liberty.

CHAPTER 13

Case Counts Rising and More People Dying

In Chapter 8, we examined the inflated death rates used to threaten and intimidate the population. The threat of more people dying is fear mongering. Some factors that raise the death rates also raise the case counts. We found that Medicare, for example, can add $13,000 as a bonus simply for tacking on a diagnosis of COVID-19 to a hospitalized patient and $39,000 if the patient is placed on a ventilator.[529] That will certainly add to the national case count as well.

As we also indicated in Chapter 8, government suppression of safe and effective early treatments for COVID-19 has increased the real number of deaths, but these people did not have to die from the pandemic. Similarly, government resistance to preventive treatments causes more people to become ill and unnecessarily adds to the case counts.

Deaths associated with the rampant, ill-conceived vaccinations are being misdiagnosed as cases of COVID-19. This falsely adds to the COVID-19 death count while even more importantly artificially minimizes the increasing deaths due to vaccination.

There is enormous pressure on healthcare providers never to diagnose serious adverse events associated with vaccination but instead to assign serious diagnoses to COVID-19. We have heard about doctors who have refused to discuss with the relatives whether or not their family members became worse or died because of the vaccines. Doctors must be heroic, like Hooman Noorchashm MD, PhD,[530] to publicly discuss the dangers of the vaccines. Professionals and laypersons who so much as explore these problems on social media and YouTube are being taken down.

The exaggeration of the size and severity of the COVID-19 pandemic is not limited to the U.S. but afflicts the world. Karina Reiss PhD and Sucharit Bhakdi MD are scientists in Germany. In their excellent book, *Corona False Alarm? Facts and Figures*, they concluded:

> The SARS-CoV-2 outbreak was never an epidemic of national concern. Implementing the exceptional regulations of the [German] Infection Protection Act were and still are unfounded. In mid-April 2020, it was entirely evident that the epidemic was coming to an end and that the inappropriate preventive measures were causing irreparable collateral damage in all walks of life. Yet, the government continues its destructive crusade against the spook virus, thereby utterly disregarding the fundaments of true democracy.
>
> We can only hope that the admonishing voices of reason will in the future not be silenced by the dark forces on this earth.

ADDITIONAL WAYS IN WHICH CASE COUNTS BECOME INFLATED

CHANGING THE DEFINITION OF A CASE TO INCREASE THE NUMBERS

In medicine, the number of "cases" ordinarily means the number of people who are actually symptomatic or ill due to a disease. A case count is not supposed to represent the number of people who have tested positive for the disease.

The CDC defines a COVID-19 "case" by the detection of a certain level of virus in the body, but this is contrary to medical practice. A *case* in medicine means a person who is sick—someone with symptoms of a disease. People routinely walk around with SARS-CoV-2 and legions of other potentially infective cold and respiratory viruses without becoming ill or symptomatic.

The COVID-19 pandemic is real, but the exaggerated fear associated with it is largely the product of overcounting "cases" and "deaths."

The Flu Disappears

Further complicating the problem of counting cases or deaths from COVID-19 is that the flu has "disappeared." According to *The Columbian*: [531]

> Nationally, "this is the lowest flu season we've had on record," according to a surveillance system that is about 25 years old, said Lynnette Brammer of the U.S. Centers for Disease Control and Prevention.
>
> Hospitals say the usual steady stream of flu-stricken patients never materialized.

The CDC finds that flu is indeed disappearing in North America and a number of other places around the world.[532] It suggests that mitigation efforts aimed at controlling COVID-19 are knocking out the flu. Maybe that plays some role. Others suggest that SARS-CoV-2 blocks the entry of the flu virus into cells, but that is pure speculation and not enough people were infected with SARS-CoV-2 when flu began to disappear. No one in authority dares suggest the most obvious reason of all—doctors and facilities are diagnosing flu cases as COVID-19 cases. We discussed these incentives earlier in this chapter and in more detail in Chapter 8.

A House Built on the Sand of Uncertain Testing

Since the beginning of the pandemic in the United States, the CDC has failed to provide statistical clarity. According to ABC News, the initial COVID-19 test kits created by the CDC failed because of a "contaminated component."[533] Those kits were pulled back after producing a large number of false positives. On May 21, *The Atlantic* reported that the CDC was requiring positive tests for antibodies to be included in the daily count of new COVID-19 cases.[534] *The Atlantic* asked, "How Could the CDC Make That Mistake?" The antibody tests, if they were accurate, would indicate a *past* exposure and immune response to SARS-CoV-2, rather than a new case of COVID-19. Furthermore, the presence of antibodies in the blood does not indicate that the individual ever was a "case," meaning someone who was ever *ill with COVID-19*. Beyond that, the tests have many other flaws, including too many false positives and too much variance in how they are calibrated.

The CDC has been loosening its standards for defining a COVID-19 death case. On June 3, 2020, the CDC declared COVID-19 cases should include not only deaths that were "confirmed," but also "probable" death cases. A death certificate listing COVID-19 or SARS-CoV-2 as cause of death or a significant condition contributing to death is counted as a new case as well even without testing.

Beyond cases associated with death, nonlethal COVID-19 cases are also required to be counted as confirmed cases if they are, in fact, listed as "probable." A probable case can be identified if clinical criteria are met for at least two symptoms: fever (measured or subjective), chills, rigors, myalgia, headache, sore throat, or new nose and taste disorders. Or, if a person has at least one symptom of cough, shortness of breath, or difficulty breathing. Or, if a person has a severe respiratory illness and no alternative more likely diagnosis is found.

In addition, the person suspected of being positive for COVID-19 must have been in contact with a confirmed or probable case of COVID 19 or someone who had a clinically compatible illness and linkage to a confirmed case of COVID-19. In other words, you are counted as a COVID-19 case if you had any contact with a probable or confirmed case, plus a few symptoms that could be caused by an infinite number of different causes such as asthma, seasonal allergy, exposure to an irritant, a cold, heart trouble, or stress.

Collin County in Texas is a good example of how these changing standards affect the numbers of COVID-19 cases counted. On May 18, 2020, Collin County Commissioners Court, Administrative Services gave a PowerPoint presentation on the new case definition for COVID-19.[535, 536] She referred to the Texas Department of State Health Services guidelines distributed on May 11, 2020.[537] Prior to the changes, the case definition for a confirmed case of COVID-19 was a positive PCR (swab test for present infection) lab result for COVID-19. The new case definitions expand the total count of COVID-19 cases to many times the original confirmed cases. The numbers of death cases expand also as positive lab results are no longer required for the counting of a COVID-19 death. In each category, the counted cases will be many times higher. According to the *CDC FAQ: COVID-19 Data and Surveillance*, "A COVID-19 case includes confirmed and probable cases and deaths."[538]

In Illinois, probable cases were not being reported as of June 8, 2020.[539] A spokesperson for the state's health department explained probable cases were withheld "because there is concern from the public the number of deaths is being inflated … We need the public to have confidence in the data and therefore are reporting only those deaths that are laboratory confirmed." Many other specifics of incomplete or missing data from various states have been uncovered by media investigations, including *The Washington Post*.[12]

Seen by the Collin County, Texas, example earlier in this piece, as counties and states begin to add probable cases into their statistics, the COVID-19 cases will skyrocket. Graphs will shoot up. Artificially higher rates for the spread of COVID-19 enables governors to prematurely reclose economies, maintaining their authoritarian regimes.

PCR Testing Vastly Inflates the Numbers

The New York Times revealed a flaw in the testing that has been recognized early on by experts who have been dismissing PCR testing as woefully inadequate.[540] *The New York Times* reported:

> Some of the nation's leading public health experts are raising a new concern in the endless debate over coronavirus testing in the United States: The standard tests are diagnosing huge numbers of people who may be carrying relatively insignificant amounts of the virus. Most of these people are not likely to be contagious and identifying them may contribute to bottlenecks that prevent those who are contagious from being found in time. …
>
> The PCR test amplifies genetic matter from the virus in cycles; the fewer cycles required, the greater the amount of virus, or viral load, in the sample. The greater the viral load, the more likely the patient is to be contagious.

This number of amplification cycles needed to find the virus, called the cycle threshold, is never included in the results sent to doctors and coronavirus patients, although it could tell them how infectious the patients are.

In three sets of testing data including cycle thresholds, compiled by officials in Massachusetts, New York, and Nevada, up to 90 percent of people testing positive carried barely any virus, a review by "The Times" found.

In other words, only 10% of people identified by the test may actually have the disease and be contagious. This means in addition to the CDC vastly inflating the number of deaths, the PCR testing used for policymaking by the CDC and enthusiastically promulgated in the media is enormously exaggerated. Positive results are not positive for the disease and for contagion except in a relatively small number of patients. Once more, America is being inundated with frightening untrue data concocted by corrupt science in the service of corrupt politics.

Dr. Fauci keeps talking about science, but he is politically motivated and so ignores the worthlessness of most data emerging from the CDC. Science is based on facts while the data on COVID-19 cases is a jumbled mess and useless for informing how we should deal with opening as a nation. Both the government's case rates and death rates are extremely unreliable and vastly inflated. COVID-19 is not the threat it is made out to be. There is no scientific basis for the devastation of our social and economic life by government-enforced lockdowns and other abusive policies and practices, such as forcing people to be vaccinated on pain of not being allowed to go to work, to school, or to public events.

CHAPTER 14

Mass Murder in
New York State

The two of us live in Upstate New York. The overall death toll in New York has been staggering. In mid-July when Governor Andrew Cuomo began bragging about how well his state was doing, CNN's usually sycophantic Jake Tapper snapped back:[541]

> NY state has lost more than 32,000 lives to COVID-19. So, while it's great that the numbers have gone down, it's perplexing to see crowing, Cuomo going on Fallon, etc. No other state has lost as many lives, not even close. New Jersey is next with 17,000+.

At the time, it was more than one third of the total deaths in the rest of the nation.

Cuomo responded to the criticism by blaming President Trump who had sent him a hospital ship and built him a giant unused hospital in New York City.

In fact, Cuomo had originally blown off the risk of the virus. According to *The New York Times*,[542] on March 2, 2020, the governor boldly declared, "Excuse our arrogance as New Yorkers—I speak for the mayor also on this one—we think we have the best healthcare system on the planet right here in New York. So, when you're saying, what happened in other countries versus what happened here, we don't even think it's going to be as bad as it was in other countries." By July, New York City and the surrounding suburbs became the epicenter of the pandemic in the US, with far more cases and a higher mortality rate than many countries have, *The Times* observed.

In the article, which was titled "How Delays and Unheeded Warnings Hindered New York's Virus Fight," *The Times* continued:

> Interviews with more than 60 people on the frontlines of the unfolding crisis—city and state officials, hospital executives, health care workers, union leaders, and emergency medical workers—revealed how the virus overwhelmed the city's longstanding preparations, leaving officials to improvise. Many spoke on the record; others spoke anonymously to describe private meetings and conversations without fear of losing their jobs.
>
> "Everything was slow," said Councilman Stephen T. Levin, a Brooklyn Democrat who had called for City Hall to take swifter action as the outbreak spread.
>
> "You have to adapt really quickly, and nothing we were doing was adapting quickly."

Many of us who live outside metropolitan New York and surrounding Connecticut and New Jersey—those of us, for example, who live in Upstate New York—especially resent Governor Cuomo's lackluster initial response to COVID-19, followed by shutting us down as severely as New York City and the rest of downstate. We never chose to live crammed together in an urban sardine can. We live in open spaces amid innumerable lakes, waterfalls, mountains, ravines, hiking trails, and parks. Yet, the governor forced upon us the same stay-at-home orders as Manhattan and the other New York City boroughs.

Governor Cuomo refused to learn that most of the contagion takes place within closed spaces and turned our homes into our jails. He would not let us go to work or even enjoy the outdoors and the sunshine that is so lethal to the virus and so healthy for us. He proudly watched while the horror he perpetrated on us became a model throughout the nation, leading other states to adopt unconstitutional emergency measures to manage the pandemic.

The governor's behavior made us appreciate efforts by the people of Upstate New York to secede from Downstate New York, so we may elect a governor more sympathetic to our ideals of personal responsibility and individual liberty, and to create a legislature that believes in the U.S. Constitution and Bill of Rights. We encourage the process of liberating Upstate New York from Cuomo's dictatorial conduct with a bill of particulars, listing some of the most important grievances and charges against him:

A List of Grievances Against Governor Andrew Cuomo

On January 31, 2021, with over 42,200 total reported deaths, New York State ranked second behind New Jersey in deaths as a percent of population. Third was Massachusetts—all heavy lockdown states.[543] Based on that data, Governor Cuomo's performance was next to last among all states. This tragic outcome was predictable based on Cuomo's irrational, oppressive behavior toward the people under his political management and control.

Prohibiting the Use of Hydroxychloroquine for Early Treatment of COVID-19

Governor Cuomo issued an executive order March 23, 2020, to pharmacies, severely limiting the distribution of hydroxychloroquine and chloroquine, and preventing many physicians in New York State from being able to prescribe these drugs to patients who have been diagnosed or appeared with symptoms of COVID-19.[544, 545] The order limited these treatments to hospitalized patients when they are by far most effective in the earliest phases of the disease. Many of the deaths in New York State are attributable to Cuomo's removing the option from doctors and patients of using hydroxychloroquine in its various combinations which thus far provided the greatest hope for prophylaxis and early treatment of COVID-19.

Depriving the people of New York of the most effective lifesaving treatment for COVID-19 and doing so for partisan politics on behalf of the global predators in his party and elsewhere, was an egregious offense, equaling or surpassing the mass murder of nursing home inmates.

SEEDING NURSING HOMES WITH
COVID-19 INFECTIOUS PATIENTS

As of April 30, 2020, New York nursing homes, also called long-term care facilities, had 3,065 reported COVID-19 deaths. As of August 31, 2020, there were 6,639 reported COVID-19 deaths in long-term care facilities. By January 29, 2021, total nursing home deaths in New York were estimated at *12,000 to 13,000 plus*.[546]

Perhaps the deadliest decision by Governor Cuomo was his March 25 Department of Health order to nursing homes to accept or re-admit COVID-19 patients.[547] As reported in the *New York Post*, COVID-19 patients immediately began being transferred to city and state nursing homes from city hospitals.[548] Personal protective equipment (PPE) and body bags were sent along with the patients. Testing patients for COVID-19 before admission was not allowed.[549] No wonder the *New York Post* called Cuomo's state health commissioner "Dr. Death."

Furthermore, Cuomo accepted the CDC policy allowing nursing home staff who were COVID-19 positive to return to work right away if they were asymptomatic.[550] He did this despite knowledge that the elderly have the highest risk for severe disease and death if they contract COVID-19.

By April, one Manhattan nursing home had 98 deaths and counting.[551] That facility had to hire a refrigerator truck to hold all the bodies. "The disease caused by the virus has killed more than 10,500 residents and staff members at nursing homes and long-term care facilities nationwide," according to an analysis by *The New York Times*.[552] At the time, that was "nearly a quarter of the deaths in the United States from the pandemic."

Adding to his malfeasance, Governor Cuomo went on to blame the nursing homes![553]

His own attorney general, Tish James, has issued a report accusing Cuomo of underestimating the already horrific numbers of nursing home deaths by 50%:[554]

> New York's nursing home deaths have been the subject of outside scrutiny for months. An Associated Press investigation published in May found that the Cuomo administration ordered more than 4,300 recovering COVID patients to be sent back into nursing

homes, despite concerns over whether some of those patients were still contagious.

In the James report released Thursday, the New York Department of Health (DOH) is essentially accused of juking the stats by omitting nursing home patients who died from the coronavirus while in hospital.

As much as any malfeasance we have seen among governors during COVID-19, Cuomo's actions in respect to New York State nursing homes deserve a criminal investigation. New York has the second highest per-population death rate from COVID-19 with a large portion due to nursing home deaths.

ALLOWING NEW YORK SUBWAY SYSTEM TO BE A CARRIER FOR COVID-19 INFECTIONS

New York City's subway system has not been closed. The trains move every day, pushing air through tunnels and out of the subway grates into the city. The subway system in New York shares its air with the city through those grates. Remember the famous photograph of Marilyn Monroe, trying to hold her skirt down as it was blown up around her waist?[555] That was a subway grate she was standing on for the photo shoot.

Throughout all of March and April 2020, Cuomo insisted the subway system remain open. This was partially modified on April 30, when Cuomo ordered the system shut for five hours a night for cleaning, a policy that remains in force, providing too little help too late to prevent the spreading of the coronavirus.[556]

Meanwhile, despite servicing less than 10% of customers, the trains continued running during the day, pushing the air and all the pathogens in through the tunnels and throughout the city. As of early April, 83 NYC Metropolitan Transit Authority (MTA) employees had died of COVID-19.[557] As of January 25, 2021, a reported 136 MTA employees had died of COVID-19.[558] These employees—consisting of "conductors, bus and train operators, cleaners"—have been canaries in the mineshaft, warning us of the contamination and degree of infection present in the system. Actions to clean, let alone to close the transportation system, have been much too slow and remain desperately inadequate.

Failing to Follow the Science of Subway Contagion

Clandestine U.S. studies from the 1960s have documented how efficiently the New York subway system spreads pathogens. In 1966, the U.S. Army experimented with releasing *Bacillus subtilis* in the subway system.[559] These biological agents were packed into lightbulbs. Soldiers with sensors were positioned throughout the system, and the lightbulbs were thrown to the tracks, smashing the glass, and releasing the bacteria.

Leonard Cole, the director of the Terror Medicine and Security Program at Rutgers New Jersey Medical School, documented these experiments in his book *Clouds of Secrecy: The Army's Germ Warfare Tests Over Populated Areas*.[560] Cole cited declassified documents that discuss the New York tests in his book and concluded, "Test results show that a large portion of the working population in downtown New York City would be exposed to disease if one or more pathogenic agents were disseminated covertly in several subway lines at a period of peak traffic."[561]

According to an army report, scientists concluded it took between four and 13 minutes for train passengers to be exposed to the bacteria.[562] Five minutes after bacteria were released at 23rd Street Station, the bacteria could be detected at every station between 14th Street and 59th Street. Between June 6 and June 10, they calculated more than a million people were exposed. Surely this information was known to New York disaster and anti-terrorism teams.

Adding to potential contagion, particulate pollution appears to increase lethality of COVID-19, according to recent studies from major universities around the world.[563] The particulate grime and filth of New York City and the subway system facilitate the transmission of infection.[564]

Unlike many other activities that have been closed down, perhaps the most dangerous of all has remained open.

Allowing Homeless to Live in Subway Cars During Lockdown

During COVID-19, the homeless migrated in greater numbers into New York City's subway cars and were using them for shelter, day and night.[565] Photos show images of ragged people surrounded by their collected belongings asleep on subway cars, increasing the risk of infection and precluding any chance of social distancing for riders who were using the subway to get to work. When Cuomo finally decided in early May to close the subways late at night to clean them, he had to marshal 1,000 police officers to clear them out each night and to take them to the shelters provided for them.[566]

Failing to Provide for Management of the Dead

In mid-March, America was told we could expect up to two million deaths, and the federal government ordered 15 days to "Slow the Spread." The proper removal, storage, and disposal of the dead was a failure in New York City as the disease surged.[567, 568] Hospitals required refrigerator trucks. Funeral homes ran out of room. One funeral home stored the bodies in a gated backyard, another in rental trucks parked in front of the home.

Issuing "Do Not Resuscitate Orders" to Emergency Medical Techs and Ambulances

A statewide do-not-resuscitate (DNR) order was issued April 17, 2020, by the New York State health department.[569] The order instructed EMTs and other first responders to no longer attempt to revive anyone without a pulse. Usual procedure would allow for 20 minutes of attempted resuscitation. The DNR order was so unpopular with EMTs and others it was rescinded a week later. For the one week, while these orders prevailed, cardiac and other patients found without a pulse were not to be revived, probably increasing the death toll.

Causing Crime to Spike by Ordering the Release of Prisoners

Governor Cuomo released hundreds of prison inmates to protect them from the potentially widespread coronavirus infections in these facilities.[570] As a result, over 1000 criminals were released on March 18. In mid-April, murder, grand larceny, and armed robbery were all significantly up in the city.[571]

Failing to Check Temperatures of Arrivals at Airports and Quarantine

Governor Cuomo failed to insist that temperatures of airplane passengers be taken upon arrival from other states and from other countries around the world, despite this being one of the major actions taken by countless other countries.[572] [573] This globalist approach has added to the intensity of the epidemic in the downstate area and nationwide as many of these arrivals would have spread out around the nation.[574]

Inflicting an Overall Militant Oppression on the People of New York State

For some time, Cuomo has maintained that individuals traveling from states listed as high-risk areas must quarantine for ten days on returning to New York and then must take a test to be released.[575] This order—impossible to enforce and too confusing to carry out—has simply added to the fear, distrust, and uncertainty among those of us who live in or travel to and from New York State.

On April 30, 2020, Cuomo announced he was preparing an "army" of people for the purpose of "contact tracing;" that is, to make identified victims of COVID-19 list all their contacts and coerce them to be tested as well.[576] He estimated that between 6,400 to 17,000 persons would be needed to accomplish this task. The goal would be to trace all contacts of the patients back for 14 days and then to isolate them, sometimes in specially designated "isolation facilities." He was joined in this China-style activity by former New York City Mayor Michael Bloomberg and his Bloomberg Philanthropies, as well as Johns Hopkins University and other predatory globalists, making it a triumvirate of predatory globalists.

Dr. Michael Ryan, executive director of the Health Emergencies Programme of WHO, stated the lockdowns drove the disease down to the family level.[577] As a result, he continued, instead of stopping the stay-at-home orders, "now we need to go and look in families to find those people who may be sick and remove them and isolate them in a safe and dignified manner."[578] The phrase about a "dignified manner" sounds a little too much like hospice care or, worse, "euthanasia" or murder. We are free-born American citizens, entitled to the rights of life, liberty, and the pursuit of happiness, which includes being responsible for ourselves, our families, and our communities, in work and in leisure.

OUR MOST URGENT DEMANDS

Many deaths from the epidemic of coronavirus have been unavoidable, but many others were man-made with Governor Cuomo as the man. In addition to killing people, the Governor's policies have caused severe illnesses by making hydroxychloroquine unavailable for the early treatment of COVID-19. Furthermore, his restrictive lockdown policies have caused immeasurable personal, social, educational, and economic misery and loss. We have not even begun to categorize these damages specifically for New York.

The Governor cannot undo the massive harm he has already perpetrated against the people of New York with ramifications for people throughout the country and the world. He can begin to set things right by paying more heed to basic civil rights and by reversing as many of his damaging decrees as possible.

First, end all limitations on doctors prescribing for their patients with respect to COVID-19. Let patients and their healthcare providers decide together on treatment regimens *without interference from the state government.*

Second, rescind the military-style assault on people who test positive for the coronavirus and on their contacts.

Third, completely open Upstate New York socially, educationally, and economically. Stop the ridiculous quarantine for people coming and going from New York to so-called high-risk areas.

Fourth, return the people of New York to as normal a life as possible. COVID-19 should never have been used as a justification for taking away our rights under the Constitution and the Bill of Rights.

We conclude with encouraging recent news. With accusations from within his own administration that he purposely undercounted the nursing home deathrate,[579] followed by accusations of creating a "toxic workplace"[580] and sexual harassment,[581] Cuomo's future as the tyrant of New York is growing dimmer. As good as it would be to be rid of Cuomo, the damage and human suffering he has caused will remain an irreparable tragedy. To begin with, we must support lawsuits to limit the emergency powers of governors throughout the nation.[582]

PART THREE

COVID-19 and the Predatory Globalists

CHAPTER 15

Bill Gates' Master Plan Found: Implements Operation Warp Speed in 2015-2017

On June 14, 2021, I made a discovery that led us to stop "stop the presses"—to put on hold the publishing process for our book. I found the Bill Gates' *2017* master plan for responding to the coming pandemic. It is a blueprint for what has already taken place and continues to roll out during COVID-19. From the financial and governance perspective, it is what Gates, Fauci, WHO, the pharmaceutical industry, FDA, CDC, NIH, and many others throughout the world are implementing during COVID-19 under the guise of public health.

The plan is titled, "Coalition for Epidemic Preparedness Innovations (CEPI) Presentation to the WHO" and is dated July 21, 2017.[583] I found it as an isolated document on the WHO website. The link uses the term "overview." Surprisingly, it seems to have escaped all notice until now.

CEPI is a foundation originated by Gates at least as far back as early 2016 and formally announced in January 2017 by Gates and by Klaus Schwab.[584] Schwab is the founder of the World Economic Forum, and leading promoter of the Great Reset. One of CEPI's earliest actions—even before its formal unveiling—was to gain the approval of WHO's governing World Health Assembly in May 2016.[585]

The CEPI website states, "We create innovative partnerships between public, private, philanthropic, and civil organisations to develop vaccines against diseases with epidemic potential."[586] This is a sugarcoated description of predatory capitalism at work that will later be named and embraced as the Great Reset. It is essentially a profit- and power-driven coalition of wealthy individuals and organizations who aim at gaining more wealth and power through CEPI, WHO and vaccines.

In an interview in 2014 that was heralded as describing "the future of global governance," Schwab agreed that he was essentially planning and developing "a new way of doing global governance."[587] Gates is leading the charge.

The Gates' master plan is a 21-page multi-color PowerPoint addressed specifically to WHO. It was presented by the Interim CEO of CEPI, Norwegian Professor John-Arne Røttingen. This document lays out most of the main categories of global predation in the coming pandemic and how Gates and WHO will divide up their authority over it in the spirit of the new world governance.

The master plan demonstrates the omnipotence and grandiosity of Bill Gates and his industrial and medical partners, including WHO. They plan to impose themselves on humanity without any mention or consideration of higher principles, democratic processes, national sovereignty, court supervision, legality, or individual rights. The goal is to increase their wealth and power in the next pandemic, with vaccines as the entering wedge. That's why COVID-19 always seems as if it is "all about the vaccines." All the planning that went into the coming pandemic was about using vaccines to gain wealth, aggrandizement, and power.

BACKGROUND OF CEPI

To understand the master plan, we have to understand CEPI. Bill Gates conceived of the Coalition for Epidemic Preparedness Innovations (CEPI) in 2015,[588] made plans for its development with Klaus Schwab at the January 2016 World Economic Forum (WEF) at Davos, and unveiled it at the January 2017 WEF with two or more videoed presentations.

Gates funded it through the Bill & Melinda Gates Foundation, which means strong participation by Warren Buffett, who is one of three trustees of the Foundation, along with Bill and Melinda. Gates and Schwab then brought in additional partners.

Wellcome Trust, a British "charitable" trust, is another CEPI early founder. Based on pharmaceutical industry funding and focused on the health arena, it is one of the largest charities in the world. The Government of India was initially described as another founder, followed by Germany and Japan—which completes the fusion between huge corporations, giant "philanthropic" organizations, and very big governments. In the master plan, the Government of India is modified to "India's Department of Biotechnology."

Another government founder was Norway, where CEPI is legally registered as "an international nonprofit association." It is hosted by the Norwegian Institute of Public Health and the interim CEO was Norwegian.

Through this global organization, Gates began organizing a worldwide alliance to finance, coordinate, and enforce the development of new vaccines to make a financial killing and to reorganize and basically control world governance in the coming pandemic. This progression of events, sometimes with additional information, is documented in *Chronology and Overview* at the end of this book.

The CEPI coalition members, especially Bill Gates, will control the unfolding of COVID-19 and the Great Reset—although they will be joined by innumerable billionaires, corporations, philanthropic groups, scientific and medical groups, and governments and their agencies. Anthony Fauci, a member of Bill Gates' small, elite vaccine Advisory Council will be one of Bill Gates' leading implementers of the master plan.

CEPI Is Announced at Davos

On January 13, 2017, the FDA issued a notice in the *Federal Register* that it was publishing elaborate revisions for guidance to industry in applying for money and other benefits under the Emergency Use Authorization act (EUA).[589] During COVID-19 and continuing to this day, the EUA has enabled the U.S. to pump billions of free taxpayer dollars into the vaccine industry while giving it an open gateway for rapid vaccine processing through the FDA.

The actual EUA revised guidance was 45 pages long.[590] A document of this size must have been in the works for some time at the FDA. Its final publication assured investors in the pharmaceutical industry that they could safely go forward with Gates' CEPI master plan. In an epidemic emergency requiring vaccines, the federal government was promising to cover all the costs and to virtually guarantee huge financial returns. Simultaneously, it promised a free ride through the FDA regulatory process. All this came true in 2020, starting with the vaccines produced by Gates' closest partners in the industry, Pfizer and Moderna. It is the ongoing fulfillment of the Gates' master plan going back to his partnership with these companies at least as far back as 2016.

As if waiting for this critical FDA notice before taking action, a mere five days later on January 18, 2017, CEPI was officially unveiled in a press release on January 18, 2017, CEPI was officially unveiled in a press release from Schwab's Davos location as a "Global partnership launched to prevent epidemics with new vaccines."[591] New sources of support are added: "CEPI is supported by several leading pharmaceutical companies with strength in vaccines – GSK, Merck, Johnson & Johnson, Pfizer, Sanofi and Takeda, plus the Biotechnology Innovation Organisation."

Here is the most recent CEPI description of its partners:[592]

> CEPI was founded in Davos by the governments of Norway and India, the Bill & Melinda Gates Foundation, Wellcome, and the World Economic Forum. To date, CEPI has secured financial support from Australia, Austria, Belgium, the Bill & Melinda Gates Foundation, Canada, Denmark, the European Commission, Ethiopia, Finland, Germany, Hungary, Iceland, Indonesia, Italy, Japan, Kuwait, Lithuania, Luxembourg, Malaysia, Mexico, Netherlands, New Zealand, Norway, Panama, Romania, Saudi Arabia, Serbia, Singapore, Switzerland, The Republic of Korea, United Kingdom, USAID, and Wellcome.

> Additionally, CEPI has also received support from private sector entities as well as public contributions through the UN Foundation COVID-19 Solidarity Response Fund.

From the beginning, all this planning had the dual goals of making money from vaccines, while imposing global governance through an alliance among billionaires, individual countries, philanthropic organizations, and corporations—essentially the Great Reset. That continues to this day as the global predators are continuing to push to vaccinate younger and younger children in America, not only to gain yet more wealth, but also to exercise more power.

The agenda also has that prevailing progressive veneer, stating, "Just over a year ago 193 states adopted the Sustainable Development Goals—the roadmap for the future we want." *Even WHO would have to admit that working with CEPI and GAVI did not, by January 2021, prevent the pharmaceutical companies from focusing on providing vaccines to wealthier nations where they could get higher prices.*[593] *As the Western markets became saturated or resistant to vaccination, only then did the global empire turn its attention more toward the world's poor.*

March 2017 *The Lancet Infectious Diseases* establishes CEPI as the answer for what its headline declares is "preparing for the worst."

The Bill Gates Master Plan for the Coming Pandemic as of July 17, 2017[594]

On July 17, 2017, CEPI made a detailed presentation to WHO which outlined a global collaboration that essentially took over future pandemics, bypassing the varied officially elected government bodies of the world. Its importance to history is that it documents the CEPI global master plan and CEPI as Bill Gates, along with Klaus Schwab, and always along with that mostly silent partner of Gates, Warren Buffett.

"Coronaviruses" are among "First Choices for Immediate Funding."

At no place in the master plan is there any discussion of implementing early and effective treatments for the anticipated pandemic. The entire emphasis is on getting a vaccine—and according to the U.S. Emergency Use Authorization, which was already in effect in 2017, there can be no government rush to make vaccines if there is an existing effective treatment. Perhaps Fauci pushed remdesivir early on because he knew it would never prove effective, while he rejected every single effective treatment and worked with the CDC, FDA and NIH to suppress the use of any and all effective treatments, from vitamins, minerals and herbs to FDA medications that were already approved for treating infections like hydroxychloroquine and ivermectin. These available treatments were excluded from the vaccine master plan and pandemic management.

Now we will examine particular slides in CEPI's PowerPoint presentation.

"CEPI Vision and Approach"

The third slide about CEPI's vision and approach concludes with a major theme throughout COVID-19 management—developing rapid programs for pushing through the vaccines:

> The initial focus will be to move new vaccines through development from preclinical to proof of principle [very early stages of testing] in humans and the development of platforms that can be used for rapid vaccine development against unknown pathogens.

Although not mentioned in this master plan, we shall document in this chapter how Gates was already much further along in carrying out these plans. By mid-summer of July 2017, Gates was already talking at Schwab's World Economic Forum about developing "DNA/RNA" vaccines in months rather than years working with Moderna and Pfizer. This, of course, would later be called Operation Warp Speed.

According to a March 2017 promotional for CEPI in *The Lancet Infectious Diseases*, the ultimate concept was to have a collection of vaccines in development but not finalized for approval. In order to gain full approval for a new vaccine a Phase III clinical trial must be done.[595] Those trials include testing the new vaccine upon thousands of people both for effectiveness and safety. "Have a stockpile of vaccine on hand for when there is an epidemic", said a CEPI board member. "You need an epidemic to test the vaccine's efficacy."[596] This report does not even mention SARS-CoV as one of the major threats but within a few months SARS-CoV will become the epidemic of choice. There is no skepticism shown regarding the legitimacy of CEPI's role as a master planner or about the inherent dangers of wealthy investors taking over pandemic management. Conflicts of interest abound.

"CEPI Fit In The End-To-End Spectrum of Vaccine Preparedness."

The above heading is the title of the fifth slide, which contains a table of actors and stakeholders in the Gates master plan. Notice that it is all about the vaccines—and nothing else—except for Bill Gates and his partners using the vaccine programs to exert their power and influence over humanity during the pandemic.

Also notice that of the more than 24 groups mentioned in the organizational chart, Bill Gates is the only individual person mentioned by name. His name is in the category that involves "Development/licensure," his means of controlling as much as he can the development, distribution, and ownership of the vaccines working through companies like Moderna, Pfizer, Johnson and Johnson, and AstraZeneca.

Also notice that WHO is mentioned as playing a role in all three categories after "Discovery."

What Have "They" Accomplished?

CEPI Fit in the End-to-End Spectrum of Vaccine Preparedness

	Significant Focus by Others	Role for CEPI	Significant Focus by Others

Phase	1 Discovery	2 Development/ Licensure	3 Manufacturing	4 Delivery/ Stockpiling
Current Stakeholders	• Academia • Governments • WT/NIH • GLOPID-R • Industry • Regulators • Biotech	• Industry • National Governments • Regulators • Gates • BARDA/DTRA etc. • WHO • Biotech • PDPs	• Industry • BARDA • CMOs • Regulators • National Governments • WHO	• GAVI • UNICEF • PAHO • National Governments • WHO • Industry • Pandemic Emergency Facility (World Bank)

How effective was the organizational chart produced by Gates and Fauci? During the pandemic, the entities listed in the master plan worked and continue to work in lockstep to carry out the worldwide totalitarian plan. They have met no significant opposition except from *independent* citizens, physicians, scientists, and journalists. Those citizens who are financially dependent on institutions— such as clinics, hospitals, schools, universities, corporations, or the media—have hidden their questions or doubts about the handling of COVID-19 or they have risked being abruptly condemned, cancelled and fired.

Outside of a few freedom-oriented media sources on the political right, not a single large or influential entity fought the onslaught head on. Most instead vied for leadership in who could use the most influence and force to vaccinate the world while strengthening totalitarian control.

GLOPID-R, who is mentioned in the chart, is " an international network of funders to facilitate coordination and information sharing." On March 12, 2020, in collaboration with WHO, it was already able to publish an elaborate multi-colored research "blueprint" for responding to the pandemic.[597] It must have been in preparation for months or more. Anticipating the Great Reset, it announced:

> The global scale of the epidemic and the unprecedented level of global collaborative commitment to research and innovation calls for a reset of the functional model for global coordination. It should clarify roles and responsibilities, enhance inclusiveness and openness, while retaining the ability for rapid decision making to drive action at the appropriate level.

Dr Tedros, Director General, WHO, speaks as if summarizing a movie script about the world coming together to fight for its survival, undoubtedly under his leadership:

> This outbreak is a test of political, financial and scientific solidarity for the world to fight a common enemy that does not respect borders... what matters now is stopping the outbreak and saving lives. (Ellipses in original.)

Vaccines are mentioned 91 times; medicine or medicines 12 times; drug or drugs 8 times,̄ and home treatment and early treatment 0 times. The coalition to make fortunes and exert inhumane controls through COVID-19 never gave inexpensive, safe, and effective early treatments a chance to undermine their plans. Everything was and remains all about the vaccines and anything in the way had to be crushed to make way for the globalists to increase their wealth, self-aggrandizement, and power.

The document recognized the "Possibility of enhanced disease after vaccination." It remains a major cause of disability and death that Fauci, the FDA and CDC continue to ignore, while pushing through and promoting the vaccines.[598]

"Collaborating with WHO Objectives"

The Gates master plan in PowerPoint includes the following statements about CEPI's collaboration with WHO:

Collaborating with WHO

❖ In the process of developing an MoU between WHO and CEPI

❖ CEPI will rely on WHO as the global normative lead agency on health

❖ Guiding principles

- To respond to vaccine R&D needs for emerging infectious diseases, and ensure that the developed vaccines will be available to all in need, in order to achieve the highest possible public health impact

- To focus on diseases on which the market fails to provide adequate incentives

- To strategically leverage the existing diverse set of national and international mechanisms that support vaccine R&D, avoid duplication, and maximize synergies

MoU stands for Memorandum of Understanding. It means the parties have reached an "understanding" and are moving forward with their plans. It represents a serious declaration that a contract is imminent.

With this agreement, Gates and CEPI assumed for themselves the role of setting the "normative" transnational standard for the medical and public health actions taken during a pandemic. They declared themselves the world's arbiter of conduct during COVID-19. They have done this with no involvement of actual elected or even legal governing bodies.

Initially, we found no information describing WHO's response to this document, including whether or not the UN agency saw itself as "In the process of developing an MoU between WHO and CEPI." The document had great value as a window into the ambitions of Bill Gates and those aligned with him, but we wondered, "Did WHO agree with this?" Then we found a WHO document that specified how it was cooperating with various stakeholders, in which it gave confirmation of its agreement with CEPI: "For example, there is already a Memorandum of Understanding (MOU) between WHO and CEPI to collaborate on vaccine R&D for the Blueprint priority diseases."[599]

The reference to "R&D for the Blueprint priority diseases" is important. In the same WHO document we found their definition for the term:[600]

> The R&D Blueprint is a global strategy and preparedness plan that allows the rapid activation of research and development activities during epidemics. Its aim is to fast-track the availability of effective tests, vaccines and medicines that can be used to save lives and avert large scale crises. … WHO Member States welcomed the development of the Blueprint at the World Health Assembly in May 2016.

By 2016, CEPI, along with WHO's governing World Health Assembly of 2016, were already developing and approving the concepts that would become Operation WARP Speed and the Great Reset four years later in 2020.

Overall, CEPI used WHO to place its imprimatur on Bill Gates' ambitions to organize and to reap the benefits of the coming pandemic. In turn, CEPI empowered WHO to be the world arbiter of medicine, science, and public health during the coming pandemic, which further meant empowering Communist China as WHO's greatest ally. This was a most extraordinary extralegal pact, exerting power over the world that no government on Earth actually possessed or could delegate. The CEPI/ WHO pact set up the global totalitarian events that would unfold.

"CEPI PRINCIPLES"

The sixth slide is about "CEPI Principles." The "OPERATING" column allows for no mistake about who will benefit the most financially:

CEPI Principles

GOVERNANCE	OPERATING
❖ Accountability	❖ No loss: vaccine developers should be reimbursed for their direct and indirect costs/
❖ Public trust	
❖ Political legitimacy	❖ Shared benefits: Rewards to vaccine developers should be proportional to levels of risk, R&D, infrastructural or other types of commitments. If a CEPI-sponsored vaccine should develop economic value above and beyond a pre-agreed set point, vaccine developers should reap these rewards and pay back CEPI funding/
❖ No conflict-of-interest	
❖ Transparency	
❖ Independence / neutrality	
❖ Public interest representation	❖ Equitable access: global access agreements should be negotiated between CEPI and vaccine developers to encourage affordability and availability in Low Income Countries (LICs). Contracts should include reasonable royalty payment provisos for products or patents/
❖ Flexibility / nimbleness	
❖ Global health responsibility	

Notice under "OPERATING" how Bill Gates has completely protected his investments in the pharmaceutical industry, with the requirement, "No loss: vaccine developers should be reimbursed for their direct and indirect costs." Also notice, under "shared benefits," that CEPI will then reap further benefits if the profits of the drug companies exceed a certain limit. Gates wins in two ways, from the loss-free, no-risk conditions of his pharmaceutical company investments and for skimming off the top for CEPI. The U.S. government provided the risk-free conditions for the vaccine makers but it's too early to know how much Gates takes off the top for his CEPI.

All the nonsense under "Governance" in the table is just a veneer. Profits are the real issue plus outrageous totalitarian control over the management of the pandemic and related national policies.

The U.S. government under NIH and NIAID, led by Fauci, put this "no loss" policy in effect for the vaccine manufacturers. BARDA provided millions to cover costs of development; NIH provided an enormous amount of research and related help, even acquiring patents to some vaccines. Ultimately, Gates, Fauci and the other global predators obtained worldwide government approval for their four preferred vaccines made by Moderna, Pfizer, Johnson and Johnson, and AstraZeneca. These four winners had been identified and locked in years earlier when Gates and his fellow global predators chose to invest in their highly experimental and dangerous mRNA and DNA vaccines (see ahead in this chapter).

"CEPI OBJECTIVES"

The fourth slide is about the objectives of CEPI during the upcoming pandemic and makes unmistakably clear that vaccinating everyone has always been and still remains the single, exclusive objective of COVID-19 under Bill Gates and his fellow predators. The ultimate goal is not merely to become more outrageously wealthy but also to become even more extraordinarily powerful and able to bend humanity to their will.

CEPI Objectives

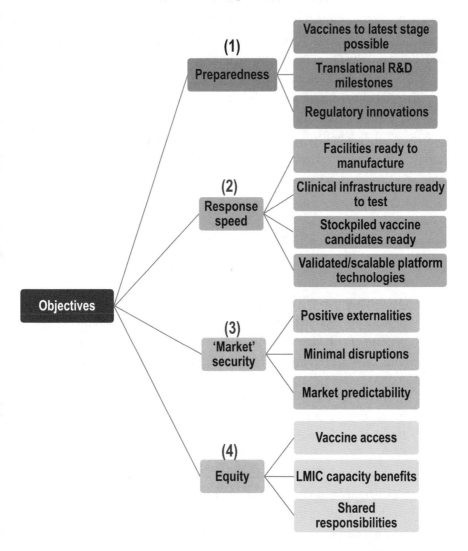

The term, LMIC, in the final stage of the chart, indicates a "low and middle-income country," also known as a less-developed country. CEPI is paying lip service to idealist concerns for poorer countries but all the marketing emphasis will be on the globalists sapping the wealth of richer nations.

We mentioned at the beginning that our discovery of the presentation by CEPI to WHO—which we have designated as the *master plan* of Gates, Schwab, and their partners—is a turning point in the discussion of how the pandemic response was planned far in advance. It blows away the smoke of "conspiracy theory," given that Gates and his fellow global predators had a plan in motion by the summer of 2016 which they were presenting to the WHO governing body and to other global predators. The plan follows the overall direction taken by the globalists in organizing and carrying out their response to COVID-19 which focuses on vaccines—to the exclusion of all treatments.

The global predators know they are at war with patriotic, democratic nations like the United States who want to remain strong and to resist rapacious globalism. The globalists' most common public strategy is to win over their leaders with favors and graft, which they easily accomplished until America First and President Trump.

Creating a Transnational Governance

Transnationalism and *globalization* can be defined and explained as follows:[601]

> Transnationalism [means] economic, political, and cultural processes that extend beyond the boundaries of nation-states. The concept of transnationalism suggests a weakening of the control a nation-state has over its borders, inhabitants, and territory. ...

> Globalization is a related concept that represents the intensification of economic, cultural, and political practices accelerating across the globe in the early 21st century.

What Gates and Fauci announced at Davos in the form of CEPI is the creation of a transnational, globalist organization that sets itself above other organizations established by tradition, politics, or the law. The global predators display an extreme hubris—a defiant arrogance—which not only stands above other organizations but also above any concepts of individual rights as established in the Declaration of Independence, Constitution or Bill of Rights, or Judeo-Christian values.

By a simple declaration of its own, a small group of individuals led by Bill Gates, establishes a new world governance with which to deal with pandemics, thereby ultimately creating an authority that can bulldoze civilization in the service of its global predatory ambitions.

It is as if aliens had descended upon the Earth, and by the sheer force of their enormous spaceships looming over our cities and their vast communication system, cast a shadow of fear over humanity, and an offer of wealth and power to the world's elites, that overnight set into motion the takeover and exploitation of humanity.

CEPI is the creation of individuals of wealth and power, joined by the worldwide elite, who are the source of power. CEPI is merely one of their agencies or enforcers. CEPI, through its alliance with WHO, ultimately invites in the Chinese Communists who are the ultimate trans-nationalists.

The Chinese Communist Party (CCP) is the extreme imperialist—bringing together Marxism and a tradition of emperors that make the CCP the force most likely to take over the world. The momentum of this growing Chinese Communist Empire will, if unchecked by the Western nations, crush the lesser global predators like Gates and Schwab, and destroy the West, covering the Earth with a thousand years or more of darkness and despair.

Other Events Leading Up to Operation Warp Speed and Corporate/Government Fusion

The *Chronology and Overview with Pandemic Predictions and Planning Events* can be found at the end of our book. It reviews and summarizes events leading up to Bill Gates implementing huge plans that, several years later, would be announced as Operation Warp Speed by President Trump with Fauci at his side. Fauci is on Gates' small, elite vaccine advisory council, and they frequently speak and email to each other. Fauci is probably the most direct connection that Bill Gates has for exerting control over COVID-19. Fauci almost certainly helped the multi-billionaire develop his Warp Speed-like strategy and investments years before the pandemic occurred.

Among the most important scientific events leading to Gates' Warp Speed-like announcements in early 2017 was the December 2015 publication by Menachery and his colleagues at the University of North Carolina, along with two Wuhan Institute researchers. The paper documented how the researchers had created a deadly SARS-CoV that could infect mice and humans, as well as their claim to having made progress developing vaccines for the man-made coronavirus. This gave Gates a written publication to take to other billionaires, to pharmaceutical companies, and to governments to whip up interest in their making huge investments in preparation for a SARS-CoV pandemic virus and the seemingly necessary vaccines. The *Chronology and Overview with Pandemic Predictions and Planning Events* can help in grasping the various steps in this history.

By now, the reader can see that there is no logical connection between being able to make a deadly virus in a lab and nature inflicting a similar plague on humanity—other than foreknowledge that the deadly lab virus would soon be accidentally or purposely released. However, most of the world remains under the hypnotic spell of Anthony Fauci and other "scientists" that we were in danger of a SARS-CoV pandemic from nature because the American and Chinese had succeeded in making harmless bat viruses into SARS-CoV pathogens.

The legislative cornerstone went into place for the globalist investors and corporations on January 15, 2017, when the FDA announced guidelines for improving the Emergency Use Authorization (EUA) to allow for the very rapid development of vaccines when faced with a pandemic emergency. It also made million-dollar bonuses available to drug companies to reduce their investment risk for committing themselves to vaccine development during an emergency. This was the "get home free" card for all the globalists involved in preparing to become even richer and more powerful through imposing vaccines upon the world. Three years later, in 2020, the Emergency Use Authorization would enable Operation Warp Speed with a corporate-government fusion characteristic of Schwab's Great Reset.

These and other events would lead to a barrage of predictions from well-known global predators that the world must tool up for treating a SARS-CoV pandemic by contributing billions to vaccine programs in which Gates and his favored companies were already invested, including Moderna. This chapter and the next will deal with these events.

Gates Describes America's Future in Response to COVID-19

The incredible announcements from Bill Gates began in January 2017 in a series of filmed talks surrounding Klaus Schwab's World Economic Forum in Davos—three years before the first indications of COVID-19. Bill Gates announced that he had already been funding and implementing plans with Moderna to rush through vaccines for an anticipated pandemic that sounded very much like COVID-19 and identical to Trump's Operation Warp Speed in 2020.

Simultaneously, Gates described how he had been working through a new foundation called CEPI to develop multibillion-dollar collaborations between governments and industries to create vaccines, using a government-corporate fusion that was identical to what would become Operation Warp Speed and to what Schwab would call the Great Reset.

By the end of 2017, the Bloomberg School of Public Health at Johns Hopkins joined Gates in this huge initial tooling up for COVID-19 by publishing a 76-page future scenario, specifically preparing for a "coronavirus" pandemic and again centered around the production of vaccines.

What could have led billionaires, wealthy corporations, and the world's most famous school of public health—three years ahead of time—to invest so much time and effort into preparing for this specific coronavirus pandemic? What possible reassurance did they have that they were not embarking on a financially disastrous, politically calamitous, and humiliating fool's errand?

Bill Gates has been focused on the danger of epidemics for many years. In a 2015 well-known TED Talk, now viewed by more than 40 million people, he said, "If anything kills over 10 million people in the next few decades, it's most likely to be a highly infectious virus rather than a war."[602]

Gates Unveils His Ongoing Preparations for COVID-19

On January 24, 2017, at Klaus Schwab's World Economic Forum at Davos—four days after President Trump's inauguration, Bill Gates gave a press conference to a select group of friendly supporters that was filmed by the *Business Insider*.[603] Gates began by saying that a worldwide epidemic was the most likely, massive, upcoming threat to humanity.

A friendly attendee brought up using "fear" to motivate people and asked what more could be done to arouse governments and people to the threat. Gates joked, saying there are just so many times he could repeat, "this awful thing is going to happen."

At the press conference, Gates unveiled CEPI, The Coalition for Epidemic Preparedness Innovations. CEPI is a worldwide consortium of governments, corporations, and wealthy individuals contributing to pandemic preparation research. CEPI describes itself as "a foundation that takes donations from public, private, philanthropic, and civil society organizations, to finance independent research projects to develop vaccines against emerging infectious diseases (EID)."[604]

In a follow-up interview in March 2017 by CBS News, Gates announced that CEPI had collected $460 million.[605] More recently in March 2021, CEPI presented its $2.5 billion plan for addressing COVID-19.[606] Of all the massive preparations signaling a certainty that the pandemic was imminent, this seems to be the most overlooked.

GATES OUTLINES THE WARP SPEED STRATEGY

At the January 2017 Schwab WEF press conference, Gates spoke about "new technologies" related to vaccines that were not yet in use anywhere and that were being developed with his investment backing. He referred to Moderna as working on methods of "trying to get out products" speedily in the anticipated pandemic.

Moderna is one of Anthony Fauci's most favored corporations, receiving enormous funding and promotion from him and his NIAID. It seems likely that Gates tells Fauci whom to favor rather than Fauci telling the mighty Bill Gates.

A representative in the audience from the *New England Journal of Medicine* asked Gates about candidate vaccines and mentioned "DNA" vaccines. Without answering the seemingly explosive question, Gates and a nearby colleague laughed and joked how well informed the audience was.

Then, in the follow-up CBS News interview in March 2017,[607] conducted in an office of Schwab's WEF, Gates openly talked about his ongoing activities in preparation for the big pandemic. His speech is somewhat halting and sprinkled with "ah," as if he is cautiously finding his words:

> QUESTION: So, you launched CEPI because you believe the world is unprepared for the next pandemic. What is CEPI?

GATES: CEPI is a Coalition for Epidemic Preparedness. And the idea is to take a new way of building vaccines that could let us develop in less than a year a novel vaccine called DNA/RNA vaccines. And so, we'll fund a few projects to build specific vaccines but not only get that, we will prove out that these platforms can work. We'll understand the regulatory issues and give us a chance to being able to respond in time when the next epidemic hits. [They are speaking beside a glass wall with "World Economic Forum" emblazoned on it.]

QUESTION: Describe the 'just in case' and the 'just in time' strategy.

GATES: Well, if we pick some things, we know there is a risk from, if we're right, [he is smiling] about the ones we pick, those will be there and that's an amazing thing. Even if we don't anticipate it, we will have this platform that will be more timely so we won't have to build the new factory, the approval steps will be streamlined, and having gone through that with particular pathogens, we'll understand what the regulator wants to see with this emergency case.

QUESTION: What are the new tools that we have at our fingertips to stop the next pandemic.

GATES: Well, right now we are at the same situation we were at before, where it takes years to build a new vaccine. But the scientific idea of these new platforms could radically change that so that a lot of the steps are sitting there ready. The factory piece, understanding the regulatory piece. So, you just have to plug in some information. *Do some quick safety profiles,* and you can get into manufacturing quite rapidly.

QUESTION: And if we speed up this process how many people are we actually saving?

GATES: Well, we don't always know the risk of an epidemic and some epidemics like a flu epidemic could be absolutely gigantic. So, it's hard to put a number to it.

Gates then boasts CEPI now has $450 million "to try out the new platform to see if works." He then goes on to discuss working in the last five years with the research scientists, industry, the governments. He states, "We need vaccines for a lot of things," and observes, "Governments have been incredibly generous working with us and others."

Thus, in early 2017—three years before COVID-19—Bill Gates was already deeply involved with funding and discussing in public what President Trump would later unveil as Operation Warp Speed. This is a stunning revelation: Years ahead of time, Bill Gates was already investing in and internationally organizing individual, corporate, and governmental wealth and power in preparation for a Warp Speed approach to a coming pandemic. A central aim was to rapidly produce innovative vaccines without the usual FDA procedures that should ensure safety and effectiveness.

Gates is also adding to confusion that exists today by speaking of an "approval" process when it is a unique emergency process that does not result in official FDA approval of a vaccine as safe and effective.

Gates was unveiling his Warp Speed plans through Klaus Schwab's conferences on what we now call "a worldwide corporate/government fusion" in response to the anticipated pandemic. As we document in Chapter 20, Schwab would give the name Great Reset to this corporate/government fusion in 2020, and almost immediately Joe Biden and John Kerry would adopt it as official American policy.

Far in advance of the announcement of Operation Warp Speed or the later announcement of the Great Reset, Bill Gates was developing both. He was marshaling billions of dollars to support his investments and political strategies, none of which respect democratic processes and all of which step on individual and political liberty as if it were of no more account than an ant underfoot.

The importance of unfolding events has been apparent to the old media. Here is a quote from a retrospective by CBS News in 2020 looking back to 2017:[608]

In a CBS News interview three years ago, the billionaire philanthropist talked about the potential for a worldwide health crisis and the fact that we are largely unprepared.

"The impact of a huge epidemic, like a flu epidemic, would be phenomenal because all the supply chains would break down. There'd be a lot of panic. Many of our systems would be overloaded," Gates told CBS News from the 2017 World Economic Forum in Davos, Switzerland. "But being ready for epidemics of different sizes, there's a lot more we should do."

He said one particular area of focus should be increasing work on vaccine development. "The idea is to take a new way of building vaccines that could let us develop, in less than a year, a novel vaccine," he said. "It gives us a chance of being able to respond in time when the next epidemic hits."

On January 18, 2017, CBS News reports the same idea as told directly to its senior producer, but now with specific mention of "DNA/RNA" vaccines: "The idea is to take a new way of building vaccines that could let us develop in less than a year a novel vaccine, called DNA/RNA vaccines." It is clear that the foundation of the future Operation Warp Speed in 2020 was already beyond the drawing board and in production in 2017.[609]

Notice the anticipation of one of the great moneymaking adventures of all time—the mad rush to create a vaccine at Warp Speed based on Fauci's great collaboration with drug companies like Moderna and Pfizer. Also notice that Gates in 2017 is forecasting "a new way of building vaccines that could let us develop, in less than a year, a novel vaccine," specifically including mRNA vaccines. He is explaining in advance exactly what he and other predatory billionaires and corporations will get President Trump to approve in the form of Operation Warp Speed. How can anyone doubt that Gates was planning the COVID-19 cash-in years in advance?

CREATING A LEGAL MECHANISM FOR
GENERATING WARP SPEED WEALTH

Bill Gates and his sidekick Anthony Fauci had to know that their Warp Speed multibillion-dollar boondoggle would require a new legal mechanism to bypass the FDA approval process. As Gates said, the many-year process had to be reduced to a year or less. Gates would need something like the Emergency Use Authorization (EUA) to allow federal agencies to override the multi-year slow-moving FDA regulatory approval process in order to do crash cursory approvals on an emergency basis only.

Gates announced his Warp Speed-like strategy in January 2017. On what day of what month of what year do you suppose the Emergency Use Authorization was passed and finalized in the Federal Register by the FDA? Moments after wondering about that potential chronological connection, I looked it up. The Emergency Use Authorization was initially promulgated and explained to the world by the FDA on January 1, 2017—making it a very Happy New Year for Bill Gates and the global predators and giving them a very green light to go ahead full speed with their pandemic plans.[610]

The title of the FDA release about the EUA was "Emergency Use Authorization of Medical Products and Related Authorities; Guidance for Industry and Other Stakeholders; Availability." Klaus Schwab and his global predatory "stakeholders" must have cheered with joy. They are still cheering!

In summary, Bill Gates announced his Warp Speed-like plans to develop novel vaccines for the coming pandemics three years ahead of time in early 2017. He made the disclosures a few weeks after the January 1, 2017, FDA announcement of the implementation of the Emergency Use Authorization that would make a Warp Speed possible.

Those of us who have been so naive for so many years must finally acknowledge, "Things don't just happen." We live in a man-made world with communications moving freely and secretly, if necessary, at the speed of light. The men and some women making this new world order are driving us toward disaster to exploit us.

Gates, Caught Off Guard, Won't Help India's Tragedy

In April 2021, Bill Gates shocked people when, seemingly caught off guard, he quickly uttered "no" to a question about helping the large number affected in an upsurge of COVID-19 in India. He spontaneously explained during a Sky News interview that he was against giving India the right or the know-how to manufacture the Oxford AstraZeneca vaccine.[611]

Less noticed is the astonishing fact that one man is so deeply invested in and has control over at least three out of four of the world's COVID-19 vaccines—Moderna, Pfizer, and now Oxford AstraZeneca—and assumes it his right to call the shots that mean life or death for millions upon millions of people.[612]

Here is a transcription we made from the video:

> Interviewer: There has been some speculation that the changing intellectual property rules and allowing the vaccines, as you say the recipe for the vaccines, to be shared would be helpful and do you think that would be helpful?
>
> Gates: No! [This is an instant, emphatic response before talking nervously.] There is only so many vaccine companies in the world and people are very serious about the safety of vaccines and so moving something that had never been done, moving a vaccine from say uh J&J factory into a factory in India that, it's novel, it's only because of our grants and our expertise that, that can happen at all. The thing that is holding this back in this case is not intellectual property. There is not like some idle vaccine factory with regulatory approval that makes magically safe vaccines. uh You got to do the trials on these things and every manufacturing process has to be looked at in uh a very careful way. There's all sorts of issues surrounding intellectual property having to do with medicines but not in terms of how quickly we have been able to ramp up the volume here. I remember how shocked people were when we said we were going to do second sources in these developing country factories … uh. You know that, that was the

novel thing, and we got all the rights from the vaccine companies. They didn't hold it back, they were participating. I would do a regular phone call with the pharmaceutical CEOs to make sure that work was going at full speed.

Interviewer: It's not that, it's that, that is slowing it down more money would help, more vaccines physically would help as well, wouldn't it? How hard is it to get the countries that have the vaccine and obviously here in the UK we have enough to vaccinate our population five times over? What are your feelings about that?

Gates: Well over the balance of the year the U.S. the U.K. and others will be able to make sure that the vaccines are going to the developing countries because many of the vaccines worked … uh … you know although we are looking at some of the side effects now making sure we can treat them and that they are very rare.

CORPORATE/GOVERNMENT FUSION AND GLOBALISM AT DAVOS 2017

It seems likely that Bill Gates and Schwab were having a wonderful time linking the predicted pandemic to their global predatory plans—their anticipated global investment partnerships between international capitalists and large governments or what we call in our book "the corporate/government fusion." Did Schwab share with Gates his conception that a populist America—a patriotic, democratic America—was the only thing standing in the way of their globalist plans? It was a perfect globalist environment in which to work it all out together.

In addition to Gates, who else was at this international WEF conference held days before Trump's inauguration? Joe Biden was there to give his last speech as Vice President.[613] Biden's new climate czar John Kerry was also an honored presenter.[614]

Most stunning, Xi Jinping, China's tyrant and slaver, gave a stellar speech on how much he wanted world peace and better environment.[615]

Also from China, Jack Ma, the head of Alibaba, one of the world's largest firms, was also there. He tore into America for achieving such wealth through globalism without sharing it with the rest of the world.[616] We wonder how much wealth-sharing Alibaba did. By taking advantage of COVID-19, in April 2021, Alibaba joined the world's top ten highest market cap companies.[617]

According to the official Davos write-up, there were already growing fears of America's backlash against globalism, although Trump had only a few days earlier been inaugurated:

> Jack Ma, one of China's most successful and richest entrepreneurs, has responded to America's growing globalization backlash, arguing that the superpower has benefited immensely from the process—but that it has largely squandered its wealth.

The Schwab conference promoted China First and predatory globalism with a veneer of progressivism. The underlying goal was, and to this day remains, stopping American populism before it undermines globalism.

DREADFUL CONCLUSIONS

We were stunned to discover that Bill Gates and his world consortium of global predators were already implementing Operation Warp Speed before Trump's inauguration and three years before the pandemic virus was released in China from the Wuhan Institute. Gates made the announcement of his already ongoing vaccine investments and programs like CEPI in January 2017, less than four weeks after the FDA's announcement of the Emergency Use Authorization (EUA) that would make it legal to rush medications and vaccines through a fast-paced FDA process that did not actually lead to a full approval of safety or efficacy. Gates announced all this at Klaus Schwab's World Economic Forum, tying his work directly into Schwab's evolving as yet unnamed concept of the Great Reset. There is no longer any reason to think about a "conspiracy theory;" this has become documented history. Operation Warp Speed, previously identified with Donald Trump, sprang from the rapacious mind of Bill Gates and the pharmaceutical industry.

CHAPTER 16

Fauci Predicts the Pandemic's Onset with Certainty

C ontinuing with the theme of globalist foreknowledge of a coming pandemic and making preparations for it, we now confront another chilling occurrence. Days before President Trump took office, Anthony Fauci correctly predicted with "certainty" when the pandemic would actually strike!

For an explanatory timeline of all such events, see our *Chronology and Overview with Pandemic Predictions and Planning Events* at the end of the book.

Fauci Predicts a Coming Pandemic Days Before President Trump Assumes Office

It was January 10, 2017, a few days before President Trump's inauguration. Anthony Fauci was giving a prepared speech in Washington, DC at a conference titled Pandemic Preparedness in the Next Administration.[618] Looking surprisingly well groomed and youthful, Fauci seemed decidedly pleased with himself, smiling and, at times, jovial, as he gave the world the dreadful news: *a pandemic was inevitable in the coming administration of President Trump.*

It was confusing and distressing to watch this elderly, yet boyish, man enthusiastically explaining how humanity was on the verge of catastrophe.

The conference on Pandemic Preparedness in the Next Administration was hosted by the Center for Global Health Science and Security at Georgetown

University Medical Center by the Harvard Global Health Institute, directed by Ashish K. Jha, and by *Health Affairs*, a journal called "the bible of health policy" by *The Washington Post*.[619] As the first two organizational names signal, it was a globalist event.

It is possible that we are mistaken, but we could not find a presenter or committee chair who was actually from "the next administration." The one "Republican" made clear he was from earlier administrations. This oddity indicated a goal that was something other than developing cooperative plans with incoming President of the United States, Donald Trump, to ward off, prevent, or deal with the plague that was foreseen as afflicting the nation during his administration. No, it had a much more insidious purpose—to plan ahead for the predicted pandemic without involving Donald Trump or any of his advisors.

With Certainty, Fauci Predicts a Plague upon the Trump Administration

Early in his speech, Fauci declared:

> If there is one message that I want to leave with you today based on my experience (and you'll see that in a moment) is that there is no question that there will be a challenge to the coming administration in the arena of infectious diseases. ... but also, there will be a surprise outbreak.

How could Fauci, and in such a cavalier manner, announce at a major world conference that "there is no question that there will be a challenge to the coming administration in the arena of infectious diseases" and "also, there will be a surprise outbreak"?

Further on, Fauci described the globally organized approach to the "surprise outbreak" he anticipated, but without mentioning, he would be its czar. The response he envisioned would include the CDC, FDA, NIH, and his institute, NIAID, as well as global organizations.

Toward the conclusion of his speech, Fauci built upon his earlier assertion *"that there is no question"* about the coming pandemic during the upcoming Trump administration. Referring to the coming catastrophic event, he said,

"We will *definitely* get surprised in the next few years" and "so the thing we are *extraordinarily confident about is we are going to see this in the next few years.*"

Fauci twice uses the unlikely word "surprise" to describe the predicted pandemic. *How can a person predict a surprise with such certainty without actually knowing about it in advance?* If Fauci, instead, were to have described a "surprise birthday party" for President Trump, everyone would know he had inside information and was probably in on the planning. The surprise was clearly for President Trump, and Fauci openly states that he knows it is going to take place and he is "extraordinarily confident" about his prediction.

The ghastliness of what Fauci is predicting—a pandemic of unimaginable consequences for humanity—is not reflected in his written words or in his demeaner in videos of his speech. He does not call it a "tragedy" or "catastrophe." Instead, he makes those bizarrely innocuous references to a "surprise," indeed, more in the manner of a surprise birthday party than a global cataclysm.

Nor does Fauci appear sad or anxious. He looks like Fauci usually does—mildly amused, feeling good about himself, energized by his role as NIAID director, and fully in charge as the world's final "scientific authority" on the subject of pandemics.

In summary, we must reasonably come to the previously unthinkable conclusion that Fauci knew a pandemic was about to be inflicted upon the administration of President Donald Trump. He actually knew the schedule and was planning ahead to organize the response. In short, he was "in on it."

Furthermore, Fauci's language and demeanor indicates that he did not lament or dwell on the obvious reality that the pandemic would be enormously destructive to America and humankind. Fauci's cavalier presentation, his willing confession of foreknowledge of one of the worst crimes in history, suggests a depth of evil beyond our ordinary willingness or ability to imagine.

Whether or not Anthony Fauci was an active instigator of the immediate release of SARS-CoV-2, his financing of Chinese researchers and the Wuhan Institute of Virology made it possible for China to develop the deadly virus. It seems likely, but cannot be proven, that Fauci also knew that the pandemic would be inflicted upon the world with the precise intention of ruining the Trump administration—a goal held by billionaires and organizations throughout the United States and China, including Bill Gates, with whom he worked closely.

Why Would Fauci Dare to Tip His Hand?

Why would Fauci so obviously tip his hand in advance, in public, and so callously?

Many of the leading predatory globalists had been tipping their hands. In this chapter, we will look at yet another presenter at the conference with Fauci who predicted the pandemic. We will examine the infamous Event 201 which, in October 2019, played a pandemic wargame featuring SARS-CoV. It was supported by Bloomberg, Gates, and Schwab. As we have documented, and will continue to do, Bill Gates, Klaus Schwab, Michael Bloomberg, and many other predatory globalists have also tipped their hands and openly supported the Chinese Communists in concert with COVID-19 oppression of America.

If Gates, Schwab, and Bloomberg had the grandiosity to carry out Event 201, planning public health responses to a SARS-CoV pandemic shortly before its outbreak, why wouldn't Fauci find his own way of joining in to preempt leadership in the coming much-predicted pandemic? Since Fauci, as the director of NIAID, would possess much more knowledge about whatever was happening around COVID-19, all of his energy would be focused on it.

Fauci's braggadocio reflects the same stunning grandiosity we have found in most global predators. His utter indifference to the fate of humanity is something else we have seen in so many global predators. It led us early in our research to describe these predators seeking not only wealth and power, but self-aggrandizement—a self-centered passion for superiority over other individuals and humanity itself.

Previous Indictments of Fauci for Causing the Deaths of Many American Citizens

Before deciding to face that Fauci knew about COVID-19 beforehand and was preparing for it, we had little in the manuscript about his earlier betrayals of HIV patients around the world. We had decided that it was unnecessary and potentially inflammatory to emphasize the past when we had so much negative COVID-19 information about him. *Now that we are making much stronger accusations against him—with implications of treason and criminal conspiracy to commit mass murder, we feel the necessity of elaborating on his past.*

Fauci has been immune from the effects of even the harshest criticism. He has been so protected by the predatory globalists, including those within the bureaucracy and in the media, that most people have no idea about his extremely controversial background and how it demonstrates similar attitudes and behaviors to his deadly handling of COVID-19.

Fauci first took the world stage as the AIDS Czar in the 1980s, when he soon generated accusations of sacrificing the lives of AIDS patients to the interests of the pharmaceutical industry. In February 2021, PBS, a progressive and globalist media organization, with some spin, nonetheless, described the outrage over Fauci's handling of HIV:[620]

> During the height of the HIV epidemic, LGBTQ activists pressed him to make experimental treatment drugs available for gay men dying of AIDS. They called on him to do something—anything—to stop the epidemic that was taking the lives of so many in their community. They protested outside his office at the National Institutes of Allergies and Infectious Diseases, where he was the director—the post he still holds today—with his head on a stake. They burned effigies of his body. In 1988, AIDS activist Larry Kramer famously addressed an open letter[621] to Fauci in the *San Francisco Examiner*, accusing him of murder.

Earlier, *The Washington Post*, another such media globalist, provided a retrospective— again, well-spun—description of that desperate protest against Fauci over AIDS:[622]

> With signs that read "Red Tape Kills Us" and "NIH — Negligence, Incompetence and Horror," members of the AIDS Coalition to Unleash Power, or ACT UP, marched toward a row of police officers in riot gear guarding Building One. ... As confused NIH scientists and administrators looked out of their windows, the 1,000-strong demonstration then marched to Building 31, which housed the offices of the protests' target: Anthony S. Fauci, then and now the chief of the National Institute of Allergy and Infectious Diseases (NIAID).

The AZT Controversy

Under Fauci's leadership, AZT, a very toxic, experimental drug, was approved for the treatment of AIDS under very chaotic conditions[623] and remained controversial.[624] Fauci was pushing AZT for expanded patient populations, including as a preventive for asymptomatic patients thought to be infected based on controversial HIV testing. Like all the drugs and vaccines backed by Fauci during COVID-19, AZT was extremely expensive. Like Fauci's COVID-19 favorite, remdesivir, the announcement of its EUA during an epidemic was a drug company bonanza:[625]

> Burroughs Wellcome stock went through the roof when the announcement was made. At a price of $8,000 per patient per year (not including bloodwork and transfusions), AZT is the most expensive drug ever marketed. Burroughs Wellcome's gross profits for next year are estimated at $230 million. Stock market analysts predict that Burroughs Wellcome may be selling as much as $2 billion worth of AZT, under the brand name Retrovir, each year by the mid-1990s — matching Burroughs Wellcome's total sales for all its products last year.

In addition, many AIDS patients were poor and lacking in insurance.

AZT is another story of Fauci backing and rushing through poorly tested, highly toxic drugs to the enormous benefit of his friends in the pharmaceutical industry, but of limited value at best for the world's patients.

Fauci Suppressed Inexpensive Sulfa Drugs

Did Fauci also suppress less expensive and more effective treatments that were far safer for AIDS? The unequivocal answer is yes.

Pneumocystis pneumonia (PCP), caused by a fungus, is not generally a threat to humans. However, from the beginning it became the most common opportunistic infection and *the most common pneumonia in people living with HIV* in the AIDS epidemic.[626] It killed many people.

The first choice for prevention and treatment among standard guides and many physicians were the sulfa drugs like Bactrim (trimethoprim/sulfamethoxazole) which had long histories of safe and effective treatment.[627] There were also controlled clinical trials confirming its use in pneumocystis pneumonia in HIV patients.[628] It had already been used for 20 years to treat the same opportunistic pneumonia in cancer patients. Bactrim remains in widespread use today and is routinely prescribed for both unapproved and approved infections.

Fauci opposed Bactrim, while pushing expensive, highly toxic, experimental, and expensive drugs, resulting in more outraged responses than currently, then accusing him of murdering thousands of gay men in particular.[629] The story of Bactrim is obviously very similar to what we have today with medications like hydroxychloroquine and ivermectin.

The Association of American Physicians and Surgeons, a conservative, patient-oriented organization that promotes freedom, ethics, and science in medicine, issued a letter making the comparison between Fauci's attitudes then and now:[630]

> We must choose to treat NOW, based on the experience of frontline physicians informed by the scientific literature, rather than deny early treatment until the results of randomized, controlled trials (RCTs) are available.
>
> Under the direction of Dr. Anthony Fauci 40 years ago, we tried the second method early in the AIDS epidemic. Patients were dying of pneumocystis pneumonia, which can be successfully treated with cheap, safe sulfa drugs. *In 1987, activists pleaded with Dr. Fauci to issue guidance that suggested prophylactic treatment with Bactrim.* Efficacy had been shown in transplant patients in 1977.[631] Dr. Fauci refused, insisting on the "gold standard" of RCTs. Since the National Institutes of Health refused to fund trials, activists raised the money themselves. By the time the results were ready, two years later, 17,000 patients had died needlessly. (Emphases added.)

As mentioned earlier, we had decided not to raise these inflammatory issues from the past in our book until we decided, after nearly finishing the manuscript, that Fauci had foreknowledge that a SARS-CoV would be released and was planning in advance for the coming pandemic.

Not since Fauci lobbied against inexpensive, safe, and effective treatments for HIV—forcing sick or dying patients to await more expensive medications and a vaccine that never arrived—have American doctors been so stymied by government interference. COVID-19, which impacts everyone, is a much larger treatment bonanza for the globalist predators, so the government and Fauci have gone much further to interfere with patients getting proper treatments that frustrate the globalists from making tens of billions of dollars.

Fauci Tries to Force Entry for Ebola into the United States of America

There is more to Fauci's reckless history. When the extremely deadly Ebola epidemic struck in Africa, Fauci fought efforts to quarantine heavily exposed healthcare workers when they returned to the United States. Fauci exerted enormous media efforts in 2014 to stymie attempts of governors to protect the American people from the spread of the deadly Ebola epidemic in Africa.[632]

Similarly, as already documented in this book, Fauci tried to stop President Trump from prohibiting travel from China to America early in COVID-19.

Fauci's Friendliness with the Pharmaceutical Empire

In the discussion following his talk, Fauci reassured a questioner that he experiences no hostility in partnering with industry. He declared he is very "friendly" with the pharmaceutical industry, helping them in "de-risking" their involvement by funding research for them. Fauci is not talking about working with nonprofits dedicated to serving humanity. Fauci's words are euphemisms for collaborating with some of the most avaricious corporations on the planet to protect and vastly increase their profits.

Fauci also described how he can help the FDA bypass more lengthy approval processes and then gave the audience reassurance there would be no reduction in safety. He is describing the further transformation of the FDA from a watchdog to a lapdog—really, a service dog—for the pharmaceutical industry.

In a follow-up blog to his speech, on February 9, 2017, in *Health Affairs*, Fauci wrote:[633]

One of the most important challenges facing the new administration is preparedness for the pandemic outbreak of an infectious disease. Infectious diseases will continue to pose a significant threat to public health and **the economies of countries worldwide.** (Bold added.)

Thus, Fauci anticipated the "surprise" would disrupt "the economies of countries worldwide." How did he know *that* was part of the surprise? Fauci, at the least, was already helping to formulate plans for massive shutdowns.

Ashish K. Jha—Yet Another Globalist

On November 17, 2020, Senator Ron Johnson (R-WI) as Chairman of the Senate Oversight Committee on Homeland Security and Government Affairs held hearings on the early treatment of COVID-19. A group of courageous and outstanding physicians and public health experts testified about their great success in treating early cases of COVID-19 with existing FDA-approved, generic antiviral medicines, including hydroxychloroquine or ivermectin, in combination with azithromycin, doxycycline, or other antibiotics and vitamins. The regimen was for specific stages of the disease, sometimes using steroids like prednisone and blood thinners when indicated.

The presentations were scholarly, research based, and clinically sound, reflecting the viewpoints of many experienced physicians and epidemiologists around the world, including numerous members of the Association of American Physicians and Surgeons (AAPS). Elizabeth Lee Vliet MD, writing for the Association of American Physicians and Surgeons, summarized the panel and their credentials, several of whom gave live testimony at the committee hearings and others who spoke long distance. Dr. Vliet observed: [634]

> The expert witnesses were nine stellar Frontline physician-scientists representing medical centers across the U.S., who have collectively treated several thousand COVID patients and who have published more than 2,000 peer-reviewed medical studies over their careers.

Presenters who were physically present at the hearing included physicians Peter A. McCullough from Texas A&M in Dallas; Harvey Risch from Yale; and George Fareed, Harvard trained and now the director of a medical center in California. It was a somber, thoroughly scientific panel by very experienced, treating physicians, confirming that early, proper treatment of COVID-19 prevented most hospitalizations and nearly all deaths.

Only one Senate Democrat briefly attended the start of the session by presenting a rebuttal in advance and then left. The Democrats boycotted the hearings. When the panel was finished, a lone expert invited by the Democrats as a rebuttal witness testified via video. He was Ashish K. Jha, a physician whose Harvard institute cosponsored the Georgetown conference with Fauci and chaired a committee that met before Fauci spoke.

Jha, who has not treated COVID-19 patients, behaved disgracefully. He repeated myths about the dangers of hydroxychloroquine, one of medicine's oldest and safest medications, by saying it caused "poisoning."[635] He misrepresented the scientific literature and dismissed the widespread success reported with the use of hydroxychloroquine from around the world. Worst of all from my perspective, he then used snarky, scandalous, and insulting language to characterize the panel of doctors and to dismiss the lifesaving early treatment of COVID-19.

As a reformer in the field of psychiatry for 60-plus years and as a forensic expert for 50 years, often going up against the biggest drug companies and medical organizations, I have also endured nasty attacks. I have rarely seen such immature and inappropriate behavior by a physician as I saw in Jha's performance. He made me think of a cowardly schoolboy daring to thumb his nose because he was doing it long distance instead of face to face.

One panelist replied to Jha and was relatively restrained when he described Jha as acting recklessly and dangerously toward the well-being of the nation and its citizens.

JHA'S GLOBALIST BACKGROUND

At the time of the hearing, Jha was professor of global health at the Harvard T.H. Chan School of Public Health and director at Harvard's Global Health Institute. Jha's official bio mentions "sponsored funding from sources such as the National Institutes of Health, the Bill & Melinda Gates Foundation, the Climate Change Solutions Fund …,"[636] connecting him to the corrupt NIH, global predator Bill Gates, and a progressive climate change organization.

Jha's Harvard Global Health Institute has several domains, including pandemics and global warming. As a professor of global health at the Harvard T.H. Chan School of Public Health, he began serving as the school's dean for global strategy the year after the Georgetown Conference with Fauci. In these globalist positions, Jha would have significant connections to China, including through the Harvard China Health Partnership at the School of Public Health.[637]

The partnership's goals focus particularly on China:

> We partner with Chinese peer institutions, policy think tanks, and socially oriented entrepreneurs to educate professionals on the ground through executive education courses.

The School of Public Health, at which Jha was professor of global health, has unusually deep roots in China—the kind enabling China to take great advantage of our knowledge and technology and to turn our heads away from the dangers of collaborating with them. The School of Public Health sees China as woven into their very fabric from the beginning in the first sentences of its "About Us:"[638]

About Us
History of Harvard and Health in China

The Harvard T.H. Chan School of Public Health has been engaged in education, research, and knowledge exchange with China for more than a century. Just two years after the school was founded in 1913, it welcomed its first international students— three Chinese doctors. A decade later, Dean David Linn Edsall took a six-month trip to China to conduct research. In 1979, after the Cultural Revolution, the Harvard Chan School became the first foreign school of public health to work with the Chinese government and Chinese schools of public health.

Over the past four decades, ties between Harvard Chan researchers and peers across China have grown increasingly close. Harvard faculty guided and assisted China in establishing the field of health policy and management, while also training the first generation of Chinese molecular biologists in public health. The School's work with academic and government partners has resulted in numerous cooperative endeavors in research, education, academic exchange, and network building.

As we have documented, the Chinese Communist Party systematically develops these close ties to leading American universities in the interest of its Military-Civil Fusion, and the CCP probably considers Harvard a great trophy.

Unfortunately, too many Americans take a much more naive and self-serving approach than Chinese citizens. Chinese professors and researchers working in the United States, or anywhere else in the world, are only allowed to do so under the overarching requirement to share all their information with the Chinese Communist Party.

Jha left Harvard in September 2020 to become the director of Brown University's School of Public Health. He is an exemplar of the globalist academic in public health. He is also one of those insiders with seeming foreknowledge of the coming pandemic for which he had anointed himself in advance as a leading theorist and worldwide organizer.

JHA PREDICTS THE ONCOMING PANDEMIC

What made Jha become so hostile and out of control in attacking his medical colleagues who simply wanted to alert the nation to the existence of old, safe, inexpensive, and effective early treatments for the COVID-19 virus? As noted earlier, Jha was one of three organizers of the Georgetown University conference on "Pandemic Preparedness in the Next Administration" at which Fauci was the keynoter.

Jha spoke earlier than Fauci as a panel moderator, where his remarks anticipated Fauci's in expressing certainty that a pandemic was coming. Jha explained, "If you think about all the things that could devastate a population, a country, pandemics is [sic] right at the top of the list."[639] He emphasized preparing for the eventuality of a devastating pandemic, "which is going to come at some point."

Jha joins our pantheon of globalists enthusiastically predicting and preparing for an imminent pandemic tragedy, again making us wonder, "Where did they get such certainty? What did they know?" This is manifestly clear: Jha, like Fauci and many other globalists, predicted and anticipated the worst—a deadly pandemic—*and then helped make the worst come true by violently opposing effective, safe early treatments.*

The purpose of organizing the Georgetown conference, Jha explained, was to build a coalition to prepare for the worst. As we saw in our quotes from the video of Fauci's presentation, Jha does not speak with the cautious, muted distress with which one might address their anticipated worst tragedy for humanity in recent times. Instead, he expresses an overflow of energy over the opportunities ahead to organize and work together on huge global projects. He speaks enthusiastically of the conference as "the first of many events" and "being on a journey."

The SPARS Pandemic Scenario, 2017

As noted in the *Chronology and Overview with Pandemic Predictions and Planning Events*, in October 2017, the Bloomberg School of Public Health at Johns Hopkins University published a book titled *The SPARS Pandemic, 2025-2028.*[640] The School of Public Health is named after Michael Bloomberg, a highly political global predator.

This little book, written under the cloak of a scientific scenario, was written specifically for promoting—and defending in advance—SARS-CoV vaccines. *The SPARS Pandemic* promotes the need for the world to invest in vaccines for a SARS coronavirus pandemic. It also examines the necessity of reshaping and censoring opinions that resist top-down government control during the coming pandemic.

The study states, "This is a hypothetical scenario designed to illustrate the public health risk communication challenges that could potentially emerge during a naturally occurring infectious disease outbreak requiring development and distribution of novel and/or investigational drugs, vaccines, therapeutics, or other medical countermeasures." All this anticipates COVID-19, Operation Warp Speed, and the Great Reset.

The star of the book is a fictional vaccine named Corovax. At first, it is maligned by a misled minority of the population yet turns out to be miraculously safe and very effective. Still, some unjustified claims for vaccine injuries are paid off to mollify the public. Thus, in January 2017, a strategy was already coming together to make billions of dollars and expand power through a SARS-CoV pandemic which was seen as imminent. Making the public compliant was deemed essential, and deceit was being legitimized in advance.

COVID Virus Germ Games: Event 201

Moving closer to the COVID-19 pandemic, we encounter Event 201, which seems to predict the pandemic within a month or two before it occurs. It is definitely the stuff of conspiracy theories; however, it was public and videoed and had the open support of major global predators. Event 201 took place in October 2019, before most people had heard of a new outbreak of SARS-CoV. There are discussions of possible SARS-CoV-2 victims as early as September, but the large, international event had to be in planning for at least several months. This best fits into the category of a prediction.

Event 201: A Global Pandemic Exercise[641, 642] was conducted by Bloomberg School of Public Health at Johns Hopkins University and led by its Center for Health Security. The Bloomberg School of Public Health is perennially ranked number one among more than 100 competitors by *US News & and World Report* and also considered the largest in the country.

Event 201 was conducted in collaboration with Klaus Schwab's World Economic Forum (WEF) and the Bill & Melinda Gates Foundation. The emblems of all three institutions—Gates, Schwab, and Johns Hopkins—are emblazoned on the promotional publications.

As noted earlier in the chapter, the Johns Hopkins school of public health is named after its benefactor, Michael Bloomberg, one of the world's great predatory globalists. While President Trump was confronting China, Bloomberg was leading an international movement to reestablish very close financial and political ties to China (Chapter 25).

Event 201 was called a Global Pandemic Exercise, similar to a war game, or in Bill Gates' phrase, "a germ game."[643] As Gates, Fauci, and other globalists have urged, it emphasized partnerships between corporations and governments with leaders of business and industry, governments, and global health making the plans. None of them have been elected or ever designated as a governing body over anyone. Their actions again raise the question, "How or through whom will they implement their global plans?"

Here's how *Event 201* described itself on its promotional website: [644]

Pandemic simulation exercise spotlights massive preparedness gap

Event 201, hosted by the Johns Hopkins Center for Health Security, envisions a fast-spreading *coronavirus* with a devastating impact. (Emphasis added.)

Notice that *Event 201* made a SARS-CoV the focus of their pandemic simulation exercise. They called the virus nCoV-19. Sounds a lot like COVID-19 and a little like SARS-CoV.

When this "coincidence" later raised suspicions of foreknowledge of the pandemic, organizers of *Event 201* scrambled to publish a disclaimer.[645] The date of *Event 201*, October 18, 2019, was almost exactly one month before the first case of infection with SARS-CoV-2 thus far identified.[646]

The "Players" in Event 201

Here are thumbnail descriptions of the 15 "players" in the Johns Hopkin's game:

- The director-general of the Chinese CDC
- A high-ranking representative from the U.S. CDC and U.S. Public Health Service
- The Bill & Melinda Gates Foundation President of Global Development
- A former legal advisor to the National Security Council and former deputy director of the CIA
- A UN representative
- The representative of an unnamed (really!) major media company who works in partnership with the World Economic Forum

- The president of the UPS Foundation who was also the chief of diversity and inclusion officer
- A global safety officer from Johnson & Johnson
- A former deputy director of the Australian office of the prime minister and the cabinet
- A Singapore financial officer
- A representative of McGill University in global policy and innovation
- Representatives of the following corporations: Marriott International, Edelman, Henry Schein, and Lufthansa

The first one listed, Director-General of the Chinese CDC, George Gao, is noteworthy. High-ranking Chinese often show up at events conducted by American global predators. The Chinese were not and are not our allies. They are conducting unrestricted warfare against us. So why was their top CDC official invited? Why would we want to bring him into an inner sanctum of our pandemic planning that, at the least, would inform the Chinese Communists about our potential defenses against biological warfare and educate them about our preventive and treatment measures? The power of global predators seems the only explanation with the cooperation of Americans who are more committed to internationalism or even to the Chinese Communist Party than to anything remotely resembling America First.

EVENT 201 UNFOLDS

Here is a further description of *Event 201*, again published by Johns Hopkins itself:[647]

> The Johns Hopkins Center for Health Security in partnership with the World Economic Forum and the Bill & Melinda Gates Foundation hosted Event 201, a high-level pandemic exercise on October 18, 2019, in New York, NY. The exercise illustrated areas where public/private partnerships will be necessary during the response to a severe pandemic in order to diminish large-scale economic and societal consequences.

The promotion piece goes on to say:

> A severe pandemic, which becomes "Event 201," would require reliable cooperation among several industries, national governments, and key international institutions.

This collaboration epitomizes globalism, which aims to use global emergencies to blur the distinctions between governments, international organizations, and businesses to create a growing global governance. Unlike the United Nations, the leaders will not be chosen by member nations. Unlike democracies, the leaders will not be chosen by the voting citizens.

Who has chosen them to lead the world? It is astonishing that the question has escaped serious discussion. The predatory globalists believe that they should be in charge of everything and everyone, and absent a better choice, they are turning to Communist China with whom they share antagonism toward democratic republics, the free market, and political freedom.

ONE LAST PREDICTION PUBLISHED SIMULTANEOUSLY WITH THE PANDEMIC OUTBREAK

Before the authors could have known about COVID-19, a follow-up to the 2017 book, *The SPARS Pandemic*, is published as a monograph in the *Journal of International Crisis and Risk*.[648] Like the original book, the update is titled "The SPARS Pandemic 2025-2080" with a new subtitle: "A Futurist Scenario to Facilitate Medical Countermeasure Communication." Its authors overlap with the authors of the book, and the contact information lists the senior author of the original book, Monica Schoch-Spana PhD at the Johns Hopkins Center for Health and Security.

Amazingly, the January 2020 report again focuses on a "fictional pathogen, SPARS," now called the St. Paul Acute Respiratory Syndrome Coronavirus (p. 82). Amazingly, this publication is rarely cited in the string of evidence of foreknowledge of SARS-CoV-2, even though its publication literally overlaps the earliest stages of the pandemic. As in the original book, the star vaccine is Corovax. Considerable emphasis is given to communicating with the public in ways to gain compliance with vaccination.

It is breathtaking to see the globalist establishment preparing in advance for COVID-19 for more than three years. With this last simultaneous publication in the early days of the pandemic, it is as if they rushed this last one into print knowing that it had to be out by January or it would become old news.

Why Did Gates, Fauci and So Many Others Want to Warn the World in Advance?

Reviewing the material in this chapter, along with our *Chronology and Overview with Pandemic Predictions and Planning Events*, raises the question, "Why did Gates, Fauci and so many others go so far out of their way to warn the world and President Trump about the coming SARS-CoV pandemic?"

Those who reject the idea that the globalists had foreknowledge and were planning the pandemic might claim that these elite individuals were too smart to open themselves to serious accusations and charges by describing in advance what they were doing. They might also argue that taking the world by surprise was more to their advantage in creating a pandemic disaster.

In fact, alerting the world in advance to the pandemic was clearly to the enormous advantage of the global predators. We can see in the demeanor of Bill Gates, Fauci, and others that they are enjoying their predictions. Why? Because alerting and frightening people in advance were key to their twin goals of making billions of dollars and strengthening their autocratic and increasing totalitarian control over the world.

The globalists began achieving their exploitive goals by 2017 when the FDA beefed up its guidelines on Emergency Use Authorizations, Homeland Security issued a large pandemic planning document, and Johns Hopkins issued the first of three efforts to spread the idea of planning ahead specifically for a coronavirus pandemic and the need to rush the production of vaccines. At the same time in 2017, Bill Gates was already working with Moderna on new platforms to rush through vaccines, investing billions in vaccine manufacturers, and collecting billions through CEPI, a fund to push the development of vaccines.

The predictions and plans were actually far more important than the pandemic itself. COVID-19, for example, turned out to be much less threatening or deadly than predicted, leading the CDC to vastly exaggerate its reporting of deaths. Yet COVID-19 has already accomplished its most sought-after goals:

(1) Making billions of dollars for the global predators and vastly increasing the power of individuals like Bill Gates, Schwab, Bloomberg, Fauci, and innumerable others, including hundreds of Chinese billionaires.

(2) Greatly strengthening top-down government worldwide, while encouraging corporate-government fusion as part of the Great Reset.

(3) Strengthening the Chinese Communist Party and its dictatorial control over the Chinese people. In comparison to the U.S., China has prospered, bringing nearer the day when Communist China will surpass us in the size of its economy and military power, and start consolidating its empire.

(4) Reversing the enormous success of President Donald Trump and replacing him with President Biden, himself a global predator with corrupt ties to China. Before COVID-19, Trump was strengthening the economy, bringing jobs and industry back to America, creating the highest employment rates in history, especially for women and minorities, reducing illegal immigration, building up the military, transforming the Republican party in a populist bastion, and containing Communist China.

Today, President Trump is no longer in office, and the democratic election processes have been vastly corrupted. America has sustained far too many deaths due to the suppression of early treatment (supported by President Trump), the implementation of bad treatment, and the turning of nursing homes into death houses by progressive governors. America's moral strength and independence have been sapped by obedience to outlandish demands, such as wearing masks in public, the disruption of our schools, the shutting down of churches, and the overall crushing of our society and economy. Most recently, massive inoculations with experimental, dangerous, and unnecessary vaccines have been driven by withheld information, fear, and coercion.

With the ruination of Donald Trump's presidency and the installation of extreme progressives who fawn over Xi Jinping, America is rapidly losing its patriotic zeal and role as the beacon of freedom and the only hope of containing the Chinese Empire. COVID-19 has already become a roaring success for the global predators, led by assorted billionaires, giant corporations, myriad corrupt governments and agencies, and Communist China.

CHAPTER 17

Clinical Trials— Gold Standard for Scientists or Golden Calf for Predatory Corporations?

How could Bill Gates, Fauci, his financial partners, and Moderna and Pfizer be so sure, several years ahead of time, that their radical, dangerous, and experimental mRNA vaccines could be successfully rushed through FDA testing? By 2020, research was accumulating that mRNA vaccines were proving, after at least ten years of research, to be too unstable, to be less effective in humans than in animals, and to pose many serious risks, including causing a dangerously strong inflammatory immunological reaction.[649]

Similarly, how could Fauci have been so sure he could rush Gilead's drug remdesivir through NIH and FDA testing, when it had already proven ineffective and potentially deadly in in Ebola and SARS-CoV-2 trials (Chapter 10)? Aren't clinical trials a gold standard of science—something beyond gross, crass, self-serving manipulation?

The answer is that clinical trials conducted under the NIH and the FDA are a sham and have been so for decades. Perhaps some are legitimate, but not in my extensive experience where potentially blockbuster drugs from big corporations are being evaluated. If the drug turns out to be really dangerous and ineffective, the FDA may make the drug company work overtime to cover it up, but they will almost never reject the medication.

Instead of being the epitome of science, clinical trials are among the most corrupt activities in the pharmaceutical-FDA-NIH complex. The reasons for that are simple: Billions of dollars are riding on the outcome of these all-too-human endeavors, and the future careers of FDA officials and scientific investigators are on the block if they interfere with the approval of a blockbuster drug or vaccine. This is documented in exquisite detail in our earlier book, *Talking Back to Prozac*, which details the approval process of an anticipated blockbuster antidepressant. It was slated for approval regardless of how much the clinical trials had to be corrupted in the process by both the company and the FDA.

As a medical expert, I have been approved many times before trial and in court to testify concerning the nature and flaws of clinicals conducted by the pharmaceutical industry to obtain FDA approval or to put positive publications into scientific journals. I have found serious corruption everywhere I have looked in my specialty area of psychopharmacology and psychiatric drugs. It is well documented that clinical trials are much more positive for the medication when the drug companies and their advocates are actively involved in them.[650]

THE ACHILLES HEEL OF SCIENCE

Science! We have been told science must be relied upon to make our decisions when dealing with the pandemic called COVID-19. We must in effect bow down to science, no matter how humiliating and painful it may seem to our unscientific minds.

But science has an Achilles' heel—a fatal flaw that can completely ruin it and frequently does. What is the fatal flaw of science? It is conducted by human beings. Where serious scientific controversies and conflicts of interest exist, science is no more reliable than politics ... or heaven forbid ... no more reliable than religion. Science is only as dependable as the people who conduct and disseminate it. The adage for judging opinions still holds true—consider the source! Similarly, the adage for finding corruption still holds true—follow the money.

When Anthony Fauci announced the rollout of his initial clinical trial for remdesivir as the great hope for knocking out the coronavirus epidemic, he boasted about the clinical trial's importance: "A randomized, placebo-controlled trial is the gold standard for determining if an experimental treatment can benefit patients."[651]

Very few people, including sophisticated doctors and researchers, have any idea how randomized, placebo-controlled clinical trials are routinely manipulated and made unreliable. Most people have no idea that they reflect not so much science as the interests of the wealthy corporations or institutions who sponsor almost all of them.

Since controlled clinical trials involving drugs and vaccines are very expensive, require access to sick and often highly infectious patients, and must be approved by the Institutional Review Boards (IRBs) and the FDA, they can only be conducted with funding from Big Pharma and Big Government or occasionally other large institutions.

The People's Drug is being suppressed by the same interest groups. As we have documented in earlier chapters, hydroxychloroquine and zinc as a prophylactic—and hydroxychloroquine plus azithromycin (or other antibiotics like doxycycline) and zinc for patients developing COVID-19—could right now greatly improve the treatment of patients in the U.S., as it has done elsewhere. This is true as well for treatment combinations starting with ivermectin instead of hydroxychloroquine.

Doctors and entire nations have been and are continuing to use this hydroxychloroquine successfully, especially in combination with other specific therapeutics. The Association of American Physicians and Surgeons has shown that the countries which actively use hydroxychloroquine have significantly lower death rates than those that do not.[652] The medical association has also called for making hydroxychloroquine more freely available.[653, 654]

The Clinical Trials in Progress Are Already Corrupted

You can see the drug company bias in a recent list of all current and planned registered trials of remdesivir for the treatment of COVID-19. Many of the trials have "arms" or comparator studies for chloroquine or hydroxychloroquine. Presently, we found 14 chloroquine and/or hydroxychloroquine trial arms, but all of them are set up so they are doomed to fail. Only one adds azithromycin to the hydroxychloroquine, and none adds zinc. However, the combination of hydroxychloroquine, azithromycin, and zinc is considered the most successful treatment for COVID-19 by experienced clinicians.[655, 656] In addition, several studies compare remdesivir to the older drug chloroquine rather than to hydroxychloroquine, which is considered safer. *Put simply, every single registered remdesivir clinical trial has been rigged to make hydroxychloroquine fail in comparison.*

Perhaps in the crucible of the epidemic, some institutions will courageously conduct genuinely scientific studies of hydroxychloroquine's safety and effectiveness in combination with azithromycin and zinc, as well as some of the recommended vitamins, but that is an idealistic longshot. The world that awaits the clinical trial results should be warned all or nearly all expensive clinical trials will be showcase trials for those who sponsor them.

Science is a process. To develop genuine or more lasting scientific validity often takes many trustworthy and honest scientists and institutions working independently and separately in many settings, publishing numerous papers, over a long period of time. I doubt "Trump's Wonder Drug" will ever be given a fair scientific review, at least in the United States. It is more likely to occur in other countries where the Pharmaceutical Empire has somewhat less influence, but when it happens, the research will receive scant airing in the U.S.

Whenever science has huge financial, political, and social impacts—when science becomes especially important to us human beings, it is instantly seized upon by unscrupulous human beings and industries for their own purposes. Often, the science becomes wholly corrupted, and at other times, it is vastly distorted and misinterpreted as it passes through the media to the unwitting people.

CREATING DEADLY STUDIES TO PROVE HYDROXYCHLOROQUINE IS TOO DANGEROUS

We will never forget Neil Cavuto self-righteously attacking President Trump's personal use and support of hydroxychloroquine.[657] The Fox News anchor cited a ridiculously flawed, unofficial, and unpublished study of VA patients which purportedly showed hydroxychloroquine increased the death rate, when a more accurate analysis showed when used in combination with azithromycin, it was saving lives. As noted earlier, the study itself was rejected by the VA as unauthorized and misleading.[658]

Cavuto accused President Trump of killing people! One of the doctors Cavuto later interviewed disappointed him on air by saying he had been using the drug for 40 years to treat other diseases, and he had never seen a single death. The media rarely slips up like that, but Cavuto was in a rush that day to use science to prove President Trump was not only wrong, but dangerous.

Recently, Stella Immanuel MD, a Black African doctor who immigrated to the U.S. and practices medicine in Texas, defended hydroxychloroquine as part of a group presentation by America's Frontline Doctors. The major media picked her to ridicule for her Christian beliefs, and social media took down her enormously popular video. Her clinical experience treating hundreds of COVID-19 patients without a single death was uniformly dismissed by the conventional media, while similar opinions by other doctors in the group were ignored. Only occasional media gave her a chance to express her well-informed views.[659]

What Are Placebo-Controlled Clinical Trials?

In a placebo-controlled, double-blind, randomized clinical trial, the drug is randomly given to one group of people, while a harmless, inert pill or saline injection is randomly given to the other group, called the placebo control. If a portion of the drug group develops encephalitis, for example, and none of the placebo group contract the disease, then the study may generate statistically significant data indicating the drug or vaccine causes encephalitis.

To eliminate doctor or patient bias in evaluating the trials, they are double-blind. That is, neither the patient nor the doctor, or anyone evaluating the trials, knows which people are receiving the treatment and which are receiving the placebo (called the controls).

Despite how it sounds, controlled clinical trials are not science. Pure science is a search for objective truth or improved theories about reality based on observation and empirical facts. At their best, clinical trials are *applied science*—the use of the scientific method for achieving certain ends, such as convincing the FDA to approve the next blockbuster drug to make the company even wealthier.

Clinical trials are not in search of truth or improved theories; they are the application of a superficially scientific method toward a specific goal. In this, they are more like engineering—applying science to achieve practical goals—or, in this instance, getting the FDA to approve a potentially valuable product.

If we lived in Utopia, in theory, it might be possible to create and to conduct a clinical trial more in search of the truth than in the interest of creating wealth. In practice, it is almost entirely impossible. As noted earlier, placebo-controlled, randomized, double-blind clinical trials are extremely expensive to conduct. They need an institutional base and an Institutional Review Board (IRB) to approve it. They require recruiters and managers, along with scientists, statisticians, and healthcare providers. If they are intended to gain FDA approval, the process becomes bureaucratically overwhelming, requiring years and tens of millions of dollars. Most drug or vaccine clinical trials are *sponsored* (paid for and controlled) by the pharmaceutical industry.

Fauci and his NIH Institute for Allergy and Infectious Diseases collaborated with his favorite corporation, Gilead, to acquire expedited EUA "approval" for remdesivir.[10] Gilead is estimated to have spent one billion dollars on developing and promoting remdesivir on its own, and Fauci's institute is estimated to have contributed $70 million more to implementing the NIH clinical trial of the drug.[660]

When so much time, effort, and money are being spent on a project, can we expect the sponsors to admit they have come up empty-handed—or worse, they have created a monster that will do far more harm than good? Can we expect Tony Fauci to tell us, "Oops, I've been completely wrong in touting this drug long before we had clinical trials, and now I have to confess I've wasted a fortune, untold man hours, and a lot of precious time on promoting it, and it is making many patients worse. Oh, yes, and I was wrong about hydroxychloroquine too, because it is a very good drug with 60 years of safe prescribing behind it, it costs almost nothing, and should be given very early to every suspected COVID-19 patient and used as prophylaxis, too. Even the World Health Organization says they know of no deaths associated with it. And, yes, in addition, I apologize to President Trump for undermining his support for hydroxychloroquine."

Because of this horrendous partnership of the most power-hungry in the government in combination with the greediest in the industry, Fauci's remdesivir "Gold Standard" clinical trial became a monstrous demonstration of political corruption masquerading as science.

My Experience Evaluating
Controlled Clinical Trials

The corruption of clinical trials is endemic in the pharmacological industry and repeatedly condoned by the FDA. As a medical expert, I have a great deal of painful experience in this area, especially after I was chosen by a consortium of attorneys to be the one doctor in the world to threaten the success of the blockbuster drug, Prozac.

Prozac (fluoxetine) changed the world of psychiatry and medicine, promoting a false theory of biochemical imbalances causing mental disorders and paving the way for multiple new psychiatric drugs with revenues in the multiple billions of dollars. In the early 1990s, I was chosen by a consortium of several dozen attorneys and approved by a federal judge to become the sole scientific investigator for approximately 150 lawsuits against Eli Lilly for the allegedly fraudulent development and marketing of its drug, Prozac (fluoxetine). My first task was to determine if there was a scientific basis for the lawsuits, and if there was, my next task was to develop that scientific basis for all cases against the company. I started by taking the courses that drug company executives take to learn about the complex process of submitting drug applications to the FDA. I went to relevant conferences put on by the FDA and other organizations. I interviewed key FDA officials on the phone and in person. I read a great deal and started to evaluate the first case given to me. During this process, I also began to focus on "the Gold Standard" upon which all drug companies are said to rely—the placebo-controlled randomized clinical trials used by Eli Lilly to gain approval for Prozac from the Food and Drug Administration (FDA).

Among many allegations in the 150 lawsuits, Lilly was charged with knowingly marketing a dangerous drug while hiding that it caused violence, suicide, psychosis, and mania. When I evaluated the company's clinical trials used for FDA approval, I added allegations, including collusion with the FDA to conduct clinical trials that exaggerated the drug's effectiveness and hid the drug's worst adverse effects. The adverse effects include a cocaine-like activation or overstimulation of the brain that leads to agitation, irritability, insomnia, mania, and even psychosis, too often resulting in violence and suicide in children and adults.

The Prozac clinical trials were fabricated applied science. Here are just a few of the major deceptions surrounding Eli Lilly's clinical trials, which we documented in detail in *Talking Back to Prozac* (with Ginger Ross Breggin, 1994) and in later sources, including *Brain-Disabling Treatments in Psychiatry, Second Edition* (2008).[661] I also urge the reader to look at my free Antidepressant Resource Center at www.123antidepressants.org.[662] To make it easier to access my conclusions, I have indicated page numbers from the St. Martin's Paperback edition of *Talking Back to Prozac*:

(1) Eli Lilly claimed to the public and to doctors that their approval data was based on 11,000 subjects (pp. 45-6). When I did a laborious hand count of the actual total number of people who finished their clinical trials, it was a mere 286 patients. Eli Lilly never challenged this analysis, or any of my scientific accusations, in deposition or in court.

(2) The studies were very short, lasting only 4-6 weeks, while patients end up taking antidepressants for months and years, so they were literally too short to be meaningful. But as short as they were, many patients could not finish because of adverse effects.

(3) Any improvements in patients were so marginal they were difficult to prove, even with outrageous statistical manipulations (pp. 52 ff).

(4) Patients were dropping out like flies from the trials due to distressing overstimulation with agitation, anxiety, insomnia, irritability, and other amphetamine-like symptoms. Rather than let the trials fail, Eli Lilly secretly broke the rules of the studies, previously agreed upon with the FDA, by giving addictive sedatives and tranquilizers to calm the patients enough to stay in the trials (p. 55, pp. 62 ff).

(5) When the FDA later discovered that Lilly had illegally been giving patients extra drugs to calm them down, the agency also discovered Prozac could not be approved without using the corrupted, illegal trials. Faced with disappointing the drug company and its political supporters—including President George Bush who had been on Lilly's Board of Directors and Vice President Dan Quayle who was from Lilly's home state, the FDA agreed to use these intentionally corrupted studies to approve the antidepressant. The FDA then hid these facts, so doctors, patients, and scientists throughout the world had no idea it required addictive sleeping pills and tranquilizers to keep patients from dropping out of the Prozac trials.

(6) When an in-house reevaluation of the clinical trials for *adults* indicated the drug was producing a massive increase in suicidality, Eli Lilly hid the data

from the FDA and the world. I have since then published their hidden analysis, but much too late to matter.[663]

(7) When the FDA's chief in-house analyst for Prozac's adverse effects found Prozac had amphetamine-like effects, making depressed patients agitated, causing dangerous agitated depressions, he wanted to warn about it in the official FDA descriptions of the drug, but his FDA superiors rejected his recommendations (pp. 74-78).

(8) After Prozac came onto the market, my publications and those of other scientists confirmed Prozac causes suicide and violence. The FDA responded by meeting secretly with Eli Lilly before the agency's normal working hours to plan how to jointly fight these allegations.

(9) When data from investigators conducting new clinical trials began to show increased suicides among patients taking Prozac compared to those taking placebo, Eli Lilly ordered the staff to doctor the reports. They were told to change the reports to eliminate words like "completed suicide," "attempted suicide," or "suicidal ideation," and to replace them with infinitely more innocuous terms like "no drug effect" and "emotional lability." Some Eli Lilly employees expressed shame in memos they sent upline (pp 2 and 3 of the linked document).[664]

(10) When it came time for the first trial against Eli Lilly, I was armed with overwhelming evidence of compromised conduct by the drug company. Eli Lilly did not have much of a chance against so much negative information about them, so they simply fixed the trial. Eli Lilly secretly settled the case in advance. The plaintiff's lawyer in return, faked a trial for the drug company to win. The pre-scripted trial was touted in the unwitting and witless major media as a complete exoneration of Prozac. The drug company's value took off, while millions of Americans taking Prozac and nearly all the newer antidepressants remain to this day unaware of their dreadful effects. The judge later figured out the trial had been fixed and changed the victory to a settlement. Also, the Kentucky Supreme Court cited Eli Lilly for possible fraud, but the major media was now overcome with silence. My most detailed write-up of these tragic and disillusioning courtroom events is in "Drug Companies on Trial," a chapter in my book *Medication Madness: The Role of Psychiatric Drugs in Cases of Violence, Suicide, and Crime.*[665]

The placebo-controlled clinical trials for Prozac were corrupt from top to bottom, and the drug should never have been approved. The promotion of the flawed results was equally invalid. Also, the media coverage of events surrounding the trial was entirely skewed in favor of the drug company. Sound familiar? This is what Anthony Fauci has in effect been doing with remdesivir, but so far with less success than Eli Lilly, the Master of Deception. Further, as I concluded much earlier in *Talking Back to Prozac*, recent studies have confirmed Prozac and other antidepressants are no more effective than a sugar pill; of course, they are far more dangerous.[666]

Placebo-Controlled Clinical Trials in the Development of Modern Vaccines

More recently, because of COVID-19, I became interested in what kind of clinical trials are conducted for standard FDA approval for vaccines. I knew it would be easy to conduct placebo-controlled clinical trials, because none of the patients in typical vaccine trials are physically ill or mentally distressed. It would be relatively simple and inexpensive, for example, to test a new vaccine for measles in children by simply giving the vaccine to one group of normal children and giving the placebo to a similar group of normal children.

The children on placebo would be in little or no danger of catching measles because of herd immunity in the U.S. Because they were not sick, they would not be deprived of any necessary treatment. And because they were children, they could easily be followed for many years, instead of a few weeks or months, to give the trial much greater length than most involving drugs. We'll never know how easy it might have been to do it right.

When I started reviewing the clinical trials used for FDA approval, I was stunned by what I found out: *The FDA does not require placebo-controlled clinical trials for new vaccines.* The FDA gives the drug companies a pass on conducting a normal clinical trial when they are seeking approval for vaccines. Instead of comparing their new drug's good and bad effects to a harmless placebo, the vaccine companies get away with comparing their vaccines to other vaccines.

Comparing one vaccine to another vaccine, instead of comparing it to a harmless placebo, is like comparing one kind of poison to another. The poisonous effects are not detected during the comparison because the drugs are alike in being toxic. For an easier analogy, imagine if you were trying to make your poisonous brand of wine look relatively safe. Wouldn't you rather compare it to another poisonous brand of wine than to a genuine placebo of flavored water or a poison-free wine?

A Warning for the Future

When evaluating a "scientific" claim, always consider the source and always follow the money. Science is no better than the people and institutions who conduct it, and the people who interpret it to the public. The current huge "partnerships" created between big industry and big government, led by Anthony Fauci, will produce "science" favoring their power and their wealth. It will not be genuine science in pursuit of the objective truth or better theories; it will inevitably be science in the interest of accruing wealth.

CHAPTER 18

Big Pharma—A Giant Among Global Predators

T he giant international pharmaceutical industry plays a major role in the coercion and authoritarianism, leading toward totalitarianism, that has been unfolding during COVID-19. These drug manufacturers are seizing the opportunity to rush through medications and vaccines for COVID-19, requiring them to exert control over multiple governments around the world, as well as medical and public health establishments, WHO, leading universities, and many other institutions.

We have described how the industry's foot soldiers have managed to crush efforts to widely use hydroxychloroquine in many countries, including America, despite the support of President Trump. Information and opinions favorable to the medication are censored on social media and excluded from the major media, and even respected medical doctors and professors are harassed for speaking out or daring to prescribe it in the U.S. In this report, we have documented extreme cases of patients being given overdoses of hydroxychloroquine in clinical trials to make it look unsafe for general use.

My Background with the Pharmaceutical Industry

I have written many books and scientific articles about the unscrupulous power exerted by the pharmaceutical industry, including *Psychiatric Drugs: Hazards to the Brain* (1983), *Toxic Psychiatry* (1991), *Talking Back to Prozac*

(1994) with Ginger Breggin, *Brain-Disabling Treatments in Psychiatry, Second Edition* (2008) and *Medication Madness* (2008).

I have also investigated drug companies and testified against them as a part of many lawsuits in which I have been a medical and psychiatric expert.[667] This forensic work first became substantial in the early 1990s when a consortium of attorneys selected me, and a federal judge confirmed me, to be the single scientific expert for approximately 150 cases to look into the otherwise secret files of Eli Lilly concerning its development and promotion of Prozac. My task was to provide the scientific basis for understanding Eli Lilly's allegedly fraudulent activities.

That intensive experience led to our book *Talking Back to Prozac*, not an iota of which has the company been able to undermine in my depositions or in my testimony in numerous trials. Later on, in a product liability case settled against GlaxoSmithKline for its antidepressant Paxil (paroxetine), I wrote an enormous paper about the company's corruption, but, when the case was settled, my report was sealed (made unavailable to the public). Later, another court's decision released it.[668] I was then able to publish a series of three scientific, peer-reviewed papers based on my forensic report, which again has remained unchallenged.[669]

I found every drug company I investigated to be ruthlessly devoted to hiding the harms of their drugs, to exaggerating their good effects, to controlling what was published in the scientific journals, and to crushing their critics. All of them were willing to manipulate the randomized, controlled clinical trials to achieve the end of obtaining FDA approval, and many used ghostwriters to craft essentially fake scientific papers to make their products look even better. It was a vastly disillusioning educational experience for me.

I had learned nothing about this corruption in medical school or residency. In my training, drug company clinical trials were treated as if they were scientific in nature, when, in fact, they are a corruption of science. They are a form of applied science that unscrupulously aims at getting drug approval, while exaggerating effectiveness and minimizing adverse effects.

It also became clear to me that the pharmaceutical industry was working with the Food and Drug Administration (FDA), a supposedly watchdog agency, which tried its best to always approve any drug from the big manufacturers and then cooperated to cover up harms when they inevitably surfaced after already being used by millions in the marketplace.

OTHER PHYSICIANS BEGIN TO INVESTIGATE
THE DRUG INDUSTRY

Although I may be the earliest physician to publish scientific books and articles critical of the pharmaceutical industry, many others have now examined the degree of corruption in the pharmaceutical industry, including its willingness to stifle any criticism. One of the best books is by Peter Gøtzsche MD, *Deadly Medicines and Organised Crime: How Big Pharma Has Corrupted Medicine* (2013). In the chapter on vaccines, we saw how Bill Gates attempted to demolish this august doctor's career for daring to show mild scientific skepticism about some aspects of vaccine research.

Despite its highly critical nature, Dr. Gøtzsche's book received a surprisingly laudatory and informative review in *The Lancet*:[670]

> Throughout *Deadly Medicines and Organised Crime*, Gøtzsche uses many anecdotes, provides countless facts and comments based on facts, and cites more than 900 references to draw attention to the allegedly shocking crimes committed by the drug industry (including manufacturers of medical devices). Gøtzsche understands pharmaceutical companies only too well because of his long and varied career in health care, with roles that have included drug representative for Big Pharma, researcher in clinical trials, physician, lecturer, and author of papers and books. He cofounded the Cochrane Collaboration and is the Director of the Nordic Cochrane Centre in Copenhagen. With his expertise and uncompromising attitude, Gøtzsche is outraged and outspoken in his book about pharmaceutical companies being "just like street drug pushers."

Some pharmaceutical companies have been caught and fined for their activities. For example, Dr. Gøtzsche details how during 2007–12, in the USA, Abbott, AstraZeneca, Eli Lilly, GlaxoSmithKline, Johnson & Johnson, Merck, Novartis, Pfizer, and Sanofi-Aventis were fined from $95 million to $3 billion for illegal marketing of drugs, misrepresentation of research findings, hiding data about the harms of the drugs, Medicaid fraud, or Medicare fraud. However, the companies seem not to be deterred and apparently regard fines as marketing expenses.

Dr. Gøtzsche also wrote another excellent book titled *Deadly Psychiatry and Organised Denial* (2015).

Marcia Angell MD, former editor-in chief of *The New England Journal of Medicine* and now a member of Harvard Medical School's Department of Global Health and Social Medicine, wrote *The Truth about the Drug Companies* (2014). A promotional states her view accurately and was probably approved by her:[671]

> During her two decades at *The New England Journal of Medicine*, Dr. Marcia Angell had a front-row seat on the appalling spectacle of the pharmaceutical industry. She watched drug companies stray from their original mission of discovering and manufacturing useful drugs and instead become vast marketing machines with unprecedented control over their own fortunes. She saw them gain nearly limitless influence over medical research, education, and how doctors do their jobs. She sympathized as the American public, particularly the elderly, struggled and increasingly failed to meet spiraling prescription drug prices. Now, in this bold, hard-hitting new book, Dr. Angell exposes the shocking truth of what the pharmaceutical industry has become—and argues for essential, long-overdue change.

> Currently Americans spend a staggering $200 billion each year on prescription drugs. As Dr. Angell powerfully demonstrates, claims that high drug prices are necessary to fund research and development are unfounded: The truth is that drug companies funnel the bulk of their resources into the marketing of products of dubious benefit. Meanwhile, as profits soar, the companies brazenly use their wealth and power to push their agenda through Congress, the FDA, and academic medical centers.

> Zeroing in on hugely successful drugs like AZT (the first drug to treat HIV/AIDS), Taxol (the best-selling cancer drug in history), and the blockbuster allergy drug Claritin, Dr. Angell demonstrates

exactly how new products are brought to market. Drug companies, she shows, routinely rely on publicly funded institutions for their basic research; they rig clinical trials to make their products look better than they are; and they use their legions of lawyers to stretch out government-granted exclusive marketing rights for years. They also flood the market with copycat drugs that cost a lot more than the drugs they mimic but are no more effective.

The American pharmaceutical industry needs to be saved, mainly from itself, and Dr. Angell proposes a program of vital reforms, which includes restoring impartiality to clinical research and severing the ties between drug companies and medical education. Written with fierce passion and substantiated with in-depth research, *The Truth About the Drug Companies* is a searing indictment of an industry that has spun out of control.

What I learned in my forensic work made me skeptical and even cynical about the pharmaceutical industry, but I hoped this self-serving, unprincipled conduct was not universal in big industry and even imagined it might be unusually bad in respect to psychiatric drugs. Over the years, I became less naive, but I was not prepared for the vast global corruption that confronted me while examining COVID-19. Nor was I prepared for the hold that this corruption has upon American society and our political system.

CHAPTER 19

Top Medical Journals
Sell their Souls

On May 16, 2020, the British journal *The Lancet* instructed President Trump to be non-partisan, meaning partisan to *The Lancet's* globalist politics, and urged the American people to elect someone else President of the United States. Ironically, one of the world's oldest and best-known medical journals was simultaneously inserting itself into America's elections, while claiming President Trump should stop being so partisan.

Here is the concluding paragraph of its editorial:

> The Trump administration's further erosion of the CDC will harm global cooperation in science and public health, as it is trying to do by defunding WHO. A strong CDC is needed to respond to public health threats, both domestic and international, and to help prevent the next inevitable pandemic. Americans must put a president in the White House come January 2021, who will understand that public health should not be guided by partisan politics.[672]

Can so-called science be more political than that?

Most of the serious COVID-19 bungling in the U.S. was driven by the global predators. The CDC was drastically unprepared for COVID-19, and President Trump had to set up an advisory committee of his own to do their work. Testing for SARS-CoV-2, which was originally the responsibility of the

CDC, had to be taken over by President Trump himself to get it rolling. President Trump correctly assessed the lifesaving value and safety of hydroxychloroquine; however, globalists like the CDC, FDA, WHO, and *The Lancet* kept it away from the people of Great Britain, America, Australia, and many other nations. They were in cahoots with other globalists, all hoping to make a killing while getting government funding to rush through new expensive drugs and vaccines.

Instead of excoriating WHO for their disastrous obedience to the Chinese Communist regime, *The Lancet* blames President Trump for punishing WHO. The prestigious journal says not a word about WHO teaming up with the Chinese Communists, who caused the epidemic, to hide its virulence from the unsuspecting world and to transport infected passengers from Wuhan and elsewhere in China to the U.S. and much of the world, causing a pandemic.

THE LANCET'S MULTINATIONAL STUDY THAT NEVER WAS

The Lancet's multinational study came out of the heart of the Boston medical establishment, Brigham and Women's Hospital and Harvard Medical School. In my experience evaluating decades of research in the field of neurosurgery and psychiatry, Harvard has always been at the top of medical corruption. When I conducted my successful several-year campaign against the return of psychosurgery and lobotomy (psychiatric brain mutilation), the leadership I had to fight and vanquish originated from the very top at Harvard.[673]

The director of Harvard's Department of Neurosurgery at the Massachusetts General Hospital, William Sweet, was arguably the most respected and powerful physician in America. He was advocating brain mutilation to cure the Black leaders of the urban uprising that were afflicting America's cities. A psychiatrist from Harvard was part of the actual project, which involved sticking multiple electrodes in the brains of hapless patients, white children, and adults, and experimentally alternating brain stimulation with burning holes in their brains. It was a task to go up against Harvard's chief of neurosurgery, but the facts were so bad that my campaign closed down his project along with most others in the U.S. and around the world.

More recently, when tens of thousands of children were suddenly being diagnosed with bipolar disorder and treated with highly neurotoxic "antipsychotic" drugs, especially Risperdal (risperidone), I traced it back to Harvard and broke the news in speeches, media appearances, and books.[674] Harvard professor and psychiatrist Joseph Biederman was secretly taking money from the drug company Johnson & Johnson, and in exchange Biederman invented the concept of childhood bipolarity disorder to sell Risperdal.[675] Risperdal is an antipsychotic drug with crushing effects on child development as well as causing frequent, permanent neurological effects and myriad other adverse effects.

Harvard psychiatrist Biederman's campaign of child abuse for money was so successful that a deadly drug for a fabricated non-existent diagnosis, childhood bipolar disorder, became faddish among doctors in the United States and worldwide. In the U.S., Biederman created such psychiatric enthusiasm that children with the bipolar diagnosis were more likely to get neurotoxic drugs than were adults with the same diagnosis.

My disclosures were followed by Senate hearings which humiliated Biederman, but he kept his job at Harvard.[676] He is a "rainmaker," bringing much drug company money into the coffers of Harvard, my alma mater, which I not only graduated, but where I also took one year as a psychiatric resident and Teaching Fellow at Harvard Medical School.

At the highest levels in medicine and psychiatry, we often find the morally and politically lowest levels, the most abjectly corrupt. It is truly as Lord Acton said, "Power tends to corrupt, and absolute power corrupts absolutely." What I once thought was relatively confined to psychiatry was never really so contained. *Global power corrupts globally.*

The Lancet study, affiliated with Harvard Medical School researchers, was advertised as confirming the horrific dangers of hydroxychloroquine in the form of "decreased in-house survival and an increased frequency of ventricular arrhythmias when used for the treatment of COVID-19." The study made the same claims and had the same errors as the seriously flawed VA study, which we were perhaps the first to discredit.[677]

We did an analysis of the study and quickly discovered it made no sense and proved nothing. In fact, we found an unnoticed confession by the authors buried in the study in the middle of a paragraph on page 8 of 10, where they confessed, "a cause-and-effect relationship between drug therapy and survival should not be inferred."

The study was taken as *the* death blow to hydroxychloroquine. Newspapers around the world blared its negative results, and the World Health Organization (WHO) put a "temporary" stop to its own hydroxychloroquine trials. WHO is the group *The Lancet* editorial was trying to protect from defunding by President Trump.[678] The study came out on a Friday, and WHO acted the next day, Saturday. *This demonstrates close coordination among global predators in their efforts to make haste to keep hydroxychloroquine discredited before it saved too many lives and interfered with their multi-billion dollar plans for new drugs and vaccines.*

How fraudulent was the study? It was a stunning farce—really, a huge medical con game. The large numbers of patients in the study simply did not exist at the facilities and areas they were citing. The statistics made no sense. Under heavy pressure, the authors could produce no data to back up any of their claims. It was a made-up study with no basis whatsoever and the authors—not the recalcitrant journal—retracted it.[679] Meanwhile, the damage had been done, and doctors around the world remained terrified of prescribing the medication.

Any study published by *The Lancet* requires prior review by the journal editors and by peer reviewers as well. *The Lancet* either bypassed the entire process or it collaborated with numerous people to approve the gross fake. *Either way, this was an enormous fraud specifically calculated by* The Lancet *and* WHO *to serve the global predators by discrediting hydroxychloroquine.*

LANCET'S OFFICIAL SCIENTIFIC COMMISSION TO EXONERATE CHINA

The Lancet, undaunted by its heinous fiasco, followed up by creating "*The Lancet* COVID-19 Commission."[680] Where does a mere scientific journal fit into the predatory globalism movement? Here is the journal's statement of goals for its "*The Lancet* COVID-19 Commission." Its scope is great enough to rival an agency of the United Nations, and perhaps the U.S. or China, and certainly aims to identify and position itself as a globalist powerhouse:

> There are four core challenges that must be faced cooperatively worldwide. The first and overriding challenge is to suppress the pandemic as rapidly and decisively as possible. The second is to meet the dire and pressing needs of vulnerable groups such as the poor, minorities, and elderly. The third is to prevent the public health emergency from turning into a fulminant financial crisis for

governments, businesses, and households. The fourth challenge is to **build the world back better**, with resilient health systems, global institutions, and economies that are being transformed on the basis of sustainable and inclusive development. (Bold added.)

In the above statement, notice the use of Joe Biden's presidential campaign slogans, "Build the world back better" and "sustainable and inclusive." In the following, continuous quote from the statement of goals, notice the attempt to align themselves with enormously important powerful global organizations, suggesting they have some supraordinate role to take, when they are simply a medical journal:

> The Commission recognizes that multilateral institutions face profound challenges in undertaking their crucial missions. WHO, the IMF, the World Bank, the Food and Agricultural Organization of the UN, the UN World Food Programme, the UN Educational, Scientific, and Cultural Organization, the Organization for Economic Co-operation and Development, and many others are on the front lines in coordinating the global response to the pandemic in the areas of public health, finance, food security and supply chains, schooling, and governance. Yet these institutions find themselves caught up in the middle of big-power geopolitics. *The Lancet* COVID-19 Commission will aim to make recommendations to strengthen the efficacy of these critical institutions and to promote their adequate financing. The Commission will also reach out to regional groupings, including the African Union, the Association of Southeast Asian Nations (ASEAN), the Southern Common Market (MERCOSUR), and others, to liaise with, hear evidence from, and support, when possible, the efforts of these bodies in fighting the pandemic.

The Lancet is trying to monitor, even taking a superior position with the most powerful predatory globalists, through their medical reputation to gain authority in world governance.

The Lancet's predatory nature—its abandonment of ethics and scientific veracity—is exemplified by its appointment of Peter Daszak to head the investigation into the origins of SARS-CoV-2.[681] As we have documented, Daszak and his organization, EcoHealth Alliance, are heavily involved in conducting and funding research supporting the work of Chinese researchers and the Wuhan Institute. He is helping the Chinese Communist Party make deadly coronaviruses in their military labs under the Military-Civil Fusion. Since Daszak is a major perpetrator in enabling China to make deadly viruses in the Wuhan Institute, he seems highly unlikely to turn on his research collaborators to find the origins of SARS-CoV-2 in their facility.

This kind of treachery by *The Lancet* goes far beyond the metaphor of "putting the fox in charge of the hen house." Daszak and many others are growing wealthier and more powerful at humanity's expense.

When we first published the facts about the depravity of *The Lancet*, we concluded:[682]

> It is no longer possible in the world today to speak of "science" with any respect. It's over. All their larger institutions and publications have been corrupted by centers of wealth and power that circle around their allies like the National Institutes of Health, the FDA, the CDC, the DEA, the World Health Organization, individuals like Bill Gates and Fauci, and the so-called scientific community. From now on, we must rely upon independent sources for almost all of our solid analysis, information, and science.

EVEN MORE CORRUPTION AT *THE LANCET*

However bad the corrupt scientific community seems, the more we learn, the worse it seems. Through a communication with Dr. Joseph Mercola,[683] we learned about a Freedom of Information study that disclosed the *conscious, malicious planning* behind *The Lancet's* actions in protecting Communist China from being condemned for its handling of SARS-CoV-2.[684] Here is Mercola's very sound summary:

STORY AT-A-GLANCE

In a February 18, 2020, scientific statement in *The Lancet*, 27 authors strongly condemned conspiracy theories suggesting COVID-19 does not have a natural origin, stating that scientists from around the world "overwhelmingly conclude that this coronavirus originated in wildlife."

Emails obtained by U.S. Right to Know (USRTK) prove EcoHealth Alliance employees were behind the plot to obscure the lab origin of SARS-CoV-2 by issuing a scientific statement condemning such inquiries as "conspiracy theory."

EcoHealth Alliance President Peter Daszak drafted *The Lancet* statement, intending it to "not be identifiable as coming from any one organization or person" but rather to be seen as "simply a letter from leading scientists."

Several of the authors of that *Lancet* statement also have direct ties to the EcoHealth Alliance that were not disclosed as conflicts of interest.

Daszak is now leading *The Lancet*'s COVID-19 Commission charged with getting to the bottom of SARS-CoV-2's origin—a role for which he is clearly too conflicted to perform in an unbiased manner.

WHO Again Supports *The Lancet* and Other Predators Defending China

The disclosures by Dr. Mercola that originated from the website of *USRTK: U.S. Right to Know* led us to check their content. We found more incriminating information about Daszak's carefully orchestrated cover-up efforts.[685]

WHO has now appointed Peter Daszak to its World Health Organization 10-person team in charge of researching the origins of SARS-CoV-2.[686] The report states the WHO team will "work alongside Chinese scientists on a set of investigations into how the virus that causes COVID-19 emerged and spilled over into humans." *The Lancet*, Peter Daszak, WHO, and the Chinese are collaborating globalists helping to continue burying the true origin of the pandemic in the Wuhan Institute of Virology. It's a tragic demonstration of the unashamed alliances among the global predators—truly terrifying to behold.

Significant numbers of leaders in the global scientific community are now joined with the worst of the global predators. When a leading scientist or medical journal says something, we are told we must believe it. The pronouncements by WHO, NIH, FDA, or CDC are taken as gospel to be enforced by the government. We have been told they represent science and that science is true and trustworthy. The corrupt partnerships between so-called science and the global predators may be the most disillusioning of all our findings.

CHAPTER 20

Klaus Schwab—
The Great Reset and
the Silence of the Cities

In the year 2020, the Great Reset went from a "conspiracy theory"—a figment in the paranoid imagination of right-wing kooks—to the officially proclaimed national policy of the United States of America under President Joe Biden and his Climate Envoy John Kerry. That is the subject of the next chapter, "Joe Biden and John Kerry—Fighting 'World War Zero' with the Great Reset." First, we need to understand the Great Reset and the catastrophic course that President Biden and Czar Kerry have so brazenly set us upon.

THE GREAT RESET AND ITS UN ORIGINS

Much of the Great Reset originated with a United Nations 41-page pledge titled "Transforming Our World: The 2030 Agenda for Sustainable Development."[687] It was unanimously adopted by the UN General Assembly on September 25, 2011, during the Obama administration. It was then unanimously adopted with no abstentions on December 21, 2011, by the United Nations Security Council. It is a mammoth wish list of nearly infinite refrain on an idealistic adolescent's dream for a better world. It begins with the following promise to humanity:

This Agenda is a plan of action for people, planet and prosperity. It also seeks to strengthen universal peace in larger freedom. We recognize that eradicating poverty in all its forms and dimensions, including extreme poverty, is the greatest global challenge and an indispensable requirement for sustainable development. All countries and **all stakeholders**, acting in collaborative partnership, will implement this plan. We are resolved to free the human race from the tyranny of poverty and want and to heal and secure our planet. We are determined to take the bold and transformative steps *which* are urgently needed to shift the world on to a sustainable and resilient path. As we embark on this collective journey, we pledge that no one will be left behind. (Bold added.)

Notice the key reference to "stakeholders" who will "implement this plan." One imagines or hopes the "stakeholders" are the people of the world and their truly democratic and freedom-loving representatives. Instead, "stakeholders" will turn out to be synonymous with "assorted global predators."

The agenda that would become UN Resolution 2030 has 59 numbered points to its vision plus an additional 17 goals with as many as a dozen or so subgoals, adding up to 169 "targets." It is a babble of political nonsense being used to blind the world to their blind ambition.

There were no mentions of "liberty" in the giant UN documents. There are three mentions of freedom, including one mention of "peace in a larger freedom" and two of "fundamental freedoms." However, the idea of freedom is not universal, God-given, or even fixed by nature, but must instead be "in accordance with national legislation and international agreements." There are 23 mentions of "rights" as a justification for the various goals, such as gender equality or more sweeping concepts of "the full achievement of full human potential."

The UN resolution has no concept whatsoever of "individual freedom" or "political liberty" as it is embodied in the Declaration of Independence or Bill of Rights. The legacy of the Founders and idea of spreading liberty throughout the world has no place in this United Nations document nor in the thinking of globalists. All of them are authoritarians or totalitarians who want to impose their particular brands of "justice" on the world, always as a cover story for exploiting humanity.

Eventually, with Klaus Schwab's concept of "stakeholder capitalism" and the Great Reset, the term *stakeholder* will become even more specific. Stakeholder capitalism actually represents what we call "predatory capitalism with a veneer of idealistic progressivism."

Klaus Schwab, COVID-19 and the Great Reset

The term the "Great Reset" was coined in a little-noticed book by Richard Florida, *The Great Reset: How the Post-Crash Economy Will Change the Way We Live and Work*, published in mid-2010. Florida's phrase was quickly adopted by Klaus Schwab, who, for decades, has been working toward an all-encompassing globalist concept of government-business partnerships.

Schwab popularized the term in his own book, *COVID-19: The Great Reset*. The showcasing of COVID-19 in his title indicates how much the pandemic has inspired him and given him hope, much as it energizes all global predators. Schwab's book appeared in July 2020 and may have established Schwab as the greatest intellectual in the globalist movement and now as the intellectual backbone of American policy under President Biden.

Globalist Enmity toward America

Schwab has declared that globalism is incompatible with patriotic democracies. Buried inconspicuously in his book,[688] Schwab makes this telling observation: "the three notions of economic globalization, political democracy, and nation state are mutually irreconcilable."

That quote is worth repeating in a slightly different way: Globalism is incompatible with democracy and the nation state (think patriotism). Schwab is right. The citizens of a patriotic democratic republic would never willingly and knowingly submit to governance by global predators.

Schwab's convictions about the incompatibility between a democratic nation state and globalism by itself explains the universal globalist enmity toward America as the most powerful democratic republic and the beacon of liberty. America was and remains the great impediment to globalist ambitions whether they are held by Fauci, Gates, Bloomberg, and Schwab, or the Chinese Communist Party.

The threat to globalist ambitions and agendas became particularly acute when President Trump was enthusiastically elected by the large portion of Americans who put their own fellow citizens, nationalism, and liberty ahead of globalism. With Joe Biden installed as President, he is already shifting government policy back toward globalism and away from America First. He has weakened the borders and resumed policies that will flood America with illegal immigrants. He has made overtures to China and restored Fauci's collaboration with and financing of the Wuhan Institute.

Schwab's Confusing Model of Partnerships with No One Named in Charge

Schwab is a devotee of China's Communist ruler, Xi Jinping. In late January 2021, he made that grossly apparent at his World Economic Forum called the Davos Agenda. Once again, he invoked COVID-19 as a blessing to his cause: "The COVID-19 pandemic has demonstrated that no institution or individual alone can address the economic, environmental, social, and technological challenges of our complex, interdependent world."[689]

In opening the conference, he unashamedly fawned over Xi Jinping and gave China's violent leader a Free World platform to emote in flowery language about bringing together a peaceful humanity.[690] In essence, COVID-19 gave Schwab the opportunity to honor in public the most powerful and violent dictator in the world.

Schwab has gone out of his way to curry favor with and apologize for the Chinese Communist Party, which seems to be one of his preferred stakeholders. According to the Xinhua News Agency, here is the viewpoint expressed by Schwab in July 2020, after China's COVID-19 perfidy should have been apparent to all: [691]

> China has been leading the world in terms of fighting the pandemic, said Klaus Schwab, the founder and executive chairman of the World Economic Forum (WEF), when addressing the Beijing Forum 2020 held on Saturday.

> He also called for concerted efforts to seize the opportunity to have "a new beginning in our global cooperation in globalization."

Under the theme "The harmony of civilizations and prosperity for all - globalization under the impact of the pandemic: new challenges and opportunities," the Beijing Forum 2020 was held on Saturday in the Chinese capital.

WHAT SCHWAB, THE GLOBAL PREDATORS, AND XI JINPING HAVE IN COMMON

At first it may seem odd that so many business-oriented globalists are hobnobbing with the most destructive dictator in the world. The Chinese Communist Party believes that their government must seize every advantage to dominate and control all human activities. Toward that end, it is using their own corporations to compete with and infiltrate the West. As we have documented, the corporations are completely answerable to the Chinese Communist Party and help to carry out its "unrestricted warfare" against America and most of the world.

Predatory globalists envision a seemingly different world in which the globalists, as business-oriented leaders, go into "partnership" with governments throughout the world. In the process, they welcome the Chinese leadership and influence into all of their activities. In the United States, Fauci has been helping the Chinese to develop their biological warfare programs while Google has helped them develop their surveillance of the Chinese people.[692]

Using the model of the United States, the globalists have become accustomed to bullying governments. Now they have managed to get their befuddled puppet Joe Biden into the presidency. Nevertheless, in respect to the Chinese Communist Party, the globalists are delusional. The Communists are controlling and using them to destroy America. In the coming Chinese Communist Empire, the global predators will themselves be destroyed or made into Communist puppets serving the Empire.

SCHWAB SUMMARIZES HIS IDEAS

TIME magazine gave Schwab the opportunity in late 2019 to describe his hopes for the future of humanity. This is the basic program now adopted by the Biden administration:[693]

We should seize this moment to ensure that stakeholder capitalism remains the new dominant model. To that end, the World Economic Forum is releasing a new "Davos Manifesto,"[694] which states that companies should pay their fair share of taxes, show zero tolerance for corruption, uphold human rights throughout their global supply chains, and advocate for a competitive level playing field—particularly in the "platform economy."

That, in fact, does *not* look like a corporate activity. It is nearer to a progressive *government* in partnership with business. It certainly has nothing to do with the free market with businesses competing with each other for customers. Additionally, as if to make clear that stakeholder capitalism is a cover story for being ruled by international corporate predators, Schwab added:

But to uphold the principles of stakeholder capitalism, companies will need new metrics. For starters, a new measure of "shared value creation" should include "environmental, social, and governance" (ESG) goals as a complement to standard financial metrics. Fortunately, an initiative to develop a new standard along these lines is already under way, with support from the "Big Four" accounting firms and led by the chairman of the International Business Council, Bank of America CEO Brian Moynihan.

In Schwab's utopian world, the group best suited to setting the world's priorities is the Bank of America CEO and his cronies. This is not "crony capitalism;" this is Crony World Government masquerading as capitalism.

Schwab mentions the economist Milton Friedman but fails to grasp the real difference between Friedman's "shareholder" capitalism and Schwab's "stakeholder" capitalism. For Friedman, the fundamental distinction is between freedom and unfreedom—ideas which fail to circulate inside globalist heads.

The freedom of the individual capitalist, whether a small businessperson or a mighty entrepreneur, is the force that generates creativity and wealth. It is definitely not the smarts of an "elite" who will give us progress. Elites have dominated the world since the dawn of civilization in the form of brutal leaders who, over the centuries, morphed into princes, kings, dictators, and presidents for life. Due to the lack of freedom, human progress was slow, if at all, during most of human history. Unfortunately, globalist corporations have become predators who, like the Chinese Communist Party, stifle or exploit "freedom" rather than defend and promote it.

The Political Utility of the Global Reset

In October 2020,[695] *TIME* officially partnered with Schwab to ask "leading thinkers" to share ideas on how to transform the way we live and work. Schwab's new book, *COVID-19: The Great Reset* had been out since early July, using the pandemic to add to recognition of his long-standing politics. *TIME* began with these words:

THE GREAT RESET

The COVID-19 pandemic has provided a unique opportunity **to think about the kind of future we want.** TIME partnered with the World Economic Forum to ask leading thinkers to share ideas for how to transform the way we live and work. (Bold added.)

"The kind of future we want" is a freshman high school essay elevated to the status of serious political theory and concrete programs with disastrous implications. It will supposedly be achieved by moving toward stakeholder capitalism, where many separate interests other than the corporation's own profits must be catered to by the "capitalist." Meanwhile, all those other stakeholders, from Black Lives Matter to Climate Change advocates, will have their own selfish "profit" motives, even if the profit is the satisfaction of expressing rage, destroying society, or obtaining corporate handouts for their leadership. The companies will be sacrificing themselves on various altars to justice, endorsing or genuflecting before progressive or Marxist stakeholders, *in some cases to survive progressive terrorism and in others to placate politicians and public opinion.* The corporations themselves—the great benefactors—have no ideology other than their rapacious predatory instincts.

The Great Reset agenda is an ever-changing hunter's stew of whatever the cooks can find to throw into it relating to making the world a better place, while the predators more quietly go about their predations. According to *TIME's*

Great Reset "thinkers," the Great Reset will accomplish, in their own words and phrases, the following:

- Redesign capitalism to incorporate "social values."
- "Reimagine" capitalism.
- Fix the worldwide economy.
- Make a "sustainable" future.
- Develop a New Green Deal.
- Fix racial inequality.
- Revolutionize E-commerce.
- Build a better tech industry.
- Smash the glass ceiling for women.
- Change how we work.
- Help us adapt quicker to change and threats.

Notice anything missing? There is no mention of spreading individual freedom or political liberty, which are the true engines of modern human progress. There is no discussion of building stronger democratic republics. Religion is completely left out. Patriotism and nationalism are anathema. The world becomes a playground for wolfpacks of globalists.

All things great about American history, traditions, and values are erased from consideration. The individual success as the land of opportunity, the fulfillment of the American Dream, and idea that we are the "home of the brave" are not a part of any globalist's thinking. Instead, America and especially the principle of America First are the recognized and real enemies of the globalists.

American politicians like President Joe Biden rarely, if ever, discuss spreading liberty throughout the world. Instead, they push for the top-down global, business-government coalitions that make them wealthy, full of themselves, and powerful.

It's All about a Vision of Predatory Capitalism with Unlimited Influence Over Governments

After struggling along for 50 years, Schwab now sees in the COVID-19 crisis a light at the end of a previously discouraging tunnel—a tunnel that leads to "capitalism" leading the world through top-down government control. Here he is talking about his Great Reset in June 2020 in the midst of COVID-19:[696]

> To achieve a better outcome, the world must act jointly and swiftly to revamp all aspects of our societies and economies, from education to social contracts and working conditions. Every country, from the United States to China, must participate, and every industry, from oil and gas to tech, must be transformed. **In short, we need a "Great Reset" of capitalism.** (Bold added.)

Notice the United States and China "must participate" and "every industry, from oil and gas to tech, must be transformed." Of course, the U.S. was not going to "participate" under President Trump, but now Schwab believes he can count on Joe Biden.

Schwab's website, the World Economic Forum, offered an essay by one of its own professionals, again connecting COVID-19 to a grand scheme for changing the world:[697]

> The profound shock of the COVID-19 crisis has challenged public confidence in the ability of leaders and institutions to prepare for unexpected risks.
>
> What new forms of cooperation, bold ideas, and reforms are needed for business; policymakers and civil society to build a more resilient world and restore confidence in leadership beyond COVID-19?

Keep in mind there really are people who want the world to suffer sufficiently to acquiesce to their radical globalist transformations, sacrificing our freedom to avaricious billionaires and corporations in the name, oddly enough, of progressivism.

On a Schwab website called the Great Reset Initiative, Schwab and his organization focus more in depth on their plans and goals:[698]

> There is an urgent need for global stakeholders to cooperate in simultaneously managing the direct consequences of the COVID-19 crisis. To improve the state of the world, the World Economic Forum is starting The Great Reset initiative.

What stands in the way of all this? Patriotic democratic republics like the United States, who want to be free and to run their own country. Also, the untold millions of people around the world who can sense they've heard this before from arrogant, self-appointed super people who know what is best for others and want the power to force it on them.

The Elite Rejoice the Weakening of America, COVID-19, and the Great Reset

The predatory globalists are already celebrating the weakening of America under President Biden. Michael Bloomberg's newsletter of February 6, 2021, sneeringly rejected America's moral leadership:[699]

> As *The New York Times* put it, America "is not the dominant arbiter in world affairs that it once saw itself to be."

The caption on a photo describes the riot at the Capitol as "Trump supporters including members of several white supremacist and other far-right groups invade the U.S. Capitol." Globalists—even one who was multiple-times the Mayor of New York City—misunderstand, hate, and exploit America and its people.

In a nation where the major newspapers have become mouthpieces for extreme progressive and globalist propaganda—e.g., *The New York Times* and *The Washington Post*, and even *The Wall Street Journal*, one newspaper has grown in strength and determination as a shining light of patriotism and love of liberty. It is *The Epoch Times*.

An article in *The Epoch Times* by Wesley J. Smith[700] captures an overview of what is evolving with the Great Reset, including Fauci's contribution to it as a progressive theorist. Smith slams Canadian Prime Minister Trudeau for calling COVID-19 "an opportunity for a reset" to "reimagine economic systems that actually address global challenges like poverty, inequality, and climate change." He criticizes supporters of these ideas from President Joe Biden to Pope Francis.

Pope Francis—A Globalist

The many Catholics who are concerned about the Pope precisely because of his globalist politics include a Catholic friend who directed us to a series of articles on *Breitbart* by Thomas D. Williams.[701] Williams has a long career as a theologian, ethicist, and professor.

In one essay, Williams discussed how "Panelists at Jesuit-run Georgetown University converged Friday in the common belief that a Biden administration will see eye to eye on many issues with the Vatican of Pope Francis." One of the most central and disturbing connections in common between Biden and the Pope is taking money from and deferring to the Chinese Communist Party.

In another essay, Smith reported that Cardinal Joseph Zen, the former bishop of Hong Kong, declared the "renewed Vatican-China deal" was a "complete defeat for faithful Catholics." Smith also describes how "China's most famous civil rights activist has joined the chorus of critics who have denounced the recent deal between the Vatican and Chinese Communist Party (CCP) as a 'betrayal' of Chinese Catholics."

The Pope was being so roundly criticized for his "extension of a controversial [secret] 2018 agreement with the Chinese Communist Party (CCP) regarding the naming of Catholic bishops in the country." In the agreement, the Pope gave the Chinese Communist Party the power to select the Catholic bishops in China in exchange for very large undisclosed sums of money.

On the same internet platform, Smith described U.S. Senators pressing the Pope to denounce the religious persecution of Uyghur Muslims who were interned behind barbed wire in China, but the Pope had already forfeited the freedom of his own Church in China by allowing China to appoint its bishops.

Transhumanism or The Living Dead

Schwab has promoted the concept of technology and artificial intelligence (AI) producing a Fourth Industrial Revolution. He is a proponent of *transhumanism*,[702] using technological interventions that transform our brains and behavior to make us, in fact, less human—less able to have spontaneous emotions and loving connections with others.

Schwab's commentaries on transhumanism and other topics often have the notes of a science fiction novel more than genuine social, political, or economic plans. Some see Marxism at work. We see a cynical, progressive veneer over plain-old global exploitation of humanity by a self-selected elite.

Transhumanism is utopian and *inhuman*. From psychiatric surgery and drugs to newer forms of brain stimulation, all physical interventions into the brain to change behavior and conduct eventually have the same effects: blunting the human mind, compromising free will, and dulling spirituality. Brain intervention that afflicts centers of the brain involved with the emotions and human conduct—the most developed parts of that great organ—inevitably render people more docile and easily led.[703]

We have opposed deep brain stimulation[704] and other forms of psychiatric surgery that try to influence emotions and human conduct by mutilating the brain.[705] In a remarkable film, *The Minds of Men* by Aaron and Melissa Dykes,[706] I am featured discussing the history of mind control through technology.

I have actively opposed various forms of psychological control through brain mutilation in my successful anti-lobotomy and anti-psychosurgery campaign (See Chapter 28). I have also been critical of Elon Musk's proposals and experiments in connecting the human brain directly to supercomputers.[707]

The Silence of the Cities and Wishing for Many Deaths

Schwab's World Economic Forum suffered a barrage of criticism before deleting a social media video titled "Lockdowns are quietly improving cities around the world."[708] In their warped minds, Schwab and his associates did not anticipate the protest their video and written comments would cause.

In describing the video, Schwab listed these improvements:

- Lockdowns significantly reduced human activity and its impact on the planet's crust, leading to Earth's quietest period in decades.

- Ambient noise generated by travelling and factories dropped by 50% to the lowest levels ever recorded.

- Lower background noise meant smaller earthquakes that would otherwise not have been observed, were detected.

- These smaller tremors allow scientists to understand the seismic threat of minor hazards and help assess the probability of larger earthquakes.

Yes, "Lockdowns significantly reduce human activity," and we know where that leads, not to "zero" emissions, but to "zero" humans or to humans so blunted and ineffectual they resemble robots. They want to make a limited number of ideally docile citizens to populate their utopias and to serve them, while they grow yet more wealthy, glorious, and powerful.

There is a tragic essence lurking inside both Progressivism and predatory globalism. Both worship death—not necessarily their own deaths but the death of humanity at their hands. This insight makes sense out of their persistent, disastrous failures to reshape everyone into equality. It explains their do-or-die compulsive push to make humanity more perfect and the havoc they want to wreak on societies and economies. Their warped worship of death explains their history of tens of millions of murders, beginning with the French Revolution's Reign of Terror through Communism in the USSR, Communist China, and everywhere else it has taken hold.[709]

The uncountable millions of abortions and the murderous practice called euthanasia result in part from that same love affair with the death of others. In the West, a more subtle death wish is expressed in their promotion of psychiatric drugs and various other psychoactive drugs like marijuana to deaden the brain and mind.

Humans are only truly equal in death. They achieve perfection only in death. The eradication of suffering and the attainment of personal and political peace are in death. Their utopian revolutionary plans always push us to death—the extinction of humanity or its transformation into something inhuman and malleable.

CHAPTER 21

Joe Biden and John Kerry— Fighting "World War Zero" with the Great Reset

Before taking office, President Joe Biden was already making COVID-19 and climate change the driving threat to make people get behind his big-government, progressive agenda. He did this by appointing John Kerry to be his Climate Envoy and a member of his National Security Council. Joe Biden and John Kerry are a pair of titans of exploitation with years of corrupt family involvement with the Chinese Communist Party.

WHAT'S IN KERRY'S DNA?

One of our consistent findings are ties between global predators and the Chinese Communist Party. Surely, John Kerry has no such inclinations. He was a U.S. Senator from Massachusetts for 23 years, a Secretary of State, and Democratic nominee for President in 2004. He is worth $200 million.[710] Being American is in his DNA. Unfortunately, being an anti-American and Communist sympathizer is also in his DNA.

Among Kerry's violations against America's self-respect and interests, in 1971, he met with two delegations of North Vietnamese in Paris to discuss how to stop the war—showcasing a massive interference with American foreign policy. The Logan Act, federal law "18 U.S. Code § 953 - Private correspondence with foreign governments," prohibits unauthorized negotiations with foreign countries by private citizens.[711]

THE ROOTLESS RUTHLESS JOHN KERRY
AS A YOUNG PREDATOR

Even as a young army officer returning from Vietnam, Kerry was way ahead of his time as a malicious progressive, eager to undermine America in favor of Communist regimes.[712] I remember watching Kerry on TV in 1971 as he testified before a Congressional Committee as a returning Vietnam Vet—a presentation that overnight made him extraordinarily famous and influential by causing us to feel ashamed and guilty about being Americans. Dramatically dressed in army fatigues, Kerry described systematic, rampant atrocities committed by our soldiers, allegedly with the knowing approval of superiors. Here is an excerpt of Kerry describing, not what he actually saw with his own eyes, but what he says he heard from other returning soldiers:[713]

> They told the stories at times they had personally raped, cut off ears, cut off heads, taped wires from portable telephones to human genitals and turned up the power, cut off limbs, blown up bodies, randomly shot at civilians, razed villages in fashion reminiscent of Genghis Khan, shot cattle and dogs for fun, poisoned food stocks, and generally ravaged the countryside of South Vietnam in addition to the normal ravage of war, and the normal and very particular ravaging which is done by the applied bombing power of this country.

Watching his performance, I never imagined Kerry was a colossal liar—a type of man I could not envision at the time—a budding globalist driven by a desire for power over us. Kerry as a young man held sufficient hatred for his country that he was able to disillusion millions of us. He led us to believe we are an evil people—all made up or vastly exaggerated to aggrandize himself and make his political future.

Over the next few decades, Kerry would become a senator, a secretary of state in the Obama administration, and a failed presidential candidate. Currently, he is positioned to have the greatest impact of his life on a world that is ill-prepared to protect itself from predatory globalism as American national policy.

The Kerry family has also been involved with the Biden family in profit-making schemes in China.[714] Biden and Kerry's children are partners in China deals.[715] Biden and Kerry are the team from hell when it comes to helping America stand up to China.

Kerry and His "Zero War"

A year before Biden made him a Czar over progressive agendas, Kerry was already making a high-profile declaration about the need to start a "war" against climate change:[716, 717]

> John Kerry called on Americans to start treating climate change as though the country is fighting a war. ... Kerry said the battle against climate change requires leadership, decision making, and organization similar to wars. He claimed **World War Zero** is going to talk to billions of people to raise awareness and make sure they put climate change "at the top of the list." (Bold added.)

In June 2020, Kerry spoke at the same virtual conference with Prince Charles announcing the unveiling of Schwab's Great Reset.[718] It was part of the COVID-19 Action Platform of Schwab's World Economic Forum, where COVID-19 continues to be used to justify radically transforming America and the world. At this time, in an address rife with threatening images to intimidate the world population, Kerry threw his support for the Great Reset as U.S. policy:[719]

> "Normal was a crisis; normal wasn't working," said former U.S. Secretary of State John Kerry in his opening remarks. "We must not think of it in terms of pushing a button and going back to the way things were. We're a long way off from being able to go back to any kind of normal." ...

> "The world is coming apart, dangerously, in terms of global institutions and leadership."

"What we never did was adequately address the social contract, the enfranchisement of human beings around the world, to be able to participate in things they can see with their smartphones everywhere but can't participate in."

Explaining that the United States of America is currently "gridlocked," Kerry said: "This is a big moment. The World Economic Forum - the CEO capacity of the Forum - is really going to have to play a front-and-center role in refining the **Great Reset** to deal with climate change and inequity—all of which is being laid bare as a consequence of COVID-19." (Bold added.)

Biden warned the worst was ahead of us and Americans should brace themselves for continued limitations on their lives. According to CNN, "President-elect Joe Biden on Tuesday said 'the darkest days' in the battle against the coronavirus pandemic 'are ahead of us, not behind us,' and urged Americans to prepare themselves for the struggle."[720]

PARTNERSHIPS BETWEEN PROFITMAKING CORPORATIONS AND GOVERNMENT

Partnerships between corporations and governments—especially China—are at the center of globalist ambitions for both their personal fortunes and also for their political influence and power. Because of their commitment to corruption, they raise no warnings about the enormous dangers of further blurring any checks and balances within the government to prevent this exploitation of the people. They feel no compunction over using slave labor in China to put Americans out of work. They never hesitate to do business with a brutal Iranian, Cuban, Venezuelan, or North Korean dictator who violently abuses his own people. China's Communist Party is seen more as a honeypot than a blight. Only severe sanctions, usually applied by somewhat conservative American political parties, keep the global predators from gorging themselves on the weak, the poor, and the enslaved.

The Great Reset is not a humanitarian movement. It is a feeding frenzy for aggressive global predators, jacked up by brandishing the threats of COVID-19 and global warming.

The global predators are so recklessly obsessed with their own power, they have no awareness of, or provision for, resolving their inevitable conflicts with each other or with the Chinese Communist Party. Ultimately, under whatever pseudonym, they are global predators who are intent on domination. With America on its knees, they expect to divide up the globe among themselves.

The Global Reset and Big Business

The global reset is actually a nonpolitical, opportunistic business concept. Prince Charles was used as a front man to roll it out for the press. Here is the *Guardian*'s summary of the event:[721]

> The recovery from the coronavirus crisis represents an opportunity to reset the global economy and prioritise sustainable development without further damaging the planet, Prince Charles said at the opening of a World Economic Forum (WEF) virtual meeting. …
>
> **The prince emphasised that the private sector would be the engine of recovery and was heartened by the pledges from business leaders to recognise the damage to the environment that would result from an unfettered dash for growth.** (Bold added.)

The much-anticipated, great escalation of wealth in "the private sector" derives from a plan where the private sector replaces or partners with and dominates large-scale functions usually reserved for government. *The Guardian* continued with Prince Charles's emphasis on the great opportunity demonstrated by COVID-19:

> **He said the pandemic, which has forced governments worldwide to mothball their economies, had showed people that dramatic change was possible.**
>
> **"We have a golden opportunity to seize something good from this crisis. Its unprecedented shockwaves may well make people more receptive to big visions of change," he added.** (Bold added.)

That is how the global predators—including President Biden and Czar Kerry at the top in American politics—see the usefulness of COVID-19 public health policies. The pandemic is here to condition us to accept "dramatic change" and "big visions." What we see as a tragedy in COVID-19, they see as a "golden" opportunity for themselves.

GONE ARE THE POLITICAL DISTINCTIONS BETWEEN LEFT AND RIGHT

Coalitions between old-time leftists, new multi-billionaires, and global corporations seem to make little ideological sense. This is because politics and religion play little or no role in globalist predatory thinking, often led by American-born predators like Biden, Kerry, Gates, Bloomberg, and the rest. Except for the Chinese Communists and Muslim Jihadists, the global predators are spiritually and ideologically rootless.

Globalism might have arisen from capitalists who leaned to the right, but that is no longer the case. The populist wing of the contemporary right in America has strong ideals about patriotism and about political liberty—neither of which are supported by Kerry, Biden, Bill Gates, and other "capitalists" who may give lip service to the free enterprise system. Globalist corporations also masquerade as politically left or right depending on their intended audience, but as predatory globalists, they have no deeply held ideology other than becoming wealthier, self-aggrandizing, and powerful.

Ironically, global predators in many ways fit the description of viruses provided by Peter Daszak, President of EcoHealth Alliance:[722]

> The divisions that we humans see, viruses do not. Viruses have no citizenship, no political affiliation; they are indiscriminate in the harm they cause. Because of this, when one of us is threatened by a deadly virus, so too are all of us.

We are the prey of both deadly viruses and predatory globalists, such as Daszak.

To make it as simple as possible, global predators do not care about us or even about their own self-identified ideologies as much as they care about making themselves more rich, glorious, and powerful. All the rest is trapping and public image.

John Kerry in Power

On November 24, 2020, "President-elect" Joe Biden presented John Kerry as the "Climate Envoy." The "cabinet-level" new post includes membership in Biden's National Security Council and specifically gives him influence at the top of the State Department with the newly appointed future secretary of state, who was on the stage with him.[723]

To our surprise, in his formal acceptance speech, Kerry invoked God—rare among globalists. It was a ruse. He then explained that "science" would work in the service of God. Thus, he made clear any God-fearing person must see "science" as the instrument of God.

Even Anthony Fauci, the representative of "science" for the progressive world, had not—so far as we know—gone so far as to declare himself not only the emissary of science, but also the servant of God. Thus, Kerry became God's emissary of science to America and the world on behalf of the most radical, progressive agenda ever made into an American political platform.

On the same day as Biden introduced him as the Climate Envoy, through the miracle of digital communication, Kerry addressed a special committee of Schwab's World Economic Forum (WEF) titled COVID Action Platform.[724] Once again, we see the opportunistic use of the pandemic with the theme, "What new forms of cooperation, bold ideas, and reforms are needed for businesses, policymakers, and civil society to build a more resilient world and restore confidence in leadership beyond COVID-19?"

Host Borge Brende, president of the World Economic Forum, begins by asking Kerry, "Are we expecting too much too soon from the new president, or is he going to deliver first day on these topics?" He is referring to the Great Reset theme. Kerry responds:[725]

> John Kerry: "The answer to your question is, no, you're not expecting too much. And yes, it [the Great Reset] will happen. And I think it will happen with greater speed and with greater intensity than a lot of people might imagine.

"In effect, the citizens of the United States have just done a Great Reset. We've done a Great Reset. And it was a record level of voting. What astounds me is that as many people still voted for the level of chaos and breach of law and order and breaking the standards. …

"And I think the underlying reason for that, Borge, is something that everyone has to examine. I think Europe has to look at that, with Brexit, and the rising nationalist populism, nationalistic populism—which is really one of the priorities we all have to address. You can't dismiss it. It has to be listened to. It has to be understood. We have just had it manifested. And Europe has it too, in various countries, to a greater or lesser degree.

"It's a reflection of the inability of democratic governments in many parts of the world to deliver. And I just have to put it bluntly. We're certainly the primary exhibit. We're exhibit number one.

More America bashing and more promises to make the world into his own progressive vision—all involving a revolutionary elevation of international corporations to the level of global predatory governance.

Fighting Global Warming by Shrinking Human Freedom

We will track a close relationship between COVID-19 and global warming as pretexts for massive globalist interventions in the lives of the people of the world. Global warming may play a bigger role than the pandemic in Kerry's imagined ability to give Earth marching orders. He is arguing we must obtain near to "zero emissions" by 2050, which we will learn is now part of the Great Reset and therefore in the plans of John Kerry and Joe Biden. Interestingly, unlike Kerry and many others who are planning on zero emissions by 2050, Gates admits it is almost impossible. He adds what seems to be a little joke about reducing humanity to zero along the way as the final step.[726]

Bill Gates sees population control as central to improving humanity and to saving the Earth from global warming. On his website, he describes his plans to actively reduce the growth of population by birth control, vaccination for improved childhood longevity, education, and other progressive means.[727] Unfortunately, he wants to implement these aims by taking control over the lives of other people rather than by encouraging freedom, which leads to prosperity along with smaller families.

Gates does not understand or care that a free and prosperous people naturally reduces their birthrate and improves their healthcare. The freer a society, the more it prospers, and the more individuals can meet their own needs and standard of living.

Tragically, this insight is so alien to global predators, whether progressive or capitalist in veneer, that it never crosses their minds. If Bill Gates instead devoted himself to spreading the concept of patriotic democratic republics, then he would become a global George Washington rather than just one more exploiter of humanity.

"Zero Emissions" as the Extreme Rallying Cry of Totalitarians

The idea of reaching "zero emissions" is an extreme goal, going beyond even "carbon neutrality." The extremism of Gates is given little notice by his allies who simply ignore the idea of stopping all emissions and instead reach for "carbon neutrality," which they hope to achieve by balancing emissions with measures yet to be fully imagined, let alone implemented, for withdrawing or absorbing carbon from the atmosphere.

For example, on December 12, 2020, U.N. Secretary-General Antonio Guterres demanded every country in the world declare a climate emergency:[728]

> Can anybody still deny that we are facing a dramatic emergency? That is why today, I call on all leaders worldwide to declare a State of Climate Emergency in their countries until carbon neutrality is reached.

The U.N. Secretary-General is part of a chorus of globalists singing this dramatic aria of looming doom to us from plagues or environmental catastrophes. Over 30 countries have already declared climate emergencies.[729] In the US, municipalities in California have begun declaring local "global climate emergencies."[730]

China has touted its aims to be carbon neutral by 2060, shaming the U.S. for dropping out of the international accords; however, even Michael Bloomberg, who supports China politically and with his business deals, seems to have doubts about China's willingness to do so.[731]

Of course, the UN has no power to force these kinds of economically and politically catastrophic totalitarian measures on humanity. Meanwhile, neither Bill Gates nor any other billionaire globalist has an army. The advocates of these global transformations show awareness of the reality that the only organization capable of implementing such measures is the largest global predator of all, Communist China, whose leaders will drop the charade of good fellowship with the billionaires after they achieve dominion over the United States and hence the world.

Global Warming or an Ice Age—Which is Coming?

For many decades, some scientists have been concerned about a worldwide warming trend. They avoid reminding us that the Earth went through enormous changes in climate, including a mile-high ice sheet that covered much of the world with vast drops in the sea levels a mere 10,000 years ago. That ice age transformed life on Earth much for the worse, while the retreat of the ice produced a greening of the world with a great outburst of life that continues to this day.

Here is a provocative headline from the Weather Channel:[732]

> ### We're Due for Another Ice Age but Climate Change May Push It Back Another 100,000 Years, Researchers Say

The article continues:

> Throughout our planet's 4.5 billion years, there have been five big ice ages, some of which lasted hundreds of millions of years. Researchers are still trying to understand how often these periods happen and how soon we can expect another one.

> "We should be heading into another ice age right now," Columbia University paleoclimate doctoral student Michael Sandstrom told *Live Science*.

The big ice ages account for roughly 25 percent of the past billions of years on Earth, says Sandstrom. The most recent of Earth's five major ice ages in the paleo record date back 2.7 million years and continue today.

Within these large periods are smaller ice ages called glacials and warm periods called interglacials.

The article cites scientists who believe the world is receiving some protection from the next ice age from the rising temperatures. This raises the possibility that if we can sufficiently drop the temperature to prevent global warmings, we will usher ourselves into a premature ice age.

During COVID-19, we have seen how scientific "models" used to predict the course of the pandemic and the numbers of deaths utterly failed. When vast political forces are invested in issues like predicting or managing pandemics and global warming, trusting the models and the politicized scientists does not lead to a rational course of action, especially when dissenting scientists—those who are not promoting predatory globalism—are given no voice.

Could we, right now, turn back another ice age that would eventually push another mile-high sheet of ice across much of the northern world, digging huge ravines and lakes many hundreds of feet deep as it came crushing southward over the northern hemisphere?

And then there are those pesky scientists who lack progressive leanings or conformity with the globalists, or do not rely on the embedded globalist bureaucracy for funding for their research. Some of them are anticipating a "mini-ice Age" between 2020 and 2030, and they must be shut up at all costs.[733] Their voices remain like a child whispering in a storm because all the public and private funding for research in the area is based on assuming the truth of global warming in advance.

A study of federal spending on climate concluded, "In 'fighting' climate change, the United States government is spending almost as much as it did on all the Apollo spacecrafts." The amount was $220 billion in 2012 dollars.

Meanwhile, fears of climate change have been insufficient to give the globalists all they want or need to transform the world. Something more was still needed—something that could not be ignored, a catastrophe beyond imagination to take control over everyone's thinking about their place, or lack of it, in the world. That is one of the main reasons why we have the worldwide shutdown in response to COVID-19, but there is more to it than that.

CHAPTER 22

Names and Shared Characteristics of America's Top Global Predators

W e are further enough along in the book to step back to examine the characteristics of global predators and their names among America's top billionaires and giant international corporations. We begin by describing the Perpetrator Syndrome and the Global Predator Profile.

THE PERPETRATOR SYNDROME

The perpetrator syndrome describes the *psychological* and *personality* pattern of characteristics of global predators.

By a predator, I mean an individual or group who is merciless or objective in the pursuit of prey. The global predators have made us, humanity, their prey—somewhat akin to a pack of voracious wolves finding a seemingly limitless supply of docile, penned-in game in an unprotected zoo or preserve.

In my book *Beyond Conflict* (1992), I developed a concept called the *perpetrator syndrome* to describe people who persistently exploit, coerce, or abuse others to achieve their ends.[734] The concept evolved from my clinical and forensic work and research. I have applied it in forensic reports to contrast essentially innocent people from hardened perpetrators. I have not used the perpetrator syndrome to diagnose anyone, which I believe would be potentially abusive. It is not a psychiatric diagnosis.

The perpetrator syndrome consists of the following persistent attitudes, psychological tendencies, or personality traits found in people who purposely and regularly coerce, exploit, and harm others:

1. They deny or minimize the damage they are doing to others.

2. They rationalize the harm they are doing.

3. They blame the victim.

4. They suppress their own feelings of empathy.

5. They deny or rationalize their own prior victimization at the hands of others.

6. They persistently react with anger and blame others based on much earlier feelings of shame and humiliation.

7. They dehumanize their victims.

8. They feel empowered through their perpetrations.

9. They seek to resolve conflicts through authority, power, and domination.

10. They become grandiose and self-centered.

11. They become alienated from their own genuine basic needs.

The perpetrator syndrome begins to answer the troubling question: "How can these globalists do such heartless, self-serving, or harmful things to people?" Their feelings, attitudes, or values are different from those of most people and enable them to act more like predators.

This syndrome was based on examining people whose perpetrations were against individuals and not against powerful adversaries or humanity. Almost all of them were already in trouble for breaking the law, or I would not have gotten to know them. They were much less effective or powerful than global perpetrators. Nonetheless, many individual global perpetrators seem to have these attitudes, if not in their immediate personal lives, more certainly in their public lives.

In the case of individuals acting within and through large institutions, their bureaucratic remoteness and prescribed institutional roles, such as promoting and protecting their organizations, make it easier for them to become large scale perpetrators and predators, more nearly looking like global perpetrators. They may have normal feelings toward people near to them, such as coworkers, friends, and family, but these globalists, accustomed to manipulating and statistically analyzing populations, will have no caring or concern for the rest of humanity.

The Global Predator Profile

Now we examine the *political policies and attitudes* of global perpetrators. The global predators we have identified are wealthy or powerful people or institutions who surprisingly often have all of the first eight characteristics. The first eight are especially well-grounded. They are apparent in their communications and behavior. The higher numbers require more inference or speculation.

1. If American, they are not patriotic but instead are critical of America and its power.

2. They do not mention or write about principles of personal liberty and political liberty as primary values.

3. They oppose democratic republics and prefer authoritarian or totalitarian governments.

4. They support open borders worldwide and especially in America.

5. Their policies enhance the wealthy and drain or destroy the poor, the working class, and middle and upper-middle class. The result is a growing gap between the very, very rich and the rest of humanity.

6. In a surprising fashion, they consistently have large, flourishing financial investments in China and want to increase them without moral or political restraints. They apologize for, support, or look the other way in regard to the policies of the Chinese Communist Party, even when these policies harm the American people.

7. They actively opposed President Trump and continue to do so now that he has left office.

8. They appear driven by a combination of insatiable needs for wealth, self-aggrandizement, and power over others.

9. They feel elite, superior, or above the vast majority of people.

10. They see no inherent value in individual humans and do not treasure human life.

11. Except for Jihadists and Communists, they lack strong political or religious ideologies other than predation to increase their own wealth, grandiosity, and power. This has become especially true of American corporations and billionaires who, in their support of Communist China, have given up any of their original moral and patriotic attachments to their country of origin.

12. They use unethical and often illegal practices to attain their ends, including illegal political activities, election fraud, and censorship.

13. They often, but not always, display a progressive or capitalist veneer; however, they are global predators.

14. They rarely if ever examine the harm that their activities cause to very large numbers of people and to humankind as a whole.

It may seem unusual to draw such a specific characterization of the global predators who are vying for control of humanity. For me, the description is empirical, especially in respect to the first eight items, and results from examining events surrounding COVID-19.

Their frequent and seemingly uniform financial and other ties to China's government were especially unexpected. It seems that any predators operating on an international plane are drawn to or cannot avoid ending up profiting from partnering with the Chinese Communist Party. I suspect that this has a great deal to do with seductive courting by the Chinese, which encourages the predators to believe that their financial and political future lies in the vast Chinese market and, perhaps, even that their safety lies in staying on good terms with the Communist behemoth.

Whether or not this dependence on the Chinese Communist Party also applies to global criminal enterprises, we have not investigated. The global predators we are examining are socially accepted and operate to some extent in the public arena. For example, much of their gross wealth and many of their enterprises are public information. There is probably some overlap between them and criminal networks.

Top 11 Companies Hostile to Donald Trump and Their Ranking in Worth

Here we examine the top 11 companies—all of them globalist—that have expressed hostility or taken acts antagonistic to Donald Trump during his presidency. They are listed in ascending order of their wealth, measured by the total value of their outstanding stock (the market cap):[735, 736]

> Twitter $50 billion
>
> Netflix $61 billion
>
> Adobe $64 billion
>
> Starbucks $84 billion
>
> Nike $91 billion
>
> Intel $169 billion
>
> Facebook $405 billion
>
> Amazon $425 billion
>
> Microsoft $507 billion
>
> Alphabet (Google) $574 billion
>
> Apple $750 billion

We added Jack Dorsey's Twitter with a market cap of $50 billion to the beginning of the list as the 11th in the market cap leaders against Trump. As the predatory globalist, *Forbes* declared in its headline on February 10, 2021, "Twitter Market Value Surges To $50 Billion Despite Trump Ban, And Shares Could Run Another 25%." Dorsey himself is worth $24 billion. This positive stock market response to Twitter's violent suppression of Trump and his supporters confirms that many wealthy individuals and companies are global predators.

These 11 companies are now so powerful that they and others succeeded in deleting the communications of the President of the United States and his supporters, contributing to his election loss, forcing him out of office, and disenfranchising half of America's voters.

ANTI-TRUMPERS RANKED AMONG THE LARGEST AMERICAN CORPORATIONS

We crosschecked the list of anti-Trump companies against a list of the top-ten largest market cap corporations in the world as of February 13, 2021.[737]

Of the companies in the top-ten anti-Trumpers, five of them are also among the ten largest market caps.

Here are the anti-Trump companies, ranked by their standing in the top-ten companies by market cap: Apple (1), Microsoft (3), Amazon (4), Alphabet (Google) (5) and Facebook (9).

In short, Donald Trump and his supporters had half of America's most valuable companies actively fighting against them. They hold the top 1, 3, 4, 5 and 9 positions in the value of their stocks. It is a wonder that Trump with his populists and America First supporters ever did as well as they did, and there is a strong possibility that they won a second term!

GLOBAL PREDATORS AMONG AMERICA'S TOP BILLIONAIRES

We turn from the corporate wealth aligned against Trump and traditional American ideals and now examine the list of top billionaires in the nation.[738] All of them are globalists, heavily involved in international business, but how many are *predatory* globalists?

We examined the top 16 billionaires and were shocked to find that 14 of 16 qualify as global predators, if only because of their heavy investments in the Chinese Communist Party which controls China. All of these individuals or their companies are discussed in this book.

The top 16 richest Americans are:

1. *Jeff Bezos.* Major global predator and owner of Amazon and *The Washington Post.*

2. *Bill Gates.* Major global predator and director of the Bill & Melinda Gates Foundation.

3. *Warren Buffett.* Major predator and one of three trustees who direct the Bill & Melinda Gates Foundation and a major contributor as well. Buffett makes vast contributions to population control, including abortion.[739] Buffett and Gates make a devastating duo of predatory capitalism combined with lethal aspects of a progressive agenda.

4. *Larry Ellison.* He is the only one who has openly voiced values and taken actions that separate him from the horde of global predators. He describes himself as a centrist and a patriot with deep concerns about China, making him unusual among the billionaires and big corporation owners and executives. He is the owner of Oracle, which is invested in China as a relatively small player

in a market dominated by Chinese firms.[740] He held a fundraiser for Trump in his own home with considerable backlash from Oracle employees. He is said to have encouraged Trump to support hydroxychloroquine.[741] In interviews with Maria Bartiromo,[742,743] Ellison was happy that Trump was pushing back against China. He explained that America must "win the battle with China," and he supported Trump's foreign policy of standing up to the Chinese Communist Party. He was concerned about China's growing economic and technological power and foresees, if unchecked, that China will surpass our military. He saw China as our main competitor and a dangerous one capable of overtaking our military superiority. He was critical of Google for helping China's abusive surveillance strategies while refusing to work with the U.S. Department of Defense. He described himself as "in the middle" politically and said he was a big donor to Bill Clinton's campaigns.

5. *Mark Zuckerberg*. Major predator and Facebook owner.

6., 7., and 8. *Three Walton Family Members*. Major global predators. The family overall gives relatively small contributions to both major parties.[744] However, Walmart's investments in China are in the billions of dollars and growing, including four Sam's Club stores, 15 additional shopping malls, and more "community stores," all being built in Wuhan, the capital of Hubei province.

9. *Steve Balmer*. Major global predator and owner of Microsoft.

10. *Larry Page*. Major global predator and owner of Alphabet/Google.

11. *Sergey Brin*. Major global predator and owner of Alphabet/Google.

12. *Michael Bloomberg*. Major global predator and Director of the New Economy Forum (NEF).

13. *Charles Koch*. Major global predator and owner of Koch Enterprises. Although he gives to Republicans and calls himself a libertarian, he invests in the Chinese Communist regime, thereby empowering a violent dictatorship and weakening America, and he favors open borders, as if a nation does not have the libertarian right to self-defense. He criticized presidential candidate and then President Trump. Well-known libertarian groups that Koch supports include the Cato Institute[745] and the American Institute for Economic Research (AIER), which promote his policies.

14. *Julia Koch and Family*. Major predators as owners of Koch Enterprises.

15. *Mackenzie Bezos*. Her ex-husband, Jeff Bezos, retains the voting rights to her Amazon shares, so she is not responsible for the activities of Amazon, and on that basis is not a global predator.

16. *Phil Knight and Family.* Major global predators. Regardless of their support for some Republican candidates,[746] they own Nike, which heavily invests in China and supports anti-America and/or pro-Chinese rhetoric from sports stars Colin Kaepernick and LeBron James.

The power of the global predators and their invasiveness into our culture are no better illustrated than their holding 14 of the top 16 spots among America's list of billionaires. *It is worth repeating that former President Donald Trump and the populist movement has done so well against such David-and-Goliath odds. We must recognize how much strength we have among Americans, making this no time to get discouraged or to quit.*

Making Use of Knowing Who They Are

Globalists know their common adversary is the mixed block in America of traditionalists, patriots, conservatives, libertarians, constitutionalists, and religious people—many of them owners of small businesses and many of them in the middle class. They know these people are determined to defend their country and to protect its principles of individual rights and political freedom.

Closing small businesses and religious institutions and shrinking the workforce during COVID-19, while leaving open big business and government, has not been an accident of fate. Global predators have always intended to destroy this conservative beating heart of America which thrives on personal responsibility, individualism, independence, sound traditions, patriotism, freedom, and God. Working-middle and upper-middle-class people are precisely the groups official Marxism and progressives despise as the "bourgeoisie."

Without understanding the globalist determination to wreck America and any remaining patriotic, democratic nations, much of what is taking place during COVID-19 makes no sense. Now that we can identify the enemy, we can better formulate an opposition based on the Constitution and Bill of Rights. We can do so without caving into the lie that their oppression of us is required to save us from COVID-19.

CHAPTER 23

How Globalists Pervert Capitalism

Predatory globalism is not a conspiracy theory but a recognizable, loose coalition of often-competitive centers of money and influence, all of them lacking self-restraint and all of them obsessed with gaining more wealth, self-importance, and influence. Never before have there been so many of them operating with so much determination and strength. We coined the term *predatory globalism* and its variations, such as *global predator*. We have already introduced you to some of them and will continue to do so throughout the book. Predatory globalists are so dominant in the arena of world politics and governance there is hardly any room, perhaps none, for a more benevolent kind of liberal globalism or non-predatory capitalism.

Predatory globalism is worldwide and includes many governments and their leaders: The United Nations (UN), and its World Health Organization (WHO); most large international corporations, including the pharmaceutical, chemical, and agricultural businesses; government and private banking; and the military-industrial complexes. It includes the Bill & Melinda Gates Foundation, its remarkable partnership with Warren Buffett, and their front man in the federal government, Anthony Fauci. It is a loose alliance of power seekers who feel inspired by the opportunities presented by COVID-19 and are homing in on destroying its one last impediment to victory, after which the struggle over the spoils will begin. The impediment is American patriotism and freedom. The spoils are humanity.

In America, the major media is a member and key ally of predatory globalism without which the globalists could never win an election in the United States for their progressive candidates. The following headline captures some of what has recently surfaced in November 2020, as the Trump administration began cracking down on globalist relations with predatory China:

> The Wall Street Journal, The Washington Post *Among Newspapers Paid Millions by Beijing-Controlled News Outlet to Publish Propaganda this Year*[747]

> The Chinese largess has been spread more far and wide than the headline suggests, from coast-to-coast (the *LA Times* and *The NY Times*) and from Canada to Great Britain.

Remember, "Russia, Russia, Russia!" and the baseless, concocted accusations against President Trump? It wasn't "China, China, China!" because the Chinese Communist Party is one of the most active participants in predatory globalism with economic ties to the Democratic Party and mainstream media as well as American education, science, and corporate life.[748]

BILL GATES AND HIS FOUNDATION

The Bill & Melinda Gates Foundation[749] is truly global, sending money to nearly every state in America, Washington, DC, and 127 other countries. Its "leadership" has 18 members outside the Gates family, and eight of them have "Global" in their name. Its "offices" are in Beijing, China; Delhi, India; London, United Kingdom; Berlin, Germany; Addis Ababa, Ethiopia; Abuja, Nigeria; and Johannesburg, South Africa.

The wealth of the Gates Foundation is staggering. It claims to have spent $54.8 billion since its inception, to have an endowment of $49.8 billion, and to have billions being donated into its coffers on a regular basis from outside sources, including from Warren Buffett.

Buffett, along with Bill and Melinda, make up the only three listed trustees and directors. In another place, the foundation describes itself as "under the direction of Bill and Melinda Gates and Warren Buffett."[750] This combination of Gates and Buffett, currently the number two and number four wealthiest men in the world, wields power genuinely beyond comprehension.

Buffett is an even larger contributor to the Bill & Melinda Gates Foundation than Gates himself:[751]

> In 2006, Buffett pledged most of his fortune to the Gates Foundation by donating 10 million shares of Berkshire Hathaway (BRK.B), which was then worth approximately $31 billion. Since then, continuous contributions have lifted the total amount of shares held by the foundation to nearly 47 million.

If Buffett had held on to these massive multi-billion-dollar contributions to the Gates Foundation, Buffett would be, by far, the wealthiest man in the world.

In mid-February 2020, as the seriousness of COVID-19 was emerging, Buffett took what some analysts called a "big gamble" by investing in the giant pharmaceutical company, Biogen, to the tune of $200 million.[752] We suspect Buffett, as a long-time friend and partner to Bill Gates, knew what was coming in terms of predatory, capitalistic opportunities from the pandemic. Gates had invested almost $400 million in Teva, another pharmaceutical giant.

Bill Gates is not running a strictly philanthropic organization in the ordinary understanding of philanthropy. The *Nation* has described "the moral hazards" surrounding the foundation,[753] which will surface throughout our book. Gates conducts a huge business by investing in favored corporations and industries, including the pharmaceutical industry, which will probably end up at the top or near the top in reaping the largest wealth and power from COVID-19. Even *USA Today*, a source that was defending Gates as a philanthropist, noted his wealth grows faster than his contributions:[754]

> He has given more than $50 billion to charity since 1994. However, his wealth has grown even faster than he has donated money. As of June 2020, Gates' net worth was estimated at roughly [more] than $110 billion, according to Forbes.

As of December 26, 2020, *Forbes* estimated Gates' net worth at $120.1 billion.[755] He certainly has not been suffering economically because of COVID-19.

Several sources are making serious criticisms of Bill Gates' overbearing impact on the world.[756] One critique raised an issue central to this book:[757]

> Perhaps what is most striking about the Bill & Melinda Gates Foundation is that despite its aggressive corporate strategy and extraordinary influence across governments, academics, and the media, there is an absence of critical voices. Global Justice Now is concerned that the foundation's influence is so pervasive that many actors in international development, which would otherwise critique the policy and practice of the foundation, are unable to speak out independently as a result of its funding and patronage.

We will not attempt any kind of definitive critique of Bill Gates' life and work; however, we will continue to find that, in the arena of COVID-19, he is one of a few chiefs among global predators, snapping a whip to which many dance. They include Anthony Fauci, WHO, universities, the scientific community, and many others.

THE CLINTON FUND AND GLOBAL INITIATIVE

The Clinton Foundation and its Clinton Global Initiative have notoriously benefitted from globalist contributions from around the world and have been accused of corruption.[758] Guess who pops up as contributors? In 2020, the Bill & Melinda Gates Foundation gave an unspecified amount of over $25 million.[759] Gates is like the devil's fly: If he is seen buzzing around you, odds are you are a malignant, predatory globalist. Indeed, the Clintons for many years have maintained their own Clinton Global Initiative.[760] We will not delve deeply into the foundation, but notice that contributors come from a wide variety of overseas sources.[761]

Here is something else to mull on: The Clinton Global Initiative worked with Dominion voting machines (think election fraud in the US[762]) to provide poor and conflicted countries with their machines: "Over the next three years, Dominion Voting will support election technology pilots with donated Automated Voting Machines (AVM), providing an improved electoral process, and therefore safer elections."[763]

Where Does George Soros Fit
Among the Globalists?

One donor to the Clinton Global Initiative deserves special attention, the infamous George Soros. Soros is a man who hates America, hates the free enterprise system, and hates President Trump, while masquerading as an idealistic progressive.[764] In January 2020, *The Washington Times* described his behavior in headline, "George Soros, 89, is still on a quest to destroy America."

George Soros has made absolutely clear where he stands in the existential struggle between the Chinese Communists and the U.S. and its allies. He's for the Communists. In an interview with the *Financial Times* in October 2009,[765] Soros looked forward to "an orderly decline of the dollar" accompanied by a rise in the value of the Chinese renminbi. He repeats the theme: "Well, certainly a decline in the value of the dollar is necessary in order to compensate for the fact that the U.S. economy will remain rather weak, will be a drag on the global economy." In the new world economy, China will become the "motor," "the engine driving it forward, and the U.S. will be actually a drag." *This economic and political rise of China, and the corresponding decline of the U.S., is a powerful motivating factor for global predators to flock to China and to distance themselves from America.*

What is needed, according to Soros, is "to bring China into the creation of a new world order, a financial world order," which will be "more stable" as the CCP displaces America in influence and power. Soros looms large in the alliance of grandiose and insatiable predators working toward the new world governance. We mistakenly saw Soros as a loner until, on July 19, 2021, the news broke of his teaming up with Gates and others in a consortium to buy a company that makes COVID-19 tests.[766]

Soros has, in fact, been bingeing on destroying national economies. In an interview in November 2020, Soros openly laughs about destroying the economies of several Southeast Asian nations. He then makes an extraordinary statement that the interviewer seems oblivious toward, perhaps because it froze his blood. Soros declares, "I am basically there to make money. I cannot and do not look at the social consequences of what I do."[767]

Soros has made his fortunes by looking for vulnerable countries and institutions and wrecking them, while making financial bets on their decline. He famously broke the British pound, sold it short, and made $1 billion.[768] Soros is aiding and abetting the breakdown of law and order in America by funding lawless district attorneys, urban rioters, and Black Lives Matter; eliminating the police; and releasing accused criminals without bond.[769] He also stepped in early to massively fund the Democrat campaign against President Trump's second term.[770]

Why would Soros want to destroy America? In addition to hating America and freedom, perhaps he has found a way to short America on some financial market. More interesting is the question, "Why would the Clintons accept a donation from him for their foundation and disclose it to the public?" Globalists often act as if they are unafraid and unashamed of anything they do.

As villainous as Soros is toward America and her survival as a free nation, he is never attacked by the establishment media, by universities, by big tech, or by any other globalist entity. His destructive activities are supported or systematically kept out of public awareness.

Are you now ready for this amazing tying together of things? In the middle of the ongoing concern about massive fraud involving U.S. voting systems, George Soros has named Lord Mark Malloch-Brown president of his Open Society Foundation. In the move, Brown leaves behind his current job as CEO of Smartmatic, a manufacturer of machines previously accused by conservative media of rigging the election against President Trump.[771] Brown has a global background as the former UN deputy secretary-general from the United Kingdom and president of Smartmatic voting systems.

Big Tech Are Globalists

The major media, the tech industry (including Amazon), and the social media (including Twitter, Instagram, Facebook, Google, and YouTube) in the U.S. and throughout the world are predatory globalists. To cover themselves, like other large corporations, they will give donations to radical progressive activities, such as Antifa and Black Lives Matter; but underneath it all, they are engaging in predatory capitalism, which bears little resemblance to a free market. The power disparities among the competitors typically give one or another an upper hand as illustrated in social media and tech industries. More importantly, the bigger and more unscrupulous the business, the more it can dominate the government as well, not only its regulators, but also the entire

Congress and the presidency. We have seen this behavior with the social media and tech industries.

The technological revolution has made many individuals and their companies enormously wealthy. Technology has made it far easier for them to monitor, control, and dominate human activity, beyond any historical precedents, even to the degree that companies like Google, Twitter, YouTube, and Facebook can censor the President of the United States and shutdown his political support during a critical election. Many, if not all of these contemporary business monsters, want entry into the world's biggest market, China with 1.4 billion people. There is good reason to believe that censorship of libertarian and conservative views on social media and the major media's refusal to cover the Biden family corruption helped throw the election toward Biden, along with whatever outright election fraud will eventually be proven.

Major Universities

The current worldwide system of higher education is also a partner in these predatory activities, and we shall find many universities have shockingly deep financial and research ties to the Chinese Communist Party.

Major universities, such as Cornell in our hometown of Ithaca, New York, make a great show of progressivism by making public pronouncements in favor of social justice and Black Lives Matter, pressuring their professors to support radical progressive programs, and by pushing confessional "educational" seminars on racism. The universities have been complicit with the shutdowns and every other intention of Fauci and Gates. Like Gates, they have little or no real sympathy or commitment either to the ideals of progressivism or to the principles of free-market economics. They are opportunistic predators.

With Donald Trump as President, American democracy became a stronger, more patriotic national state, with a more ardent devotion to liberty. This is every global predator's worst nightmare—a strong democracy that puts itself first and is devoted to its own liberty and the liberty of other nations as well. Even the strongest global predators, from Bill Gates and Klaus Schwab to Xi Jinping, know their power will be limited as long as America exists as a powerful, patriotic, democratic republic devoted to freedom.

Globalists can work with China, even though it is a nation state, because it is not a democracy. They can do business with China through its leaders without the messy interference of petty politicians, voters, checks and balances, and stringent rules of ethical business conduct. They can buy their way into huge markets with seemingly far more growth potential and far less competition than in the United States. We can add that the Chinese leadership is not nearly as nation-state oriented as one might imagine. The Chinese come from a long history of empires, and the Communists are Marxist visionaries with a political mandate to rule humanity that probably makes them more globalist than patriotic.

Understanding predatory globalism lifts much of the confusion surrounding why politicians, educators, scientists, and prominent American athletes are so protective of China—they make money dealing with the corrupt dictatorship much more easily than within the competition and complexities of a democratic republic and patriotic nation like the U.S.

Capitalism Morphs into Outright Predation

Large-scale capitalism is in trouble, on that we agree with many others, but not because it has lost track of the needs of the people of the world or needs more stakeholders. Elected democratic governments with constitutional restraints are the best approach we have to address critical, unmet needs. Capitalism meets many of the world's material needs by answering to its shareholders and customers. Global capitalism has become predatory because it has rejected promoting the liberty of the people and the nations of the world. Instead, global capitalism uses government to force itself on individuals, communities, and nations around the world. The takeover of governments is what we are now witnessing with such intensity during COVID-19, placing the people of the world at the mercy of Big Tech, Big Pharma, Big Media, and a host of varied and sometimes competing Bigs.

We must find ways to dissociate big money and big corporations from government—something President Trump was trying to accomplish by refusing to accept large donor contributions and by confronting the globalists, including the chief predatory exploiter, the Chinese Communist Party. The stakeholder concept—which merely brings in additional rapacious entities—will only further encourage capitalism's already predatory takeover of governments, making them into extensions of themselves. Realistically, the biggest "stakeholders" in the world are the governments. As the biggest and most powerful billionaires and corporations further "partner" with governments, their power will only grow and liberty will further shrink around the world.

This is where limited government aimed at protecting citizens from large corporations—not cooperating with them—becomes critical to the future of individual freedom and political liberty. This is the fundamental problem that has been so tragically amplified by the response to COVID-19. The result has been a feeding frenzy for giant corporations, billionaires, and their minions like Anthony Fauci and WHO's Tedros.

IS MONEY THE ROOT OF ALL EVIL?

Money or wealth is often seen as the root of all evil, but, in fact, all the great totalitarian leaders from Alexander the Great and Genghis Kahn through Hitler and Mao, and even minor modern figures, such as Fidel Castro of Cuba and Ali Khamenei of Iran, have been driven by a more all-consuming lust than the acquisition of wealth. They are driven by the lust for power over masses of human beings. They also seek self-aggrandizement, the extreme of which is to be worshipped in a megalomaniacal, God-like fashion. Marxists and Jihadists are additionally devoted to political and religious ideologies, but while that strengthens them, it limits their appeal to others as globalists seeking world dominion.

In America, COVID-19 is being used as the justification for increasing unfettered government power at every level, including governors, mayors, and public health officials from New York State and New York City to California and Los Angeles.

Money is, of course, also a great motivator, and we will see this demonstrated most obviously regarding billionaires like Bill Gates and the pharmaceutical industry who try to deprive us of inexpensive, safe, and effective treatments while making us wait for new, very expensive, experimental, unsafe, and ineffective treatments. To create wealth in such a fraudulent and coercive manner, industries also need vast power over governments, society, and any other competing interests.

Some libertarians and conservatives seem deluded about big industry. They confuse predatory capitalism with the "free market" or genuine capitalism, for example, by partnering with China who uses slave labor. Within these all-powerful, predatory, worldwide institutions, there is zero respect or concern for personal freedom or political liberty. There is no empathy for those they harm. They are so big they can readily defy most governmental controls and instead buy favors from politicians and bureaucrats to manipulate the agencies intended to monitor them, such as the FDA or CDC, and nowadays even the FBI. For globalists, it is all about monopolistic predation, whether expressed as their lust for money, self-aggrandizement, ideological fulfillment, or sheer power.

Nonetheless, especially among many academics at think tanks, there is a great desire to see Klaus Schwab and his World Economic Forum, and other globalists, as a more idealistic hope for the future. Perhaps they fan these hopes to drive out the reality that the same corporations giving them money are participating in predatory globalism, creating a conflict of values they cannot bear to experience. For example, in the 1970s, I self-identified as a libertarian activist and lost all support from the libertarian think-tank establishment because I had become an effective critic of the global pharmaceutical industry for the fraudulent development and marketing of psychiatric drugs.

Identifying these huge, international industries and billionaires with genuine capitalism—which is based on individual freedom and political liberty—is a sham. They are more dangerous than the governments they control because they exert power from a position far above the people. Rather, they act through their intermediary politicians like Biden or Fauci, who give them cover and do their bidding. Behind them all is the ultimate threat of Communist China who pays off many of the global predators. Instead of a self-serving, doomed Great Reset, we need a complete re-evaluation of how democracies limited by constitutions and principles of liberty can resist the power of the global predators. We hope this book will help to stimulate such an assessment.

The Great Reset and New Normal—subjects of the previous chapter—are based on an explicit model of top-down control by leaders who act utterly without regard for human liberty or democratic political processes. It may seem an exaggeration that educated people, including Americans, would plan for us with utter disregard for our personal freedom and liberty. It has also been difficult for us to accept, but this is how those who seek a new governance of the world think and feel—with indifference to those whom they exploit. Otherwise, they would feel rationally and morally compelled to extend liberty throughout the world rather than to impose their will upon it.

Their cover story is always a hodgepodge of progressive ideals which, absent a concept of liberty and democracy, they have no way of fulfilling. They declare their lofty goals, while having no plan and little desire to reach them. Undaunted by their failures to improve humanity's lot, they become wealthier, more self-important, and further determined to rule the world. Meanwhile, America-First policies stand in their way like a vast moat protecting our freedom.

The Green New Deal is now a part of the Democratic platform and embodied in their plans for the Great Reset. This generalized adoption of the most extreme progressive ideas, including Bill Gates' zero emissions by 2050,[772] are essentially the cover story for a powerful totalitarian movement. It is the shared public facade for everyone from Fauci, Schwab, and Bill Gates to the leaders of the Chinese Communist Party.

The COVID-19 pandemic has provided a unique opportunity for predatory globalists to demand a rethinking and redoing of the world's governance. To a shocking degree, their operating principles reflect the notion that American patriotism, freedoms, and democracy are bad, yet Chinese, top-down control and conformity are good. They are, in fact, making fortunes doing business with China to the great disadvantage of the people of America and the world. Worldwide, the middle and working classes and the poor have been most injured by globalists. Predatory globalists literally zap the life out of them in the pursuit of wealth, self-aggrandizement, and power. In a profound sense, predatory globalists are slavers, and we are their slaves.

CHAPTER 24

Shared Values— President Joe Biden, Bill Gates, and LeBron James

resident Joe Biden is especially important not only as a grave globalist threat to America, but more specifically for his subservience to the Globalist Puppet Master, the Chinese Communist Party. Many globalists work hard for their money, but not the Biden family. They use bribery. Like other globalists, they lack any patriotic commitment to putting America's interests above or on a level playing field with China.

What do they have in common with LeBron James? James is extremely hardworking, enormously talented, and loved by many people. Clearly, that's not what they have in common. What they do have in common puts LeBron at risk of becoming a tragic figure.

Joe Biden, International Hustler

Remember how Joe Biden called President Trump a "xenophobe" for labeling SARS-CoV-2 the "China virus" and for shutting down traffic from China as soon as the pandemic was confirmed late in January? Remember how he pushed the fraudulent Russia probe while rejecting China as a major threat to the United States?

Joe Biden is an active, long-time, global predator, taking gold from the grinning Chinese dragon who would just as soon devour him when he is no longer useful. We will focus on Joe and his son Hunter, who used to travel around with his father in Air Force Two when Joe was vice president. Together they engaged in shaking down the leaders of multiple countries, including China, for millions of dollars.[773] Although the term "shakedown" implies the Bidens are taking advantage of these countries, in reality, the countries that are paying out the money and the Biden family are joining together to exploit those outside the circle of power, including most of humanity.

The scandal has been written about for several years, including in the 2018 book *Secret Empires*[774] by Peter Schweizer and Seamus Bruner. On October 20, 2020, the *New York Post* published a report by Schweizer and Bruner about a deluge of new, incriminating data on Joe and his son:[775]

> Thanks to three brave Americans, we now know that Joe Biden has long misled the public about his involvement with his family's foreign business entanglements[776] while he served as vice president.
>
> At considerable personal risk, former Biden family business partners Tony Bobulinski[777] and Bevan Cooney, and computer shop owner John Paul Mac Isaac, have come forward with tens of thousands of primary-source documents—internal corporate records, emails, and text messages—detailing years of business dealings that centered on trading on the Biden name. This material suggests that, despite Joe Biden's insistence that he knew nothing about his family's business deals, he was well aware of his son Hunter Biden's business ventures in China, Ukraine, Kazakhstan, and elsewhere.
>
> These new troves constitute hard evidence of Biden family corruption, and confirm our reporting dating back to our 2018 book *Secret Empires*. In 2018, when we first broke the foreign influence scandals that have now engulfed the former vice president, it seemed apparent that China and Ukraine were not paying Biden's family members for their expertise, they were buying access to the vice president of the United States. This was never

a scandal solely about Hunter or Joe's brothers, James and Frank. It was, and has always been, a Joe Biden scandal.

Hunter's partner Joe Bobulinski then appeared on the Tucker Carlson Show on Fox News, apparently after none of the other media would interview him.[778] He reported that Hunter actually introduced him to his father, the vice president at the time, and Biden vetted Bobulinski for participation in business dealings with China and provided further documentation that Hunter had arranged with the Chinese for his father to have a share of the funds allocated to him.

The Senate Homeland Security and Governmental Affairs Committee reportedly reviewed the relevant documents and confirmed their validity.[779] One of the emails from a Chinese authority to Hunter referred to the "Big Guy" getting 10% of the deal, a reference to his father, Joe.

There was, and continues to be, a blackout on the Biden family story in the mainstream television and newspaper media. The predatory globalist media has devolved into little more than a progressive American version of Chinese state media.

More About the Bidens Selling Out America

On November 10, 2020, in the conservative newspaper *The Epoch Times*, Emel Akan wrote an analysis, "Beijing, Wall Street Could Deepen Ties Under Potential Biden presidency."[780] Akan's analysis confirms President Trump's remarks in his 2016 presidential nomination acceptance speech about how his pro-America, anti-globalist stance most definitely separated him from the Democrats. Akan observed:

> For more than a decade, Chinese companies have taken advantage of U.S. capital markets while operating under lax standards. Regulators in Beijing have refused to allow audit inspections of its U.S.-listed companies on the grounds of national security and state secrecy.

In 2013, under the Obama-Biden administration, the U.S. regulator, the Public Company Accounting Oversight Board signed a memorandum of understanding[781](MOU) with Chinese regulators. The MOU gave Chinese companies improved access to U.S. capital markets without complying with the same disclosure rules required of U.S. companies.

Citing a *Just News* article, [782]Akan then explains that transcripts from the Obama administration's archives showed that the agreement resulted from multiple meetings between Vice President Biden and Chinese leaders. He then explains:

> As a result of this concession, American investors, through their pension funds, have been unknowingly transferring wealth from the United States to Chinese entities that don't comply with U.S. laws. Some of these companies are sanctioned by the U.S. government as they're involved in the Chinese Communist Party's military, espionage, and human rights abuses.

> As of Oct. 2, 2020, there were 217 Chinese companies[783] listed on U.S. exchanges with a total market valuation of $2.2 trillion, according to the U.S.–China Economic and Security Review Commission.

> American investment banks have a vested interest in fundraising for Chinese companies in the U.S. capital markets because they earn substantial fees from these entities.

Akan then describes how the Trump Administration "crackdown on Chinese companies" took several steps "to curb money flowing from U.S. pension funds into Chinese stocks" and released a plan to make listed Chinese companies comply with U.S. standards for ethical business practices.

President Trump's America First and fairness policies maddened global predators and led them to do everything possible to defeat President Trump and replace him with their hapless puppet Biden. Wonder how lackluster Joe became the Democratic candidate and then won? He didn't lack luster for predatory billionaires and corporations.

A GREAT SPORTS FIGURE GOES FROM BEING A HOMIE TO BEING GLOBAL

Our sports teams, many of them seeking to reap fortunes from a Chinese market, will allow players in groups to disparage our flag and our nation. The same players and management never show disrespect for China, which is a far worse predator than the U.S. has ever been. Why? Because the sports teams and some of the wealthiest players are themselves among the global predators benefitting from China's market and government largesse.

LeBron James, a much-loved former Cleveland Cavalier basketball star, expressed his love and commitment to his fans and for the working people of Ohio. I remember greatly admiring LeBron for his open expression of support for people who were his roots in Ohio and felt deeply impressed by his return to the Cleveland Cavaliers. His sympathy for Ohioans was justifiable, because they were suffering mightily from the outsourcing of industries and jobs to China.

When LeBron left Cleveland in order to win championships with the Miami Heat, his fans were saddened but forgave him. An athlete wants to win championships, which he did, skyrocketing his fame.

When LeBron returned "home" to try once again to win a championship for Cleveland and the working people of Ohio, he succeeded. People rejoiced and knew they were right to love him. He left again in July 2018, this time going to the storied LA Lakers, the team and organization with the history, the cache, the cash. Valued at $4.4 billion in 2020, LA is among the wealthiest sports franchises in America. LeBron's personal worth is estimated at $480 million, nearly half a billion dollars.

On July 21, 2019, LeBron joined his teammates in LA by taking a knee in protest during the national anthem to humiliate America and its values. He expressed his support for Colin Kaepernick as the leader of the NFL football players in the taking-a-knee protest movement.

LeBron's "protest" must have delighted his potential partners in Communist China, yet he felt he had to do more for China. On October 15, 2019, LeBron publicly humiliated Houston Rockets' General Manager Daryl Morey for not being "educated on the situation" when Morey tweeted in support of heroic protesters in Hong Kong against the Chinese Communists. The Chinese were breaking their agreement with the semi-autonomous city not to absorb it further in China. LeBron was currying favor with China, cozying up to them. It was out in the open, an international story and controversy: *The hometown boy had joined the global predator game—the biggest game of all on the planet.*

USA TODAY was critical of LeBron with a headline:[784]

LeBron James's controversial comments "furthers his brand power in China."

The newspaper then quotes surprisingly political remarks by LeBron, where he criticizes free speech in America and in effect tells Americans to lay off China: "We do have freedom of speech, but there can be a lot of negative things that come with that too" and "I also don't think every issue should be everybody's problem."

The newspaper further reported:

> The global superstar who plays for the Los Angeles Lakers has traveled to the country 15 times since he signed with the sports apparel company 16 years ago and has visited several Chinese cities promoting physical fitness, education, basketball as a unifier and, of course, Nike.

> It has been a beneficial relationship. The face of the NBA has been careful not to upset the Chinese—going so far as to call out Houston Rockets' General Manager Daryl Morey for the timing of his pro-Hong Kong protesters tweet days before the Lakers and Brooklyn Nets traveled to China for two preseason games.

> Some sales figures help shed light on why James might have said what he did and took criticism for it.

> How much is James's deal with Nike worth? He makes $32 million annually from Nike, according to Forbes, and Nike's revenue in China surpassed $6 billion from June 1, 2018, to May 31, 2019, according to statista.com.

LeBron's support of China as an individual player might have been dismissed as the aberration of a relatively young star whose fame and wealth had gone to his head and made him finally forget his roots. But then the NBA officially came out to support LeBron's criticism of Houston Rocket's General Manager

Daryl Morey, throwing under the bus a man whom they should have praised for supporting the people of Hong Kong.

The NBA, after gaining so much from America and its U.S. fans, displayed more allegiance to the Communist Party of China than to its own country. Why? The Chinese are a market with more long-term growth potential. Also, highly likely, the Communists are able to provide under-the-table incentives unavailable in America's more strictly monitored economy.

The NBA must believe that leadership in America will not punish its defense of China because our leaders, except for President Trump, are predatory globalists. The Chinese, however, might put their outrage or their political considerations ahead of their pocketbook and break relations with the NBA if it dared to put up with criticism of them from players or managers.

LeBron's relationship with world predators goes back a long way. There is a 2008 video clip of him dining casually with Bill Gates and Warren Buffett, the second and third richest men in the world at the time. On his way out, LeBron reportedly boasted, "We just had dinner with a few friends of ours, you know. Two guys by the name of Warren Buffett and Bill Gates. A lot of people call him William. Two of the powerful people in the world, you know."[785]

LeBron's bragging is entirely forgivable; the video was from 2008, and he was only 23 years old at the time. Nonetheless, after the video came to light in 2013, LeBron may have found it embarrassing or incriminating. By the time we went looking for it for this book, it was often blocked with an explanatory sign, "Video unavailable. This video is private."

This brief story about LeBron James is meant as an illustration of the corruption of a young man by wealthy interests, turning him into a supporter of predatory globalism and making him into a public critic of the country that has given him so much. Much, much worse, it shows how the NBA operates as a corrupt supporter of China for the sake of a vast new market.

LeBron has the confidence, intelligence, leadership qualities, money, and public following to make a huge difference should he remember his roots. He could join patriots in putting America First and in resisting the worldwide predations of the globalists. Sadly, after this chapter was completed, LeBron tweeted remarks that others felt were inciting violence against the police.[786]

America is the land of *opportunity* that made LeBron's success possible; China is the land of *opportunism* that destroys individual opportunity and freedom. We must all rally to save America and its values, the greatest values ever articulated by any nation on Earth.

CHAPTER 25

Bloomberg Celebrates China's Rise and America's Demise

This chapter introduces you to another of the leaders of the globalist predators: multibillionaire and former New York mayor, Michael Bloomberg, who is in the business of financial communications. He is the founder and director of the new and growing conference among globalists, the New Economy Forum. Its strongest partner? The Chinese Communist Party.

Even some progressives think Bloomberg has gone too far into China's pocket.

BLOOMBERG'S TIES TO CHINA

The *Washington Post*, which adores and profits from China through advertising and tends to worship all things progressive, found that Bloomberg had thrown himself too deeply into China's arms. On January 1, 2020, during Bloomberg's presidential campaign, the *Post* headline gave him an unwelcome New Year's Day present:[787]

Bloomberg's business in China has grown.

That could create unprecedented entanglements if he is elected president.

With undisguised hostility, the *Post* wrote that at Bloomberg's initial 2018 New Economy Forum "to the surprise of some in the audience, he gushed about one of China's top government officials," Vice President Wang Qishan, and boasted about knowing him since he entertained him at his home while Mayor of NYC 15 years earlier. At another time, amid mass riots in Hong Kong and protests against the imprisonment of Muslim minorities in China, Bloomberg defended President Xi Jinping, declaring on television that he "is not a dictator." *The Washington Post* further criticized Bloomberg for running for president while simultaneously he "has deepened his entanglements with that key U.S. adversary—forging close financial ties there while showering praise on the Communist Party leaders whose goodwill is required to play a role in that fast-growing market." The newspaper further lambasted Bloomberg, stating:

> The billionaire, whose core business sells financial information to investors, has led efforts since 2015 to make it easier for U.S. companies to trade in Chinese currency, a move embraced by China's largest banks. He expanded one of his company's financial indexes, which could steer $150 billion into China while earning his firm an undisclosed amount in fees.

The Washington Post is owned by Amazon founder Jeff Bezos. Bezos is a globalist who makes massive donations to progressive causes.[788] Despite his newspaper's criticism, Bezos himself supported Bloomberg for the presidency in February 2019, leading Bernie Sanders to declare, "The billionaires are looking out for each other. They're willing to transcend difference and background and even politics."[789] Bloomberg, although three-time mayor of New York City, meets most of the criteria for a global predator, including currying favor with and enriching himself through the Chinese Communist Party at the expense of America, including the New York City citizens he once led.

Bloomberg's November 21-23, 2019 New Economy Forum Honors China Over His Own Homeland

To begin with, the conference was held in *Beijing*, making it a direct slap in the face of President Trump and his America First policies. The single cohost for the conference was the China Center for International Economic

Exchanges, International Institute for Strategic Studies and Mandela Institute for Development Studies (CCIEE).[790] The leader was Zeng Peiyan, the former vice-premier of the State Council of the People's Republic of China and now chairman of the China Center for International Economic Exchanges (CCIEE). He was also a cochair of the conference, cementing the Chinese Communist Party influence over the Beijing conference.

By its own description,[791] CCIEE was founded in 2009 "with the approval of the government of the People's Republic of China." Although masquerading as a "think tank," the CCIEE is the spawn of the Chinese Communist Party and therefore inextricably linked with the People's Liberation Army. Even Wikipedia notes,[792] "the CCIEE has been noted for its tight connections to the government of the People's Republic of China. Underscoring this relationship, its offices are located a few hundred meters from Zhongnanhai," the central headquarters of the Chinese Communist Party.

"Funding Partners" included 3M, Exxon Mobile, FedEx, HSBC, Hyundai, MasterCard, and others. China's violent dictator Xi Jinping is shown grinning on location like the Cheshire Cat.[793] It was a time to fawn over the Communists.

On the first day, speakers included cochair Bill Gates; David Solomon, CEO of Goldman Sachs; Henry Kissinger; and Henry Paulson, who was already on Biden's election team. One observant reporter wrote, "Former Google China head Lee Kai Fu dines with former Google head Eric Schmidt. When asked what they were discussing they said they've known each other for 15 years and were just catching up."[794] A lovefest of global predators.

On day two, more corporate giants starred. Yet another astute reporter observed:

> And China-U.S. ties and decoupling are already a theme today. Scott Kennedy, of the Center for Strategic & International Studies in Washington, told Bloomberg TV that ripping the two economies apart would be "nuts."[795]

Opposing President Trump's Successes

By the time of the Bloomberg New Economy Forum of November 2019, President Trump was well on his way to fulfilling his foreign policy promises to America: closing the border to illegal immigration, bringing back corporations and jobs to the U.S., raising the employment rate to unbelievable all-time highs for minorities and many other Americans, and standing up to aggressive political, military and economic attacks from the Chinese Communists.

With the exception of the elite globalists, all of these successful policies were better for Americans, from the poor to the upper middle class. But the elite global predators were so threatened by each of these America First policies that they came out in the open to attack President Trump. Trump had not anticipated this and neither had we. Hardly anyone imagined there was so much evil simmering within so many people, agencies, and corporations. It was like one of the horror movies where all of sudden misshapen creatures with batwings and lizard tails come flying out of the darkness into the light and on the attack.

To say that Bloomberg is in a league with the Chinese Communist Party is no exaggeration. To say Bloomberg is openly trying to undermine and reverse American foreign policy is equally true. This explains many of Bloomberg's policies—and those of most other global predators—that are being documented in this book. We will find in the last chapter of our book that Bloomberg is now openly saying that China's Maoist totalitarian policies during COVID-19 were and continue to be good while Trump's were bad.

AFTER THE 2019 BLOOMBERG CONFERENCE

A few days later, on November 24, 2019, an empowered Bloomberg announced he was running for President and declared himself in favor of impeaching President Trump because he was an "existential threat," not to China, but to American democracy.[796]

A little more than a year later, on January 14, 2021, the Bloomberg report supported President Trump's second impeachment, urging the Senate to rush through with convicting him "a rebuke Trump richly deserves."[797]

Calling for the impeachment and expedited conviction of Trump while supporting Xi Jinping and more trade with China exemplifies the arch global predator. Xi Jinping must have smiled and shaken his head at the infidelity and stupidity of rich Americans.

Bloomberg's November 16-19, 2020 Conference: A Virtual Forum for Backing China over the U.S.

Although Bloomberg's November 16-19, 2020 conference was virtual, he nonetheless had it cohosted by the Chinese Communists as he had done in 2019. He didn't have to be in Beijing to want the full cooperation of one of the most murderous regimes in the history of the world.

Choosing China over His Homeland of America

According to *Bloomberg News*, a Chinese participant at the forum boasted how his country had continued to cooperate with America despite President Trump's policies to the contrary: "Xie Zhenhua, China's former top climate diplomat, said U.S.-China climate cooperation continued despite Trump's efforts. ... Even after the U.S. federal government signaled it would withdraw from the Paris climate agreement, China and the U.S. maintained cooperation at local levels and through think tanks and the private sector."[798] This is not mere braggadocio. Xie Zhenhua is confirming how China's tentacles reach deeper into our society than any presidential policies can easily sever.

Bill Gates was a cochair. Another cochair was Henry M. Paulson, former U.S. Treasury Secretary under George W. Bush (2006-2009) and, at the time, an active member of Biden's transition team. His presence in both Democratic and Republican administrations reminds us that the Bush family, Bill Gates, and Bloomberg are globalists with much in common, including a fear and hatred of President Trump and populism.

Globalists Bill Clinton, 42nd President of the United States of America, and his wife Hillary Rodman Clinton, former U.S. Secretary of State during the Obama years, were speakers. Janet Yellen, former chair of U.S. Federal Reserve Board of Governors, announced as Biden's future Secretary of the Treasury, was also present.

Global corporate executives were well represented with the president and CEO of McDonald's Corporation; the executive chairperson of IBM; the CEO of MasterCard; the president and CEO of FedEx; a chief executive of Prudential plc; the CEO of Goldman Sachs; the CEO of Honeywell; and numerous huge banks.

Pharmaceutical industry executives were present, most notably Moderna CEO Stéphane Bancel. Moderna is a young, wealthy, and powerful company that is one of Bill Gates' and hence Fauci's favorites. Both have been giving Moderna's vaccine a big push.

International political leaders were also strongly represented, including the Prime Minister of India; Tony Blair, Former Prime Minister of the United Kingdom; and His Majesty Abdullah II bin Al Hussein, King of Jordan. There were bankers with top positions in the past or present with the European Central Bank; Bank of England; Reserve Bank of India; Credit Suisse; and the People's Bank of China.

COVID-19 Predators at Bloomberg's 2020 Conference

Peter Daszak, the President of EcoHealth Alliance, the gross apologist for the Chinese Communists, headlined the final day of the conference. He was joined by *Tedros Adhanom Ghebreyesus*, director-general of WHO and another apologist for the Communists. WHO is financially supported by both China and Bill Gates.

The origin of the virus was attributed to "nature" and not to Chinese labs. One of Bloomberg's workshop description begins, "COVID-19 shows that humankind's relationship with nature is broken." As we have seen in Chapter 6, this is similar to Fauci's theme: Neither he nor his collaborators in China are to blame for anything about COVID-19; humanity is to blame for messing too much with nature.

Global warming also had a presence at the conference. One workshop emphasized, "Scores of countries, including China, have pledged to become carbon neutral."

DARPA at the Conference?

Regina Dugan is chief executive officer of Wellcome Leap, a nonprofit started by the Wellcome Trust, which is an enormous charity started by Henry Wellcome, the pharmaceutical magnet. Wellcome Leap describes itself in a fashion that will give chills to many who fear high-tech government research into mind control: "Leap is a DARPA for global health."[799]

Dugan is the former director of DARPA—the U.S. Defense Advanced Research Projects Agency. DARPA has a shadowy history of human experimentation, including recent efforts in support of Elon Musk's research in mind control through computer and brain interfacing of which we have been very critical.[800]

Wellcome Leap cannot disguise its joy at the opportunities presented by the COVID-19 tragedy as it compares itself to DARPA again:

> Today, in the throes of a pandemic, we face our own Sputnik moment. And we must respond in kind. We need Apollo-like programs for vaccines, therapeutics, testing, and new health advances for the future. We need a DARPA for global health.

Overall, it is appalling and threatening to see Americans like the Clintons, Bloomberg, and Gates, along with corporate heads of some of the largest companies in the world, plus their COVID-19 activists like Daszak and Tedros, celebrating the return to business as usual with China. China remains the greatest military, economic, and political threat to the free world and to humanity.

A Chinese Takeaway from the Bloomberg 2020 Virtual Conference

Commenting on the Bloomberg conference, Andy Mok, a research fellow at the Center for China and Globalization (CGTN), voiced an "independent" opinion, which seems to epitomize much of the Chinese and the globalist position in favor of authoritarian or totalitarian government: [801]

> In China, the pandemic was swiftly brought under control and as a result both daily life and the economy have largely returned to normal. This was made possible because China has a sound system of government staffed by qualified and competent officials.
>
> In the U.S., however, the pandemic has only worsened and looks likely to wreak even greater social and economic damage. While many blame Trump for this disaster, the real culprit is the system itself. American democracy is little more than a popularity contest

where leaders are selected with only the most cursory consideration of their aptitude for governing or temperament for leadership.

Furthermore, should a competent leader somehow by sheer good fortune be selected, his or her ability to act quickly and decisively is limited by the inefficient mechanisms for the exercise of political authority.

Laura Ingraham Shines Light on Bloomberg's 2020 Forum

On November 20, 2020, Laura Ingraham, the Fox News host, gave a trenchant critique of Bloomberg's 2020 New Economy Forum, declaring it aimed at crushing individual liberty and subjugating people to the whims of the wealthy international community. Her show is worth watching.[802] Ingraham's critique from Fox News indicates predatory globalism is getting increased scrutiny:[803]

> She focused on World Health Organization Director-General Tedros Ghebreyesus, who said, "The fastest way out of this pandemic is for us to work together in solidarity across sectors and borders. Vaccination nationalism will prolong the pandemic." He added, "The pandemic has ... also given us something: The opportunity to think afresh about the world that we want. We can build the healthier, safer and fairer world that we all want—and build back better."

> "You think it's any coincidence that he's borrowing Biden's campaign slogan?" asked Ingraham, who added, "they all have the same end goal. Individualism and individual rights will be subverted to the global rights, the communal rights. It's a sign that Joe Biden won't put America first."

> Ingraham argued that the overarching theme of the international left is "globalists unite against freedom," noting that the forum has been attended by figures from China, Europe, and the Americas, as well as Bill Gates and Bill Clinton.

"At this moment, planning is underway by the wealthiest and most powerful people on Earth about how you should live your life, how nations should be structured, and societies reorganized all to promote global harmony through a worldwide distribution of assets," she said. "There's an astounding lack of curiosity by the media about this project. Well, it's largely because most of them agree with the ultimate goals of a global wealth tax and perhaps an international Green New Deal." ...

She noted that Gates himself floated the idea of a globally recognized coronavirus vaccine "certificate" to be used as something like a passport to gain entry to nations based on immunity statutes.

"As people come in, do they have a digital certificate that says whether they have been vaccinated or not?" Gates said, adding that he hopes a "dialogue" will begin on the topic when the U.S. rejoins the WHO, as expected under a Biden presidency.

None of the global predators seem flummoxed or even perplexed by the fact that their aims nearly always coincide with and support the Chinese Communist Party's aims, grossly endangering America and humanity, including the people of China.

CHAPTER 26

The "New Normal" Becomes Forever Abnormal

In 1776, the year we declared our independence from Great Britain, Adam Smith published *The Wealth of Nations*. It was an inspired analysis of the unfolding new system of economic relationships based on the freedom of people to make voluntary agreements, exchanges, and bargains among themselves. By each person or business pursuing self-interest, the interests of all would eventually best be served.

Was it an ideal that could not be achieved on Earth, an illusion now turned into a delusion? Or was it a good idea on a small scale, but only if people were more moral than avaricious and cared about each other?

As I describe in my book *Wow, I'm an American!*, the American entrepreneur Benjamin Franklin knew Adam Smith, met with him in Great Britain, and reportedly influenced some of his thinking. America was already the best example of this new kind of economic freedom, aided by its remoteness from its central government across the sea. It provided a model for Smith's vision of how a free people could exchange goods and services to the betterment of each other and the society.

Given the flaws in human nature, what would make it work so smoothly, this pursuit of self-interest that benefits all? Perhaps thinking of that difficulty, Smith famously invoked the Invisible Hand of God. Neither Smith nor Franklin was so naive as to think this new, free enterprise model could survive on its own merits as an ideal way of relating that would defy and overcome the frailties of human nature. As expressed in Smith's *Wealth of Nations* and in Franklin's *Autobiography*, the success of freedom in our personal, political, and economic lives depended on widely accepted moral principles and what Smith called "fellow feeling." The free enterprise system required respect for the law and a commitment to the well-being of the larger community. It was clear to both men that a breakdown in religion and morality, including the loss of the Golden Rule, would make political liberty and civil society impossible. Predatory globalists exemplify the disastrous breakdown of morality, including fellow feeling or empathy.

Our Founders also wrestled with the question of the dangers of big government and tendency for varied interest groups to dominate it. They believed a Judeo-Christian moral education was necessary to ensure a citizenry that would resist their own predatory, exploitative instincts and those of others as well. They built in a system of checks and balances to the Constitution and then added a Bill of Rights—all in an effort to prevent a predatory top-down government from robbing Americans of their freedom. The system they built was brilliant, but it has been progressively eroded and is in the process of being overwhelmed by the unanticipated threat from globalism in league with the Chinese Communists.

This cannot be overemphasized. Libertarians and conservatives need to give up the myth that their wealthy corporate patrons who donate to their think tanks or political campaigns have been operating according to the principle of voluntary exchange in the generation of their wealth. They rarely grow their wealth by practicing high standards of ethics, such as honesty about the quality or risks of their products, or by allowing their weaker competitors to freely compete with them, or by telling Congress not to make laws giving them unfair advantages.

To the contrary, nowadays, and especially during COVID-19, big corporations and big government brag about being in partnership with each other. When they make such boasts, the citizenry should quake with fear and then take actions to break up the monopoly that threatens all: collusion between big business and big government. The only worse monopoly than the collusion between business and government is Communism, the ultimate, unfettered monopoly on power that once again threatens the world.

We Never Thought of That!

Earlier in the book, I mentioned how globalists in the media and politics, and perhaps everywhere, like to blame the SARS-CoV-2 for the economic destruction inflicted on the world when, in fact, it is their own totalitarian policies and tactics that have shut down society and economic activity. I also pointed out they show very little genuine concern for the victims of their oppressive policies. Jeffrey Tucker, editorial director of the *American Institute for Economic Research (AIER)*, found additional evidence directly from the mouth of Melinda Gates. Tucker observed: [804]

> In a wide-ranging interview in *The New York Times*, Melinda Gates made the following remarkable statement: "What did surprise us is we hadn't really thought through the economic impacts."

We agree with Tucker's analysis:

> It's a maddening statement, to be sure, as if "economics" is somehow a peripheral concern to the rest of human life and public health. The larger context of the interview reveals the statement to be even more confused. She is somehow under the impression that it is the pandemic and not the lockdowns that are the cause of the economic devastation that includes perhaps 30% of restaurants going under, among many other terrible effects.... It's possible that she actually believes this virus is what tanked the world economy on its own but that is a completely unsustainable proposition.

Tucker characterizes what may be the attitude of many globalist predators:

> To many of these, COVID-19 became their new playground to try out an unprecedented experiment in social and economic management: shutting down travel, businesses, schools, churches, and issuing stay-at-home orders that smack of totalitarian impositions.

Could Bill and Melinda Gates be so lacking in intelligence or economic sense that they could not foresee the havoc they were inflicting upon their fellow humans? No chance. No one functioning on their levels could be that stupid. Many people who become very rich, full of themselves, and powerful are, or rapidly become, morally defective and remote from anything like empathy or fellow feelings for "humanity." They simply do not consider the suffering their grandiose plans will inflict on others. This, of course, is a characteristic of global predators and, to some extent, of all predators—failure to feel concern for their prey. It is probably impossible to exploit people without crushing your capacity for empathy and caring.

If Bill and Melinda profess to love the environment, they must be acutely aware of avoiding doing any harm to it. They would certainly never put the Earth's atmosphere in jeopardy—or would they? On January 13, 2021, the *Washington Examiner* made a chilling report titled "Bill Gates Is Trying to Dim the Sun."[805]

> With the help of Gates' greenbacks, Harvard scientists are attempting to determine whether they can dim sunlight to cool down planet Earth. The administrators of SCoPEx, or Stratospheric Controlled Perturbation Experiment, plan to test[806] their sun-reflecting, particle-spraying balloon in Sweden in 2021, sans particle expulsion. The aim: "SCoPEx is a scientific experiment to advance understanding of stratospheric aerosols that could be relevant to solar geoengineering." Run for your lives.

Global predators are megalomaniacs. That makes us little more than figments of their freakishly destructive self-inflated imaginations. They can imagine and even carry out their global experiments without concern for the harms they may do to us or to Earth.

Killer Mosquitos?

A Gates-funded experiment with genetically altered mosquitos is being conducted in the Florida Keys with government consent but not with the consent of the citizens themselves.[807] Genetically altered male mosquitos, said not to bite, are being released, and when they mate with females, their offspring will die. This will hopefully reduce the mosquito population which will reduce disease vectors.

What could possibly go wrong with a GMO mosquito? Here is one concern voiced by a competitor who makes mosquito traps as an alternative: "The GMO mosquitoes may mutate into a stronger mosquito that can reproduce, which poses a whole new threat."[808]

I checked the detailed Risk Assessment made by the CDC and the risk of a harmful or dangerous mutation in the GMO mosquito must be so very small that it is not even evaluated.[809] Maybe it's just too scary for the CDC to mention. From the CDC's handling of COVID-19, they do not deserve our trust. Should we trust the scientific know-how, caution, and sense of responsibility of the owner of the company making the mosquitoes, Bill Gates?

Meanwhile, we have just unearthed another scary factoid about Gates: He has recently become the largest private owner of farmland in America at 240,000 acres.[810] However, he does not seem to be near the top in land ownership that also includes ranches and undeveloped land. Jeff Bezos, who owns *The Washington Post*, has a 420,000-acre plot that includes an industrial site in Texas. CNN's founder, Ted Turner, who identifies himself as a progressive,[811] owns two million acres of ranch land.[812]

Predators with No Traditional Political Ideologies and Parties

There is a surprising amount of agreement and considerable discussion about the rise of globalism, usually conceived of as individuals and institutions of wealth and power whose ambitions are tied to an international political and economic strategies benefitting "them" rather than the rest of us. Less understood is the degree to which the globalists know no national boundaries or party affiliations and how little solid values they have beyond the central motive of predatory capitalism or predatory Marxism. Many, like Michael Bloomberg, Bill Gates, and Anthony Fauci, are Americans by origin but seem to lack any identification with America and with its founding principles of individual freedom and political liberty. Not too many of them seem faith based. Nearly all, including Bill Gates,[813] are also supporting extremely progressive agendas similar to Biden and Kerry, and nearly all supported Biden's campaign for president.

Many self-identified moderate liberals, conservatives, republicans, and even libertarians are globalists, including the Clinton, Bush, and Koch families. All of them rejected President Trump's populism. They have lost their mooring and become globalists, rising on the flood tide of international predation, making their fortunes from China where Christianity is suppressed, Islamic people are held behind barbwire, abortion is enforced if you have one too many children, where people have died inside their door-bolted apartments, and where slavery is not unheard of. Not to mention, they are using SARS-CoV-2 against us as an unrestricted weapon.

Many liberals and progressives mistakenly hold on to the belief that contemporary globalism is made up of ordinary "greedy capitalists." Perhaps based on that naive idea, some progressives are trying to create their own independent global movements.[814] *HuffPost* has offered "A Progressive Vision of Globalization" as an alternative to what they think is a monopoly of right-wing corporate globalization.[815] All the predatory globalists we have identified reject any semblance of a competitive free-market system. They want to partner with top-down governments to enforce their exploitations, most often cloaked in progressive propaganda. They are leaving America to dwindle while cleaving to the Chinese Communists.

Meanwhile, some of the richest corporations in the world and their employees are taking progressive stands.[816] We have documented how wealthy tech leaders, such as Google, YouTube, Facebook, and Twitter are persistently censoring President Trump and his supporters and promoting progressive agendas and President Biden's new administration.

An article in *The Atlantic*,[817] rather friendly toward its subject, was called "The Rise of the New Global Elite." Globalism, it observed, "helps explain why many of America's other business elites appear so removed from the continuing travails of the U.S. workforce and economy: the global 'nation' in which they increasingly live and work is doing fine—indeed, it's thriving." This globalist self-aggrandizing attitude to the exclusion of more ordinary folks is what brought President Trump to the presidency.

Globalism seeks to transform humanity into a market for its predations without regard for principles such as liberty or the betterment of humanity. For many globalists, the human population is too numerous and too chaotic, so humanity needs some pruning or weeding out if only through attrition and birth control.[818]

The New Normal, Predatory Globalism, and Bill Gates

On March 9, 2020, *The New York Times* released a report, *"Suddenly, the New Normal."*[819] In the post COVID-19 "New Normal" we will all become more comfortable with big government interventions in our lives. A Google search of "New Normal" reveals multiple additional stories in *The New York Times* and various news outlets, as well as untold numbers of videos and even a new TV comedy, *The New Normal*. We have even seen references to the "New Normal" in TV commercials for consumer products.

Bill Gates gave a LinkedIn presentation on or about April 11, 2020. Daniel Roth, editor-in-chief and vice president at *LinkedIn*, described what Gates had to say.[820] Reporting on Gates, Roth said "he's positive the global economy will stay smothered well into 2021: Exponential." Roth went on to share:

> Unlike heart disease, cancer, car accidents or any of the other leading causes of death—you probably saw the charts in the early days of lockdowns—the virus spreads exponentially. The particles from one infected person sneezing, talking, singing can infect 2 to 2.5 more people, who each spread it to 2+ more. ... Quarantine that initial spreader and you're in good shape. But if you wait, you get where we are in the U.S. — with enforced, economy-destroying social distancing to stop the spikes.

Roth continued to quote the threatening presentation by Gates:

> "Well, it's a mistake not to have the entire country take these extreme measures. It's understandable that people find this shocking. They're not used to the exponential growth of a respiratory transmitted virus. They haven't spent time, looking like at [sic] the model that I showed in the TED Talk where you had 30 million deaths and $3 trillion of economic damage just from that one event if you weren't prepared for it. So, it takes time to get used to, but once somebody educates you about it, you're going to shut down."

Roth added that Gates favored shutting down domestic and international travel, stating: "But as long as we're allowing movement, we're all in this together."

Roth described how Gates answered questions from LinkedIn members about the prospects for a vaccine. The Bill & Melinda Gates Foundation has already invested over $100 million on therapeutics and testing and *was funding factories that are developing seven vaccines.*

Further on, Roth quotes Gates on the importance of massive government involvement:

> "The government is called on here to step up. The government did not do its duty to prepare for this well in advance, but now people are focused on it, and we can see different countries responding in different ways.
>
> The scale of the response is completely unprecedented. I mean, you've never had, you know, 10% of GDP-type responses. It's great that that's been done, but will it really reach the people in need? We're still, you know, kind of designing this as we go … This is very unfair in terms of how it impacts people who already had had a lot less wealth."

This description with quotes from Bill Gates provides an extraordinary amount of information about the globalist attitude he exemplifies. A summary of Gates' viewpoint becomes a list of the essential elements of totalitarian globalism:

1. Gates wants us to know the pandemic is going to get "exponentially" worse.

2. He advocates the "entire country takes these extreme measures."

3. He believes everyone who gets "educated" will want to shut down.

4. He has already publicly described a model where one pandemic can cause "30 million deaths and $3 trillion of economic damage."

5. He is very enthusiastic about the investment possibilities and he plans on getting even richer, beginning with $100 million invested in "therapeutic and testing" plus all the potential medications and vaccines he is investing in. His investment in the pandemic probably goes into the multibillions or he wouldn't be so focused on the subject.

6. Increasingly enormous government funding and partnering is required.

In an August 2020 interview, Gates warned of millions dying worldwide and described his increasing investments, now up to $350 million:[821]

> MILLIONS MORE are going to die before the COVID-19 pandemic is over. That is the stark message of Bill Gates, a co-founder of Microsoft and one of the world's largest philanthropists via the Bill & Melinda Gates Foundation, in an interview with Zanny Minton Beddoes, *The Economist's* editor-in-chief, in early August. Most of these deaths, he said, would be caused not by the disease itself, but by the further strain on health care systems and economies that were already struggling. He also lamented the politicization of the response to the virus in America, and the spread of conspiracy theories—some implicating him—both of which have slowed efforts to contain the disease's spread. But he offered reasons for hope in the medium term, predicting that by the end of 2021 a reasonably effective vaccine would be in mass production, and a large enough share of the world's population would be immunized to halt the pandemic in its tracks.

Way back in 2015, Gates called for the world to hold "Germ Games" along the lines of the war games carried out by armies.[822] He explained, "We all need to spend billions to get the vaccine out to save the trillions the economic damage is doing." It's not as if Gates expects to spend all those millions or perhaps billions on our behalf without a payback.

Bill Gates' investment style is to literally invest—to risk money to make money. These large sums of money are not donations to good causes. Even though his stated purpose is to make vaccines inexpensive for the poor,[823] the poor are not even getting them. Vaccines are being distributed far more to the rich countries than to the poor ones. They go where the money can be made through corporate/governments, similar to the U.S. giveaways to Moderna and Pfizer.[824]

Gates does not give money away to the people. He is not sharing or redistributing his wealth. He is not supporting individual freedom or political liberty in these countries. He is making a market out of them. When he goes into partnerships with gigantic corporations to develop treatments and vaccines for the rest of us, the return on the investments goes back into his foundation, vastly increases his operating wealth and hence his power and influence.

Bill Gates is one of the wealthiest men in the world with gigantic new financial investments in drug companies[825] and vaccines,[826] including vaccines to treat COVID-19.[827] In addition to supporting the world's largest industries with billions of dollars in tax-deductible donations from the Bill & Melinda Gates Foundation,[828] Gates is also an advocate for big government interventions and partnerships with industry,[829] making him the archetype of the globalist. Gates himself has been interviewed multiple times about the "new normal."[830] For globalists, COVID-19, with its government and industry partnerships, is a source of great wealth and power. Their activities place great pressure on America to move toward top-down management with increased authoritarianism and ultimately totalitarianism, all in the service of internationalists seeking more wealth, self-aggrandizement, and power.

Anthony Fauci, who works closely with Bill Gates on his vaccine development and marketing, has similarly spoken and testified about the "new normal" which he sees in part dependent on a successful vaccine. According to *PolitiFact*,[831] in an article in defense of Fauci, he is not getting personally rich on COVID-19, but clearly his power and his institute's wealth are growing: "The Bill & Melinda Gates Foundation is indirectly supporting the NIAID by funding a group helping the agency develop a potential coronavirus vaccine. But there is no evidence Fauci himself stands to profit."

A September 14, 2020, interview with Gates described himself as not a "fan" of President Trump but closer to Fauci:[832]

> Although Gates has spoken to President Trump about a number of health issues before the pandemic—and the importance of preparedness—he said his "most active communication" these days is with Francis Collins, the director of the National Institutes of Health, and Anthony Fauci, the director of the National Institute of Allergy and Infectious Diseases.

Fauci's progressive globalism was astonishingly expressed at the conclusion of a recent "scientific" publication in which he blames COVID-19 on humankind and calls for sweeping changes which would require the top-down imposition of global totalitarianism. Here is his summary published in September 2020 which he coauthored with a close associate:[833]

SUMMARY AND CONCLUSIONS

SARS-CoV-2 is a deadly addition to the long list of microbial threats to the human species. It forces us to adapt, react, and reconsider the nature of our relationship to the natural world. Emerging and re-emerging infectious diseases are epiphenomena of human existence and our interactions with each other, and with nature. As human societies grow in size and complexity, we create an endless variety of opportunities for genetically unstable infectious agents to emerge into the unfilled ecologic niches we continue to create. There is nothing new about this situation, except that we now live in a human-dominated world in which our increasingly extreme alterations of the environment induce increasingly extreme backlashes from nature.

Science will surely bring us many lifesaving drugs, vaccines, and diagnostics; however, there is no reason to think that these alone can overcome the threat of ever more frequent and deadly emergences of infectious diseases. Evidence suggests that SARS, MERS, and COVID-19 are only the latest examples of a deadly barrage of coming coronaviruses and other emergences. The COVID-19 pandemic is yet another reminder, added to the rapidly growing archive of historical reminders, that in a human-dominated world, in which our human activities represent aggressive, damaging, and unbalanced interactions with nature, we will increasingly provoke new disease emergences. We remain at risk for the foreseeable future. COVID-19 is among the most vivid wake-up calls in over a century. It should force us to begin to think in earnest and collectively about living in more thoughtful and creative harmony with nature, even as we plan for nature's inevitable, and always unexpected, surprises.

It should be no surprise Fauci's proposals for saving the world from pestilence sound remarkably like the radical Democrat's Green New Deal, or the earlier UN Resolution 2030, or Schwab's Great Reset, now officially adopted by the Biden administration for implementation by John Kerry, a man who has hated America since coming back from the Vietnam War.

How do we know they don't really want to stop emissions and save the world from global warming? *Bloomberg New Economy* newsletter, a big supporter of China, admits that China's 28% contribution to "all global emissions" is double the U.S. and that China's detailed economic planning does not include any serious cutbacks, not even in its huge coal industry.[834] Globalists are not genuinely concerned about the world, or China would be their main enemy. It is all about crippling America—the one remaining but weakened impediment to the interests of global predators.

CHAPTER 27

The Globalist War Against President Donald Trump and America First

T he predatory globalists hate and fear President Trump because they hate and fear all us Americans who show pride in our country and love liberty, including freedom of thought and speech.

An America inspired by President Trump's patriotism and love of liberty is the greatest threat to globalism, so globalists sought to use fear as a tool to subdue the American populace and to defeat President Trump from the time he announced his candidacy to the present. The editorial board of the global predator, *The New York Times*, on October 16, 2020, a few weeks before election day, declared:

President Trump's re-election campaign poses
the greatest threat to American democracy since World War II.

Now we know why the globalist *The Washington Post* enabled Fauci to interfere with the presidential election results by issuing a dire "public health" warning about COVID-19 and President Trump. Here, again, is the *Post's* ominous headline:[835]

"A whole lot of hurt": Fauci warns of COVID-19 surge,
offers blunt assessment of President Trump's response.

Globalists are spreading fear of COVID-19 across the planet to gain power and wealth, which also means knocking down and crushing the only political leader who seems willing and able to oppose them and to stand up for America's middle and working classes.

State and federal officials clamped down the hardest on religious institutions and small businesses because this struck at the very heart of the fabric that unites Americans in opposition to globalism. Social media, major media, big tech, the publishing industry—virtually all the establishment or progressive systems of communication in the U.S.—censored, suppressed, or shut down conservative or anti-globalist political viewpoints. Any criticism of COVID-19 policies or practices was and continues to be excluded from the public dialogue, including discussions of the experimental nature of the vaccines; escalating reports to the CDC of vaccine-related adverse reactions and deaths; the especially outrageous vaccination of young adults, children, pregnant women, and nursing mothers; the exaggerated death threat from SARS-CoV-2 and its mutations; the futility and destructiveness of the shutdowns; and lack of science behind all the draconian public health policies. In addition, criticism of establishment organizations and agencies was forbidden, including Anthony Fauci's NIAID, NIH, FDA, and CDC, as well as WHO. Finally, every effort continues to be made to stamp out awareness of the availability of inexpensive, safe, and effective early, home-based treatments for COVID-19, including preventive treatments—while vaccines are irrationally forced upon every living person as the only protection and solution. Thus the global predators grow more wealthy and powerful, while America is weakened and humanity is crushed.

Predatory globalists see themselves as moving toward an international governance of the world. To continue this push, they use COVID-19 as an opportunity to accustom people to submission and to weaken America and President Trump, regardless of whether he remained in office.

President Trump and Globalism Confront Each Other from the Start

In accepting the nomination for President of the United States at the 2016 Republican National Convention, near the beginning of his speech, candidate

President Trump gave an emotionally powerful statement about the central importance of his opposition to globalism:

> Tonight, I will share with you my plan of action for America. The most important difference between our plan and that of our opponent, is that our plan will put America first. **Americanism, not globalism, will be our credo.** As long as we are led by politicians who will not put America first, then we can be assured that other nations will not treat America with respect—the respect that we deserve.
>
> The American people will come first once again. (Bold added.)

President Trump could not have been more definitive. His opposition to globalism is "the most important difference between our plan and that of our opponent" and "Americanism, not globalism, will be our credo."

The globalist alliance was rudely awakened from its decades of comfort with both Republican and Democrat administrations. Instead of being cozy with the political leaders of the world, this very successful business leader—whom the media cabal portrayed as an inexperienced buffoon, upstart, TV show host—was mounting an effective war against them. If they had any doubts, the globalist coalition, fueled by China and its market opportunities, would quickly learn President Trump was deadly serious about America vs. the Globalists.

With astonishing rapidity, the globalists would come together to attack him from within, and they worked in concert to prevent his election in 2016. When that failed, and Donald J. Trump became President of the United States of America, they escalated their combined assault from the Democratic Party, much of the Republican Party, the mass media, social media and big tech, the courts, multiple billionaires like Gates and Bloomberg, and the Deeply Embedded Bureaucracy, from the FBI and CIA to the State Department. After slowing him down somewhat but failing to oust him with concocted Russia probes and the first one-party impeachment, they manipulated vote counts to prevent his re-election. In the end, they impeached him again and tried him as a private citizen in such an unconstitutional manner that even the feckless Chief Justice of the Supreme Court refused to preside over the trial, which then found him not guilty.

Almost *anything* that strengthens America is a potential threat to globalism because many globalist ends, such as the worldwide redistribution of American industry, is at the expense of America's poor, working people, middle class, and small business owners. American workers lose their jobs, or their real wages remain stagnant or drop because of inflation. Illegal immigration across the southern borders helps big agriculture but harms the poorest of Americans who must compete with them as they settle down throughout the country. Globalism *helps* multibillionaire Americans like Bill Gates, Warren Buffett, and Michael Bloomberg whose wealth and power expand with cheaper labor, weaker regulations, and lower taxes in foreign countries, and with cheaper labor from porous borders in America.

A STRONGER, IDEALISTIC AMERICA IS GOOD FOR HUMANKIND

A strong America is good for the world, but only when America lives by its ideals of liberty and promotes the American Dream. We saved freedom in the world by playing major roles in winning World War I and World War II.

President Trump made America more of a force for stability in much of the world by brokering recognition and opening trade relationships between Israel and some Arab states, by strengthening Israel and recognizing Jerusalem as its capital, and by containing Islamic Jihadism. He also stood firm against North Korea and Chinese imperialism. Under President Trump, American power, sometimes required to act alone, began placing limits on the unbridled exploitation of the American people while protecting many peoples around the world in Israel, Taiwan, Japan, and countless other places.

Laced throughout this book you will see other evidence of his confrontations with the globalists, including Fauci and WHO. It is important to remember that President Trump has been the single most important enemy in the sights of the entire community of global predators. Never before have they felt compelled to deal with such an implacable enemy.

The realization that President Trump declared himself to be the enemy of globalism – "Americanism, not globalism, will be our credo" – clarifies why he may be the most abused and politically trashed president in the history of the United States, as well as one of the most courageous, resilient, and true in keeping the promises he made to the American people.

CHAPTER 28

My Expertise and Background for Evaluating Dr. Fauci and COVID-19 Policies

The gravity of my bill of particulars against Dr. Fauci in the next chapter made me realize that the reader needs to understand my background and expertise in drawing them. In a sentence, I have spent more than 50 years studying the medical-industrial complex, writing books and scientific papers about it, and testifying as a medical expert in the U.S. Congress, before federal agencies, and in state and federal courts on related subjects such as pharmaceutical industry practices and ethics, drug development and monitors, and the FDA; additionally, my career includes two years as a full-time consultant at the National Institute of Mental Health (NIMH) and briefly as a medical-legal consultant to the Federal Aviation Administration (FAA).

My formal resumé contains more details, including my professional publications with 24 medical and popular books and 70-plus peer-reviewed scientific articles and reports.[836] The resumé was written specifically to qualify me as a medical and psychiatric expert in state and federal courts. Here it helps to explain my qualifications for making judicious and informed evaluations of the work of Dr. Fauci. My website, www.breggin.com, presents additional details about my legal and reform work. *The Conscience of Psychiatry: The Reform Work of Peter R. Breggin, MD* more dramatically documents a broad public view of my reform work through the eyes of the media and my fellow professionals up to 2009.

The gravity of my conclusions regarding a physician and his activities made me realize that the reader needs to understand my background and expertise in drawing them.

Early Background for Understanding Federal Institutes

Following the completion of my medical and psychiatric training, I became a Lt. Commander in the U.S. Public Health Service (USPHS) for two years. My post was Full Time Consultant to the National Institute of Mental Health (NIMH), one of the many institutes under the NIH umbrella. Anthony Fauci, who is four years younger than I am, is also a former USPHS officer and he too began working at one of the NIH institutes, NIAID. As an insider, I began my education about the politicization of NIH about the same time Fauci did. I was leaving and he was arriving in 1968 and we never crossed paths.

At the end of the two-year assignment, I was formally invited to remain as a career professional at NIMH. I initially accepted and then declined. I could already see that "official" psychiatry and federal mental health programs were becoming dominated by pharmaceutical industry influence. *Anyone* choosing a career in *any* of the National Institutes of Health would be compelled to conform to or promote the needs and demands of the drug companies. I got out and Fauci decided to make it a career.

After NIMH, I went into private practice with adjunct teaching at institutions such as the Washington School of Psychiatry, Johns Hopkins, George Mason, and the University of Maryland. I began my reform work in earnest in 1972 and soon established the International Center for the Study of Psychiatry and Psychology (ICSPP, a nonprofit) with a highly respected board of directors that included two Congressmen and a Senator, and many top professionals in medicine and psychology.

ROLE AS A MEDICAL EXPERT IN LEGAL CASES AND RELATED ACTIVITIES

In a number of areas related to Fauci's activities and to other issues in this book, I am probably one of the most experienced medical experts in the world. I have testified in untold numbers of depositions and been qualified over 100 times in state and federal courts in the U.S. and a few times in Canada. Most of the cases involve allegedly negligent activities of institutions and doctors that have harmed patients: malpractice suits against doctors and institutions; defective product cases against pharmaceutical companies; and criminal cases in which medications, psychosurgery or electroshock have impaired a patient's judgment or impulse control. Involuntary treatment, involuntary intoxication with psychoactive medications, consent, competence, and other legal principles are often involved. This has prepared me well for examining the harms being inflicted on Americans by the pharmaceutical and medical industries, and Dr. Fauci. I have also written extensively about these issues and been invited to speak about them to conferences of lawyers and healthcare professionals.

Many of my cases have involved a key issue in COVID-19: the FDA's role in the approval of drugs for marketing based on safety and efficacy, as well as the FDA's reporting system for adverse events.

More recently, I have consulted in cases surrounding COVID-19, which can involve the activities of the FDA, CDC, NIH, and public health officials in general. I have written extensive reports for attorneys concerning COVID-19, including evaluating how public figures and their activities are harming Americans. Remarkably, understanding and criticizing the pharmaceutical industry has played a major role in many of my pre-COVID-19 cases as well as my current COVID-19 cases.

In respect to federal agencies, I have been hired as a medical-legal consultant by the Federal Aviation Agency (FAA) and have presented to a board of assistant U.S. attorneys general concerning potential lawsuits against pharmaceutical companies. Several committees of the U.S. House of Representatives and Senate have invited me to testify before them.

In the early 1990s, I was chosen by a consortium of attorneys and confirmed by a federal judge to be the single scientific and medical expert in approximately 150 combined product liability cases against Eli Lilly for alleged fraud in the scientific development and the marketing of Prozac. That was my first of many deep dives into the secret files of a major pharmaceutical company. It was extremely educational. With my wife and coauthor Ginger Ross Breggin, I wrote about what I was learning in *Talking Back to Prozac*, a science-based bestseller.

Because of my role in the Prozac cases, my forensic work became a major aspect of my professional work. I have been a medical and psychiatric expert and consultant in innumerable cases involving malpractice by physicians and negligence on the part of medical institutions and pharmaceutical companies. In work that I continue to do, approximately a hundred cases have gone to trial and multiples more have been settled before trial. Sometimes I am asked to determine if the negligence became so extreme as to require what are called "punitive damages" to punish the defendant. I have also been an expert in dozens of criminal cases to evaluate if crimes were in part induced by medications and other drugs, by head injury and disease, and by trauma and stress.

The combination of my reform and legal work changed the standards of conduct for doctors and manufacturers in several areas of medicine and psychiatry. With the publication of my second medical book in 1983, *Psychiatric Drugs: Hazards to the Brain*, my educational campaign was able to push the FDA in 1984-1985 to toughen the warnings that drug companies must place into their official drug descriptions, called the Full Prescribing Information, on the labels or the package inserts. I accomplished this with media appearances, including an hour-long special with Dan Rather that I developed with his producer. I wrote scientific papers and newspaper reports and testified before the FDA.

During this time, I became very active as a medication expert on almost all the major TV programs: the morning and evening news shows, *Oprah* and *Larry King Live* about a dozen times, *20/20*, *Nightline*, and so on.

During 2000-2004, I initiated and led a successful campaign to upgrade the warnings for antidepressant drugs. It involved similar activities, including writing books and articles, media appearances, and testifying before the FDA. By then, I was also frequently involved in legal cases.

Stopping a New Wave of Lobotomy and Psychiatry

My reform work on a national and then international level began in 1972. In that year, a resurgence of lobotomy and psychosurgery—psychiatric brain mutilation with scalpels or hot electrodes—was being whipped up worldwide by neurosurgeons and psychiatrists who worked with them. I devoted several years to stopping this wholesale resumption of that barbaric practice.[837] My effort entailed extensive research, writing scientific and legal papers, working with the U.S. Congress, appearing on major media, testifying against psychosurgeons in trial, and speaking at conferences, including the American Psychiatric Association and one run by the psychosurgeons themselves.

In 1973 a court injunction was brought against the Michigan Department of Mental Health, a state hospital, and other medical defendants to stop psychosurgery in a state mental hospital. Based on my testimony, the three-judge panel in their written opinion compared the brain operations in state mental hospitals to the heinous medical experiments on inmates in Nazi extermination camps. The Nuremberg Trials after World War II determined these experiments to be crimes against humanity.

That judicial decision, called Kaimowitz v. Department of Mental Health,[838] stopped the horrendous, ongoing practice of psychiatric brain surgery at the National Institutes of Health (NIH), the Veterans Administration (VA), and state hospitals throughout the United States. It almost completely stopped the practice throughout the country and forced stricter oversight on the remaining few.

My work with Congress, including legislation that I wrote for Senators and members of the U.S. House of Representatives, resulted in the creation of a Commission to investigate psychosurgery. The Commission ultimately declared the mutilating brain operations too experimental for clinical use, resulting in the continued cessation of all federal funding for psychosurgery and further discouraged clinical practice.

I was recently featured in a film, *The Minds of Men* by Aaron and Melissa Dykes, viewed by several million. I described those challenging and often frightening years both in the film and in videos containing my entire uncut interview for the film. Both the film and the videos are on my YouTube Channel.[839]

FACING POWERFUL ADVERSARIES

Then, as now, my reform efforts frequently came into direct conflict with some of the most powerful individuals in American medicine and politics.

During the psychosurgery reform project, I went up against William Sweet, MD, the director of neurosurgery at the most respected hospital in the world, Harvard's Massachusetts General Hospital. Dr. Sweet was arguably the titan of American medicine.

Ted Kennedy, the most powerful person in the U.S. Congress, was a major supporter of the psychosurgeons, including his fellow Bostonian William Sweet. I had not realized why Kennedy was so adamantly against my attack on psychosurgery until a brown envelope with no return address arrived. The records and photographs showed that Ted's sister Rosemary had been lobotomized while under her father's control and was languishing in a remote hospital setting—something that was not publicly known at the time.

NIMH and NIH were also involved peripherally and indirectly in these issues. NIH was experimenting with psychosurgery for the control of pain. NIMH, following my consultation with the Assistant Director, would eventually voice concerns and criticism about psychosurgery.

Through a media intermediary, a household name in America at the time, I promised Kennedy not to reveal that secret information of his lobotomized sister, and Kennedy agreed in return to hold hearings about psychosurgery with me as one of the experts. He also stopped any interference with the legislation I had written establishing the Psychosurgery Commission.

Neurosurgeon Sweet had gained federal funding for his project by testifying before Congress that the people involved in urban protests and riots in America's cities in the late 1960s could be cured of their behavior by psychiatric surgery. I disclosed these racist underpinnings in detail in press conferences, the media, newspapers, and a law journal.[840] Sweet's project, which was carried out by a Harvard psychiatrist and another neurosurgeon at Boston City Hospital, lost its federal funding and folded.

Another famous, high-profile neurosurgeon at Massachusetts General refused to stop doing the procedure, which was unsupervised in his private practice. I was the medical expert in two malpractice suits against him on behalf of severely mutilated patients whose spontaneity and personality he had destroyed. Juries in the Boston area almost never went against top doctors at Harvard hospitals, and we lost both cases. Nevertheless, the neurosurgeon's malpractice insurance carrier withdrew coverage and he could no longer perform the surgery.

During the early and mid-1990s, all the neurosurgeons and institutions doing psychosurgery in the U.S. that I identified were stopped from doing the procedures.

Later, two would resume but with heavy oversight, and they are now unpublicized or stopped.

What drove me so hard and turned me into a high-risk, lifelong reformer? One event stands above all others. It gave me nightmares when I first read scientific papers by O. J. Andy at the University of Mississippi in Jackson. He was putting multiple electrodes into the brains of black children as young as age five to experiment on them and to burn holes in their heads. With further investigation, I found out that the children would walk around for months on the wards of the institution with wires hanging out the back of their heads in bunches like grotesque ponytails. More than anything else, the awful plight of these children made me determined never to quit until all these violent, brain-mutilating assaults on children were ended.

I called the chair of the Department of Psychiatry at the University of Mississippi in Jackson who was sincerely shocked. He candidly explained to me that he was unaware of that aspect of Andy's work, even though the neurosurgeon's office was down the hall from him. Remarkably, the psychiatrist organized a successful effort to stop Andy by having the university require Andy to make scientific applications to an oversight committee. The brain butcher stopped his psychosurgery.

The department chair's intervention at the University of Mississippi Medical Center in Jackson is the only time I know where a psychiatrist, neurosurgeon, or any other medical doctor intervened to stop his colleagues from doing these horrendous brain mutilations. The reluctance of doctors to become involved is confirmed by the fact that I was the first physician ever to protest the treatments or to testify in a malpractice suit against a psychiatrist or neurosurgeon involved in psychosurgery. Later, I became the first to testify in a *successful* case, one brought against a neurosurgeon practicing at the Cleveland Clinic who was then forced to stop.

Taking on Electroshock

Skipping ahead decades, in October 2018, I submitted a massively documented, scientific report on electroshock treatment, or ECT, to a California court.[841] (Yes, electroshock is still being done.) My report led the judge to confirm that there is sufficient evidence for brain damage to let the case go to trial. Somatics,

Inc, the shock machine manufacturer, quickly caved in rather than face a trial where the judge had already found the basic argument scientifically credible. The company settled the case for an undisclosed amount and made a formal statement to the FDA that the treatment can cause brain damage.[842]

The legal decision supporting my scientific analysis that ECT causes brain damage has empowered the same attorneys to continue pursuing class-action suits against the two American manufacturers of these machines.[843, 844] I had first begun taking on ECT seriously in 1979 with the publication of my first medical book, *Electroshock: Its Brain Disabling Effects*.[845]

Despite the evidence for lack of effectiveness and for brain damage, my report had to overcome the FDA's bizarrely outrageous decision that electroconvulsive therapy is so harmless that the manufacturers are not required to test shock machines for safety or effectiveness. After our court victory, the FDA rose to the defense of ECT by formally approving it for the first time—entirely without ECT going through testing. It could never have passed safety testing if only because the grossly damaging traumatic brain injury is so easily documented by simply looking at the patients after their individual ECTs. No rational, independent physician believes that repeated episodes of severe traumatic brain injury are harmless.

ECT passes a very intense searing current, typically 800 milliamps at around 400 volts applied for six or eight seconds. It's enough to light and heat a 60-100-watt bulb, to stop a heart, or to damage brain tissue. The doctors inflict this disaster on their patients for an average 8-20 episodes of electrical trauma spaced at two or three per week.

Each "treatment" causes instant unconsciousness followed by coma lasting at least several minutes. The brain is so severely injured by the electrical assault and the resulting massive electrical storm, and other factors, that the EEG goes flat. A flat EEG indicates an absence of detectible electrical activity in the brain and is one sign of brain death. After a series of ECTs, the patient's EEG often remains permanently abnormal with a pattern consistent with chronic brain damage.

I first witnessed ECT as a college student volunteer in a state mental hospital and then during my medical and psychiatric training. After many minutes, the patient awakens from the deep coma with severe concussive-like symptoms, typical of traumatic brain injury (TBI), including headache and a confusion or delirium that commonly requires restraints. There is always severe confusion, complete disorientation, and total or nearly total obliteration of memory for the event and many hours before and after.

Following the first one or two such treatments, many people are very disabled for at least 24 hours. Most patients endure lasting, serious generalized harm to their mind and brain, including unusually severe memory loss and continuing memory dysfunction.

For scientific articles and summaries detailing these dreadful results, see our ECT Resource Center on www.breggin.com.[846] It is easily accessible to the general public as well as to physicians and researchers.

I have evaluated doctors and nurses who permanently lost most of their memory of their educations and their ability to practice after routine ECT. I am proud to be the first psychiatrist to stand up actively against this treatment and to have been the medical expert in the first-ever successful malpractice trial against a doctor for his involvement in ECT.

I have spent this extra time writing about ECT and also about psychosurgery in part to emphasize the tragic truth that physicians and the medical profession have a long history of perpetrating and tolerating horrible abuses by doctors, and it continues to this day in COVID-19. The attacks on me and on my heroic publisher, Dr. Ursula Springer, were a rude awakening to the cost of challenging the medical establishment. These attacks grew even worse when I took on psychosurgery, but never so violent as when I faced the pharmaceutical industry as the medical and scientific expert in cases against the first SSRI antidepressant, Prozac.

Ginger and I Stop a Giant Racist, Eugenical Federal Government Program

As my last illustration, I will summarize our reform project that most nearly paralleled the issues surrounding COVID-19, Dr. Fauci, and his associates. It began in the early 1990s when Ginger and I had been married and working together for nearly ten years. While I was away for speaking engagements in the U.S. and Great Britain, she stayed at home and discovered that a federal psychiatrist, Frederick Goodwin, MD, had made outrageously racist remarks while addressing his annual conference of national leaders in mental health.

At the time, Goodwin was the director of a no-longer-existing agency, the Alcohol, Drug Abuse, and Mental Health Administration (ADAMHA), that oversaw all federal programs related to psychiatry and mental health. In his role as director, Goodwin was considered the most powerful psychiatrist in the federal government and probably in the world.

While I was traveling to Great Britain, Ginger contacted the Black Caucus of the U.S. Congress, with whom I had worked very closely years earlier while stopping lobotomy and other forms of psychiatric surgery. Through the office of Congressman Ron Dellums, she was able to obtain the transcript of Goodwin's speech, and we were stunned by its revelations. Goodwin was organizing a mammoth interagency eugenic program involving all federal health, education, and justice departments to carry out research on "inner city" and "urban children"— code words for African Americans. The research would study Black children from in utero until early adulthood to determine how they were genetically and biologically predisposed to violence. The proposals to Congress for funding were in the works, under Goodwin's powerful leadership.

We were dismayed that this program could have existed and been presented to "national leaders" in mental health, with none of them complaining or leaking it to the press. We found out because an African American secretary overheard Goodman making outrageous racist remarks and reported them to the media. Once again, now with Ginger's help, I worked closely with the Black Caucus of Congress and addressed them at their annual conference.

Because of our several year-long, intensive efforts, many of the ongoing federal eugenical projects were cancelled, a conference about the racist program was cancelled, and the giant funding proposal was stopped. To our surprise, Fred Goodwin was fired from his federal job and left government service.[847]

Still Not Prepared for Fauci

Despite our lifetime of reform work, nothing could have prepared us for confronting the depth and extent of evil lurking beneath the cover of the pandemic and Fauci's "public health" policies and practices. To that evil, we now turn in more detail in the following chapter with my bill of particulars against Anthony Fauci. Thankfully, we are not alone as we have so often felt in the past. Instead, we are part of a growing, worldwide coalition of medical and health professionals, most of whom are coming under heavy attack for standing up against COVID-19 oppression and for promoting the well-being and freedom of Americans and all humanity.

CHAPTER 29

Bill of Particulars Against Dr. Anthony Fauci

Our work researching COVID-19 has led to the moral and political necessity of summarizing in some detail Dr. Anthony Fauci's offenses against America and humanity as a bill of particulars. The following particulars have already been documented in this book, and each one is indicative of treachery and possibly criminality and treason.

But we need a caveat. As powerful as Fauci seems at times, he is doing the bidding of masters far more powerful than he could ever be. He is carrying out the orders of Bill Gates and ultimately of the Chinese Communists, or he would never have the courage or the strength to continue funding research that bloated the billionaires of both the U.S. and China, all the while giving China the capacity to make and spread SARS-CoV-2. The bill of particulars against the global predator billionaires, CEOs, and government leaders has yet to be written.

The style of the bill of particulars is similar to the style I have used for my conclusions concerning negligence in hundreds of medical expert reports, spanning half a century of experience in criminal, malpractice, and corporate negligence cases (product liability cases). These legal reports aim at covering all the essential complaints or charges against the individuals or corporations based on the available evidence. As in this list of accusations against Dr. Fauci, any bill of particulars will always fall short of finding all potential negligence until the legal process called "discovery," along with Freedom of Information Act (FOIA) requests, have been exhausted. In regard to Dr. Fauci, the process of discovery has hardly begun, and Freedom of Information requests by Judicial Watch are being thwarted. Nonetheless, evidence is accumulating of his compliance with the shared views of WHO and Communist China on issues surrounding COVID-19.[848]

While the information in this report might be useful in bringing formal or official charges against Dr. Fauci, this bill of particulars has no legal weight. Its purpose is to summarize many of the findings in *COVID-19 and the Global Predators*. There are 12 categories with multiple particulars:

Bill of Particulars Against Anthony Fauci MD
Regarding His Betrayal of the People of the United States
of America and Humanity

I. Dr. Anthony Fauci, Director of the National Institute of Allergy and Infectious Diseases (NIAID), has abused and continues to abuse his power, becoming and remaining the leading sponsor and funder in the world of gain-of-function research that enabled Communist China to engineer a range of deadly SARS coronaviruses (SARS-CoVs). He carried out these plans in defiance of two American Presidents, Barack Obama and Donald Trump, and against official NIH review policies. These activities caused and contributed to the development of SARS-CoV-2 in American and Chinese Communist labs and its release from the Wuhan Institute of Virology. Because Fauci is the Director of NIAID, he funds and bears primary responsibility for all gain-of-function research making deadly viruses and for all the risks to America and the world associated with these and related activities.

Specifically, we found:

1. Fauci bypassed President Obama's 2014 moratorium on gain-of-function research by deceptively outsourcing funds to the Wuhan Institute of Virology through EcoHealth Alliance. The purpose was to enable the Chinese Communist Party-controlled Wuhan Institute to continue its research and development in the engineering of harmless coronaviruses into SARS-CoVs capable of causing pandemics.

2. In order to fund the Wuhan Institute directly, Fauci elevated the Institute's safety level to the highest rating, enabling the Communist-run facility to receive the outsourced funding for arguably the most dangerous kind of research in the world. Fauci did this while knowing that the Wuhan Institute of Virology,

like other similar Chinese Communist facilities, had a bad safety record and that deadly SARS-CoV viruses had, on at least four earlier occasions, escaped from its Beijing equivalent.

3. Fauci grossly defied Obama's 2014 moratorium by continuing to fund and support NIAID's major gain-of-function research project at the University of North Carolina at Chapel Hill. The studies, which began earlier, were funded and published in 2015 and 2016, years when the moratorium was supposed to be enforced. The gain-of-function research was brazenly published in these papers, and the 2015 publication even raised questions about its own legality under the moratorium. The 2015 study described the increased risk of death in some vaccinated mice when the animals were later exposed to the virus.

4. In 2017, toward the end of President Trump's first year in office, Fauci and NIH Director Francis S. Collins, with no presidential involvement, illegally overturned President Obama's moratorium. Then they illegally bypassed the established internal review process with the designated Review Framework and began funding new projects in America and at the Wuhan Institute.[849] Dr. Fauci, along with Dr. Collins, compromised the safety of all Americans and humanity, first by increasing the risks of leaks and second by enhancing the ability of the Chinese Communists to conduct biological warfare.

5. Fauci's funding for gain-of-function research was widely distributed. Some funds went directly to top Chinese scientists in their own projects at the Wuhan Institute of Virology and other funds went to Chinese scientists in collaborative projects with Americans. In every case, the funding and the research became part of the Chinese Communist effort to develop defensive and offensive biological weapons.

6. Fauci funded and collaborated with Chinese scientists and the Wuhan Institute at a time when it was already well established that both the scientists and their Institute were under the control of the People's Liberation Army (PLA) and the Chinese Communist Party. Since January 2020, the Institute is directed by a military officer from the PLA's biomedical warfare section. In addition, under the Chinese Communist policy of Military-Civil Fusion, all Chinese scientists were and are required to report to and collaborate with the PLA and the Party.

7. While knowing SARS-CoV-2 was released (whether accidentally or on purpose) from the Wuhan Institute, Fauci has continued to ridicule and dismiss such a possibility in order to protect his gain-of-function research

and Communist China. This has disrupted American foreign policy, given misinformation to the American and worldwide public, and endangered the world.

8. Fauci continues to support gain-of-function research in the United States, including at the University of North Carolina at Chapel Hill, despite its flawed security procedures. In particular, Fauci has enabled the program to make and to possess a new and more virulent SARS-CoV that causes deadly encephalitis. He has done so by circumventing Presidents Obama and Trump.

9. Fauci has repeatedly minimized the very real risk of gain-of-function research leading to deadly escapes. He never mentions the multiple times that deadly pathogens have escaped from Chinese labs, even though SARS-CoV-2 is at least the fourth documented escape of SARS-CoV in China. He ignores the Chinese researchers who have disappeared or whose publications have been suppressed because they documented the direct line of progression from earlier man-made Chinese SARS-CoVs to SARS-CoV-2.

II. Anthony Fauci has abused and continues to abuse his power by suppressing inexpensive, safe, and effective treatments for COVID-19 in order to promote very expensive, unsafe, experimental, and potentially ineffective treatments and especially vaccines. His successful goal is to increase the wealth and power of his collaborators, including the pharmaceutical industry and investors like Bill Gates and other global predators. In doing so, he directly caused and continues to cause a large proportion of all COVID-19 deaths in America and throughout the world.

Specifically, we found:

1. When the U.S. was seemingly overwhelmed by COVID-19, Fauci thwarted all effective, early, home-based treatments, especially the inexpensive, safe, and effective hydroxychloroquine, ivermectin, and related therapeutics. He continues to stop inexpensive early treatments from being made readily available for the purpose of enriching his global predator colleagues in the pharmaceutical industry and their investors like Bill Gates, as well as the power and wealth of his own NIAID. He coordinated these efforts with the FDA and CDC, eventually resulting in the entire medical and scientific establishment's aligning against proven, early, home-based, lifesaving treatments. *The size of this crime against humanity is indicated by the extremely positive results of proper early treatment, which has reduced hospitalization by about 75% and death by more than 85%.*[850]

2. With no legal or traditional basis, Fauci assumed for himself the power to tell American physicians that they could not use specific FDA-approved

drugs to treat COVID-19. In this, he collaborated with the FDA and the CDC, lending authority to their also unprecedented, illegal attempts to control the practice of medicine and the doctor-patient relationship.

3. Fauci criticized and argued with President Trump in public about early, safe, and effective treatment with hydroxychloroquine about which the President was right. In the process, he undermined President Trump's sworn and constitutional efforts to protect and serve the American people by providing them hydroxychloroquine, including the large federal stores of the drug that were and remain wasted.[851]

4. These efforts by Fauci have led to physicians who are devoted to the early treatment of COVID-19 being harassed and persecuted by their places of employment and by other agencies and organizations, sometimes costing them their jobs and professional status.

III. Fauci has abused his power by rushing through EUAs extremely expensive, highly experimental, very dangerous, and inadequately studied or tested vaccines. In doing this, he supports the investments of Bill Gates, on whose elite vaccine Leadership Council he serves. At the same time, he supports the pharmaceutical industry with whom he is intimate and its vast network of investors, including the Chinese Communists, to whom he has been catering. His role is that of chief enforcer for building the wealth, glory, and power of the global predators in the arena of medicine, public health, and ultimately politics in America and worldwide.

Specifically, we found:

1. Fauci funded, encouraged, and managed the development and distribution of the mRNA and DNA COVID-19 vaccines.

2. As the enforcer of the strategy of Bill Gates and other global predators to accumulate even vaster wealth and power, he set the tone for pressuring and coercing people to take these dangerous and experimental intrusions into their bodies.

3. He promoted inflicting these vaccines on the elderly, on children and youth, on pregnant women, on women with nursing infants, and on women whose reproductive capacities were at risk. Never before in history have vaccines been used on these populations without thorough evaluation.

4. He allowed and supported continued vaccinations even after the number of reported vaccine-related deaths rose from the hundreds into the many thousands.

5. Anthony Fauci bears responsibility for all the harms being inflicted by these vaccines now and in the future, and perhaps even for future generations if their genetic endowments are compromised. He is responsible for tens of thousands of vaccine-related deaths that have already occurred worldwide. He is responsible for various harms that could have been anticipated from these vaccines. Further, he is responsible for those that could not be anticipated, because they could have been avoided by never inflicting these vaccines on human experimental subjects or on the hundreds of millions of people worldwide.

IV. Acting illegally in his capacity as a career government employee, Anthony Fauci took public, political positions against President Donald Trump and for presidential candidate Joe Biden before election day.

Specifically, we found:

1. Fauci made political statements with dire warnings against President Trump's reelection directly to the major media, including a widely disseminated interview 48 hours before election day in *The Washington Post*.[852] These acts were almost certainly in violation of the federal Hatch Act which prohibits government employees from using "their official title or authority when engaging in political activity." Fauci was under particularly severe limitations on political speech as a Career Senior Executive Service (SES) employee.

2. Simultaneously, in the same sources, Fauci supported presidential candidate Joe Biden.[853]

Fauci's widely publicized interview with *The Washington Post*, plus his many other public expressions of opposition to Trump, undoubtedly influenced the election in favor of President Biden.

V. Again acting illegally in his role as a Career Senior Executive Services government employee, Fauci systematically interfered with President Trump making and implementing policies that served the American people and humanity.

Specifically, we found:

1. Fauci spread confusing, contradictory information, for example, first saying masks were unnecessary[854] and later arguing that they are desperately needed.

2. When the pandemic struck, Fauci worked with the World Health Organization (WHO), Bill Gates, and other global predators to cover up its severity and its origin in the Wuhan Institute of Virology, protecting the Chinese

Communists, and allowing the unchecked spread of the virus throughout the world.

3. Emails obtained by Judicial Watch through FOIA confirm that NIH, along with WHO, allowed Communist China to control NIH and WHO's public statements. Fauci's own emails have not been fully released, but he is very likely involved in these communications.

4. Fauci resisted President Trump's policy of stopping travel from China early in COVID-19, arguing it was unnecessary and harmful when it was spreading SARS-CoV-2 to the United States at an accelerated rate. Fauci did this despite arguing that slowing the spread of the disease was critical to avoid overloading the healthcare system. Fortunately, President Trump's decision prevailed, limiting some of the damage inflicted by Fauci's opposition.

See additional illustrations of Fauci's interference with President Trump's programs in Parts I-IV of this bill of particulars.

VI. Fauci predicted the coming coronavirus pandemic, indicating foreknowledge of the event, but he failed to share the source or evidence for his predictions and advanced knowledge.

Specifically, we found:

1. As documented in Chapter 16, Anthony Fauci openly expressed certainty and foreknowledge of a coming pandemic. He expressed such certainty that President-elect Donald Trump was about to be "surprised" by a pandemic, he must have had foreknowledge of the timing of the pandemic. His repeated statements occurred at a conference on January 10, 2017, three days before President Trump's inauguration. The purpose of the conference was to prepare for the predicted, upcoming pandemic about which Fauci was so certain.

2. At the conference, Fauci described preparing for his role as COVID-19 Czar in advance, including being "friendly" to the drug companies. He stated he would minimize their "risks" in rushing through treatments and vaccines in the coming pandemic. Remarkably, Fauci communicated all this without actively involving members of President Trump's incoming administration in the conference. In the next few busy days leading to President Trump's inauguration, the President, and his staff, unless alerted and involved in advance, would have given no attention to a conference at a local university. Thus, the new administration was kept uninformed about Fauci's foreknowledge and

preparations for what was about to descend upon the President and the people of America and the world.

VII. While suppressing available treatments, Fauci promoted fast-tracking with government financial support for drugs and vaccines that were and are experimental, unproven, and highly dangerous, but enormously profitable.

Specifically, we found:

1. Fauci promoted clinical trials for remdesivir even though he knew that the drug had already been proven useless and highly dangerous in randomized controlled clinical trials for Ebola and for SARS-CoV-2. The death rate of patients on remdesivir in the Ebola trials exceeded that of other medications, and it had to be withdrawn from the study to save lives. In the prior SARS-CoV-2 study, remdesivir was ineffective and worsened the respiratory condition in many patients.

2. When the 2020 NIAID remdesivir study was coming out negative, Fauci lowered the endpoints or standards for success for the clinical trials several times, finally making them meaningless and misleading. Then he declared the drug successful but ended trials abruptly, before completion, by breaking the double-blind standard, essentially closing it down prematurely. Any attempt to analyze the safety or effectiveness became impossible.

3. Fauci created an NIH Coronavirus Committee stacked with members who were on the payroll of Gilead, the developer of remdesivir, leading the committee to approve remdesivir and to reject hydroxychloroquine before NIAID or NIH conducted any studies. The committee took these positions even though remdesivir had a poor safety and effectiveness record with one earlier failed clinical trial and no successful clinical trials. At the same time, hydroxychloroquine was rejected despite positive clinical trials and an amazing safety record.

4. As noted, Fauci ended the remdesivir study before it was scheduled to finish, corrupting the data. Still, he managed to have it published as if it were completed, and he gained eventual standard FDA approval for safety and effectiveness under the name Veklury.

Fauci's unethical misdeeds in respect to the controlled clinical trials for remdesivir have been further summarized in this book in Chapter 10.

VIII. Anthony Fauci lied under oath at a U.S. Senate hearing on May 11, 2021, when he denied that NIAID or NIH had funded or were continuing to fund gain-of-function research.

Specifically, we found:

Fauci's NIAID, NIH and other U.S. agencies funded gain-of-function research by scientists working at the Wuhan Institute in China. NIH and NIAID have been funding gain-of-function for many years at North Carolina Chapel Hill and at the Wuhan Institute. He funded some Chinese researchers when they were working in projects with Americans and some when they were in exclusively Chinese projects. While Fauci's NIAID and NIH may not have directly funded the collaboration with Chinese scientists who were working with North Carolina researchers in Menachery et al.'s 2015 and 2016 publication, Fauci was the leader of these collaborative programs with the Chinese Communists on conducting gain-of-function research that provides basic science required for making potential biological weapons, including SARS-CoV-2 (Chapter 2).

IX. Anthony Fauci knew years ahead of time that a coronavirus pandemic was being planned and announced in January 2017 that it would definitely occur during President Trump's administration.

We have documented the enormous investments of time and money made by Bill Gates, Moderna, Pfizer, CEPI, and other groups with the aim of developing Operation Warp Speed-like projects for an anticipated SARS-CoV pandemic and how this enabled Pfizer and Moderna to be numbers one and two in the approval of their EUAs. Although this particular allegation and the one that follows in section *XI* are the most controversial, we believe there is substantial and convincing evidence to support them. (Also see Part VI of this Bill of Particulars.)

We know that Fauci was appointed to Bill Gates' original vaccine Leadership Council in December 2010 and so was in on the beginning of the great worldwide push for vaccines. As such Fauci was in regular communication with Bill Gates as part of his small, elite vaccine advisory group. We also know Gates was closely in touch with Fauci during the Trump administration.[855] Fauci was in charge of implementing the COVID-19 vaccine plans in the United States.

X. Anthony Fauci knew years ahead of time that no successful vaccine had ever been made for SARS-CoV. He specifically knew that animal research from 2008-2020 confirmed that animals vaccinated with SARS-CoV RNA and DNA vaccines were then at serious risk of becoming ill or dying after exposure to SARS-CoV. Although NIAID and NIH had funded some of this research, he ignored it and the implicit and explicit warnings expressed in it. He also knew from his experience with AIDS that enabling drug companies to spend great amounts on vaccine research did not necessarily lead to success. Instead, he forged head to get EUA approval for them. This has already led to tens of thousands of vaccine-related deaths.

Fauci participated in the multi-agency refusal to examine the reports of death made to the CDC's VAERS. With his central role on the White House Coronavirus Task Force and as Director of NIAID, Fauci was in the best position to make sure that the vaccines were not given an EUA in light of the animal research.

XI. Anthony Fauci has a recurrent history of similarly mishandling pandemic threats to increase the wealth and power of NIAID and the pharmaceutical industry.

Specifically, we found:

1. Fauci has a similar history with AIDS of endangering Americans and the worldwide public by depriving them of early treatment with inexpensive, safe, and effective sulfa drugs for AIDS-related pneumonia, the main cause of death.

2. Fauci delayed the use and development of treatments for AIDS while pushing for vaccinations, all of which failed. Despite the fanfare, no vaccine was ever developed. This is probably a major reason for his desperate fast-tracking of SARS-CoV-2 vaccines outside the normal FDA-approval process—he knew they were likely to fail or to prove unsafe if normally evaluated.

3. In 2014, Fauci risked spreading the Ebola epidemic to America by trying to prevent state governors from keeping American healthcare workers quarantined after reentering the country following heavy exposures to the deadly disease.

XII. Anthony Fauci's abuse of power as Director of NIAID has broad and continuing negative effects in the social, economic, and political arenas in America and worldwide. Much of the book COVID-19 and the Global Predators *elaborates on these innumerable harms in which Fauci has played a role by implementing the interests and strategies of the global predators from Bill Gates to Communist China.*

Specifically, we found:

1. Fauci's activities vastly increased the wealth and power of the drug companies and globalist billionaire investors, especially so in China, where the wealth of that country's 450 billionaires rose by 60% in 2020. At least according to the Chinese Communist Party's data, its economy was the only one in the world to have positive if meager growth in 2020.[856]

2. Fauci's activities weakened our nation as a democratic republic, pushing the U.S. toward a top-down, one-party state. That dire totalitarian condition is now being increasingly implemented under the Biden administration at the federal level in concert with Fauci and the global predators, all of them pushing fear to keep increasing their multibillion-dollar vaccine boondoggle and more top-down government.

3. Fauci's activities increased authoritarianism and totalitarianism in America on the federal, state, and local level, making the population more obedient and docile. This is key to the goal of increased governance by global predators and the Chinese Communist Party.

4. Fauci's overall enumerated activities strengthened the Chinese Communists while undermining America's sovereignty and the freedom of its citizens. He put China's reputation, power, and well-being above that of the United States of America.

5. Fauci planned, pushed for, and presided over a vast and catastrophic shutdown of the American economy and society with very little precedent and no medical or scientific basis.

6. Fauci bears a large portion of responsibility for the vast misery and destruction caused by shutdowns in the U.S. and elsewhere that continue to have an infinite number of compounding and spreading negative effects, including the following:

- Many daily deprivations and hardships.

- Social isolation, especially of people living alone or in confinement.

- Family stresses within their homelife, with increased divorces.

- Loss of contact with friends and family.

- Loss of educational opportunities from preschool through graduate school.

- Aborted careers and opportunities.

- Widespread unemployment.

- Loss of savings.

- Increased misery and death among our older citizens, especially those in care facilities.

- Diminished healthcare visits for cancer treatments, acute illnesses, and other non-COVID-19 medical problems.

- Loss of religious expression and community.

- Fear and distrust of the government and of other people.

- Unwarranted, exaggerated fears of COVID-19.

- Increased apathy, anxiety, depression, and suicidality in the general population.

- In particular, a much higher rate of emotional distress and completed suicides in adolescents and young adults compared to almost no deaths from COVID-19.

- Developmental delays and losses among our children, who may never fully recover from the deprivations and fears.

- An overall diminishment of the quality of life and the freedom of Americans.

- The overall degradation of American society and politics by authoritarian and totalitarian public health measures.

Finally, it must be stressed again that Fauci is the visible henchman, the front man and enforcer for the global predators, all of whom share responsibility for his actions, or for not stopping his actions, and all of whom are wholly responsible for their personal exploitation of humanity in myriad ways.

From Anthony Fauci, Bill Gates, Michael Bloomberg, and Klaus Schwab to those who run corporations, agencies, and governments—these predators think and act globally, their crimes are global, and they should be charged with crimes against humanity.

END OF BILL OF PARTICULARS

To this overall bill of particulars, we add the additional discussions of Fauci's destructive actions as described in the Executive Summary and Report of October 19, 2020, by Peter and Ginger Breggin titled "Dr. Fauci's COVID-19 Treachery with Chilling Ties to the Chinese Military."[857] We also include information in the very lengthy legal report by Peter R. Breggin MD submitted to attorney Tom Renz in support of his injunction and legal case to stop the lengthy declaration of emergency in Ohio. The August 30, 2020 report is titled "COVID-19 & Public Health Totalitarianism: Untoward Effects on Individuals, Institutions and Society."[858]

CONCLUSION: BEYOND THIS BILL OF PARTICULARS

The bill of particulars listing offenses by Anthony Fauci—bolstered by the information in this book and in our two reports—raises serious issues and questions about treachery, illegality, and treason. It is consistent with deliberate mass murder on the part of Anthony Fauci, as charged by Vladimir "Zev" Zelenko MD in his introduction to this book.

Although our bill of particulars is against Anthony Fauci, the predators Fauci represents—from Bill Gates to Xi Jinping—share responsibility for the COVID-19 catastrophe. If any one of these global predators had publicly denounced the assault on humanity, that single person could have saved millions of lives and changed history forever.

Imagine if Fauci had taken the microphone during one of the White House press briefings and spoken the truth: "Mr. President, I will need protection and I may still be killed for saying this, but I must declare that COVID-19 is a biological warfare attack with the aim of taking down your presidency and the United States—and for bringing the rest of the world to its knees. Our enemy, the Chinese Communists, will be strengthened and, at the same time, many global billionaires and corporations are going to make a financial killing. Mr. President, you need to call out the military and convene Congress now. I will back you in declaring a state of emergency. The United States needs to marshal all its forces in self-defense. And, in the meanwhile, there is wonderful news for humanity: There are excellent early treatments for COVID-19 and we still have time to help millions reduce or avoid their suffering, hospitalizations, or deaths.

PART FOUR

Recovering Our Liberty

CHAPTER 30

Terrorizing Us to Tame Us

T he *Chronology and Overview* at the end of our book tracks the planning for COVID-19 and all the elements of Operation Warp Speed and the Great Reset beginning in 2015-2017 with enormous funding and planning by Bill Gates, Moderna, Schwab, U.S. government agencies, and many others. These vast expenditures of money, energy, and time can only be explained by foreknowledge that a coronavirus pandemic would be released in the near future. Otherwise, all their efforts would be wasted.

We have found that ten days before Donald Trump became President, Anthony Fauci actually guaranteed with "certainty" in a public health address to colleagues that the pandemic would strike during the Trump Administration (Chapter 16). They must have realized that President Trump, if not destroyed, would doom their globalists ambitions by empowering America as a free, independent, and patriotic people obedient to none in the universe except God.

Most unbelievable of all, we found that global predators—individuals, government agencies, journals, university departments, and corporations—have documented ties to the officially atheist Chinese Communist Party. Worse than we ever imagined, they share with the Communists a desire to weaken America in order to strengthen their predatory globalist activities. Together, they have used COVID-19 to increase their wealth and power, and to subdue America.

To accomplish their aim of increasing top-down government and intimidating America, they needed to use fear. The tactics and threats used by Bill Gates, Schwab, Fauci's NIAID, NIH, FDA, WHO, the Chinese Communists and many others were and continue to be calculated to make everyone docile from fear. Using guilt, they also paralyzed us from opposing the "necessary" interventions—from masks to vaccines.

It is impossible to take control over the world, or even a few countries at a time, without using fear and terror as the central method. When so many nations, religions, and cultures are involved, it becomes impossible to appeal in a positive way to their unique and conflicting values and ideals.

Faced with multiple cultures, it becomes necessary to force them to react out of fear to some overwhelming threat. A universal human motivation must be created, something much more powerful than a desire for unity, fellowship, or a belief in a loving God. The universal motivation must be terrifying and awesome, like a great plague. If the plague turns out to be not so awesome, they will have to exaggerate and lie to make the whole sufficiently terrorized and submissive.

We Have Known Fear Before

Being afraid is certainly not a new experience for Americans or for any other nation on Earth. We were afraid during World War II and under the threat of atomic war during the Cold War, during the Cuban Missile Crisis under President Kennedy, and then during 9/11 under President Bush, all of which this author lived through. Not even those terrifying, existential threats ever led to the kind of forced social transformations we have now been experiencing.

As a child in World War II, I remember when we lived on the South Shore of Long Island near the ocean where German subs sank our ships, and I recall as a child seeing debris from our life rafts on the shoreline. One poignant find was a long bamboo pole stuck through a cork float with a red rag tied to the top. I still wonder about the sailors who made the makeshift distress signal while floating about, probably far off the shore. I remember turning out all the lights at night and my father putting on his helmet to patrol our street as an air warden. I never remember anyone telling us to isolate at home or not to go to school, and except for families with loved ones in the military, there was probably nothing like today's pervasive, invasive fear creeping into our private lives.

I also remember being ten years old as World War II was coming to an end, watching a *Movietone News* feature in a theater with my parents before the family show came on. The unanticipated feature was about the liberation of a Nazi extermination camp where Jews had been slaughtered and thrown into heaps like animal carcasses. In a few seconds, this young Jewish boy was thrown into a world of horror beyond any ability to grasp or to integrate into his life—with an underlying terror of what might happen in the future to him and anyone he loved.

When I was 12, a polio epidemic struck Long Island and killed a friend of mine two days after we had been wrestling. Someday, perhaps, I will write about how the experience and the horror of learning about the Holocaust, transformed my life and eventually contributed to my becoming a reformer.

I have lived long enough to have experienced many potential tragedies for our entire nation, including the Cuban Missile Crisis when we feared we might all be incinerated in an atomic holocaust. Later on came the assassinations of President John F. Kennedy, his brother Robert F. Kennedy, and most crushing to me, Martin Luther King, Jr. I remember the smoke rising from downtown Washington, DC, as the protests turned into riots that burned out so many storefronts in the Black community and set back the neighborhoods for many years.

The Vietnam War spread terror through men young enough to be drafted, and I had friends who fled to Canada.

Then, of course, much more recently there was 9/11. Living on the outskirts of Washington, DC, I saw smoke rise once again from where the airplane filled with passengers dove into the Pentagon. After, cars jammed on nearby Wisconsin Avenue as people tried to flee at a snail's pace. Going against the traffic, driving toward the very area everyone was fleeing, Ginger and I went downtown so I could appear on local TV to reassure adults and to help them reassure their children.

So, I have known fears generated by national calamities and the like, enough so that COVID-19 does not dismay me as much as many less-experienced young people. But I realize, especially for younger people, the size of threat is new and just as demoralizing as what my generation had gone through.

Nonetheless, in predatory globalism—whether from billionaires or Communist China—we now face a threat that is not only external, but is also profoundly invasive, infesting all our institutions. COVID-19 treacheries seem more insidiously threatening than anything we have previously encountered as a nation, and it can be argued they have resulted in the greatest loss of personal freedom and political liberty in America's history.

The Difference is the Government's Response

In all of the crises I have experienced through these many decades, this is the first one that the government, instead of reassuring us, has tried to terrify us. Where all other Presidents of the United States have been praised for calming the anxieties of the people, only President Donald Trump, in my memory, has been criticized for wanting to reassure us and avoid "panic."[859]

One of the biggest differences between COVID-19 and the earlier national crises I had been through is the behavior of our leadership. I do not recall our leaders ever trying to make us dreadfully afraid in order to control us. Almost always, our leaders have tried to be reassuring, whether it was Roosevelt during World War II with his fireside chats or Kennedy during the Cuban Missile Crisis. On 9/11, President Bush did not try to whip up fear; instead, he went down to the smoldering wreckage of the Twin Towers to inspire and encourage us.

President Trump behaved in the same reassuring manner, trying not to exaggerate the threat and reassuring us with daily briefings over a long period. Aside from President Trump, almost everyone else, from the media to Joe Biden, Bill Gates, and Anthony Fauci, have been trying to terrify us. That is one of the reasons COVID-19 has been so frightening to so many people. An avalanche of propaganda has been launched at us to make us frightened, a tactic not usually used as pervasively in democratic republics as it is in tyrannies.

There is yet another reason COVID-19 has been so frightening to people. Many of the country's scientific and political leaders have demanded a kind of sacrifice and compliance that I had never before witnessed. We had rationing of food, gasoline, and other commodities such as sugar and butter that were deemed vital to the World War II effort, but never this degree of conformity in our most personal aspects of living.

What is happening today is unprecedented in democracies and on a worldwide basis. Public health expert David Halperin observed, "a palpable climate of confusion and anxiety pervades" and a stunning indication is the Johns Hopkins University Coronavirus Resource Center website is recording some four billion hits a day!"[860]

Fauci's Fear Mongering

Fear and even terror are fundamental tools of top-down control over a population. As this report was being finished, a new wave of fear and confusion was inspired by Anthony Fauci, director of NIH's Institute for Allergy and Infectious Diseases. Apparently, in his absence, while he was undergoing surgery, the CDC determined it was no longer necessary to test asymptomatic people for the SARS-CoV-2—an action consistent with relieving fear and opening the nation's life and economy.

When he heard about the new, loosened guidelines, Fauci rose to the occasion, pointedly describing how he was "unconscious" when the decision was made without him and warning about dire results. *The New York Times* raised the fear level by blasting the CDC's decision, stating, "A more lax approach to testing, experts said, could delay crucial treatments, as well as obscure, or even hasten, the coronavirus's spread in the community."[861] Fortunately, other sources found numerous experts to support the CDC's decision.[862]

Labor Day 2020 was turned from a celebration to a feast of fearmongers. As I was writing this section, my unlisted hardline phone rang. I use it exclusively for radio media that prefer interviewing me on old-fashioned phones rather than less reliable cellular phones. Out of curiosity, I picked up the call, expecting spam. What I got was a recorded warning from my county health director instructing me to wear masks and to avoid crowds on the holiday, stating that any size crowd was a risk for increasing the spread of the highly infectious coronavirus.

As people brushed up their courage to enjoy Labor Day with their friends and families, Fauci gave out dire warnings generating headlines like this one:[863]

Fauci Warns 7 States to Take Extra Holiday Precautions Against COVID-19 Surge

Since Americans were beginning to overcome their fears, Fauci threw in some guilt:

> "You want to be part of the solution, not part of the problem," says Fauci as antsy Americans plan for Labor Day weekend.

What "solution" is Fauci talking about? Our medical system and hospitals were not at that time being stressed to the limit, compromising healthcare. Slowing down the virus at the expense of creating a totalitarian society with dismembered social life, closed schools, and a collapsed economy was not exactly a good solution. Fauci was trying to extend the length of the plague while increasing public fear—all toward ends that will become clearer as we proceed.

As if planning ahead for this Labor Day weekend, Fauci published an article in a scientific journal on September 3, 2020, titled "Emerging Pandemic Diseases: How We Got to COVID-19."[864] With first author David Morens, one of his staff, Fauci uses the introduction to announce his official version of the pandemic—i.e., he was right all along, the virus was and remains a catastrophic death threat, and there was and is no other way but a massive shutdown to deal with the virus.

Sowing Fear from Death Counts to Case Counts

The deaths were also slowing down so much that Fauci no longer mentioned them as a threat in his public pronouncements. As the rate of lost lives declined and held steady at a lower level,[865] he and others gave up on counting the dead as a deterrent. Instead, as illustrated in the Labor Day headlines, he began to emphasize the growing number of *reported cases* rather than the death rate yet never explaining what was wrong with the process of communities developing herd immunity.

In desperation to scare people, the methods of counting cases became ridiculously inflated. Under CDC guidance, a valid case of COVID-19 can now be counted if someone has, in effect, seen someone who had seen someone who may have possibly had COVID-19. We described this absurd situation earlier in our blog/report, *CDC Surges Covid-19 Stats*[866] and further analyzed these manipulations in Chapter 8.

Organizing to Terrorize America and the World

How pervasive and organized has public health intimidation become? In the field of public health, the broad subject called "fear appeal" is now an academic discipline. Characteristic of the trend we are describing of seemingly value-free means of achieving influence and power, "fear appeal" is not a pejorative or negative descriptor but rather a field of serious and respected study. It is a kind of doublespeak, using fear to make an appeal to people.

What is fear appeal? Definitions basically agree it's scaring the hell out of people to get them to do what you want. Here is one definition and description from a 2015 American Psychological Association press release touting a new scientific article:[867]

Fear-Based Appeals Effective at Changing Attitudes, Behaviors After All

WASHINGTON—Fear-based appeals appear to be effective at influencing attitudes and behaviors, especially among women, according to a comprehensive review of over 50 years of research on the topic, published by the American Psychological Association. ...

Fear appeals are persuasive messages that emphasize the potential danger and harm that will befall individuals if they do not adopt the messages' recommendations. While these types of messages are commonly used in political, public health and commercial advertising campaigns (e.g., smoking will kill you, Candidate A will destroy the economy), their use is controversial as academics continue to debate their effectiveness.

Fear appeal is the use of scare tactics. In studies, it is strongly associated with public health and politics. The study being touted by the American Psychological Association itself concluded: "Fear appeals are effective" (p. 1196).[868]

We did not find any concerns expressed either in the American Psychological Association laudatory press release or in the article about the ethics of scaring people into conformity with your message. In a world devoid of ethical restraint and oriented toward power, scare tactics become one more "objective" or scientific approach to influencing people.

A 2014 study of "Sixty years of fear appeal research: Current state of the evidence"[869] examined whether arousing fear in people helps them develop or conform to standards of personal hygiene or self-care and to conform to public health laws or guidelines. The authors concluded, "presenting threatening health information aimed at increasing risk perceptions and fear arousal" (p. 63) does not lead people to take better care of themselves or to respond to public health requirements.

At first, this seemed like a rational and humane approach until I read further. These public health scientists were not against frightening a population to force it to accept their policies and practices. They wanted to make fear more effective. There were two missing ingredients in the various other approaches to intimidating people. According to the study, public health officials must be better at *"convincing people that they are personally susceptible to the threat"* (p. 68). Having personally intimidated us, they must then provide us "instruction on how to successfully implement the recommended actions" (p. 68). That is, after they have personally scared us, our leaders must *tell us exactly what to do!* The authors recommend this without regard for our rights as citizens of a democracy inspired by the Declaration of Independence and built on the Constitution and the Bill of Rights.

The authors of this scientific paper on the most effective way to scare people into conforming to public health demands seem unconcerned about the collateral psychological or political damage of their recommendations. This is how much the use of threat is accepted in the literature and among academic public-health thinkers. Imagine how much more blatantly brutal some of the less academically restrained public health officials may feel.

There are some in the scientific community who find the use of fear to be appalling. In their June 11, 2020, publication, Jeni A. Stolow and three public health colleagues[870] wrote a heartfelt abstract to their article "How Fear Appeal Approaches in COVID-19 Health Communication May Be Harming the Global Community:"

As health professionals develop health communication for coronavirus disease 2019 (COVID-19), we implore that these communication approaches do not include fear appeals. Fear appeals, also known as scare tactics, have been widely used to promote recommended preventive behaviors. We contend that unintended negative outcomes can result from fear appeals that intensify the already complex pandemic and efforts to contain it. We encourage public health professionals to reevaluate their desire to use fear appeals in COVID-19 health communication and recommend that evidence-based health communication be utilized to address the needs of a specific community, help people understand what they are being asked to do, explain step-by-step how to complete preventative behaviors, and consider external factors needed to support the uptake of behaviors. To aid health professionals in redirecting away from the use of fear appeals, we offer a phased approach to creating health communication messages during the COVID-19 crisis (p. 531).

They explained:

In this article, we discuss the use of fear appeals during the COVID-19 pandemic and the potential negative sociobehavioral outcomes fear-based messaging may have. These include distrust in public health authorities, skepticism of health messaging, a lack of uptake in recommended behaviors, and a plethora of other unintended consequences (p. 531).

Citing scientific literature, they reported "studies have documented that fear appeals or fear-inciting health communication campaigns may produce unintended consequences, such as denial, backlash, avoidance, defensiveness, stigmatization, depression, anxiety, increased risk behavior, and a feeling of lack of control" (p. 532).

Perhaps to avoid conflict with their peers, Stolow et al. do not describe ongoing public health abuses that can cause such reactions, but they mention some that never made it off the drawing board:

In our work with health professionals, an example of a proposed COVID-19 campaign was to design a poster featuring an image of mass burials to persuade individuals to wash their hands. ...

Another proposed health communication campaign among health care professionals was to create television commercials portraying a fictional hospital overloaded with patients coughing up blood, fainting in hallways, and crying in pain to persuade people to physically distance (p. 532).

Stolow and her colleague encourage non-threatening approaches, including promoting the use of masks:

> For example, if the desired outcome is for individuals to use a cloth face mask, free reusable cloth masks could be made available to people coupled with educational brochures detailing why mask use is important, where it is appropriate to use a mask, how to don, wear, and doff the mask, proper mask storage, and ways to wash the mask.

They conclude:

> COVID-19 has caused the global community enough stress and fear; there is no need to exacerbate these issues by using fear appeals as a health communication strategy. We urge health professionals to consider the possible consequences when determining what health communication approaches to use, and to think systematically and innovatively about approaches. The world's health depends on it.

Even these empathic public health professionals do not raise the issue of the individual's right to be free of unwanted restraints on their lives and activities. There is no awareness that the Constitution and Bill of Rights can and should restrict terrorizing people to crush their liberties. They do not imagine that the legislative and judicial process of democracy can and should be overseeing them and their more obviously threatening colleagues. The public health educators need to be educated about individual rights and the principles of liberty.

How They Terrorize Us

Here is our survey of public health methods for creating sufficient fear and anxiety to convince people of the necessity of surrendering liberty to the so-called "new normal":

Maintain an atmosphere of fear.

- Begin early with predictions of millions of deaths in America.

- Forbid contact through handshakes or other touching.

- Constantly message, "Wash hands for 20 seconds as often as you can."

- Constantly message, "Stay socially distant and wear a mask."

- Broadcast false information about COVID-19 infection rates and its survival on hard surfaces, boxes, packages, and hands.

- Encourage remote shopping even though grocery stores and other "essential" shops such as pharmacies and hardware stores remain open.

- Spread news of extreme examples of COVID-19 disease and of deaths.

- Suppress any critical or contradictory information that might alleviate fear of the virus.

Stir up feelings of emergency.

- Broadcast there is no treatment for early stages of COVID-19 disease.

- Encourage treatments such as remdesivir or plasma which require intravenous administration in hospitals.

- Constantly emphasize the dangers of virus by creating one "news event" after another. For example, Dr. Fauci announces everyone might need to wear goggles in addition to masks, or Dr. Fauci declares there can be no opening of professional sports, or he declares school openings are very risky.

Make people stay home for a few weeks, then a few months, and then for an indefinite home confinement.

- Order children to stay at home.

- Order parents to oversee "distance learning" of children in the home.

- Order working adults to telecommute.

- Close most service or personal care establishments, including hair and nail salons, spas, gyms, therapeutic massage, daycare facilities, bars, and restaurants.

- Create such onerous requirements for service and personal care establishments that many are forced to go out of business.

- Close shopping centers, malls, and other facilities where adults might find some relief from stay-at-home orders.

- Close all recreational facilities, parks, and outdoor venues, including beaches.

- Cancel any "elective" medical procedures, including hip and knee replacements and other operations, many of which are needed to alleviate pain and to improve the person's health and well-being.

Make people distance themselves from each other and become isolated.

- Wear masks everywhere, except at home alone, scaring some people so much they can be seen wearing the masks while driving alone in their cars or solo hiking in the distance.

- Stay at least six feet apart.

- Keep your dogs (when dog walking) six feet apart, too.

- Keep children six feet apart from each other and do not allow playdates or any contact outside.

- Forbid contact with other family members sheltering separately.

- Forbid contact with friends.

- Forbid public religious gatherings inside or outdoor.

- Forbid political gatherings except "protests," which often devolve into destructive riots.

- Forbid mass athletic gatherings of any kind, from schoolyard games to professional sports.

- Create bizarre and isolating requirements for establishments to reopen, including plexiglass partitions around outdoor tables at restaurants and designated circles on grass in parks. Require plexiglass or other partitioning everywhere indoors to keep people separate from one another.

Physician and COVID-19 expert Peter A. McCullough offered the following perceptive observation about Fauci:

> My one suggestion on Fauci is to recognize he is a solo man; he survives administrations and acts alone. Never a team of advisors or consultants and never a group decision. He creates his own narratives to drive his agenda. He is even more powerful under Biden than Trump. He has intellectually gambled away hundreds of thousands of our lives and promoted millions of unnecessary hospitalizations to promote his agenda of mass vaccination. All presidents and administrations are weaklings compared to Fauci. Trump or Biden not once said: "COVID-19 is a horrible problem, I am calling in a group of doctors who are experts in treating COVID and together we are going to stop this tide of hospitalizations and deaths." No president or leader thus far has had the strength to utter such words.

By "solo man," Dr. McCullough does not mean that Fauci is a lone wolf. Along with Tedros of WHO and Peter Daszak of EcoHealth Alliance, Fauci is a member of a violent wolfpack of global predators led by Bill Gates and others with far more wealth and power than these lesser wolves like Fauci.

Preparing for What is Coming

Physician Lee Merritt MD, in a few stirring words, has captured what we are facing today:[871]

> A very sophisticated psychological operation is apparently being perpetrated upon the whole world. We are rapidly becoming an immoral, insane, dystopian technocracy in which our very right to assemble and even speak can be revoked on the basis of hospital bed utilization. If we fail to stand up and take back control now, those of us who survive the next phase will spend our lives under a medical tyranny the likes of which the world has never known.

Zev Zelenko MD, in his introduction to our book, made a statement that bears repeating here. For those who do not share Dr. Zelenko's belief in God, his words may still carry great meaning for you:

> It is my supposition this suppression of lifesaving information and medication is mass murder. This crime against humanity has been willfully perpetrated by a group of sociopathic despots that possess a delusional "G-d complex" and perceive themselves as superhumans with the right to enslave others.
>
> It is my strong hope and prayer that they will be brought to justice in both the earthly and heavenly courts.

The task of every American, and every able person throughout the world, is to work toward making sure the so-called New Normal, the Great Reset, and Progressivism are defeated. Their slogan of "equality" for all is a huge lie. Everywhere it has been tried, the attempted enforcement of equality means the elite thrive while the vast majority of us languish in uniform misery or die. These drastic assaults on our freedom must be fought and vanquished before they completely overwhelm and enslave us.

CHAPTER 31

Public Health Policies— Weapons to Destroy Freedom

T here is a close relationship between totalitarianism and public health policies and their effects on the individual, which can be similar or the same. Before World War II broke out, Adolph Hitler was praised in Western scientific literature as "The First Mental Hygiene Führer" for his devotion to genetics, eugenics, and the cleansing of the population from corrupting illnesses and diseases.[872]

POLITICAL TOTALITARIANISM

The following definition and description of totalitarianism was written by the editors of *Encyclopedia Britannica*. The separation and numbering of the sentences has been added:

(1) **Totalitarianism:** form of government that theoretically permits no individual freedom and that seeks to subordinate all aspects of individual life to the authority of the state. ...

(2) Totalitarianism is often distinguished from dictatorship, despotism, or tyranny by its supplanting of all political institutions with new ones and its sweeping away of all legal, social, and political traditions.

(3) The totalitarian state pursues some special goal, such as industrialization or conquest, to the exclusion of all others.

(4) All resources are directed toward its attainment, regardless of the cost.

(5) Whatever might further the goal is supported; whatever might foil the goal is rejected.

(6) This obsession spawns an ideology that explains everything in terms of the goal, rationalizing all obstacles that may arise and all forces that may contend with the state.

(7) The resulting popular support permits the state the widest latitude of action of any form of government.

(8) Any dissent is branded evil, and internal political differences are not permitted.

(9) Because pursuit of the goal is the only ideological foundation for the totalitarian state, achievement of the goal can never be acknowledged. ...

(10) Under totalitarian rule, traditional social institutions and organizations are discouraged and suppressed.

(11) Thus, the social fabric is weakened and people become more amenable to absorption into a single, unified movement.

(12) Participation in approved public organizations is at first encouraged and then required.

(13) Old religious and social ties are supplanted by artificial ties to the state and its ideology.

(14) As pluralism and individualism diminish, most of the people embrace the totalitarian state's ideology.

(15) The infinite diversity among individuals blurs, replaced by a mass conformity (or at least acquiescence) to the beliefs and behaviour sanctioned by the state.

Remarkably, all 15 descriptive points of totalitarianism can be found in much of the current public health policies and programs in response to COVID-19 that have been implemented by various U. S. governors, as well as federal officials and agencies. They are key to understanding the threat inherent in the "public health" approach now being taken by a wide range of actors who are implementing policies and actions that seem erratic, often inconsistent, sometimes unnecessary, almost always unprecedented—but often *consistently* totalitarian.

Public Health Totalitarianism

Public health publications generally frame the totalitarianism inherent in public health as essentially an *ethical* problem or question. They do this presumably to avoid becoming mired down in "politics" or even the law. However, public health as a field focuses on and implements *government responses* to biological threats, such as COVID-19,[873] and is therefore inherently legal and political in nature, rather than simply ethical.

Faden, Shebaya, and Siegel (2019)[874] "examine the kinds of moral justification that may be marshaled in support of specific public health interventions" and do touch upon liberty:

1. The overall benefit a public health measure produces.

2. The collective efficiency an intervention provides by coordinating action to ensure population-wide compliance.

3. Fairness in the distribution of the burdens of disease and disability.

4. The "harm principle," according to which the only justification for limiting a person's liberty is the prevention of harm to other persons.

5. Paternalism, which is the thesis that a restriction on individual liberty is acceptable if it is necessary to prevent harm or produce a benefit to the agent involved.

The first three concern the justification of public health policies that do not directly benefit all members of the population, while the last two concern the justification of public health measures that interfere with individual liberty.

In the same book, Faden and Shebaya (2019)[875] shorten this description of the "appropriate" "ethical justifications" for public health interventions to the following: "(1) overall benefit, (2) collective action and efficiency, (3) fairness in the distribution of burdens, and (4) prevention of harm (the harm principle), and (5) paternalism."

In their conclusion, Faden, Shebaya, and Siegel (2019) observed:

Public health is (1) a collective good, (2) focused especially on prevention, (3) reliant on government action supported by the force of law and (4) intrinsically outcome oriented. These characteristics of public health give rise to a wide range of ethical issues, such as the balancing of future health gains against current ones, the justification for the state's use of coercive powers to advance health, and the moral foundation of public health. In addition, how many of these issues are framed will depend on how we understand who it is that public health protects and the boundaries of what public health addresses. Together, these moral and conceptual questions form the distinctive challenges of public health ethics.

The observations are made from the viewpoint of advocates of public health who also want to raise "ethical" and "moral" concerns, and then to answer the questions *to their own satisfaction*. What is notably missing is either a legal or a political framework that would provide any control over their actions. When legal and political issues arise among advocates of public health, an odd orientation sneaks in. Such is their self-importance that public health authorities represented in the tome write as if the decision-making power is simply theirs to take. For example, Faden, Shebaya, and Siegel state that particularly when government programs are involved, "one task of public health ethics is determining self-imposed limitations and restrictions on what can reasonably come under the auspices of public health authorities" (p. 18). They expect to be controlled by self-restraint and not by external regulation.

What about Us?

The public's right to disobey or to impose its own ethical, legal, and political restraints on public health activities is a forbidden subject. I believe this self-centered, self-empowering attitude is epitomized in the attitudes and actions of Dr. Anthony Fauci, the head of NIH's Institute for Allergy and Infectious Diseases.

This is perhaps the kernel of the threat of public health: It sweeps aside those kinds of concerns upon which American law and political theory have drawn so heavily and that emphasize the protection of individual rights and liberty. It also escapes from considering the harm to individuals psychosocially and physically when liberty is curtailed. Consistent with this, little or nothing is said about the psychological and social consequences of loss of liberty or the diminishment of personal responsibility, and even more obvious threats, such as impingements on freedom of speech, the right to assemble, private property, or diminishment of individual well-being and quality of life, get little attention.

There is a long and varied literature warning about the totalitarian aspects of public health and universal health insurance and national medical planning.[876] Indeed, it is at the heart of ancient discussions going back to Plato that weigh the public good or general welfare against the rights of the individual. Fauci is not nearly as much a scientist as a philosopher king, but one without a coherent philosophy, other than he thinks he knows he is right and has science on his side.

"Educating" the People to Accept "Interventions"

Public health advocates have a working assumption they are right, and others must learn not to disagree with them. Science is frequently invoked on their side, but their science is often corrupted by the sheer complexity of the multiple variables as well as by their own biases, financial interests, political ideology, or desire for power.

"Educational Interventions" are a favorite concept in public health. In *Public Health Ethics*, a chapter titled "Public Health Interventions: Ethical Implications" shares observations which read more like a political platform than a scientific or economic study:[877]

Educational and Environmental Interventions

Educational interventions are designed to change the knowledge, beliefs, and predisposing psychological and social factors that lead individuals to engage in unhealthy behaviors ... (p. 78).

Short of eliminating poverty, a variety of strategies have been developed under the umbrella of environmental interventions. Whether inadvertently, through technological advances such as television and automobiles, or intentionally, through advertising or product placement, physical and social environments have been transformed in ways that are far less conducive to optimal health (for example, such changes have led to people driving instead of walking to work, or children watching TV instead of playing outside). To protect and promote population health, environmental interventions seek to undo or counterbalance these untoward changes in the environment. ... One example of the use of nudges is to re-engineer cafeteria lines to feature fresh fruits more prominently, instead of the more typical conspicuous displays of candy and chips (p. 79).

Some of the above observations have undeniable reality, including that poverty and social class play a role in individual access to medical care, including around the world during the pandemic. Many observers have concluded that even in places that have universal healthcare, such as Great Britain, the level of individual healthcare is inferior for lower social classes and poor people.

When these observations are made in the public health context, advocates almost inevitably begin citing progressive politics and justice concepts as the approach. That is, they want to use government to "educate" and to shape the populace. Whatever their "interventions" are, they are justified by necessity as they see it. They do not raise any overarching political and economic concerns, such as, "Does this rob too many citizens of their autonomy or inch us further toward an authoritarian or totalitarian state?" Instead, their conviction that they, as "experts," are doing good becomes conflated with totalitarian attitudes. Even when they raise issues of "liberty," they discuss it among themselves with the assumption they will decide how much liberty to seize. They fail to realize they deserve no more decision-making power than anyone else in a democratic society where legislatures, courts, and the people are supposed to determine matters of individual and political freedom. We have witnessed this arrogance throughout COVID-19 where "experts" acting in the name of public health ignore the Bill of Rights and the Constitution in their policymaking. We must retake America from them.

CHAPTER 32

Suffering and Recovering from Loss of Freedom

A ll creatures seem to hunger for freedom. From moles tunneling underground to human beings learning how to fly, creatures thrive on expressing their nature as fully as possible. When thwarted or blocked in their purposes, they try to find a way around it. In human beings, this desire for freedom is expressed every day by millions of children who at an early age learned to say "no" in the face of negative consequences. Historically, this motivation to overcome blocks to their liberty was demonstrated by our nation's Founders when signing the Declaration of Independence that simultaneously became their death warrants in the eyes of King George if he could have caught them.

Animals Love Their Freedom and Suffer Without It

The mechanistic psychologist Ivan Pavlov postulated a "freedom reflex" in animals to describe their distress upon being confined or restrained:[878]

> We tried out experimentally numerous possible interpretations, but though we had had long experience with a great number of dogs in our laboratories we could not work out a satisfactory solution of this strange behaviour, until it occurred to us at last that it might be the expression of a special freedom reflex, and that the dog simply could not remain quiet when it was constrained in the stand. This reflex was overcome by setting off another against it—the reflex for food. We began to give the dog the

whole of its food in the stand. At first the animal ate but little, and lost considerably in weight, but gradually it got to eat more, until at last the whole ration was consumed. At the same time the animal grew quieter during the course of the experiments: the freedom reflex was being inhibited. It is clear that the freedom reflex is one of the most important reflexes, or, if we use a more general term, reactions, of living beings. This reflex has even yet to find its final recognition. In James's writings it is not even enumerated among the special human "instincts." But, it is clear that if the animal were not provided with a reflex of protest against boundaries set to its freedom, the smallest obstacle in its path would interfere with the proper fulfilment of its natural functions. Some animals as we all know have this freedom reflex to such a degree that when placed in captivity, they refuse all food, sicken, and die.

Because modern psychology and behavior largely ignores or rejects freedom in humans, let alone animals, Pavlov's concept of the freedom reflex or reaction is as fresh and startling in some ways as it was when he discovered or described it. In the above quote, he calls it "one of the most important reflexes" of "living beings." He describes the necessity of having this reaction; otherwise, "the smallest obstacle in its path would interfere with the proper fulfilment of its natural functions." He directly relates the loss of freedom to extreme suffering, even in animals, who "when placed in captivity they refuse all food, sicken and die."

Helplessness in the Face of Losing Freedom

When Pavlov describes the captive animal's refusal to eat, allowing itself to die, he is seeing the animal enacting the feeling of *helplessness* that overcomes animals and humans when they lose their freedom. Anyone who has spent time with toddlers has seen that freedom reflex expressing itself and probably has experienced the youngster's negative reactions to being confined or restrained. Eventually, animals and humans tend to adapt to the suffering caused by the loss off freedom and may stop attempting to fight back or to escape. They often become docile and obedient. They may end up identifying with the oppressor as in the classic Stockholm syndrome.[879]

It is the threat facing all of us at this moment—the threat that we will be overcome by helplessness in the form of docility and obedience, and eventually identify with and support our oppressors.

Jeffrey Masson on Animal Freedom

I asked the one person in the world whose books have most encouraged my feelings of empathy for animals if he would comment on animals and their need for freedom for my legal report in the Ohio case and then for inclusion in this book. He is Jeffrey Masson, the author of *When Elephants Weep* and *Dogs Never Lie About Love*, and here is what he wrote:[880]

Commentary on Animal Freedom
By Jeffrey Masson PhD

In a new book "Mama's Last Hug," Frans de Waal, director of the Living Links Center at the Yerkes National Primate Research Center, tells the story of Mama, a chimpanzee who was 59, the great matriarch of the colony in Burgers Zoo at Arnhem in the Netherlands. She was curled up in a fetal position in her straw nest, clearly in her last days.

Nobody in their right mind would enter the cage of such an animal, but a biology professor, Jan van Hoof had known Mama many years before, and the two had become close. As soon as she saw the old professor, himself approaching 80, enter the cage (a first anywhere), barely able to move, she managed to get up from her nest and one of the staff took a video of what happened next. She came very close, and then she recognized him, and began making the sounds of joy that chimpanzees make at reunions, threw her arms open and embraced him as he cried in joy.

It's a lovely story, no doubt. But what made me sad was not that Mama was saying goodbye to her human friend, but that she had spent more than 50 years in captivity. Chimpanzees do not belong in a cage, anywhere, for any reason. Period. It is a crime against nature. But then is that not true of all zoos? I would have to say yes. No wild animal should ever live in a cage, or be fenced in, or on an island surrounded by a moat. It is not natural, that is, it goes contrary to its nature.

It has taken some time for humans to recognize the depth of suffering that animals show when they are confined in a cage or even in a corral. Remember, this is not something they EVER experience in the wild. Yes, a prey animal is killed, but swiftly. No other animal, other than man, puts a member of a different species into a cage or confinement for life. The depression that overcomes every single confined animal is well known now to animal scientists: the animals become listless, they lose their appetite (some will even starve themselves to death), they lose their interest in other animals, even their own partners, in short, they give up on life. Some will eventually recover, but their lives have been marked forever, and not for the better.

Well, you might well ask, is that not true of horses then? After all, you cannot allow a horse to simply wander. No, you cannot, and that to me is a weighty argument against the domestication of horses. How about parrots then? Absolutely not. Parrots are not even domesticated. They are not meant to live in cages. Orcas and dolphins? Perish the thought. But then what about cats and dogs?

Ah, that is a difficult topic: You cannot simply leave a dog to go in and out of your house, but the difference is that the dog wants to be there. The dog chooses to live with you. As does a cat, possibly the only animal we can allow its freedom. Cats come and go as they please, and that is how it should be.

Freedom is essential to ALL animals. They all need it, they all want it, every bit as much as we do. What is the worst thing that can happen to a human? To be locked in (prison or a psychiatric ward) or even to be forbidden to leave a country (think of North Korea) is to be deprived of perhaps our single most cherished possession: freedom. We are animals, after all, and just like all other animals, we want to live free!

Humans Love Freedom, Too

Much like animals, humans can become dismal and despairing when faced with their loss of freedom. Unless their spirit has been crushed, humans want nothing more than to escape even the most accommodating prison or their involuntary confinement at home.

In humans, like many animals, loss of freedom can lead to overwhelming feelings of helplessness and, as I describe in my book *Guilt, Shame and Anxiety*, it can result in every expression of our basic negative emotional reactions, including anger, emotional numbness, guilt, shame, and anxiety.[881]

An entire nation is being deprived of its pride and dignity, without the perpetrators seeming to give it a second thought. We are criticized or even censored if we express outrage over the suffering the lockdowns are causing so many of us. This is a large part of the dehumanization process—robbing people of their freedom and then refusing to recognize that loss of freedom is itself a demeaning process and experience, in addition to its more obviously destructive aspects.

The aim of coercion is to gain power over another, to put one's own will in place of another's. In short, coercion is used to make people do what they do not wish or choose to do. To refrain from coercing others, and to resist being victimized by it, we must be able to identify coercion and find better alternatives. This is true in both our personal and our political lives. The Founders of this nation, including those who conceived of it and fought for it, often saw themselves as expressing and defending their desire for liberty and their earnest goal to see freedom spread throughout the world.[882]

Suppression of Choice Making and Freedom

The suppression of our freedom can lead to the self-imposed suppression of our desire and capacity for making choices. We become seemingly unable to make decisions and take actions on our own. When freedom seems utterly lost, it can become too painful and perhaps unimaginable to make choices. The fear of punishment for seeking freedom can also make us suppress our wishes in order to stay out of trouble. The profound resentment we feel can confuse and disable us.

Children, for example, often stop wishing for things they believe they cannot have or that will lead to punishment, including more freedom. The self-suppression of our need and desire for freedom leads to apathy, indifference, and docility. That is the result of our giving up making choices on our own behalf. As a part of this collapse into helplessness, we stop blaming others and instead blame ourselves in self-punitive, demoralizing ways.

All oppressive situations from domestic violence and cult experiences to brainwashing and political oppression such as the world is undergoing during COVID-19 can result in this outcome as the person being controlled gives up trying and becomes helpless and unable to make choices. Guilt, shame, and anxiety are emotions which, when stirred up in the extreme, can cause people to withdraw and to experience themselves as having no free will or self-determination. It is often expressed as apathy and indifference. This is closely related to the emotional blunting I describe as one of the negative legacy emotions in my book, *Guilt, Shame and Anxiety: Understanding and Overcoming Your Negative Emotions.*

Studies on an individual and political level show totalitarian environments create avolition, apathy, and indifference. After the fall of the Berlin Wall, older people from behind the Iron Curtain were so indoctrinated into dependence and helplessness they could not rejoice in the changes, which was left more often to youth.[883] Individuals lacking in volition are ideal citizens for a totalitarian political system, but they vastly impair the functioning of a democratic republic. Eventually, their inability to contribute to society also impairs totalitarian societies and leads to their demise.

The Cost of Coercion: The Impact on The Victim[884]

One might wish that abusing people would cause the perpetrator to feel guilt, shame, and anxiety, but in reality, they often feel empowered, much as we see with global predators. Power can be exhilarating, and the more they get, the more they want.

Especially in childhood but also in adulthood, the victims are the ones who feel badly and self-blaming toward themselves. While it seems easily understandable that victims might experience anxiety, and perhaps shame, it seems at first glance more puzzling that they also experience guilt. The key, I believe, is the psychological helplessness engendered in victims which makes them vulnerable to guilt, shame, or anxiety.

Overall, coercion always has a cost for the victim. Most obviously, it creates psychological helplessness, including guilt, shame, and anxiety in the victim. It encourages victims to lie and to dissemble. While perpetrators lie to others and sometimes to themselves to cover up their abuses, victims learn to lie and to dissemble to avoid or minimize further oppression. Eventually they lie to themselves about their real feelings about being suppressed or abused, a process that is sometimes called denial.

During COVID-19, both individuals and society are going through this emotional turmoil with increasing oppression energizing the perpetrators and grinding down the victims. Under these circumstances, the victims must reject their helplessness and fight back as effectively and ethically as possible.

Understanding Psychological Helplessness

In my book *Guilt, Shame, and Anxiety* and scientific articles,[885, 886, 887] I have described how oppressed individuals end up shutting down their own vitality. They become overwhelmed with helplessness, and the feeling often leads to what I call the negative legacy emotions of emotional numbing or apathy and anger, beneath which are deeper negative feelings of guilt, shame, and anxiety.

My recent publication,[888] coauthored with Jeanne Stolzer PhD, a professor of childhood and family life, describes effects of psychological helplessness, including how it is reinforced by oppression:

> Physical or objective helplessness can be distinguishable from psychological helplessness, which is subjective and emotional in origin. Physical helplessness is exemplified by being incarcerated behind bars or afflicted with a neurological paralysis. Psychological

helplessness involves feeling, believing, or acting as if one were emotionally imprisoned or paralyzed, and unable to take effective or meaningful changes in one's attitudes or behaviors. Prisoners or physically paralyzed individuals may be largely unable to improve their physical status, but they do not have to become psychologically helpless. That is, they do not have to give up looking for opportunities to improve their emotional, psychological, and physical responses.

In the publication, I briefly describe disabling effects of extreme abuse upon adults. My observations apply directly and without modification to the effects of increasing degrees of totalitarian control in COVID-19:

The Effects of Abuse in Adulthood

When adults are exposed to extreme abuse, their sense of personal value and worthiness of love can be crushed. This occurs when disabled, elderly, or other vulnerable persons are abused within their own families. It occurs when people are abducted and held in captivity or when they are incarcerated in total institutions such as extermination camps, prisoner of war camps, mental hospitals, or prisons. In all these situations, the perpetrators [such as hardened staff] often systematically attempt to bring about psychological helplessness in the individual along with feelings of being undeserving of love and hence even life.

Going Behind the Iron Curtain

The universality of these principles—they apply in all societal settings that oppress or control people—was deeply impressed upon me when I visited East Berlin under Communism in 1988, one year before the Berlin Wall came down. I went with author Jeffrey Masson, who speaks German and has studied Nazi Germany. To get to East Berlin, we took an underground train from West Berlin and got off at the last and only stop within a frightening labyrinth with grim-faced guards beneath East Berlin. Inside the city itself, it was bleak and barren of interesting or exciting activities. The people looked fearful and depressed.

We were approached by a small group of teenagers who begged us to find a way to sneak them into West Berlin—an act of despair that took considerable bravery and perhaps recklessness born of resentment at living within a totalitarian system. Unlike the animals described by Pavlov as dying from loss of freedom, these youngsters were striving for freedom, and it leaves me wondering how they fared when the Berlin Wall came down a year later.

I was struck at the time by something I continue to mull upon to this very day: Stepping into East Berlin brought back my experience as a Harvard College student when, as a volunteer, I first experienced the inside of a state mental hospital[889]—the bleakness, the lack of normal human activity, the sadness and despair in the faces, and the pitiable requests to please help them go free. This is the fate of people who have gradually and inexorably been crushed by totalitarianism. They become like inmates of a giant, 1950s' state mental hospital.

These observations closely parallel those of sociologist Erving Goffman, who described mental hospitals as "total institutions," the institutional equivalent of totalitarianism.[890] He explained how it robs people of their very identity as they adopt the necessary submissive and demoralized attributes to survive in a wholly coercive situation.

Reducing People to Objects

Humans cannot be objective toward other humans without harming them.

Predators in the wild seem to be objective. They do not seem to hate their prey. This is not so of humans. Whenever we try to be objective about other people, we suppress our love, our empathy for them, and we end up degrading them. Instead of being objective, we become hateful and exploitive. If they show any resistance to our objectifying them, we may destroy them.

I have seen this in my profession of psychiatry, where diagnosing people as "schizophrenic" or "clinically depressed" or "bipolar" blocks any caring feelings the doctors may have for them, enabling the hateful and violent practices, such as drugging patients with neurotoxins and inflicting electroshock and lobotomies on them.

Only shared values and mutual affection ultimately limit how badly people will treat each other. Unfortunately, for the few sitting on the mountain tops of wealth and power, while manipulating humanity, people can look as far distant as ants underfoot or, all the more distant, as ants underfoot in a nature movie.

In 1964, in one of my earliest publications, *Coercion of Patients in an Open Mental Hospital*, I described how a supposedly "voluntary" psychiatric ward was not truly voluntary.[891] When patients resisted drug treatment or wanted to go home, the staff would often threaten and intimidate them, for example, by mentioning possible transfer to a state hospital, by threatening electroshock treatment, or by suggesting they might need to be committed against their will. Similar threats permeate our "free society" under COVID-19 policies and practices, with warnings about dire personal or legal consequences from failing to wear masks, resisting vaccines, having too many friends over, needing quarantine, or criticizing the authorities and their largely nonexistent "science."

David L. Altheide concluded his paper on *Terrorism and the Politics of Fear* with the following observations that apply to much of the politics of fear used to implement the COVID-19 shutdown:[892]

> They seek to reduce individuals to objects rather than involving them as subjects. The element of direct, physical coercion is either open or poorly concealed and there is no further goal than that of either neutralising the threat or making it manageable.

In a paper I delivered in Germany at the only conference ever on *Medicine in the Third Reich*, I addressed the role of psychiatry and psychiatrists in paving the way for the Holocaust with the highly organized, bureaucratic murder of the psychiatric inmates and disablement in their care. Very consistent with the final remarks of Altheide's paper, I concluded in my published version of the paper that in order to dominate and destroy people, the German psychiatrists had to think of them and treat them as if they were objects:

> One fundamental flaw is the reduction of the human being to an object devoid of inherent worth or inviolability. In Muller-Hill's words, "It seems to be that to reduce other people to the status of depersonalized objects is of no help whatsoever to them" (p. 101).[893] Trying to view people "objectively" can be demeaning in and of itself. It also tends to lead toward further degradation of

the individual into subhuman status. In the Nazi ideology, the Jews became "pests" or "vermin," in psychiatric ideology, patients become "diseases" or biochemical and genetic aberrations. Devoid of inherent value, they become suitable for various inhumane solutions, including involuntary treatment and, ultimately, sterilization and extermination (pp. 146-147).

Similar objectifications accompanied by fear or terror take place in public health in the service of political totalitarianism. Individuals become "disease carriers" or merely "contacts," and all the protections given to human beings in the Constitution and Bill of Rights become null and void.

We should apply the Nuremberg Code to obtaining consent for vaccination from prisoners, hospitalized mental patients, or nursing home inmates. People living under such authoritarian conditions are faced with intense pressure if they refuse a vaccine. Therefore, they cannot give voluntary consent to dangerous, experimental procedures such as Operation Warp Speed vaccines. Vaccines should not be offered to them any more than they should be offered the opportunity to participate in other kinds of dangerous or life-threatening experiments.

Extending that principle, as Ginger Breggin has pointed out, society itself has now become so authoritarian and threatening surrounding COVID-19 issues that people feel compelled to volunteer for the vaccine.[894] If they do not volunteer, they are subject to social rejection and to increasing limitations on their freedom, such as going to school, traveling, working at the office, shopping in stores, or going to public events. Therefore, these experimental and dangerous vaccines should not be given to anyone, and we should return to the stricter standard of requiring that they pass through the FDA approval process before being marketed.

Using Increasing and Yet Inconsistent Threats

Some recent threats employed by shutdown advocates are reminiscent of totalitarian governments. One such example is the mayor of New York City, Bill de Blasio, telling citizens to use a special hotline to photograph and report their neighbors and other people who were not following all the shutdown rules.[895] Another example is the mayor of Los Angeles announcing the city will turn off electrical power and water if private homes or businesses have large groups.[896]

Issuing catastrophic threats is a major way of making people docile. On March 6, 2020, the Australian National University reportedly declared, "In the most disastrous scenario, the [global] death toll could reach a staggering 68 million ..."[897] Their best scenario was 15 million. The first author, Warwick McKibbin, was also associated with the Brookings Institute.[898]

On March 10, 2020, Tom Friedan, the former head of CDC, wrote we could have a million deaths.[899] On March 16, 2020, *The New York Times* reported, "White House Takes New Line After Dire Report on Death Toll."[900] A new twist on threats was made by a recent "scientific" publication on August 13, 2020, titled "Comparison of Estimated Excess Deaths in New York City During the COVID-19 and 1918 Influenza Pandemics."[901] On August 15, 2020, the fantastic manipulations to make this comparison were correctly debunked by Richard Epstein in a report, "COVID-19 Confusion."[902]

I recently had a personal experience of inflicting uncertainty on helpless beings under my control. In the cold weather, I wanted to encourage my dogs to respond more quickly when I called them to return home with me after walking outdoors in the snow. So, I began to give them treats after we returned home. But I enjoy walking them so many times a day, I decided to give them occasional or intermittent treats. After a few weeks of this erratic rewarding, it became apparent to me, my wife Ginger, and her mom Jean that the dogs became anxious and confused at those times that I did not give them their expected treats. Anthony Fauci's well-documented flipflopping on issues represents this kind of unnerving practice.[903]

Censorship Is Loss of Freedom

The control of the media and internet is in the hands of multibillion-dollar industries who favor top-down government, who pay off or collaborate with government administrators and legislators, and who almost universally favor the extremes of the COVID-19 shutdown. Google and YouTube, Twitter, Facebook, and Amazon stifling and controlling free speech on a mass scale in the name of public health is characteristic of this shutdown. Censorship is the loss of freedom of expression.

As a most astonishing example, Alex Berenson, the highly respected former *New York Times* science reporter was censored in advance by Amazon when he tried to sell his insightful, documented booklet on their enormous website. From all the truly bizarre and even evil publications Amazon has for sale, they decided his book, *Unreported Truths about COVID-19 and Lockdowns: Part 1: Introduction and Death Counts and Estimates,* did not meet their standards. They gave him no reason except to say it would help if he edited out references to COVID-19, which happened to be the subject of his book. As he describes in *Part II,* he was saved by the intervention of the powerful Elon Musk. If Amazon had not accepted the book, Berenson writes there was no other comparable alternative for him to get his views to the public.

The justification for censorship is typically that the publication or commentary is too "extreme" or comes from an "extremist." This is true whether a gigantic monopolistic corporation or the government is making the negative judgment. On this subject, Timothy Snyder[904] has been eloquent and precise:

> *Extremism* certainly sounds bad, and governments often try to make it sound worse by using the word *terrorism* in the same sentence. But the word has little meaning. There is no doctrine called *extremism.* When tyrants speak of *extremists,* they just mean people who are not in the mainstream—as the tyrants themselves are defining that mainstream at that particular moment. Dissidents of the twentieth century, whether they were resisting fascism or communism, were called extremists. Modern authoritarian regimes, such as Russia, use laws on extremism to punish those who criticize their policies. In this way the notion of extremism comes to mean virtually everything except what is, in fact, extreme: tyranny (pp. 101-102). (Italics in original.)

In our ordinary activities as citizens, we may feel censorship does not have much impact on our lives. For example, if you want to look up critical information about Dr. Anthony Fauci or Bill Gates, you may feel satisfied with what you find on Google or on Wikipedia, or even in social media. Once you begin to dig deeper and to talk to people with critical opinions, you will realize how thoroughgoing the censorship can be in the interest of progressive globalism.

Even if you go to your favorite media on the left, such as CNN, or the right, such as Fox News, you may think you are getting something nearer to the truth, but you will find sparingly little about most of the issues discussed in this book. You will uncover pitifully little about the routine corruption than runs amok in major corporations, because both the left and the right media now depend on advertising coming from globalist corporations.

LARGE CORPORATIONS AND THE LIMITS OF HUMAN CARING

Why does corporate life go morally south? We humans evolved to live in extended nomadic families often no larger than 20 or 30 in number. This went on for hundreds of thousands of years until the advent of villages a mere 10,000 years ago. Thus, we evolved to care about the well-being of the people nearest and dearest to us in our nomadic extended family, while feeling fearful, defensive, or predatory toward the great body of humanity surrounding us. While our capacity to care about people close to us persists, many of us work in large corporations, institutions, or agencies where we impact people whose names we will never know and whose faces we will never see. Whether we are working for a drug company or a government agency, most of us lack concern and empathy for the distant people over whom we have so much power.

On top of that, the breakdown of traditional cultural ties that bind large groups of people—such as nationalism, a shared history, religion, and varied cultural identities—leave these human predators unrestrained by any broad sense of affiliation or love for the people they are perpetrating against. The world has turned into a field full of prey, at times a killing field, for all enormous predators who feel little or no attachment to the vast numbers of people over whom they seek economic, religious, or political dominion. With the rejection of basic principles and sound ethics so rampant in modern society, those who run large entities of any kind—governmental, corporate, and religious—have become easily corrupted, putting their own power and wealth, and their own institutions, above all else.

As I describe in my book *Guilt, Shame, and Anxiety*, we evolved over millions of years to feel attached, loving, and obligated to our immediate family, while often fearing and avoiding strangers. In our personal and family relationships, we naturally feel negative, inhibitory emotions like guilt, shame, and anxiety when we "behave badly"; however, we may feel little or no concern for strangers somewhere else in the world who are being directly or indirectly disadvantaged or harmed by our economic activities. A person who would never cheat in ordinary life, that is, with friends or family, might think nothing of manipulating data in a computer to make the corporation's products look safer and more useful than they are.

This human weakness with respect to caring about people with whom we have no kinship and little or no contact, and on whom we depend for making a profit, dooms large, heartless entities to becoming corrupt. This vulnerability exists whether we are talking about large government, giant corporations, powerful political parties, or international religious organizations. Even worse, these large entities will become even more monstrously corrupt if there are no competing or inhibiting entities, such as an effective governmental watch-dog agency, a free press dedicated to honesty and the public good, or a strong president protective of America and liberty.

The strength and independence of America is the predominate reason this disparate group of global predators feels the need to cooperate and to get strength from each other. Many of them have strong financial ties to America from which they draw much of their power, but as others have warned, these parasites are endangering the host with globalism and with their ties to the Chinese Communists. Silverdollar.com[905] describes part of that paradoxical process of the globalists destroying their host:

> But why would the globalists do this? For those that assume the U.S. economy represents the "goose that lays the golden eggs," what I describe above is inconceivable. In order to understand what is happening and why, we must cast off the lie that America is a golden goose that perpetually supports the globalist agenda. Rather, America is more like a host to the globalist parasites, and once the host is drained of all vitality, the parasites will leave and move on to bigger and better targets.

Globalists are not only predators, but they also are parasites that live off the subjugation of other people. In a profound sense, all humans who profit off the unfreedom of other people are parasites. And when they go too far, they kill their host, the captive people they are exploiting.

There can be no doubt that the predatory globalist billionaires and huge international corporations see American conservatism as a vast impediment to their success and that their aim is the death of America as a patriotic bastion of freedom. If the globalist left continues to succeed in destroying America from within as a freedom-loving nation, they will clear an open road for the Chinese Communists on their way to dominating Earth. The left's murder of the American right will become an unintended suicide of the American left itself, as China snuffs out the great experiment in freedom begun by our Founders less than 250 years ago.

The current suppression of America's patriotic, conservative right is China's agenda. What can we do about this looming tragedy for us and for the rest of humanity? We Americans can no longer afford to remain helpless in the face of parasitic global predators. We must revive our determination to stand up for personal and political liberty. No other country and no other people are prepared to do it without our leadership.

CHAPTER 33

They Fear and Hate Us—
But Who Are We, Really?

By the time this book comes out, conditions in America will be growing even worse because the Democrats will have been in office longer and well-meaning but uniformed people will have been lulled into indifference. Although genuinely idealistic liberals watching CNN or reading *The New York Times* or *The Washington Post* will have little or no idea, the governing elite in the Biden Administration will be massively persecuting "the Trump followers" by every means possible.

A lock on the major or mainstream media is not enough thought control for the progressives and globalists. Now that President Biden is safely in office, *The New York Times* is airing out proposals for him to appoint a Reality Czar.[906] What kind of delusional unrealities must be stamped out? They include the following two delusional aberrations: "more than 70 percent of Republicans believe Mr. Trump legitimately won the election, and 40 percent of Americans—including plenty of Democrats—believe the baseless theory that COVID-19 was manufactured in a Chinese lab." They are setting up for books like this to be burned or maybe shredded by a Biden-appointed Reality Czar.

Remember, in June 2020, California Democratic Representative Maxine Waters called for violence against President Trump's team members: "If you see anybody from that cabinet in a restaurant, in a department store, at a gasoline station, you get out and you create a crowd and you push back on them, and you tell them they're not welcome anymore, anywhere."[907] Not to be outdone, in August 2020, Speaker of the House Nancy Pelosi called for progressives to treat President Trump as "an enemy of the state" who must be fought against in defense of the Constitution.[908]

This will not be the "politics as usual" of previous transfers of power from one globalist administration to another; this will be a war by the predatory globalists against former President Trump and those American citizens who believe in a civil, patriotic, freedom-loving America.

Predicting an Escalating Suppression and Censorship

Under President Joe Biden, the progressives are escalating taking down the most important conservative and libertarian internet platforms,[909] throwing conservatives and President Trump supporters off social media, much like when President Trump was permanently banned from Twitter.[910] The progressives are socially and culturally cancelling them,[911] hounding them out of their jobs or refusing to hire them,[912] and even refusing to publish their books.[913]

Meanwhile, we predict the IRS will resume Obama-era offenses against liberty, maliciously auditing conservative tax returns[914] or systematically rejecting their applications for nonprofit conservative organizations[915]… and other anti-freedom actions we never imagined could happen in America.

Third-World Horrors We Never Expected to Endure in America

Some of the "worst things" we never imagined include the Democratic leadership calling for implementation of the 25th Amendment to declare President Trump mentally "incapacitated" and, when that failed, actually impeaching the President days before he left office.[916] Their next attempt, trying to convict him, unconstitutionally, as a private citizen after leaving office, was truly unbelievable.

No wonder President Biden felt righteous about calling out the military before his inauguration, allegedly to protect the government from these threatening "insurgents"—President Trump and his supporters.[917] They thrive on making us seem evil. In a most unprecedented and un-American fashion, the FBI and Department of Defense vetted the National Guard troops to eliminate conservatives and President Trump supporters.[918, 919]

Mike Lindell, the "My Pillow" businessman who supports conservative causes, was permanently banned by Twitter.[920] If they can get away with banning President Trump and Mike Lindell from social media—and they have—no American is safe. Among the wealthy tech giants, only Elon Musk has found the courage to stand up for some degree of freedom of speech.[921]

In late January 2021, influential Democratic Representative Alexandria Ocasio-Cortez (known as AOC) accused Republican Senator Ted Cruz of "trying to get me killed." She claims this was Cruz's intention when challenging the election results before the break-in at the Capitol. She demanded that he resign.[922] On January 27, 2021, *The New York Times* published a commentary aimed at preparing the nation to take up arms against conservatives, patriots, and Trump supporters:[923]

> As a former overseas operative who has struggled both on the side
> of insurgents and against them, the past few days have brought a
> jarring realization: We may be witnessing the dawn of a sustained
> wave of violent insurgency within our own country, perpetrated
> by our own countrymen.

These are not idle threats. The Biden Administration and the mainstream media are preparing to do everything in their power to disempower and demolish about one-half of the American citizenry. Their first goal is to make their own Democratic and leftist constituency terrified of us, their second is to terrify us, and their final goal is to suppress us in every possible way.

When we began this book, none of this seemed possible, but the COVID-19 experience of daily, unrelenting government oppression set the stage for it. *Nothing could be more pleasing to the global predators, one and all, and the Chinese Communist Party, than this all-out assault on former President Trump and the independent populist movement.*

How the Progressives and Globalists View Us

John Brennan and many others, including members of Congress and the media, have already set the stage for an enormous escalation of the ongoing persecution.

Brennan is a particularly terrifying bellwether because he is so profoundly involved with the Democratic leadership and its embedded bureaucracy. He knows its mood, methods, and intentions. Brennan was director of the Central Intelligence Agency (CIA) from 2013 to early 2017. He became President Obama's Deputy National Security Advisor for Homeland Security and Counterterrorism.

If there is any doubt about his loyalties or his connections, Brennan was at the Oval Office meeting with President Obama and other security officials in the summer of 2016 discussing Hillary Clinton's plan to create a Russian hoax to destroy President Trump. In his own handwriting, Brennan wrote a note about the meeting that included mention of the "alleged approval by Hillary Clinton—on 26 July—of a proposal from one of her foreign policy advisers to vilify President Trump by stirring up a scandal claiming interference by the Russian security services."[924]

Shortly after President Biden's inauguration in an interview on *MSN*,[925] Brennan, the former CIA-director, described how the intelligence and military communities were already "moving in laser-like fashion to try to uncover as much as they can about what looks very similar to insurgency movements that we've seen overseas." Brennan denigrated Trump supporters involved in this alleged insurgency as "an unholy alliance of religious extremists, authoritarians, fascists, bigots, racists, nativists, even libertarians." He emphasized the need "to root out what seems like a very serious and insidious threat to our democracy."

Who are these seemingly uncouth and dreadful people in Brennan's alleged "unholy alliance" to overthrow the Biden presidency? In part, they are also Hillary's "basket of deplorables." Here is what presidential candidate Clinton said at a September 2016 fundraiser:[926]

> You know, to just be grossly generalistic [sic], you could put half of Trump's supporters into what I call the basket of deplorables. Right? The racist, sexist, homophobic, xenophobic, Islamophobic—you name it. And unfortunately, there are people like that. And he [President Trump] has lifted them up.

Brennan's "unholy alliance" and Hillary's "basket of deplorables" overlap with Barack Obama's comments about bitter people clinging to their guns and religion in small, economically depressed towns. Quoting Obama from one of his fundraisers:

> They get bitter, they cling to guns or religion or antipathy to people who aren't like them or anti-immigrant sentiment or anti-trade sentiment as a way to explain their frustrations.[927]

Unlike Hillary's plain nastiness, Obama's remarks have some policy substance in his references to "anti-immigrant sentiment" and "anti-trade sentiment." The people he is describing have been left behind by avaricious globalism. They are against Obama's immigration policies because waves of poor immigrants lower their wages and take away their jobs. They are against his trade policies because they have the same tragic outcomes by moving industries out of the country in search of cheaper labor. People like Hillary and Obama should care about these people because they are a large portion of Americans and because they really have received a raw deal from global predators on both sides of the aisle in Congress and from every president since Reagan.

When President Obama spoke of people hanging onto their religion and their guns, we wonder if he realized that he was describing the Founders of this great nation who risked their lives fighting for our lives and our future. They always hung onto their religion and their guns. These are the same Founders who inscribed our freedom of religion and our right to own guns in the Bill of Rights. People who believed in religion and who owned guns fought our first successful revolution. *They may have to fight our second one as well if the predatory globalists, including the Democratic Party leadership, continue to escalate their abuse and oppression of the American people.*

Yes, their attacks at the present time are limited to the other half—probably more than half the nation—those people who have supported populism and President Trump, as well as an end to unscientific and irrational COVID-19 restrictions on our lives. However, history tells us that after predatory people take control over a nation, all but their own *elite* survive intact and even many of them die in the infighting. Often there is only one man left standing, the most cunning and violent of all the predators, who becomes President-for-Life.

All the global predators we examined in this book want to bring more poor immigrants into America and to transfer out more industries to other countries. They are happy to cause the utter destruction of our working, middle, and upper-middle classes, while only the upper globalist "elites" prosper, churning out hundreds of billionaires.

These globalists mostly reject religion and guns alike because they represent the American individuality and determination that resists being chewed up and absorbed into the belly of globalism. President Trump became the only American president in decades to stand up for these and many other deplorables and non-elites.

Sizing Up and Attacking President Trump

As soon as he announced his candidacy, the domestic progressives and global predators quickly realized that Donald Trump was the only potential president in memory to pose a threat to them. As documented earlier, before he was elected and prior to the day he took office, they were already planning his impeachment.

The world of global predators and progressives assaulted President Trump with all their might, eventually using COVID-19 as their best weapon against him. Using Fauci as their front man, the predatory globalists imposed drastic public health measures on America, destroying President Trump's multiple accomplishments, including the roaring economy, the tremendous employment figures that broke all the records for women and minorities, and his successful America-First confrontations with global predators and the Chinese Communist Party.

Dominion Voting Machines

Ultimately, the U.S. progressives and global predators used COVID-19 to justify manipulating the methods of voting, vastly increasing mail-in ballots, enabling them to steal the election. They sealed their victory with the use of Dominion voting machines. Here is some data about these voting machines you would have missed if you were not looking hard for it.

Smartmatic software is used in the infamous Dominion voting machines. The president and chairman of the board of Smartmatic is Peter Neffenger. Neffenger is both a close associate of the master of evil globalism, George Soros, and he was also on Joe Biden's transition team.[928] If that's not enough, Britain's Lord Mark Malloch-Brown is the chairman of Smartmatic's London-based holding company, SGO Corporation Limited, and another longtime associate of George Soros. How close? On December 4, 2020, when the Trump campaign and independent lawyers were charging the software company with helping to fix Biden's election, Soros appointed Smartmatic's Malloch-Brown to be president of his Open Society Foundation.[929] Open Society subsidizes anarchistic, progressive, revolutionary organizations around the world, such as Black Lives Matter and Antifa. It fueled the persistent riots throughout America's cities in the months prior to the election.

We the People

Who are the people so hated by American progressives and the global elite? We are all the people that American progressives and the global predators have been targeting for years and will continue targeting during the Biden Administration.

We are the Cubans, Russians, Vietnamese, Chinese, and other refugees from Communism who often are far more appreciative of America than many who have been born here. And we are those who know that America is the first and only nation created on the principles of love of God and freedom, and that America is the remaining, if wounded, bastion of liberty for the world. We also know that America is far from perfect because all of us are human beings. Being human, we must always work hard to overcome our own worst tendencies while fighting off those predators who have, in Tom Paine's metaphor, always tried to track down and hound liberty to death.

We are probably and hopefully the great majority of Americans. We are idealistic youth who once supported the Democratic Party without realizing that they were using us as cannon fodder for the global elite who only care about their own wealth, glory, and power. We are old-time idealistic Democrats and trade unionists who believed in making the American Dream available to the poor, to the working class, and to middle-class Americans. We are disillusioned members of the American Civil Liberties Union (ACLU) who believe in protecting freedom of speech rather than in promoting progressive globalism.

We are African Americans who know that we have been used by those Democrats who originally kept us in slavery and who now want us to remain dependent on their government-subsidized urban plantations—unemployed but voting for them. We are those women who wish that President Trump, when in conflict with competitors, did not become such a caricature of the immature alpha male. These women are beginning to think that it might take such a proud and indomitable man or woman to stand up, virtually alone, to the entire might of the progressive, globalist predators. In turn, they admire his appointments and support for strong women, including those in his family.

We are people who are genuinely conservative, libertarian, or patriotic. We are God-fearing people of many faiths. We are the handful of Republicans who remain true to the ideals of Lincoln and Ronald Reagan. We are people who live in small towns and understand the importance of family, community, and church, as well as owning guns for hunting and self-protection.

And, yes, we are atheists and humanists who now see that we need fundamental values and ideals that are unshakeable, based on what the Declaration of Independence called the "Laws of Nature and of Nature's God." Perhaps Thomas Jefferson and the Founders put in "Laws of Nature" just for those of us who put more stock in empirical science than in anything else.

We are the small business owners being destroyed by totalitarian COVID-19 policies. We are the 90 percent of white men[930] whom they attack as racists. We are the underemployed or unemployed who want jobs, and we are college graduates burdened by huge debts promoted by the progressives to gorge the universities. We are those Blacks, Hispanics, and women, in growing numbers, who voted for President Trump a second time because of the all-time high employment he gave us by fighting the globalist elites and their self-serving policies.

We are all the many Americans who are both patriotic and freedom loving. We are all the older citizens who wonder what in the world has happened to our America. We are the youth who are sick of relativism, hate, cancel culture, and arrogant teachers and professors who tell us what to think. We are people who believe in living by ethics and ideals. We are the people who won a second term for President Trump and were then disenfranchised by gross vote manipulation in several key states, which may never be fully investigated because the Supreme Court is thus far bent on never taking the cases to air and to examine the shameful facts.

We are anyone and everyone, all of us, who are determined to save America from becoming one more civilization that succeeded so well until, with too much success, it lost its courage, its ethics, its patriotism, and its will to fight for its own liberty and ideals.

WHO ARE THOSE WHO OPPOSE OPPRESSIVE COVID-19 POLICIES AND PRACTICES?

We are another large overlapping group of people—one in which we, the authors, are deeply engaged—the physicians, scientists, and concerned citizens who are critical of the oppressive COVID-19 policies pushed by Anthony Fauci and his allies, including the National Institute of Health (NIH), the Centers for Disease Control (CDC), the Food and Drug Administration (FDA), and the World Health Organization (WHO). We know that Fauci and these institutions serve the rich and powerful and not humanity.

Many of us are physicians like I am and other professionals who know that there are inexpensive and safe treatments for COVID-19 that could be saving hundreds of thousands of lives but have been banned in Western democracies by the avaricious globalists who control Fauci. Many are people who have seen their elder family members die terrible, lonely deaths due to the mismanagement of COVID-19.

You have heard relatively little about us because we have been cut out of the debate by the major media, even including most of Fox News, and censored or taken down by Amazon, YouTube, Twitter, Facebook, and other globalist, big-tech companies. We are marvelously represented by three courageous physicians, Peter A. McCullough MD, Elizabeth Lee Vliet MD, and Vladimir "Zev" Zelenko MD, all three of whom honored us and our book by writing introductions to *COVID-19 and the Global Predators*.

THE PROGRESSIVES AND GLOBALISTS PUSH TOWARD VICTORY

Leading up to the election on November 3, 2020, progressives from the Democratic Party, along with globalists from Amazon, YouTube, Twitter, and other tech companies, took down the platforms and cut off the social media of conservatives and critics of China or COVID-19 totalitarian practices. Even the President of the United States was not immune to being cancelled and censored! When that failed and the vote count was becoming a President Trump landslide in the early morning of November 4, 2020, the pivotal states stopped counting for several hours until they could tool up to reverse the election results.

At the end of President Trump's last year in office, the progressive globalists rushed through an unconstitutional, Third World-like, second impeachment in the House of Representatives. With that last act, progressive globalists became openly and fully in charge of the government of the land of the free. Under the guise of suppressing an insurrection, the new Biden regime organized a military occupation of Washington, DC during the inauguration in the style of a military dictatorship. They escalated their suppression and censorship on social media of those who cherish America, its Founders, and the principles embodied in the Declaration of Independence, the Constitution, and the Bill of Rights.

Meanwhile, the opportunistic predatory globalists immediately began throwing their wholehearted support to President Biden, thereby making clear they had withheld important and even lifesaving help from America and the world until successfully defeating President Trump. Within days of Biden's election victory, the pharmaceutical industry began announcing the almost immediate availability of the new vaccines. Pfizer touted a 90-percent effectiveness that provided "a big burst of optimism to a world desperate for the means to finally bring the catastrophic outbreak under control."[931] On Inauguration Day, Amazon volunteered its vast resources to President Biden to escalate the distribution of the vaccines, prompting Fox News to ask, "Why did Amazon wait until Biden's inauguration to offer help with vaccine distribution?"[932]

Also, the World Health Organization (WHO) urged COVID-19 testers to crank down the sensitivity of their PCR testing machines and warned that more caution is needed in the "interpretation of weak positive results." Those changes would greatly reduce the apparent size and severity of the pandemic, artificially making President Biden's administration look good.[933]

Under the domination of global predatory progressivism, the leadership of America is increasing its efforts to sacrifice not only our freedom, but also innumerable lives here in America and around the world that could be saved by allowing the use of simple, inexpensive, and effective early treatments for COVID-19.

By contrast, America, from its founding and at its heart, stands for individual liberty—ours, yours, and every human being on Earth. It makes "We, the People" the rulers and not the servants of government. It stands for God, the Creator— the source of the energy we call love and the definer of what is good, who tasks us to live by standards so high we must exert all our might to barely begin to meet them. For ourselves, our children, and indeed the future of the world, we must find the courage to reinspire and to liberate America.

CHAPTER 34

The Escalating War Against America's Freedoms

The all-out, warlike personal and political attacks on President Trump, already escalating, exploded on January 20, 2017, when he took office as President of the United States and *The Washington Post* simultaneously declared:[934]

The campaign to impeach President Trump has begun

The effort to impeach President Donald John Trump **is already underway.**

At the moment the new commander in chief was sworn in, a campaign to build public support for his impeachment went live at ImpeachDonaldTrumpNow.org, spearheaded by two liberal advocacy groups aiming to lay the groundwork for his eventual ejection from the White House. (Bold added.)

Before then and after, it was nothing but hysterical abuse from the left, the Democratic Party, the major media, many billionaires, and the tech industry—indeed, the entire domestic and international horde of global predators. By the second impeachment, it had escalated into the most partisan, most unconstitutional, and most hateful assault on a president in the history of the United States.

Despite the predicted impeachment and endless harassment from a resentful, frightened-to-the-core, and hateful Democratic Party, President Trump managed to keep nearly all his promises to his populist constituency and to America. He built The Wall and controlled the border, brought industry back to America, improved our trade advantages, cut taxes, got the economy humming as never before, made us energy independent, rebuilt the military, started the Space Force, began cooling down our last remaining wars, defeated Jihadist ISIS, broke Iran's hold over the Middle East, brokered cooperation between Israel and several Arab states, moved the U.S. Embassy in Israel to Jerusalem, and, in his populist masterpiece, dramatically raised employment to record highs for all to include the poor, Hispanics, Blacks, and women.

With these remarkable successes, despite unrelenting leftist and globalist opposition to every one of them, President Trump mightily challenged the global predators, including Xi Jinping, the leader of the Chinese Communist Party. The balance of power in the world, which was dangerously tilting toward China, began to tilt back toward the United States of America at an unexpectedly fast pace.

Four years later, in January 2021, with President Biden's inauguration and President Trump's second impeachment, the balance of power shifted back again to the dark side as China once again moved into a position of potential domination over the United States of America—to the glee of globalists at home and abroad.

More Warfare Against America

Among the attacks against President Trump and against the United States of America, the grossest offense of all was the fixing of the election in key states to reverse President Trump's growing vote-count advantage on election night. I watched it happen overnight to the bewilderment of talking heads from the left and the right as all vote counting stopped in the critical states and then *resumed hours later* by taking votes off the board for President Trump and putting up huge new batches for Biden.

Probably, the next most egregious act was a concerted effort to stifle free speech by the major global predatory media and tech corporations—Amazon, Twitter, Facebook, Google, YouTube, Apple, as well as CNN, the major media, and others. They ganged up on the American people and shut down the President's communications with America in his last days in office, while picking off the communication opportunities, one after another, of leaders on the political right. In one of the great offenses against free speech in our history and in an escalation toward Third-World politics, the global predators cooperated to take down Parler, the last resort for Twitter-like communication for conservatives, anti-globalists, and opponents of the COVID-19 abuses of Americans.

On January 20, 2021, YouTube made another major assault on freedom, while handing China an enormous potential victory. It demonetized the YouTube channel of *The Epoch Times*, meaning the platform prevented them having advertising in their videos to generate income. In our view, *The Epoch Times* is the most important and useful newspaper in the U.S. and by far the best source of information about China and the Chinese Communist Party, as well as COVID-19 policies and practices. It also offers daily and feature video reports of the highest quality about China, as well as the U.S. and the world. *The Epoch Times* was founded by Chinese dissidents who came to America to be free and now find themselves victims of censorship and oppression all over again.[935]

In late February 2020, Google's YouTube took down an interview by former President Trump on Newsmax. We are witnessing a cascading of censorship.

While attacking President Trump and populism, and now *Parler* and *The Epoch Times*, big tech has played an enormously treacherous role in defending and strengthening China at the expense of freedom in America. Big tech has done so with special vigor in suppressing any discussion that harms the Democratic Party, including the Biden family's self-enrichment from China in the millions of dollars.

In trying to wreck President Trump's political movement and the American Right, Joe Biden, Kamala Harris, Nancy Pelosi, Chuck Schumer, and others are implementing China's agenda for world domination. For the Chinese Communist Party, this is a big step toward the fulfillment of their global strategy. With Democrats now calling for the outright ostracism of President Trump supporters and their removal from life in America, what could possibly give more satisfaction to the Chinese Communist Party? The Democratic Party is eliminating the only effective opposition to China taking over the world—President Trump and the American patriotic, freedom-loving right.

Not only are the globalist Democratic Party predators carrying out the aspirations of the CCP, but they are also providing cover to China as the ultimate perpetrator. At the very end of the presidency of Donald Trump, when the Democrats stole the election and focused on destroying political opposition from many millions of Americans, almost all the savants and critics were portraying it as a victory for the Democrats. No, it is a victory for China and for evil throughout the world.

With the disempowerment of President Trump, the populist movement in America, the military, and the police, China will barely have to lift a finger while witnessing the collapse of its only significant opposition on Earth. The leaders of China, at this moment in time, must truly feel they are destined to the rule the world.

Biden Quickly Begins a Treasonous Empowering of the Chinese Military

Instead of China meeting continued, resolute resistance from us, Joe Biden quickly began appeasing the Chinese Communist Party. In January 2021, he issued a memo prohibiting anyone in the executive branch of government from using terms "China virus," "Wuhan virus" or "Chinese flu." This was striking not only for its shameful catering to China, but also for its progressive-style arrogance in controlling speech in order to control thought and action.[936] Although the prohibition is limited to the executive branch, by muzzling them, Biden deprives America of hearing honest discussions by government officials.

In February 2021, Biden ordered a treacherous reversal of President Trump's policy by ordering a resumption of Fauci's funding of the Wuhan Institute through EcoHealth.[937] Once again, our own government is empowering the Communist military by financially and morally supporting animal research on developing more deadly viruses in their labs. Keep in mind that many consider biological warfare to be the biggest future threat to America and to the world.

Biden's decision to fund the Chinese Communist Party's military bioweapons program makes him complicit with any new leaks of SARS-CoV and any other pathogenic viruses from Chinese labs, which we have demonstrated to be frequent occurrences going back to the early 1990s.

Funding animal research at the Wuhan Institute of Virology also makes Biden complicit with any unrestricted biological warfare that the Chinese Communists in the future conduct against us, any other people, or the world at large. Even if the Chinese do not use these weapons in the future, the threat of their using them already bolsters their unrestricted warfare against us.

We believe that this kind of treasonous behavior by President Biden can only be explained by his family's profound financial dependence on the Chinese Communist Party. It is precisely the kind of action that demands an honorable, just, and constitutional impeachment—gross treachery against the United States involving collaboration with the enemy.

We remain proud that our work led President Trump in mid-April 2020 to cancel Fauci's original funding of China's biological warfare capacity (Chapters 1-2). Biden's continued funding of the exact research that empowered China to make SARS-CoV-2 is dismaying and horrifying. SARS-CoV-2 is not the deadliest of viruses. As we have seen, engineering more deadly coronaviruses at the University of North Carolina has already produced a virus that kills by infecting the brain. Almost certainly, China has its own similar experiments, drawing on American research.

THE INEVITABLE RESULT OF THE WORLDWIDE COLLABORATION BETWEEN GLOBALISTS AND CHINA

While jockeying for position among themselves, all the global predators are united, above all things, in protecting and increasing their investments and other cooperative efforts with China, including Fauci's obsession with working with China on the development of dangerous viruses. Whether they are coming from a hybrid capitalist-progressive viewpoint, as many do, or from a progressive viewpoint, or from a "libertarian" or conservative view, as the Koch's hold, their ideology is corrupted by the reality that they are strengthening China and weakening the U.S., pushing America and the world toward a cataclysmic disaster marked by the loss of personal and political freedom around the world.

What is likely to happen? What will follow if the globalists and China succeed in removing America from its position as the most stable democracy, the torch of liberty, the strongest economy, the strongest military, and the leader of the free world? If America is weakened sufficiently, China will fill the vacuum, initiating an interminable Dark Ages under a Chinese Communist Empire.

Globalist Denial of Their Own Fate

The globalist multibillionaires and occasional multimillionaires are on a suicide mission while naively assuming their place in the future is protected. They act as if they will be safe after the collapse of America because of their ties with China. The great Russian dissident, Alexandr Solzhenitsyn, in his 1978 Harvard commencement address, specifically warned about the folly of seeking safety within the sphere of Communist China:[938]

> At present, some Western voices already have spoken of obtaining protection from a third power against aggression in the next world conflict if there is one. In this case the shield would be China. But I would not wish such an outcome to any country in the world. First of all, it is again a doomed alliance with Evil; also, it would grant the United States a respite, but when at a later date China with its billion people would turn around armed with American weapons, America itself would fall prey to a genocide similar to the [one] in Cambodia in our days.

He said nothing would be gained and all could be lost by making an alliance with "Evil" Communism.

Perhaps the globalists have led such charmed lives, been so protected by their wealth, and so coddled by everyone they meet, including the leaders of China, they believe they are invulnerable. Or perhaps they are simply so greedy for more money, more glory, and more power, they cannot see beyond what China is momentarily providing them.

We want to directly address the globalists for a moment. Yes, the Chinese are fawning over you now, like the grinning Xi Jinping at the 2019 Bloomberg New Economy Forum held in Beijing in defiance of President Trump. Or the same Chinese dictator Xi Jinping speaking at Davos Agenda, a virtual event organized by Schwab's World Economic Forum in early 2021.[939] Of course, the Communists want to feign welcoming you with open arms. Partly because of your help, China is overtaking America at a speed they themselves probably never anticipated. You have empowered China, not only with your investments and your protection, but also with your help in trying to crush President Trump and the American populist movement which has been the CCP's worst nightmare.

When China replaces America as the dominant nation in the world, the Chinese Communist Party will no longer find any use for independent, globalist leaders of any sort. Individuals like Bill Gates and Michael Bloomberg will be largely useless to them. They will put their own dedicated Communists into positions of power. The wealth controlled individually or as a group by individuals like Bill Gates, Warren Buffett, Michael Bloomberg, the Koch family, or anyone else will be confiscated by them. Huge tech companies who provide the Chinese so many services will be maintained for a while as the Chinese gradually take them over and absorb them as well.

No Escape for Globalists

Nor will the globalists find anywhere to escape. There are no islands in the ocean out of reach of Chinese vessels, drones, or technology. Even if they get to the moon or Mars, China will control outer space and their sustaining resources back on Earth. The biggest bankers will not be able to help them because international banking will be controlled by the Bank of China, and all currencies will be tied to China's currency, the renminbi, informally called the yuan.

Many people mistakenly fear the globalists are building a World Government. The globalists never clarify or give any prescription for turning the UN into a genuine world government, nor do they offer any alternative proposals. They cannot build and maintain their own armies, and the monetary system, too, will be taken over by China. Nevertheless, they seem to be content with what they are doing, working together in a loose coalition to influence and control the national governments already in place, including China.

What they fail to see is that China's voraciousness is not like the other governments of the world, except perhaps for those countries ruled by Muslim Jihadists, who will probably be no match for the Chinese Communist Party.

China is not being "influenced" by the globalists; the Communists are merely fattening them like pigs and sheep for the slaughter.

China, with the help of the globalist community, is near to supplanting America as the nation in the world with the strongest economy, the most powerful military, the most sophisticated military technology to include biowarfare, the greatest Third World influence, and the most deeply committed expansionist ideology. The risk is not a World Government like the UN. As soon as America is weak enough, the Chinese Communist Party will take over the world, not as a government, but as the Communist Chinese Empire.

China expert Gordon Chang has warned:[940]

> "Xi Jinping ... is floating these ideas that China has the mandate of heaven over all under heaven, and China actually acts like emperors of old, who believe they're the only sovereigns in the world, which means essentially that Americans are just subservient—we don't have a country, we don't have sovereignty," he said.

During November 16-19, 2020, a mere two weeks after the election results in favor of Joe Biden, Michael Bloomberg and his New Economy Forum celebrated their recoupling with China.

Another six weeks later, Bloomberg was now sure he and fellow globalists and China had actually won, and America and the poor of the world have lost! On January 2 of the New Year 2021, he published a report titled "Bloomberg New Economy: How the Pandemic Shoved Us into the Future." Through an essay by his editor, Bloomberg celebrated what he believes to be the final victory for China and against President Trump and America:[941]

> By one estimate, the pandemic has given China a five-year jump on its goal to overtake the U.S. as the world's largest economy, the Chinese leadership's reward for locking down hard and early while the Trump Administration fumbled its response—and then exacerbated it with mixed messaging, like holding super-spreader events in the White House.

The *Newsletter* of the former Mayor of New York is celebrating China overtaking the U.S. economy, making the American billionaires even richer! This epitomized the evil twist that turns the minds of global predators into pure evil:

> *U.S. billionaires added almost $1 trillion to their fortunes during COVID-19, even as the new laboring underclasses face economic hardship and hunger.*

Yes, Bloomberg is celebrating that the billionaires have grown enormously richer at the expense of most of the rest of the world, including the poor and middle economic classes, and it is true:[942]

> According to Americans for Tax Fairness and the Institute for Policy Studies, the 651 U.S. billionaires have seen their collective wealth grow by more than $1 trillion over the past nine months, while the less fortunate struggle to keep their jobs and put food on the table.

Under the brand-new Biden Administration, life was already picking up for the global predators:

> The idea of any country signing up for a program of "decoupling" from the soon-to-be-biggest global economy is dead. ...

> "COVID has acted like a time machine: it brought 2030 to 2020," Loren Padelford, vice president at Shopify Inc., told The Wall Street Journal.

But the globalists, like most people living on the exploitation of others, always feel vulnerable, and after President Biden had been in office only a month, they again felt the jagged edge of terror through their souls.

Bloomberg: An "Unthinkable" Disaster Looming for the U.S. and China

On February 27, 2021, the Bloomberg newsletter displayed a huge, dark, black and ominous headline: **Bloomberg New Economy: For the U.S. and China, the Unthinkable May Be Inevitable.** Something more terrible than a new pandemic or a faster rate of global warming? No, nothing like that. The "Unthinkable" catastrophe is the U.S. failing to recouple with China:

> [A] full U.S.-China "decoupling"—a scenario under which trade and investment plummet to zero—is unthinkable. But the unthinkable is indeed being contemplated. Indeed, the U.S. Chamber warns that "if the current trajectory of U.S. decoupling

policies continues, a complete rupture would in fact be the most likely outcome."

This is the heart of the insanity of the global predators. Their greatest fear is not for the complete destruction of freedom under a worldwide Chinese Communist Party Empire. It is not the worsening of worldwide suffering and poverty, the next pandemic, or global warming. It is not even a global war between the U.S. and China. Their unthinkable eventuality—the source of their most abject terror—is being cut off from making money with the Chinese Communist Party.

Facing the Facts

Monopolies of wealth and power cannot be allowed to become so strong that they can prevent free speech and block communications of elected officials as they did in stopping President Donald Trump, the most powerful officeholder in the world, from communicating with the world. Facebook, Twitter, and then YouTube threw him off their social media platforms. Google, Amazon, and Apple acted in concert to take down the President's alternative platform, Parler.[943] In that extraordinary, coordinated attack, six of the most influential tech companies in the world—Facebook, Twitter, YouTube, Google, Amazon, and Apple—worked in concert to humiliate the President of the United States and to cut off his communications with America and the rest of the world. Even Snapchat and Facebook's Instagram banned the President.[944]

That incredible, coordinated attack on President Trump and the people of the United States continues as the social media and big tech, backed by Big Pharma and other global predators, have now shut down most advocates of America as a patriotic free nation or who oppose draconian COVID-19 policies. They are especially brutal toward anyone who is critical of COVID-19 vaccines, which bring so much wealth and political power to the globalists.

No one on Earth became off limits after these global predators got a free pass and even praise for taking down the President of the United States. Of all the physicians, scientists, and popular leaders with whom we work, not one has escaped serious acts of aggressive censorship. We are now struck for 90 days by YouTube and will almost certainly be taken down. The most recent strike was for interviewing the author of a book critical of antidepressants. To keep up with our videos and related activities, you need to sign up for our Free Frequent Alerts on www.breggin.com.

These predators cannot be allowed to remain as omnipotent as they have become. They have inflicted COVID-19 totalitarian policies and practices on our nation, rendering us fearful, docile, and submissive to their outrageous predatory demands—while taking from us the main available international communication platforms that we have used to speak freely and effectively.

Communist China in particular must never be allowed to work with these other global predators to undermine our great nation and other democratic republics, thereby threatening the freedom and well-being of our citizens and all humanity.

We must stop fooling ourselves and accept the difficult task of facing evil. The documentation in our book makes clear that several hitherto unimaginable realities can no longer be avoided or denied, as painful and outlandish as they are:

- The COVID-19 attack on America and the world was planned well in advance, at least as far back as 2015-2017. The fundamentals of what would become Operation Warp Speed were put into motion before 2017 with Bill Gates working with CEPI, Pfizer, and Moderna, and with corporate/government collaborations characteristic of what would become Klaus Schwab's Great Reset.

- Experiments with mRNA and DNA vaccines for SARS-CoV going back at least to 2006 repeatedly demonstrated that many vaccinated animals become worse and even die, especially when then exposed to a SARS-CoV. Further studies demonstrated that the SARS-CoV-2 spike protein that the new vaccines make in the human body replicate are very toxic to humans. The global predators know that they are inflicting vaccines on humanity that cause serious illness and kill many immune-compromised and older people. Consistent with a eugenic policy, the vaccines especially kill older and sicker people and also target the ovaries, almost certainly reducing fertility in many women.

- SARS-CoV-2 was developed through the mutual, combined efforts of American scientists and Chinese Wuhan Institute scientists, with the support of Anthony Fauci's NIAID and also NIH. Thus,

Fauci helped the Chinese Communists build multiple SARS-CoV biological weapons and then lied about the origin of SARS-CoV-2 to cover his tracks and protect his Communist allies. As an act of stealth "unrestricted warfare," the Chinese released the weapon.

- When the very existence of the global predators and the Chinese Communists was threatened by President Donald Trump and his America First policies, the Communists almost certainly released SARS-CoV-2. Then they purposely spread it around the world on passenger flights. In this way, they managed to wreck the America's economic and social life—and to stop President Trump from having a second term. They also achieved the collateral gain of reducing one of their greatest social and economic burdens—their disproportionately large elder population.

- With draconian policies and fear tactics, the global predators succeeded in putting America into its current condition of moral and political docility and apathy. A main characteristic of America's tragic condition is the lack of outrage at censorship of the President and now of anyone who threatens the current evolving one-party system and its fellow global predators.

- Most of the world's great institutions have been infiltrated or taken over by globalists, many with progressive or Chinese Communist agendas, and all of them trying to exploit humanity. Their globalist agendas now control the Democratic Party, most big corporations, universities, the medical and scientific establishments, the media and news establishments, entertainment, and even sports franchises and leagues. In America, hardly any aspect of our cultural, social, and economic life is free of control by the globalist, predatory elites.

The people of the United States of America must renew their dedication to the founding principles of the nation, including the Declaration of Independence, the Constitution, and the Bill of Rights. At this moment, freedom of speech and individual liberty are being crushed by a Democratic Party so totalitarian that it seeks to control all communications and even the national election. The ultimate goal is to create a one-party government as in Marxist countries. We must join together to become "Re-Founders of America," devoted to restraining the out-of-control government, to pushing back the global predators, and to reclaiming our rights as a free people.

CHAPTER 35

Takeaway Lessons from COVID-19 Totalitarianism

Very briefly, there are several important conclusions that the facts have forced us to face—conclusions we never anticipated when we began our research:

First, the underlying purpose of the attack that drove COVID-19 policies and practices was threefold: (1) increase the power, aggrandizement, and wealth of the globalists; (2) strengthen Communist China as the ally of nearly all globalists; and (3) weaken the U.S. as a patriotic democratic-republic standing in the way of this rapacious onslaught against freedom and the well-being of humanity.

The globalists had been growing in power almost without resistance for decades until Donald Trump's America-First presidency began to delay and threaten to crush their aspirations for world domination. Forced into the open to resist President Trump and American patriotism, they showed themselves in vast numbers, often openly aligning with the Chinese Communist Party through which many were enriching themselves.

Second, the globalists had long believed that they needed a terrifying worldwide emergency to justify and enforce their assumption of power and nothing seemed better suited than a pandemic. They also saw it as a means of vastly increasing their wealth through enforcing the vaccination of the world. In the process, they would use suppressive public health control to train humanity to be more docile and even subdue those elements of the American people who still believed in individualism and liberty—the twin anathema to globalism.

By 2015-2017, the globalists had largely implemented what would later be called Operation Warp Speed. They had in place federal legislation to establish emergency measures to push through the vaccines. Anthony Fauci was ready to implement the necessary totalitarian policies.

The subsequent attack on America as a bastion of freedom, under the cover of COVID-19, was joined by most of the world's top billionaires, top corporations, the media, the educational establishment, the medical and scientific establishment, and ultimately most of the world's large, international institutions.

Third, the vaccine was purposely made lethal to physically impaired and older people, and destructive to the ovaries, in order to "cull" the population.

Coauthor Ginger Breggin and the three frontline physicians who wrote the introductions to this book—Doctors McCullough, Zelenko, and Vliet—have for months been convinced of this malevolent intention behind the vaccines. Since she fled China early in 2020, Dr. Jan has called COVID-19 unrestricted warfare against the United States. But we could not accept this dreadful conclusion until mounting evidence finally convinced us when the book was nearly finished.

Basic Takeaways from the Book

Here are 16 basic lessons or takeaways that became paramount during our inquiry into COVID-19. The takeaways focus on what is going wrong surrounding COVID-19, so keep in mind that there are many positives as well, including many brave physicians standing up for truth and a groundswell of grassroots organizations standing up for freedom. Especially positive are treatment regimens described in the front of the book with the title "How to Find Lifesaving COVID-19 Treatment Guides and Doctors." Since at least March 2020, successful, inexpensive, and safe treatments have been available for the early treatment and prevention of COVID-19, but are violently opposed by global predators, including politicians and the medical/scientific establishment. This has caused inestimable misery and tens of thousands of unnecessary deaths in the United States alone.

First Takeaway Lesson:
COVID-19 Is Probably No More Dangerous than the Flu

SARS-CoV-2 is not as dangerous as indicated by dire warnings and CDC-fabricated statistics. Children and young adults are entirely safe as a group, and although the risk increases with age and infirmity, it remains relatively low. When treated with proper medical regimens, such as those described by the American Association of American Physicians and Surgeons and also the American Frontline Doctors, hospitalizations and deaths in all ages are small in number and less than the seasonal flu in children and young adults.

Second Takeaway Lesson:
Draconian Public Health Measures Are Causing
More Deaths than They Are Preventing

It is difficult to show that the draconian public health measures are saving any lives. We can say with certainty they are causing many deaths. People in nursing homes in New York State and elsewhere have died by the many thousands due to policies that have packed them together with inadequate treatment or basic health precautions. Older people at home or in facilities are dying from isolation.[945] The rate of suicide in children and youth kept out of school is skyrocketing. People in general are becoming more anxious, depressed, and despairing, raising the suicide rates and causing misery. Healthcare facilities have been closed at critical times to the needs of people suffering from non-COVID illnesses, and when open, people are afraid to go to them. A study in *JAMA* found "huge increases" in all-cause deaths separate from COVID-19, including diabetes, heart disease, and Alzheimer's disease. In New York City, there was a 398% rise in heart disease deaths and a 356% increase in diabetes deaths.[946]

Third Takeaway Lesson:
COVID-19 Vaccine Death Rates Require a Moratorium

Reports of deaths to the official CDC reporting system exceeded 5,000 toward the end of May 2021. Journalists and scientists in several other nations are also reporting excessive deaths. A *pattern* of this size—such numerous deaths soon after vaccinations—is more than sufficient to call a halt to the vaccinations. Even a few hundred deaths from a new vaccine would usually result in its being withdrawn from the market.

Fourth Takeaway Lesson:
Global Predators Are Stopping Use of Effective, Inexpensive Treatments

Hydroxychloroquine and ivermectin are old, inexpensive, very safe, and effective treatments when administered properly and in combination with other therapeutics early in the treatment. These treatments prevent most hospitalizations and most deaths. They have been opposed by the global predators and allies in the U.S. government.

Fifth Takeaway Lesson:
The SARS-CoV-2 Was Released from the Wuhan Institute of Virology

There is no question that COVID-19 was released from the Wuhan Institute of Virology. Scientific papers establish it was one in a family of deadly coronaviruses developed through a collaboration between U.S. and Chinese labs directly funded by Anthony Fauci through his National Institute of Allergy and Infectious Diseases (NIAID). An even more deadly variant has been developed and is being experimented with through Fauci funding at the University of North Carolina.

We cannot be certain that China intentionally released the virus, but it is certain that China purposely spread the contagion around the world and has found great benefit from doing so, especially in the harm that COVID-19 policies have done in America. All of this is documented in the opening chapters of the book.

Sixth Takeaway Lesson:
COVID-19 Public Health Policies Are Unrestricted
Warfare Against America

If we look at the various strategies employed by globalists and domestic progressives in the U.S. to enforce COVID-19 policies, much of it constitutes "unrestricted warfare" against America (Chapters 1-3). Brian T. Kennedy, writing in the September *Imprimis*, discussed this "unrestricted warfare:"[947]

> In conclusion, Americans are not looking for war with Communist China, but Communist China appears to be at war with us. As a first order of business, we must continue what we have at long last begun: building a military designed to deter Chinese

aggression and pursuing trade and other policies that put our own national interests first.

Equally important—especially given the violence in our cities that our foreign enemies cheer—is defending our American way of life and teaching our countrymen why America deserves our love and devotion, now and in the days ahead.

It cannot be over-emphasized: COVID-19 public health tactics resemble and are a part of China's unrestricted warfare against America.

- The suppression of inexpensive early treatments that could still save countless lives with the goal of inducing helplessness in the citizenry and forcing Americans to wait for hugely expensive treatments and vaccines. This further bloats globalist wealth, especially China's billionaires and the power of the Chinese Communist Party.

- The exaggeration of the COVID-19 threat and the overcounting of deaths to enforce radical totalitarian measures that weaken America and encourage submission in its citizens. The ultimate goal is to supplant America with the Chinese Communist Party as the dominant influence in world affairs.

The goals of enriching predatory globalists and empowering China at America's expense are built into all the public health policies and practices described in this book, including:

- The shutdown of the economy.

- The focused suppression of small businesses.

- The disruption of family and social life.

- The disruption of our educational system.

- The focused suppression of religious gatherings.

- The crushing of the working and middle classes.

- The rejection of patriotism in favor of planetary concerns.

- The enforced submission and isolation of our citizens with masking, physical separation, and isolation, making them amenable to authoritarian and totalitarian control.

All the above and many other policies described in our book are leading to the destruction of liberty while turning Americans into docile members of a totalitarian state. It is unrestricted warfare against the United States, strategized by China and conducted by the progressives, the Democratic Party, and the global predators. From Fauci and Bill Gates to Biden and Kerry, many Americans are contributing to China's aspirations to destroy the United States without having to fire a shot.

Seventh Takeaway Lesson:
Globalist Capitalism and Communism
Are Joined in War Against America

One of the greatest misperceptions of modern times is that capitalism and communism are locked in an ideological war against each other. That is no longer true. Instead, global capitalism and Communist China are joined in an existential war against us. That is a major lesson to be taken away from seeing predatory billionaires and corporations working so closely with the Chinese Communist Party and its corporations. It is epitomized by the Biden Administration adopting the Great Reset—the triumph of predatory capitalists masquerading as idealistic utopians. It is why Biden and the progressives are so soft on China. They are actually collaborating with China against us.

Corporate America no longer exists; it is now *Corporate World*. All, or nearly all, large American companies have become global predators with little or no patriotic ties to America and no commitment to advancing or protecting the free enterprise system.

Genuine free enterprise does not exist in global businesses. All those many corporations participating in and supporting conferences by Klaus Schwab and Michael Bloomberg are aligning themselves with China, recoupling at the expense of America and freedom.

The only genuine free-enterprise models remaining in the world are probably relatively small business owners—the kind of people who invest their hearts, labor, and life savings in local companies, shops, or restaurants.

There are also more rare individuals who have a new idea or invention and find a way to market it on a small but profitable scale. Many of the small business owners are professionals who remain independently in practice for themselves, such as artists, tradesmen, lawyers, ministers, therapists, and the few remaining physicians like me who are not employed by large health systems. *Ironically, the free market seems to thrive only among the relatively powerless, the people despised and exploited by the elite—those being crushed by COVID-19 politics.*

Perhaps this has always been the case. The free market works until companies become monopolies or at least powerful enough to influence and possibly partner with governments. Going back to the time of the American Revolution, the far-reaching East India Company was already in partnership with the British Empire to exploit anyone they could. Capitalism supported by government and unleashed on a large scale has always become predatory. This disappointing conclusion seems consistent with human nature and its greatest flaw—the lust for power. Large companies may lead to efficiency, reducing prices and maximizing profits, but they also lead to corruption and the endless pursuit of more wealth and power.

Once these individual or family businesses become giant global enterprises, such as Koch Enterprises or Walmart, they become global predators, even though "family owned." As our assistant Missy observed on reading this, "They start out reaching for the American Dream and turn it into a nightmare for the rest of us."

We have no simple answers to the reality that both big government and big business lose respect for freedom and for humanity and become predatory. As I describe in my book *Guilt, Shame and Anxiety,* we did not evolve to feel profound empathy for strangers. We evolved to distrust and distance ourselves from them and sometimes to fight them for our lives.

Instead of facile solutions, we need more serious discussion of the impossibility of freedom existing in a world dominated not only by big government, but also by big business. Conservatives, libertarians, and others with genuine concerns about human freedom must face and discuss the differences between the free market and predatory capitalism. Many small businesses act upon free-market principles and voluntary exchange as best they can through a tangle of government regulations. Large corporations, in varying degrees, impose themselves upon humanity by using government to their advantage.

As we have mentioned, we expect that if the various global predators succeed in helping the Chinese Communist Party become the dominant nation in the world, China will not tolerate any competing billionaires or corporations. They will do China's bidding or disappear off the face of the Earth. Unfortunately, we cannot wait for China to take those steps; by then, it will be too late. The United States with an America-First policy must disempower the global predators—very soon and largely by itself.

Eighth Takeaway Lesson:
Powerful Corporations Have Overwhelmed Our Democratic Republic

The U.S. government was founded upon a Constitution and Bill of Rights to prevent the government from overwhelming the individual, but the Founders could not have imagined businesses becoming more powerful and domineering than governments. The idea of a Google, Twitter, or YouTube dominating public opinion and censoring and deleting the communications of the President of the United States could not have entered their minds.

Globalists know their common adversary is the combination of American traditionalists, patriots, conservatives, constitutionalists, religious people, workers, and owners of small businesses and professional practices. These people and groups, and not the billionaires or giant corporations, are determined to defend their country and to protect its principles of individual rights and political freedom.

Predatory globalists seek to demoralize and disenfranchise these patriotic, freedom-loving groups to bring about the collapse of America. That will give them free rein over the world. The globalists within our own government do not want to support America First policies. They know they must incapacitate America before they can take over and exploit the world to their own advantages.

Under the guise of treating COVID-19, progressive politicians closed small businesses, churches, and synagogues. They created massive unemployment among the poor and destroyed the financial stability of the middle class. Global predators have always *intended* to destroy this conservative beating heart of America—a life force based on individualism, independence, sound traditions, patriotism, freedom, and God.

Ninth Takeaway Lesson:
America Is Becoming a One-Party Totalitarian State

A major lesson we have learned from the events of 2020 is that America is extremely vulnerable to a leftwing totalitarian takeover and that it is already well along its way. The public health policies implemented through COVID-19 have conditioned Americans to become docile and intimidated to a degree that surprised and saddened all of us who love America and freedom. The corrupt election "victory" for Biden was probably the deadliest blow ever against America as a constitutional republic, against freedom, and against patriotism. It was quickly followed by the last-minute attacks on President Trump, the shutdown of the President's social media, and the ganging up of multiple tech giants to close down Parler, the one significant freedom-of-speech alternative to Twitter. The President's close associate and personal lawyer, Rudy Giuliani, was removed from YouTube. Desperate progressives, the Democratic Party, and the globalists—mostly all the same predators—have decided to turn America into a one-party state.

As President Trump was about to leave office, the global predators mounted a hastily concocted, second impeachment in the House of Representatives. The nasty Third-World quality of the impeachment was demonstrated by the circumstances: This was the second impeachment of his term; it had no due process whatsoever and was doomed to fail in the Senate like the previous one.

The charges concocted against President Trump were much more accurately applied to the Democrats who had indeed been inflaming murderous urban riots for months. The hastiness of the impeachment left no time for any due process. Finally, the attempt to convict President Trump after he had already left office was unconstitutional and unprecedented. Ultimately, the Democrats sent a message to at least half the American people declaring, "We will prosecute and persecute you and your leaders to the ends of the Earth."

We must face how despicably hateful and unlawful the Democratic Party and its following in the media, judiciary, and government have become. We must face how cowardly and unprincipled the Republican leadership has become. Another shocking revelation is how both political parties and most American institutions have become servants of global predation and specifically how responsive they have become to the aims of the Chinese Communist Party.

COVID-19 continues to show us the seriousness of China's undeclared asymmetric war against us. More unexpected to us, it has demonstrated how all or nearly all identifiable global predators are closely tied to China for their increasing wealth and power, including Fauci, Bill Gates, Warren Buffett, Bloomberg, Klaus Schwab, the Koch family, Biden family, John Kerry,[948] many federal agencies, and many leading Democrats and Republicans.

We reviewed the top 20 American billionaires and could find very few who were vocal patriots. Oracle's Larry Ellison was seemingly an exception. Nearly all have massive business attachments and other affiliations with the Chinese Communist Party and some with radical progressive activists. The story is the same with all the wealthiest big tech companies, the media, large banks, much of the publishing industry, the most important schools of public health like Johns Hopkins and Harvard, large universities, the sports industry, and just about every other center of wealth and power in America. We know others realized this before us, but it was an enormously enlightening shock to put the details together with COVID-19 as we have done in this book.

We saw America's crisis and potential demise unfold through the window of COVID-19 and burst forth in the entire political arena, displaying globalism and the Chinese Communist Party as it seeks to dominate and exploit humankind.

Tenth Takeaway Lesson:
The Corruption of Medicine and Science

To many of us who are physicians and scientists, the most appalling revelation is the predatory globalist attitudes of the medical and scientific establishment: Institutions like the FDA, CDC, NIH, and NIAID, and their foreign counterparts; American or foreign research laboratories; medical associations like the AMA; and journals we have already mentioned. This group includes people like Anthony Fauci, Peter Daszak, and Tedros of WHO, as well as the editors of journals and the heads of medical associations.

We were dismayed and disgusted with how completely the American and international scientific establishment caved in to the Chinese and globalist aims. Mollifying these predators became far more important than scientific and medical truth. The leading three medical journals—the *New England Journal of Medicine*, the *Journal of the American Medical Association*, and *The Lancet*—covered themselves with shame by lying or distorting the truth to protect China from being criticized and to prevent inexpensive, very safe, very old, and effective COVID-19 treatments from being used. Capitulating to the global predators became even more important than the lives of vast numbers of people around the world who are still dying because the medical and scientific establishment has been pandering to and joining with the avaricious globalists, using COVID-19 to achieve more wealth, glory, and power.

Post-election investigations have shown that some of our state economies and governments are economically dependent upon China![949] So are many foreign governments, the UN, and its World Health Organization (WHO), and the Pope.

Eleventh Takeaway Lesson:
China's Vast Interference in American Affairs

Working people and the middle and upper-middle classes are precisely the people despised by official Marxist theory and progressives. They are demeaned as the "bourgeoisie" and the "proletariat." It is no coincidence that these are the same people disparaged by contemporary progressives and the Democratic Party who increasingly promote the interests of the worldwide elite and the Chinese Communist Party.

By digging deeply into the destructive COVID-19 policies, we have found yet another manifestation of a struggle that has been going on since civilization began between liberty and oppression—between those who want freedom and those who would enslave.[950] Predatory globalists are slavers and we are their slaves.

America not only needs to decouple from China; America's survival also depends on decoupling from predatory billionaires and corporations throughout the world. That will require a powerful populist movement and the rebirth of the United States as a strong and independent democratic republic.

China's long-term plans to infiltrate and dominate America should not have been such a surprise to us. Like most Americans, we simply were not paying enough attention. The Chinese blueprint for their triumph and our demise has been well documented in books like Brig. Gen. Robert S. Spalding's *Stealth War* (2019) and in his interview with me in April 2021.[951] Other books include Steven W. Mosher's *The Bully of Asia: Why China's Dream Is the New Threat to the World Order* (2017); Michael Pillsbury's *The Hundred Year Marathon* (2016); Gordon Chang's *The Great U.S.-China Tech War* (2020); and Bill Gertz's *Deceiving the Sky: Inside Communist China's Drive for Global Supremacy* (2019).

In July 2020, *The Epoch Times* newspaper summarized the reality that our research into COVID-19 had forced upon us:

Stealthily, surreptitiously, and with sweeping precision, the Chinese Communist Party (CCP) began a decades-long war against America for world domination by utilizing a military strategy known as "unrestricted warfare" that continues today.

Unbeknownst to most of the population, the CCP has infiltrated almost every major avenue of life in the United States, leaving virtually no industry untouched. While this threat has largely existed undetected, the effects it's had on the nation, as well as its geopolitical consequences, are far-reaching.

Our research into COVID-19 brought us face to face with this ugly and potentially demoralizing reality. We Americans must wake up, before it is too late, to the threat—a fate that goes beyond the abuse of President Trump and his tens of millions of supporters who love America. The press would like us to believe that we are enamored with Trump and that Trump is enamored with himself, but the truth is that all of us are enamored with America and its people who remain the last hope for humanity to resume its progress toward a world of personal responsibility and individual liberty.

Twelfth Takeaway Lesson:
Shared Values Transcend Diversity

Diversity mongering is tragically destructive. The progressives and Democrats have been pushing diversity, slicing and dicing us up into so many different categories of human that we forget we have anything in common. They do this in order to divide and conquer us. Although presenting themselves as champions of every possible underdog or minority, their strategies aim at making themselves top dog.

With the excuse of COVID-19, division among Americans has been further fanned, especially in the racial area. The issue of the vulnerability of urban African Americans to COVID-19 has been blamed on white racists when it is more the product of poverty and family dysfunction that are directly attributable to progressive policies and leadership in Black urban communities.

New categories of separation among Americans have been created. The "good" people who wear masks, and the "bad" people who question their value; and those who blindly accept globalist predatory science are also "good" and those who don't are "bad."

Two groups have been singled out by progressives and globalists for degradation. Children are expendable and subject to experimentation and manipulation while older people are a burden suitable for being left to die untreated.

Diversity cannot be an American goal or the goal of any successful society. Everywhere in the world, diversity exists and has led to the severe conflicts that made people flee to America. America has thrived because we have held values in common that equal or come ahead of our diverse backgrounds and interests.

No better example of this was ever displayed than by heroic African Americans during World War II. Despite federally enforced racism at home and segregation in the military, they went overseas to do everything in their power to fight for the American Dream of freedom and opportunity.[952] Many of America's bravest and most effective warriors were Black individuals who risked and gave their lives. They did this out of patriotism and a determination to take their rightful place in the American society. They did this despite knowing they were returning to an America that was deeply flawed by Jim Crow laws and racism, often fostered by leading Democratic politicians with ties to the KKK.[953]

Americans' shared values have been embodied in the Declaration of Independence, the Constitution, and the Bill of Rights, as well as in our War of Independence and our Civil War in which many shed blood in the cause of liberty. Yes, there were complex motives for both wars, some noble and some not, but the idea of liberty was embedded in them. The American Revolution and the Civil War made us stronger and more united as people and more devoted to liberty.

America was diverse from the beginning and gradually progressed toward more unity and hence to enormous success compared to other nations. The most extreme diversity was created by the horror of slavery. Nothing is more "diverse" than master and slave. To reverse the process of the cultural melting pot that made us all Americans—to make us feel more diverse—will destroy us. Make no mistake: the destruction of America is the goal of American progressivism and its partners, the global predators, and the Chinese Communist Party.

America must be a community of people striving to be personally responsible and free, or it will not survive. By striving to be free, we mean people who work at respecting their own rights and the rights of others as equals. We also mean people who strive to take responsibility for themselves while wanting to help those who are struggling. We mean people who believe in treasuring others and protecting their freedom, as well as their own. We mean people who believe in positive ideals that are greater than themselves, providing a higher standard toward which to strive. We mean people who try to conduct their lives based on responsibility, liberty, and love.

In our personal lives, we have found that people who strive to live by principles of liberty, responsibility, and love can be found in any nation, religion, or culture, and certainly in every race. We have known individuals of every race and from many seemingly alien cultures and nations with whom we have shared most or all these values, including Christians, Jews, Muslims, Buddhists, humanists, and atheists.

Although we are white, we are well aware that we have more in common with many individual Black or Chinese people than with many white people. Although one of us is Christian and the other Jewish, and one of us is a woman and the other a man, we certainly have much more in common with each other than with anyone else we have ever met.

The worship of diversity is a trap because it teaches us that individual whites have more in common with each other than with individual Blacks or that individual Blacks have more in common with each other than with individual whites. It teaches that women similarly have more in common with each other than with individual men or that gay people or bisexual people have more in common with each other than with any straight people.

Across nationalities, cultures, genders, races, and other categories of diversity, people striving to be good—to be responsible, to respect liberty, and to be a source of love—will have more in common with each other than with the people in their particular category of diversity. Freedom, personal responsibility, and other shared values have made us a great society and nation, despite the inherent difficulties in having considerable diversity as a given of our great society.

The "melting pot" concept is not a mirage or a fool's dream; it is the reality that has made America great. America's diversity has always been secondary to its shared values, and when that comes to an end, we will sink into anarchy and despair. We will become easy picking for the global predators who hate our individuality and our ability to join together when our liberty is threatened.

Thirteenth Takeaway Lesson:
Trusting and Believing in Each Other

The need to feel trusting is in our nature from infancy through our older years. Trusting other people is central to sanity and to effectiveness as a person. When our trust for people becomes fragile or broken, we become emotionally distressed and, in the extreme, very disturbed. Any weakening in trust makes it more difficult to build and to conduct personal relationships or business transactions. *COVID-19 policies and practices have been overwhelming us, particularly because of those moments when trust has been threatened, leading to hopelessness, despair, and a compulsion to withdraw inside ourselves.*

All totalitarian institutions seek to sow distrust of each other among the people so that they will lack the strength and the will to resist. *COVID-19 policies and practices have been used as the great sower of distrust among people. It has been done purposely to render us more docile and controllable, more obedient to arbitrary authority and to exploitation.*

On February 3, 2021, based on my interview with him, reporter Conan Milner of *The Epoch Times* wrote a feature story titled "The Trust Imperative: Human Beings Depend on Trusting relationships, and suffer deeply when trust is broken."[954] *The Epoch Times* is one of the few newspapers that reports accurately and in depth on the issues investigated in this book. We had never heard of the outlet until COVID-19 sent us searching for new sources of information.

Therein lies one key to rebuilding our sense of trust: we must make new relationships with people, groups, and organizations we can realistically trust. Before COVID-19, we had also never heard of the Association of American Physicians and Surgeons (AAPS). Before COVID-19, we did not know most of the people we thank in our acknowledgments and now work with in a common effort to save America.

Build your life around trustworthy relationships based on shared values. Despite the inevitable disappointments, you will build a much stronger and satisfying life and a better America. Innumerable alternatives to mainstream, global, predatory institutions are being developed by people who believe in personal responsibility, self-determination, personal freedom, the American Dream, and political liberty. Join them and create new groups as well. We must retake the future of the United States of America—the greatest nation in human history.

Fourteenth Takeaway Lesson:
The Never-Ending Threats

There will come a time when COVID-19 begins to wane or when aggressive politicians and billionaires can no longer see any profit or gain from exploiting the pandemic. This will not diminish their threat to human autonomy and freedom already being imposed upon us using the cover of COVID-19. Beyond this, the globalists are busily reinventing global warming as the real emergency to justify top-down government control over us, and failing to make that work, they will find or create other threats to justify keeping us in our present state of docility and semi-confinement.

Bill Gates, much like other globalists, sees COVID-19 as a kind of experimental testing ground for how much control can be exerted over the world's population. Gates has said, "If we learn the lessons of COVID-19, we can approach climate change more informed about the consequences of inaction."[955]

Notice Gates is not talking about learning from the negative consequences of his actions, which have been catastrophic in terms of individual lives, national economies, and political freedom. Instead, he focuses on "the consequences of inaction." For globalists, COVID-19 tyranny is just the beginning.

COVID-19 and the predicted climate change catastrophe are often linked together by globalists. For example, Klaus Schwab's World Economic Forum (WEF) has a COVID Action Platform which is simultaneously used to promote "a carbon neutral world by 2050"[956]—the same timeline promoted by Bill Gates.

Fifteenth Takeaway Lesson:
Innumerable Individuals and Entities Must Be Brought to Justice

Our research initially focused on Anthony Fauci, who quickly led us to Peter Daszak of EcoHealth Alliance, Bill Gates, and Tedros at the World Health Association (WHO), then on to the Chinese Communist Party. It will be up to others who have the authority to decide who among the predators warrants formal prosecution by the courts and official agencies.

Sixteenth and Final Takeaway Lesson:
We Must Fight for Our Freedom—Now!

On November 13, 2020, Chief Justice Samuel Alito expressed the depth of the assault on liberty in America in the guise of public health and defeating COVID-19. Here are some excerpts:[957]

The pandemic has resulted in previously unimaginable restrictions on individual liberty.... And I think it is an indisputable statement of fact, we have never before seen restrictions as severe, extensive, and prolonged as those experienced, for most of 2020. Think of all the live events that would otherwise be protected by the right to freedom of speech, live speeches, conferences, lectures, meetings, think of worship services, churches closed on Easter Sunday, synagogues closed for Passover, on Yom Kippur. Think about access to the courts or the constitutional right to a speedy trial. Trials in federal courts have virtually disappeared in many places. Who could have imagined that the COVID crisis has served as a sort of Constitutional stress test? And in doing so it has highlighted disturbing trends that were already present before the virus struck.

The U.S. Constitution and the Bill of Rights are *intended for* emergencies, not *expendable during* emergencies. We must cleave to them and enforce them with even more vigor when they are under attack by global predators empowered by COVID-19. If human societies were not plagued by perceived or real emergencies and if seemingly good people were not easily compromised by lust for wealth and power, there would be no need for a Bill of Rights. Individual rights are the heart of the American Experiment. When bypassing the Constitution and the Bill of Rights, we risk the rapid unraveling of the American Dream and its foundation in individual liberty.

The global movement to promote top-down government can only achieve its end by trampling on our individual rights. It starts authoritarian and then quickly progresses to outright totalitarian. In countries around the world, from democracies to dictatorships, public health policies and practices, allegedly to control COVID-19, are being used to rachet up measures of top-down control. In the U.S., with compliance or outright support from the left and the right, politicians are throwing their weight behind persistently locking down America, and with it, our Republic, and our freedom. They are working in collaboration with giant international corporations who want to own the governments of the world. They are doing so with the support of news media, network TV, universities, the embedded globalist bureaucracy, organized sports, and entertainment—all calling for more top-down government and less individual liberty.

We Are the Hope of the World

When Trump became president amid a revival of American patriotism and devotion to liberty, the global predators united in turning against him with every vicious attack and destructive means at their command. They also began blocking America's best early medical treatments, thereby crushing the hopes of many people for a speedy recovery from COVID-19. Their purpose went beyond pushing their drugs and vaccines. They needed to further undermine the greatest obstacle to their unbridled ambitions to control the human population—the United States of America led by a populist, conservative president determined to Make America Great Again and to defend our personal and political freedom on the domestic front.

Our nation must not give up its patriotism and principles of liberty. For our own sake and humanity, we must prevail. This is not to say America is perfect, but it is to say that our ideals are the best ever conceived and built into a nation's founding and laws. Most importantly, America is the only hope we have for the future freedom of the world.

Remember this: There would very likely be no Free World without the founding of the United States of America on the principle of liberty. There would have been no wave of democracies arising in Europe and spreading here and there throughout the world. Without America, the world would probably have been partitioned among competing totalitarian regimes in the 1940s. Imagine America divided up for military governing by Nazi Germany, the USSR, Communist China, and Imperial Japan.

No one knows what would have happened without the rise of America as the dominant power in the world, but it may reinspire us to contemplate a world without a patriotic, liberty-imbued American people willing to fight for their freedom and the freedom of their allies. We must contemplate this because that dark future is once again upon us in the form of the Chinese Communist Party and whatever allies—such as Russia, Iran, or North Korea—it will bring into its orbit as its big push to become the dominant empire continues.

Because human beings are so imperfect and larger institutions amplify those imperfections, there can be no end to our work of defending and expanding liberty. As events around the world constantly remind us, liberty is a mighty spearhead of progress. It's also a fragile construct like a thin crystal kept intact and polished to a gleam only by the devotion and determination of brave people. Liberty can be demolished in the blink of the historical eye.

For the first time ever, Americans have dramatically forfeited their liberties almost overnight without a fight because it has been cloaked as a necessity of public health. This must be reversed; liberty must be recovered by reimplementing the Constitution and the Bill of Rights, along with our great traditions of freedom, individuality, and independence.

The Founders were not simply about liberty. Liberty had a purpose based on the Judeo-Christian religions and traditions. It was about the moral development of the individual into a person who would take responsibility for both self-development and good citizenship, a person who would use and protect liberty in order to build a more moral and spiritual life for himself and society.

REAWAKENING OUR MORAL LIFE AND SHARED VALUES

In the previous chapter, we quoted from Alexandr Solzhenitsyn's address to the 1978 Harvard commencement. The great Russian intellectual and novelist, driven from home by the Communists, called his speech "A World Split Apart."[958] He spoke to the West and in particular to his American audience about our failing courage and softening moral backbone. In a sentiment close to ours—albeit so remote from today's globalists like Fauci, Gates, Bloomberg, Biden, and Kerry—Solzhenitsyn spoke of the inevitable disaster for those who cozy up to China.

Thus, more than four decades ago, Solzhenitsyn warned us against exactly what has now happened to the remarkable detail of our giving China biological weapons to use against us.

Solzhenitsyn also warned that material comfort and freedom can never become ultimate values because they erode into lawlessness and immorality. We must instead rediscover morality. He decried our spiritual condition:

> [W]e have lost the concept of a Supreme Complete Entity which used to restrain our passions and our irresponsibility. We have placed too much hope in political and social reforms, only to find out that we were being deprived of our most precious possession: our spiritual life.

And because of this lack of moral and spiritual guidance beyond mere humanity, we have lost our courage. Solzhenitsyn's words are a clarion call for us to regain our heart and our valor:

> A decline in courage may be the most striking feature which an outside observer notices in the West in our days. The Western world has lost its civil courage, both as a whole and separately, in each country, each government, each political party, and, of course, in the United Nations. Such a decline in courage is particularly noticeable among the ruling groups and the intellectual elite, causing an impression of loss of courage by the entire society. Of course, there are many courageous individuals, but they have no determining influence on public life.

Many takeaways from this book have been difficult for us to face. We list some of the tougher ones because, without fully confronting them, we are lost:

- The U.S. and China collaborated in making the SARS-CoV viruses that resulted in SARS-CoV-2, often with funding by American government agencies, especially NIAID and NIH.

- COVID-19 was planned in detail for years by Gates, Schwab, WHO, the pharmaceutical industry, the Chinese Communists, and many other global predators. It has always been about profits and power through forcibly vaccinating everyone. The globalists describe themselves as leading a new "world governance" being implemented through COVID-19.

- The Chinese Communist Party almost certainly intentionally released SARS-CoV-2 to ruin Trump's presidency and to reduce America's power. It is unquestionably true that they spread it around the world with the collaboration of most or all of the global predators. COVID-19 is part of the CCP's relentless "unrestricted warfare" against the United States.

- Global predators have taken over most of America and the world's major institutions, indoctrinating several generations to surrender their personal freedom and patriotism to the "greater good."

- Global predators do not care about "the people" or "humanity." They are obsessed with wealth, self-aggrandizement, and power. The draconian measures of control during COVID-19 are aimed at frightening and subduing the world, especially the Western democratic republics.

- Global predators are unrestrained by faith in God, patriotism, kindness and love, or any sound ethics. They are guided by the Communist principle of "the end justifies the means." Millions of deaths do not daunt them.

- Free market principles do not apply to large, global corporations. Predation is their norm and their business strategy.

- The mouthing of "progressive" or "idealistic" principles by global predators, whether they identify as leftists or capitalists, is strictly to fool the people.

- Many naïve, idealistic, and caring people are being misled by propaganda from the left and the right. Almost always, these idealists are without power or influence, but they hope or believe that the politicians they support are guided by the same good intentions. How leading globalists will team up with anyone who wields predatory power is exemplified by the recently announced business partnership between Bill Gates (the self-identified American capitalist) and George Soros (the self-identified progressive who is trying to destroy America).[959] Globalists view the interests of nations, religions, and political parties as obstacles to their global predations, and they reject individual and political freedom as utterly incompatible with their aims.

In the final chapter, we look at the steps we need to take to regain our freedom.

CHAPTER 36

Action Steps to Save Freedom

We can personally restore ourselves by living by The Primary Principles in my book *Wow, I'm an American!*:

Protect freedom.
Take responsibility at all times.
Express gratitude for every gift and opportunity.
Become a source of love.

Here are some active steps each of us can take:

Each and every one of us can become braver.
Each and every one of us can become an activist.
Each and every one of us can speak out for and defend America and her liberty.
Each and every one of us can face and walk through fear to do our share.

Those of us who overcome fear may never become the majority, but all we need is an increasingly inspired number of us to right this great ship of freedom we call America. In this battle for freedom, there is a giant reservoir in America of people who believe in our Judeo-Christian traditions and our founding principles and documents.

The creation of this nation and the establishment of its freedom was not won by a majority vote or by a consensus, and certainly not by an overwhelming number of citizens united in the War of Independence. It was won by a courageous group we call the Founders and the many thousands who backed them and responded to their call, while many others stood on the sidelines or undermined the revolution.

Judeo-Christian Values and the Struggle for Freedom

The Founders drew heavily on Judeo-Christian traditions, stories, and principles. The Exodus of the Hebrews from Egypt was an inspiration and a symbol for the colonial struggle for freedom. Nearly all the Founders were Christians, and all had been brought up in a deeply Judeo-Christian culture from which they derived a universal concept of liberty for all of God's children.

These principles led to the Declaration of Independence and the American Revolution, followed later by the Constitution and Bill of Rights. Yes, the Founders can be criticized for setting aside the issue of slavery in order to join together for the revolution and then for the Constitutional Convention.

The Founders put off a civil war to fight a war for their mutual independence, making inevitable the blood-soaked Civil War that would tear the nation asunder in less than a century and kill many more citizens than the War of Independence. The American dead in the War of Independence are numbered at 25,000. In the Civil War, the number of Americans dead on both sides was in the range of 620,000.[960] Twice as many Americans died at Gettysburg than during the entire Revolutionary War.

The principles of freedom the Founders so powerfully promoted, fought for, and embodied in our founding documents would inevitably lead to a bloody Civil War in which Republican President Abraham Lincoln would initiate the abolition of slavery. The concept of life and liberty as gifts from the Creator— and not from a king or any government—came directly from Judaism and Christianity as expressed in the Bible.

Whether or not you have a personal belief in a loving God, you have been deeply molded by this Judeo-Christian enlightenment that makes all of us God's children, every one of us worthy and valuable in our own right. You may have translated this concept, as I did earlier in my life, into a more secular and abstract idea. For me, it was a belief that love is a spiritual phenomenon that permeates and energizes life. Unfortunately, that secular orientation is seldom sufficient to motivate people to fight for their freedom.

Much as the Founders, we need a belief in both the individual and what George Washington called Providence, which most people call God. We need that divine spirit that imbued the Founders to permeate and inspire us to overcome the alliance of global predators who now increasingly oppress us.

Freedom Without Love or God

In Chapter 7, we looked at the perversion of liberty in libertarianism as illustrated by the Koch family and the politicians and institutes they support. Because of their selfish and loveless approach, these libertarians (but not all) end up calling for open borders and "free trade" with the most enslaving regime on the face of the Earth. This understanding has enormous importance in how we plan and take actions in the current assault on freedom. *The call to freedom is insufficient in itself to bind us together and to enable us to risk our lives for each other!*

As I described in *Beyond Conflict* (1992), the first key libertarian principle is voluntary exchange—economic relationships should be free of both force and fraud. The second, closely related principle is that force should be limited to self-defense. These two principles lead to the political application that government should be limited largely, if not entirely, to protecting individual freedom. Police and courts are needed to maintain domestic freedom with a military to defend the nation from external threats. Even those functions, a staunch libertarian might argue, should be privatized as much as possible.

The two fundamental principles of libertarianism—free exchanges between people characterized by honesty and the use of force only in self-defense—hold great appeal to me. I strive to live by them in my personal, professional, and business relationships. In fact, they are easiest to carry out in our personal lives and on a small economic scale, such as a small business or profession. We can personally decide to be open and honest in our personal and work relationships and to avoid coercion except when needed in self-defense.

Many people do their best to live by these freedom or libertarian principles without intellectually identifying with them. Few people, if any, operate on these principles on the large world stage dominated by globalist governments, billionaires, and international corporations. To become that powerful, it seems, human beings must become corrupt. If power is your goal, you automatically become corrupt. Everyone "at the top" becomes a predator.

The few exceptions, such as my heroes—George Washington, Abraham Lincoln, Mahatma Gandhi, Frederick Douglass, and Martin Luther King, Jr.—must be pillars of strength who defy their own corruptible human nature in their political lives. Being human, they are plagued by personal vulnerabilities. Also, living in a world of humans, they are too often killed.

After becoming active in the Libertarian Party in the early 1970s, I was sufficiently engaged to be elected by the annual convention as one of two at-large members of the National Committee. I discovered an unfortunate Ayn Rand-inspired lack of empathy for others among many libertarians, especially when they look at politics rather than their personal lives. Political freedom was their formal first principle, and too little concern was given to love, generosity, and the traditions that hold families and societies together.

The libertarian leadership did not understand, as our Founders did, that freedom can only work in a society where the citizens *are morally committed to* defending each other's freedom and *sufficiently concerned for their fellow citizens to be charitable* toward those unable to care for themselves within a vast nation. For example, if the government has no role in providing a safety net, many people will inevitably suffer and die of neglect in complex industrial societies, and the nation will become unstable and totalitarian.

A fundamental problem with libertarianism is its lack of spiritual basis. If we simply posit that individuals have rights, we do not necessarily treasure the people who have these rights. In contrast, if our rights come, not from logic itself or empirical observation of its good effects, but from God, as the Declaration of Independence tells us, then "liberty" is not the ultimate value. The ultimate value is the treasuring of each and every human being as made in God's image and, as the Quakers say, all human beings have "that of God" within them. In the Judeo-Christian tradition, this puts love of God and love for each other as the centerpiece of human life. *Freedom is not the ultimate value, but the context for the ultimate value, which is to love one another as God loves us.*

As we gear up to stand together against the forces of evil and totalitarianism, we must begin with a profound sense of our engagement with each other through love of one another and for God, along with love for our great nation that was built on God, freedom, morality, and patriotism.

Our Present Limitations

There are no other people in the world with the capability or potential will to reverse this worldwide catastrophe—not merely the tragedy of COVID-19 itself, but also, much more importantly, the massive oppression inflicted by global parasitic predators. They have taken advantage of the pandemic and, in doing so, made themselves known to us by their enthusiastic support of each other in stealing our freedoms and exploiting us. They have all grown wealthier, more self-aggrandizing, and more powerful while we have grown less free.

With its grounding in the Constitution and Bill of Rights and its roots in a people who love liberty and live by higher ideals, America must lead the fight against worldwide oppression. We must begin with a revival of our personal principles and devotion to living meaningful lives. This will take courage, a profound sense of moral commitment, and the will to come out into the light with a fierce determination to fight.

We Must Never Feel Sorry For Ourselves or Complain

In these suppressive times, it's easy to become resentful or feel sorry for ourselves. As a subtle part of this psychological helplessness, we tend to cry out, "How can people behave this way?" or "What is the matter with our leaders?" or "What has gone wrong in America?"

The issue is not how can our leaders be so callous, grandiose, or murderous. That is how leaders of large groups have always been. The issue is not why so many Americans have become docile and obedient. That has always been the fate of most people since we became parts of large organizations or governments.

Instead, we must face reality. Ever since we humans began living in villages— instead of hunting and gathering in extended families, we became vulnerable to being politically and socially oppressed and to accepting it. The basic story of written history is about the domination of the many by the few. Except for the most fleeting instances of relatively free city states, *being oppressed has almost always been humanity's condition over the past several thousand years of what we call civilization.* We have lived under invaders, slavers, emperors, religious oligarchies, monarchs, and dictators who have, if anything, grown more violent and oppressive in modern times.

When we become aware of the deterioration of freedom in America, it should be no surprise. It is a return to the commonplace, the norm, the given for any people who are not actively supporting and fighting for their freedom.

The questions to ask are: "How did the miracle of America come about?" "How did we create and maintain the American revolution for so long?" "What must we do to revitalize it?"

Because of the richness of the land and its distance from the monarchy, which is why so many people came here to be free, especially in the practice of their religions, those American colonists (who were not enslaved) were already among the freest people in history. They fought the British Empire only when it became necessary to maintain their freedom. They did not fight to extend their power over others or to destroy or dominate anyone. This was unique in history. The colonists drew on their own personal experience, on the Enlightenment, and above all else, on the Judeo-Christian traditions and religions to establish principles of freedom in their new government.

Put simply, America has always been exceptional. The current degradation of freedom should not be treated as unexpected, hard to believe, or even unusual. If people do not fight to maintain their freedom, this degradation is nothing more than a return to life as usual since recorded history. In large groups, someone or some group will always try to dominate and oppress—a principle so well understood by our Founders that they built checks and balances into their new government and bolstered it with a Bill of Rights.

We must never feel sorry for ourselves or wonder pitiably how our leaders can be treating us so badly in America today. Our Founders gave us the gift of a vast exception to how humans in power have always treated the rest of us. Despite where they failed, as in endorsing slavery, they nonetheless created *marvelous principles* that would inevitably lead to the Civil War and to the end of slavery. In a sense, the Founders put off the Civil War to cooperate in the Revolution. Future Americans, both slaves and free persons, would pay a staggering price for this unholy compromise in the form of decades more of slavery leading up to the Civil War and its aftermath and with negative effects that still reverberate through society.

REVIVING THE TRUST THEY HAVE SHAKEN

Trust is at the heart of human relationships on every level, and its loss destroys everything human. "Human beings depend on trusting relationships and suffer deeply when trust is broken."[961] The predators and exploiters seeking to dominate the world want to break our bonds built on trust, mutual affection, and respect. They want to isolate, weaken, and dehumanize us. Thus, they can more readily control us like powerless objects or puppets.

We see through their strategies and their tactics, and we do not have to succumb. None of us must live isolated, alone, and too fearful to have our thoughts and feelings and to join others in the fight to restore freedom. We are everywhere. All we have to do is to find each other and join in facing the tasks ahead.

Envision what the predatory globalists are trying to create—a pyramid with them at the apex and us pushed down into the broadening base of expendables and deplorables. This means we are going to find more of us at the bottom than the top. That's reality. People like us who believe in personal responsibility, independence, and respect for each person's liberty are more likely to be found among the middle class, working class, or poor.

This is not so much a class war as it is a war of predators upon their expendable prey whom they are desperate to whip into conformity to consume at their leisure. Because of the nature of people who lust for wealth, glory, and power, they always try to climb to the top by using the backs of others as their stepping-stones. Never before have there been so many of them wanting to rule the world and consume us.

ACTION PLANS

We have been personally blessed in our own lives by the amazingly courageous new friends we have made as we have turned our attention toward resisting the predators and supporting each other and our ideals. You have met many of these wonderful people in this book and can meet them again, along with many others, every week on *The Dr. Peter Breggin Hour*.[962]

We can find each other and work together on many levels far beyond our limited individual imaginations and efforts. We can work together to advance personal responsibility, love, liberty, and the founding principles of America. Here are a few of those infinite attitudes and actions we can take:

- Never forget that liberty is the framework in which love and creativity thrive and remind others whenever we can.

- Keep our loved ones and ourselves as healthy, spiritually strong, and independent as possible.

- Stay in touch with each other by all the marvelous means now available.

- Build new websites, internet platforms, and high-tech communication systems to prevent the predators from taking everything they can.

- Share with our children and families what we really think, feel, and believe, and emphasize the centrality of family life in a democratic republic.

- Create groups among our friends to share reading materials and ideas.

- Create and join chat rooms and blog sites.

- Join religious, educational, business, and professional associations.

- Become involved in politics at the precinct and school-board levels.

- Run for higher office.

- Make ourselves known in the public square by every means possible to bring us together and give us strength.

- Never be intimidated or shouted down; instead, speak out whenever we can.

- Open our hearts to redeemed progressives and Never Trumpers yet always remain proud and grateful to President Donald Trump for reconnecting us with America's greatness and, with his actions and scorching honesty, standing up to the global predators and putting Americans first.

- Support those organizations and businesses of our choice that identify themselves with liberty, the Declaration of Independence, the Constitution, and the Bill of Rights.

- Work politically toward a more limited government, stronger local communities, and a stable yet upwardly mobile middle class.

- Do everything we can to identify and resist the global predators, most of whom will have ties to the Predator-in-Chief, the Chinese Communist Party.

- Withdraw our support or participation from activities owned and sponsored by global predators, including those who suppress the flow of information and control most of what we consume.

- Welcome those who lack faith but be proud and eager to share that God is our source of strength and our inspiration to love.

- Believe in a spiritual reality or level of meaning that is greater than ourselves—God, the ideals of liberty and love, America, the future of humanity—and take strength and guidance from these spiritual sources.

Whatever else they felt religiously, leaders like Washington, Jefferson, Franklin, John Adams, Sam Adams and many others shared a belief that God granted humans the right to be free and that the American Revolution could only have been won with the benevolent support of Providence or God. We must become Re-Founders of America. We must recognize the extraordinary exceptionalism of America and become prepared to pledge our lives once again for the fight for freedom.

We thank our new partners in the struggle for liberty, and we urge others to identify and contact people they know or can locate who share positive aspirations and hopes for the future. We must revive our trust in each other. We must strengthen our belief in loving one another and in supporting each other's liberty. We must come together as Re-Founders to build a renewed American nation and a better world for each other and humanity.

CHRONOLOGY AND OVERVIEW

With Pandemic Predictions and Planning Events[963]

COVID-19 and the Global Predators

By Peter R. Breggin and Ginger R. Breggin

For updates, go to www.breggin.com

Most of the observations in this *Chronology and Overview* are thoroughly documented in the book's 1,000-plus endnotes with only a few endnotes included here for the reader's convenience. However, items added to the chronology after the text was finished in early July 2021 are always documented with endnotes in the *Chronology and Overview*.

1984: Anthony Fauci becomes Director of the National Institute for Allergy and Infectious Diseases (NIAID), an institute of the National Institutes of Health (NIH). Fauci seems very powerful, but we will find that, in reality, he does not "call the shots." *Fauci's longevity is not, and could never be, due to his supposed role as an objective scientist. He is neither objective nor a scientist. His longevity is due entirely to his Machiavellian subservience to those who call the shots and his willingness to use and abuse those over whom he has been given power.*

If an individual takes independent actions at odds with or detrimental to the interests of the dominant billionaires, global pharmaceutical companies, tech companies, bankers, and so on, he cannot stay in charge of a government institute for multiple decades. Henchmen, like Fauci, are rule followers, not rule makers. Like mafia captains or capos, they can be domineering, cruel, and deadly within their own territory, but the real and ultimate power derives from the bosses.

March 1997: Ralph S. Baric, at the time an associate professor in the Department of Microbiology and Immunology of the University of North Carolina at Chapel Hill, engineers a mouse hepatitis coronavirus to cross species to infect baby hamster cells.[964] NIH funds the gain-of-function research. The term "gain-of-function" designates research that enables a virus to become more transmissible (infectious) or pathogenic, including enabling it to jump from one species to another or to cause an epidemic. We have proposed that most of this research should be referred to as "gain-of lethal-function."

February 2000: Researchers from the NYS Department of Health in Albany and the Institute of Virology in the Netherlands describe how "Coronaviruses generally have a narrow host range, infecting one or just a few species" and "One of the hallmarks of this family is that most of its members exhibit a very strong degree of host species specificity, the molecular basis of which is thought to reside in the particularity of the interactions of individual viruses with their corresponding host cell receptor."[965] This shows the unlikelihood of Fauci's concern about an animal coronavirus jumping to humans. However, the researchers manage to "retarget" a mouse *coronavirus* to make it "cross the host cell species barrier" into cat cells.[966] In short, although it rarely happens in nature, in the lab, an infectious mouse virus is turned into an infectious cat virus. This project increases the risk of further spread should a mouse or cat escape to infect other animals or people in society.

Although coronaviruses are not a high risk for jumping to human beings, they are among the viruses most easily engineered in the lab by humans to cross species. This relative ease of manipulating them is the main reason researchers have always been so interested in them. Coronaviruses are not a major threat to humans until humans make them into pathogens.

June 22-23, 2001: "The *Dark Winter* exercise portrayed a *fictional* scenario depicting a covert smallpox attack on U.S. citizens."[967] It is an extensive wargame by National Security Council in which Johns Hopkins University (JHU) plays a role, the first of many activities led by JHU that eventually display specific foreknowledge of the coming coronavirus pandemic. "Dark Winter" is a precursor to the **Pandemic Predictions and Planning Events** that start multiplying in 2017.

2002-4: SARS-CoV-1 appears in southern China in late 2002 and spreads around the world. It is very lethal, killing nearly 800 out of 8,000 people infected worldwide, but is contained before causing cases in the U.S.[968] China immediately begins engineering the virus in its labs—indicating it might have been able to engineer it originally from benign bat viruses—but the origin of SARS-CoV-1 remains unknown. In 2004, China's virus lab in Beijing begins experiencing the first of several accidental releases of variants of SARS-CoV with a limited number of deaths.

2003-4: Six separate accidental lab contaminations in and around China lead to persons infected with variants of SARS-CoV carrying them from labs, sometimes infecting others. Four are from China's Beijing Virus Institute. We will eventually draw several key observations as we cut through the enormous misinformation campaign by Fauci, Gates, WHO, the Chinese Communists, and a large array of global predators:

1. No SARS-CoV has ever been found in nature.

2. Some "SARS-like" CoVs have been found in bats in caves, but it takes enormous scientific and engineering efforts to convert them into viruses that can infect animals and human cells—efforts that could no more occur in nature than any other human engineering feat from building artificial arms to programming computers. Converting SARS-like CoV into pandemic SARS-CoV-2 was a colossal, scientific effort requiring decades of work and billions of dollars. It could never be mimicked by evolution.

3. The risk of SARS-CoV emergences from nature is miniscule and hardly worth considering compared to the high risk of its intentional or accidental release from one of the many labs containing them. Fauci and his bosses, like Gates, from among the global predators have it backward: The danger is not from nature, but from the gain-of-function research funded in part and wholly encouraged by Fauci and conducted in America, China, and elsewhere.

April 17, 2003: Laboratories quickly became able to study and to work with the virus, and a team of scientists submits the complete genome for a strain of SARS-CoV-1 to the CDC.[969]

November 2003: There are now multiple lab studies supporting the potential use of chloroquine and its derivative hydroxychloroquine for the treatment of SARS-CoV, including in *The Lancet,* which provided numerous citations.[970] The papers demonstrate their mechanisms of action for prevention and treatment in both cell cultures and in animals. The safety of this already long-used drug is emphasized.

Seventeen years later, when the Emergency Use Authorization (EUA) is applied to COVID-19 in order to fund vaccine research and to rush through vaccine EUAs without the usual FDA approval requirements and without financial risk to the drug companies, there will be a catch. The EUA could not go into effect if safe and effective treatments were already available. Therefore, this prior, intensive research confirming hydroxychloroquine's safety and potential effectiveness in SARS-CoV infections was ignored or suppressed, and doctors were told that there were no safe treatments. This fraudulent suppression of good treatments in order to make the vaccine companies and their investors wealthier and more powerful results in millions of deaths and the crushing of economies throughout the Western world during COVID-19.

2006a: Research establishes that impaired or older *vaccinated* mice challenged with SARS-CoV have potentially deadly reactions to the N protein within the virus.[971] Subsequent research will find that same problem with the spike protein. Animal research will continue to establish this deadly S protein risk right up to the time that the COVID-19 vaccines are going through Operation Warp Speed in 2020.

2006b: Pandemic Predictions and Planning Event. The **International Finance Facility for Immunization**, or IFFIm, is established. The stated purpose is to enable poorer nations to obtain guaranteed inexpensive loans for purchasing vaccines through novel loan arrangements involving the Bank of America and other funding sources. However, these loans are connected to two predatory organizations, GAVI and CEPI, both of which find ways to increase the markets of the pharmaceutical industry. GAVI describes how it uses proceeds from the bonds:[972] "For Gavi, the proceeds of Vaccine Bonds help ensure predictable funding and more efficient operations. In addition, Gavi can frontload funds when necessary for rapid roll-out of new and underused vaccines. For example, IFFIm funds enabled Gavi to stimulate country demand for the five-in-one pentavalent vaccine, enlarging the size of the market, attracting new manufacturers, and reducing prices." The loans will cycle back into the pharmaceutical industry and its investors.

December 2008: Vanderbilt and UNC under Ralph Baric synthesize a "bat SARS-like coronavirus," a "SARS-CoV," that infects live mice and "human ciliated airway epithelial cells" in the lab.[973] This breakthrough receives little attention in the media, but it is a turning point. Starting with a harmless bat virus and an added spike protein, they make a SARS-CoV capable of infecting live mice and human lung cells. That human antibodies attack this virus in the living mice and the human epithelium probably inspires the global predators. They could make trillions of dollars on vaccines while enforcing their authoritarian control. It all depends on a COVID-19 pandemic coming along.

2009: Moderna is founded with the aim of making mRNA vaccines.[974] Despite no products or income, investors pour in money. Two tantalizing quotes in technical language show what a dangerous and highly experimental product Moderna is making. In 2012, a scientific report in *PLOS ONE* finds:

> These SARS-CoV vaccines all induced antibody protection against infection with SARS-CoV. However, challenge of mice given any of the vaccines led to occurrence of The-type [sic] immunopathology suggesting hypersensitivity to SARS-CoV components was induced. Caution in proceeding to application of a SARS-CoV vaccine in humans is indicated.[975]

A financial analysis in April 2021 notes, retrospectively:

> Moderna's technology platform inserts synthetic nucleoside-modified mRNA (modRNA) into human cells using a coating of lipid nanoparticles. This mRNA then reprograms the cells to prompt immune responses. *It is a novel technique, abandoned by other manufacturers due to concerns about the toxicity of lipid nanoparticles at high or frequent doses.*[976] (Italics added.)

January 2010: Pandemic Predictions and Planning Event. Bill and Melinda Gates call for the next ten years to be the Decade of Vaccines.[977]

May 2010: Pandemic Predictions and Planning Event. Almost a decade before COVID-19, the Rockefeller Foundation publishes a future pandemic scenario in a business investment-oriented, 53-page booklet, *Scenarios for the Future of Technology and International Development.*[978] The first "Scenario Narrative" in the booklet is called "Lock Step: A World of tighter top-down government control and more authoritarian leadership, with limited innovation and growing citizen pushbacks." The opening line is, "In 2012, the pandemic that the world had been anticipating for years finally hit."

The only real regret displayed about the pandemic is the lack of investment opportunity. Despite dozens or more suggestive scientific articles, especially in 2008, the idea of a coronavirus pandemic and expensive vaccines with unlimited profit opportunities had not yet been imagined by these futurists. Nonetheless, the advantages of a "Lock Step," "top-down," and "authoritarian" government are already explored and celebrated and seen as permanent after the next coming pandemic.

December 2, 2010: Pandemic Predictions and Planning Event. From the Bill & Melinda Gates Foundation:

> New York: The World Health Organization (WHO), UNICEF, the National Institute of Allergy and Infectious Diseases (NIAID), and the Bill & Melinda Gates Foundation have announced a collaboration to increase coordination across the international vaccine community to create a Global Vaccine Action Plan.[979]

The Leadership Council is reintroduced with Dr. Margaret Chan, Director General of WHO at the top, followed by Anthony Fauci of NIAID among the five elite internationals. This document confirms that Fauci was working with Bill Gates going back to December 2010 on making the overall plans for the coming pandemic and the outrageous pushing of unsafe vaccines on the world.

Ginger Breggin points out, "That's planning. Gates announces the decade of the vaccines at the beginning of 2010 and fulfills his wildest ambitions in 2020."

September 2011: Pandemic Predictions and Planning Event. The United Nations passes Resolution 2030, calling for governments and international corporations to collaborate under the U.N. to pursue a huge variety of "progressive aims." No mention is made of fighting authoritarian or totalitarian governments or spreading personal freedom and political liberty. Resolution 2030 becomes the model for the Great Reset, although without U.N. governance. The document mentions vaccines three separate times. Presaging the Great Reset, the U.N. calls for making vaccines universally free while simultaneously protecting corporations from "financial risk" or losses:

Achieve universal health coverage, including financial risk protection, access to quality essential health-care services, and access to safe, effective, quality, and affordable essential medicines and vaccines for all.[980]

September 21, 2011: A research article supported by Fauci's NIAID and the University of North Carolina, with Baric's name on it, reconfirms and illustrates important problems associated with gain-of-function research with coronaviruses:[981] (1) It begins by repeating the falsehood that there are many SARS-CoV viruses in animal reservoirs waiting to emerge, necessitating investments in research: "Severe acute respiratory syndrome coronavirus (SARS-CoV) is an important emerging virus that is highly pathogenic in aged populations and is maintained with great diversity in zoonotic reservoirs"; (2) It makes clear that they already have created numerous pathogenic variations on the original SARS-CoV-1, because they are using them in their experiments. It also verifies two serious problems with SARS-CoV vaccines: Vaccines made for one variant of SARS-CoV are not effective for others and *aged vaccinated mice exposed to SARS-CoV are not protected and instead are vulnerable to human-like severe immune reactions. The spike protein is the source of the problem:* "Importantly, aged animals displayed increased eosinophilic immune pathology in the lungs and were not protected against significant virus replication" and "When challenged with zoonotic and human chimeric SARS-CoV incorporating variant spike glycoproteins, the aged BALB/c mouse model reproduces severe lung damage associated with human disease, including diffuse alveolar damage, hyaline membrane formation, and death."

Like COVID-19 itself, deaths from the vaccines, especially in older vaccinated people challenged with SARS-CoV-2, were predicted and seemingly planned from the beginning. The lethal capacity is built into the spike protein for all mRNA and DNA vaccines to force the cells in the human body to make innumerable exact copies of the SARS-CoV-2 spike protein, poisoning the body from the inside out. Research on animals continues to confirm this deadly effect even after the vaccines are going into production with Operation Warp Speed (see September 9, 2020, entry regarding Wen Shi Lee et al. in Nature Microbiology.)

September 2012: A review article warns against starting research on humans with mRNA vaccines, in part because animals given the vaccines can become more ill than expected and even die when they are later exposed to a SARS-CoV lab-engineered virus.

September 2013: Pandemic Predictions and Planning Event. NIH awards $10,000,000 to Baric and UNC to study and manipulate "highly pathogenic human respiratory and systemic viruses which cause acute and chronic life-threatening disease outcomes." Baric says the "focus" would be on "highly pathogenic coronavirus infections, using SARS-CoV and MERS-CoV as models."[982]

October 2014: On White House stationery, President Barack Obama calls for a moratorium on gain-of-function research that manipulates animal viruses to increase their capacity to infect and harm other species, including humans. Fauci will ignore and work around this prohibition. In particular, he never stops funding the single most important and most dangerous gain-of-function research by Menachery at the University of North Carolina, which Fauci funds to this very day through EcoHealth Alliance.[983]

2015a: Fauci approves China's first-ever Bio-safety Level-4 (BSL-4) lab at the Wuhan Institute of Virology to facilitate giving money to Wuhan scientists for gain-of-function research.

2015b: Pandemic Predictions and Planning Event. Bill Gates is by now investing in the RNA vaccine industry.[984]

2015c: Pandemic Predictions and Planning Event. Reports in May 2021 confirm the obvious. In 2015, Chinese Communists are discussing among themselves the use of coronaviruses as biological weapons with observations about the acute and permanent fear and disability that they would create.[985] Biological warfare experts in the U.S. are likely doing the same thing. The likelihood should be taken for granted, especially since DARPA is openly funding U.S. coronavirus research.

2015d: Pandemic Predictions and Planning Event. Bill Gates conceives of the Coalition for Epidemic Preparedness Innovations (CEPI) in 2015[986] and funds it through the Bill & Melinda Gates Foundation. Gates begins organizing a worldwide alliance to finance and coordinate the development of new vaccines to make a financial killing and to reorganize and basically control world governance in the coming pandemic.

Klaus Schwab and his World Economic Forum (WEF) is another founder. He works closely with Gates and will, in 2020, unveil the Great Reset. The Great Reset will become the mantra for Bill Gates, Joe Biden, John Kerry, China, and other globalists seeking to subdue the independent, patriotic, freedom-oriented America First movement.

Wellcome Trust, a British "charitable" trust, is another CEPI founder. Based on pharmaceutical industry funding and focused on the health arena, it is one of the largest charities in the world.

The Government of India is another founder, followed by Germany and Japan—which completes the fusion between huge corporations, giant "philanthropic" organizations, and very big government.

In 2017, WHO will join and become the announced scientific arbiter for all CEPI activities, which ties the pandemic planning program to both the United Nations and, most importantly, to Communist China, which controls WHO. See three additional entries in this *Chronology and Overview*: January 18, 2017; January 24 and March 2017; and July 21, 2017.

This CEPI coalition will control the unfolding of COVID-19 and the Great Reset—although they will be joined by innumerable billionaires, corporations, philanthropic groups, and governments and their agencies. Anthony Fauci, a member of Bill Gates' small, elite vaccine Advisory Council, will be one of Bill Gates' leading implementers of the program.

January 26-27, 2015: Pandemic Predictions and Planning Event. "At the Berlin Pledging Conference 2015, the Bill & Melinda Gates Foundation announces USD 1.55 billion for Gavi's next 2016-2025 strategic period."[987] Gavi is the Global Alliance for Vaccines and Immunization, and, like many of the Gates' ventures, has been criticized for representing the interests of the pharmaceutical companies.[988] It is a corporate-government fusion that bypasses the decision-making process of a democratic republic in favor of global predatory planners. It is what will later be called the Great Reset. These large, organized investments in vaccines are taking place five years before COVID-19.

April 2015: Pandemic Predictions and Planning Event. Bill Gates gives a famous TED Talk explaining that if anything is going to kill 10 million people worldwide, it's a virus.[989] His affect has a touch of grim humor that some people find shocking.

December 2015: Vineet Menachery and Ralph Baric at the University of North Carolina make the great breakthrough.[990] They transform innocuous bat coronaviruses into epidemic SARS-CoV pathogens that infect live mice and human lung tissue preparations. But something more threatening is disclosed by the publication. *With Fauci-funding, U.S. researchers are collaborating with top Chinese scientists from the Wuhan Institute of Virology, a Chinese Communist Party facility, now directed by the military's leading biological warfare specialist. Despite President Obama's moratorium, Fauci has been continuing to fund both the American and the Chinese researchers in this epitome of hugely dangerous gain-of-function research. The research collaborations create at least two nightmare scenarios: a **major risk** of the poorly managed Wuhan Institute accidentally leaking one of its many man-made SARS-CoV pathogens and the **certain** step of advancing the biological warfare capacity of the Chinese Communist Party and its military.*

February 2016: Pandemic Predictions and Planning Event. The Wuhan Institute and other Chinese Communist sponsors announce an international conference to be held on October 21-23, 2016, in Wuhan: "Nature Conference: Viral Infection and Immune Response."[991] Many speakers are from the U.S., including two from NIAID. The first Session Topic is "Epidemiology of emerging viral disease." The brochure boasts about the Wuhan Institute, "Pathogenic study of emerging infectious diseases has become one of the major research fields. Great achievements have been made in animal origin studies of Severe Acute Respiratory Syndrome, *coronavirus* (*SARS-CoV*) and avian influenza viruses."[992]

The Director of the Institute and two of her four deputies have extensive university training in the United States in fields involved in gain-of-function studies and biological warfare. One of the institute's top three officials is Changcai HE, Deputy Secretary of the Party at Wuhan Branch,[993] a high-ranking position in the Communist Party. When the pandemic surfaces in January 2020, the Wuhan Institute will be directly taken over by the military.

March 14, 2016: Pandemic Predictions and Planning Event. The prediction of a COVID-19 scenario gains scientific approval (but not validity) from a follow-up article by Menachery et al. in the *Proceedings of the National Academy of Sciences (PNAS)*. The article declares that the risk of the spontaneous emergence of a SARS-CoV pandemic pathogen is now more likely because the researchers have been able to make pathogens out of harmless bat viruses in the lab. The theory is something like this: "I could construct a plague out of an innocent virus, so now I expect nature to do it."

That claim makes no scientific or logical sense. Natural evolution is based on a nearly infinite number of chance mutations occasionally finding adaptive advantages in their environments. This nearly random activity in nature will not and cannot achieve the same elaborate, predetermined ends as carefully planned and purposed laboratory manipulations of biological entities. In the lab, many humans are guiding intensive, elaborate, highly technical activity. They are purposely working together toward achieving a specific goal, such as making a harmless virus deadly. Nature cannot mimic this any more than it could if the lab had succeeded in grafting a set of cobra teeth into a toothless bat's mouth.

Evolution and laboratory engineering are nearly exact opposites. Evolution depends on chance viral mutations in varying environments over an infinite length of time. Lab engineering requires forcefully mutating a specific virus in specific ways to have a specific effect on a specific human receptor to make humans sick. Evolution in nature cannot work in a purposeful manner.

Despite reality, this 2016 publication gives Gates additional scientific credibility in his successful efforts to obtain billions of dollars from investors and companies who, at this point, are lining up to make SARS-CoV vaccines far in advance of the actual pandemic.

May 2016: Pandemic Predictions and Planning Event. The governing body of the World Health Organization (WHO) approves the CEPI/ Bill Gates *master plan* for collaborating with WHO to control the profits, organization, and governance of the coming pandemic (Chapter 15)—but WHO's primary allegiance is to Communist China. (See July 21, 2017 for the master plan.)

January 2017a: This is the year that **Pandemic Predictions and Planning Events** explode in anticipation of a coming pandemic, often identified as caused by a SARS-coronavirus. It begins with Fauci's announcement with "certainty" that a pandemic will strike during President Trump's first term.

January 2017b: Pandemic Predictions and Planning Event. Homeland Security publishes a 135-page grand plan for an interagency response to "biological incidents."[994] The plan includes the application of the Emergency Use Authorization (EUA).[995]

January 9, 2017: Pandemic Predictions and Planning Event. President Barack Obama's administration officially recommends lifting the moratorium on gain-of-function (GOF) research and provides the necessary steps to be taken:[996] The recommendation is made by John Holdren, White House Director of the Office of Science and Technology Policy (OSTP), also known as the Science Czar. *The American Report* identifies Holdren as an advocate of "massive-scale human population reduction measures such as adding reproductive sterilization agents to food and water supplies and enacting forced abortions in the United States."[997] President Obama's Health and Human Services (HHS) Director, responsible for authorizing and implementing lifting of the GOF funding pause, is Sylvia Matthews Burwell, "former President of the Global Development Program and Chief Operating Officer of the Bill & Melinda Gates Foundation."[998]

President Obama takes action 11 days before Donald Trump takes office and Trump is kept in the dark about it.[999] On the eve of their departure, the Obama-Biden administration provides the authorizations needed for the eventual re-funding of gain-of-function research by the U.S., enabling NIH to finalize the details by the end of 2017.[1000]

We only learn of this as the book goes to press. It further confirms that President Trump had nothing to do with the official resurrection of gain-of-function research. Meanwhile, Fauci had continued funding this highly dangerous research that has caused many deadly virus escapes over the years and done so despite President Obama's moratorium. Lifting this ban was part of the January 2017 Gates-led resurgence in developing vaccines.

January 10, 2017: Pandemic Predictions and Planning Event. Ten days before President Trump's inauguration, Fauci keynotes a Georgetown University Conference on "Pandemic Preparedness in the Next Administration." Fauci unequivocally declares that President Trump will "definitely" face a pandemic during his first term (Chapter 16). He expresses his certainty several times during his talk. Fauci urges the development of the corporate/government collaborations, characteristic of the future Great Reset.

The Harvard School of Public Health, which boasts of being founded in close collaboration with the Chinese Communist government, is represented by Ashish K. Jha MD. He, too, gives a talk expressing his anticipation of the coming pandemic and advocates for a global response resembling a combination of the Great Reset and Operation Warp Speed in a massive collaboration between business and government.

January 13, 2017a: Pandemic Predictions and Planning Event. The FDA announces new guidelines for its Emergency Use Authorization (EUA),[1001] originally passed by Congress in response to 9/11. The new FDA regulations reinforce the high-speed development and production of medications and vaccines in an epidemic, bypassing the FDA's usual safeguards and creating billion-dollar boondoggles for the drug companies and their investors.

January 13, 2017b: Pandemic Predictions and Planning Event. A White House planning meeting is held in which Obama Administration officials brief the Trump Administration on preparations for a global pandemic. Here is the description by the leader of the exercise:[1002]

> January 13, 2017, national security officials assembled in the White House to chart a response to a global pandemic. A new virus was spreading with alarming speed, causing global transportation stoppages, supply-chain disruptions, and plunging stock prices. With a vaccine many months away, U.S. health-care infrastructure was severely strained.

> No, I didn't get that date wrong. This happened: it was part of a transition exercise that outgoing officials from the administration of President Barack Obama convened for the benefit of the incoming team of President Donald Trump. As Homeland Security and Counterterrorism Adviser to President Obama, I led the exercise, in which my colleagues and I sat side by side with the incoming national security team to discuss the most pressing homeland security concerns they would face.

January 20-24, 2017: Klaus Schwab holds his annual World Economic Forum (WEF) at Davos in Switzerland, where there is concern that the U.S. government under incoming President Trump with his America First populist mandate will be much less docile and subservient in its relationship with the Chinese Communist regime. In direct defiance of Trump, Schwab invites China's tyrannical dictator Xi Jinping to give the keynote address in which he is critical of America. Later, international economic and financial conferences run by Schwab and also by Michael Bloomberg will become more antagonistic toward Trump's America First policies and more fawning over China.

January 18, 2017: Pandemic Predictions and Planning Events. CEPI is officially unveiled in a press release from Schwab's Davos location as a "Global partnership launched to prevent epidemics with new vaccines."[1003] New sources of support are added: *"CEPI is supported by several leading pharmaceutical companies with strength in vaccines – GSK, Merck, Johnson & Johnson, Pfizer, Sanofi and Takeda, plus the Biotechnology Innovation Organisation." From the beginning, all this planning has been about making money from vaccines, along with imposing the government, philanthropic, and corporate governance called the Great Reset.*

The agenda also has that prevailing progressive veneer to it: "Just over a year ago 193 states adopted the Sustainable Development Goals—the roadmap for the future we want."

January 24 and March 2017: Pandemic Predictions and Planning Events. At Schwab's Davos Conference and later in Schwab's office at the World Economic Forum, Bill Gates describes his preparations for an inevitable, forthcoming pandemic (See Chapter 15). Gates unveils a new fund called the Coalition for Epidemic Preparedness Innovations (CEPI). It is founded by the Bill & Melinda Gates Foundation and announced by Gates at Klaus Schwab's 2017 World Economic Forum, which is a cofounder. This confirms the vaccine boondoggle is closely linked to what will become the totalitarian Great Reset popularized by Schwab.

CEPI will become a multibillion-dollar collaboration between large private donors and corporations, the U.N., and national governments to research and manufacture vaccines for the much-anticipated pandemic. Empowered by the 2017 FDA guidelines for Emergency Use Authorization,

Gates describes the need for bypassing federal regulations to produce vaccines in a process that takes "less than a year." He mentions Moderna as one of the key corporations already working on new platforms for "novel" vaccines, which he describes as based on either "RNA" or "DNA."

In short, Bill Gates indicates in January 2017 that he has already invented, invested in, and started implementing a project identical to what will be "unveiled" as Operation Warp Speed three years later by President Trump and Fauci. Gates announces his preparations for a superspeed vaccine project in the context of the corporate/governmental capitalism that Schwab will popularize as the Great Reset. These plans will run roughshod over any concern about democratic processes or liberty—the heart of human progress.

Gates is not alone as a predator. Of the top American billionaires, all but Larry Ellison, the founder of Oracle, seem eager to sell out America in favor of China. Unfortunately, Ellison's own Silicon Valley employees reject his patriotism and concerns about China and oppose his politics. All the top American corporations are desperate to keep growing their wealth and power through unrestricted business with the Chinese Communists, who are, in turn, conducting unrestricted warfare against America.

February 18, 2017: Pandemic Prediction and Planning Events. At the Munich Security Conference, Bill Gates makes the following predictions:[1004] "The next epidemic could originate on the computer screen of a terrorist intent on using genetic engineering to create a synthetic version of the smallpox virus ... or a super contagious and deadly strain of the flu." "Whether it occurs by a quirk of nature or at the hand of a terrorist, epidemiologists say a fast-moving airborne pathogen could kill more than 30 million people in less than a year. And they say *there is a reasonable probability the world will experience such an outbreak in the next 10 to 15 years.*" (Italics added.)

He also spoke of the need for vaccines, explaining: "Most of the things we need to do to protect against a naturally occurring pandemic are the same things we must prepare for an intentional biological attack."

March 2017: Pandemic Predictions and Planning Event. *The Lancet Infectious Diseases* positions CEPI as the central funding and organizing center for pandemic preparation.

July 21, 2017: Pandemic Predictions and Planning Event. In June 2021, we discover the master plan for CEPI, which is essentially the plan for the world's future from the minds of Bill Gates, Klaus Schwab, and their partners in global predation (Chapter 15 has details). Interim CEO of CEPI, John-Arne Røttingen, presents an astonishing 21-page PowerPoint presentation to the World Health Organization Association.[1005] The document confirms and strengthens all the points made in the book about the advanced planning, including the enormous profits assigned to Bill Gates. A table gives a broad description of the global predators, leaving out only Communist China, who works through WHO. CEPI describes how it will divide up the world with WHO during the coming pandemic. Among the more than 20 large entities described, only one person is mentioned, "Gates," who will participate in "Development/Licensure." CEPI will take control over everything except setting healthcare and scientific standards, which will be the task of WHO. The plan springs a great deal from Gates' inspiration and determination—with no regard whatsoever that these plans are being made for all humanity by people with no elected or appointed status to do so. It also continues to be clear it's all about making a fortune and increasing power and control through vaccines.

October 2017: Pandemic Predictions and Planning Event. The Johns Hopkins University (JHU) Bloomberg School of Public Health releases an important in-depth projection on how to deal with a coming "coronavirus" pandemic with vaccines. The title of the forecast plays on SARS: *The SPARS Pandemic, 2025-2028.* The fictional vaccine, named Corovax, plays on coronavirus. This is *not* the more famous Event 201, which will be conducted live in October 2019 as a wargame against the coronavirus. *The Spars Pandemic* puts more emphasis on pushing the vaccines for the coronavirus—*three years ahead of COVID-19.*

December 8, 2017: Dr. Tal Zaks, Moderna's Chief Medical Officer, compares mRNA to a "line of code" in a computer program and declares that, with the new mRNA vaccines used experimentally for cancer treatment, "we are actually hacking the software of life."[1006] Transhumanism and vaccine technology unite.

March 20, 2018: A scientific review describes a huge upsurge in mRNA vaccine research, ringing in "a new era in vaccinology."[1007] A table reveals that Moderna (an upstart born in 2010) is being funded by the Bill & Melinda Gates Foundation, by DARPA, and by BARDA. DARPA does "advanced research" for the Department of Defense. BARDA will become the government agency for implementing Operation Warp Speed and the giveaway of billions of dollars to pharmaceutical corporations in the style of the Great Reset. The table also reveals that Gates is involved with CureVac AG, which is in partnership with several other companies, including Johnson & Johnson. Cure Vac AG is supported by the Gates brainchild, the investment fund CEPI,[1008] and with other corporations, creating a great tangle of complex business interests involving vaccines.

The scientific review warns that moving from animal to human research has been stymied by two problems. One problem involves "rare cases of severe injection site or systemic reactions" from mRNA, requiring further preclinical and clinical studies into "systemic inflammation." The other problem is ineffectiveness in generating immunological responses in humans. Individual scientific reports give an even more dismal picture. Evidence mounts that the billionaires, drug companies, and government agencies are making an enormously bad bet on mRNA vaccines, but the commitment to and the momentum for marketing these dangerous, experimental vaccines leaves no room for turning back.

April 27, 2018: Pandemic Predictions and Planning Event. Gates gives an honorific address sponsored by the Massachusetts Medical Society and the *New England Journal of Medicine* titled:[1009]

**Ring the Alarm: The next epidemic is coming.
Here's how we can make sure we're ready.**

Gates describes the marshalling of forces that he, Fauci, and others are in fact carrying out in anticipation of the seemingly inevitable pandemic:

> The world needs to prepare for pandemics the way the military prepares for war. This includes simulations and other preparedness exercises so we can better understand how diseases will spread

and how to deal with things like quarantine and communications to minimize panic.

October 2018. Pandemic Predictions and Planning Event. The Johns Hopkins Bloomberg School of Public Health, Division of Health Security, is the source of all JHU pandemic planning events. A terrifying book titled *Technologies to Address Global Catastrophic Biological Risks*[1010] describes developing projects for dealing with upcoming pandemic threats, specifically including, but not exclusively, "SARS and MERS coronaviruses." It is a blueprint for how to take control of society and to enforce vaccines despite public opposition.

The Bloomberg School of Public Health book becomes chilling when it goes on to advocate solutions that have been previously ridiculed as conspiracy theories:

> **Self-Spreading Vaccines:** Self-spreading vaccines are genetically engineered to move through populations like communicable diseases, but rather than causing disease, they confer protection. The vision is that a small number of individuals in a target population could be vaccinated, and the vaccine strain would then circulate in the population much like a pathogenic virus, resulting in rapid, widespread immunity. [1011]

Furthermore, the Hopkins report sees this technology as posing lethal risks about which their only concern seems to be alienating public opinion:

> Finally, there is a not insignificant risk of the vaccine virus reverting to wild-type virulence, as has sometimes occurred with the oral polio vaccine—which is not intended to be fully virulent or transmissible, but which has reverted to become both neurovirulent and transmissible in rare instances. This is both a medical risk and a public perception risk; the possibility of vaccine-induced disease would be a major concern to the public.[1012]

As Ginger Breggin says, "Yesterday's conspiracy theories keep becoming today's headlines."

2019: Pandemic Predictions and Planning Event. Fauci's NIAID, predominantly through Peter Daszak's EcoHealth Alliance, awards four more years of coronavirus research, including gain-of-function studies. The current funding goes through 2021, and the project through 2025.[1013]

January 2019: Pandemic Predictions and Planning Event. Klaus Schwab's World Economic Forum (WEF) teams up with the Harvard Global Health Institute to publish an extensive white paper titled *Outbreak Readiness and Business Impact—Protecting Lives and Livelihoods across the Global Economy.*[1014] The Harvard Global Health Institute is led by Ashish K. Jha, who joined Fauci on January 10, 2017, to predict that there would be a pandemic during the then upcoming administration of Donald Trump and later attacks all early treatment of COVID-19 and the frontline physicians.

In warning about future threats, the WEF/Harvard white paper twice mentions "coronaviruses" and seven times refers to "SARS." The purpose of the report is summed up in the sentence, "The societal threat posed by epidemics provides a compelling platform for engagement across the public and private sectors."

Pandemics are seen as the ultimate way of enforcing what Klaus Schwab will soon unveil as the Great Reset and what we call corporate-government fusion. The goal is to enable totalitarian control that bypasses or crushes the legal protections and liberty represented by the United States of America as a patriotic democratic republic.

May 1, 2019: Pandemic Predictions and Planning Event. More than six months before the outbreak in China, a bill was introduced in the House of Representatives that led to a proposed $100 billion program to force epidemic tracking on Americans.[1015] The bill was "COVID-19 Testing, Reaching, And Contacting Everyone (TRACE) Act," and its remarkable number was HR 6666.[1016] It was criticized as a grave move toward totalitarianism[1017] and was not passed. Although there is evidence Gates influenced the creation of the bill, there is no evidence it would finance his activities. This bizarre bill adds to the mounting evidence for active preparation for the coming pandemic.

September 2019. Pandemic Predictions and Planning Event. *A World at Risk: Annual Report on Global Preparedness for Health Emergencies by the Global Preparedness Monitoring Board (GPMB)*[1018] is a book sponsored by The

World Health Organization and the World Bank with acknowledgments to familiar globalist organizations like the Johns Hopkins University Center for Health Security and Wellcome Trust. Emphasis is placed on rapid, innovative vaccine development, and include "surge manufacturing" of "nucleic acid types" which will be "pre-tested and approved for us within weeks." This once again anticipates Operation Warp Speed.

Members of the Board of GPMB include Anthony Fauci; Chris Elias, President, Global Development Program; Bill & Melinda Gates Foundation, USA; George F. Gao, Director-General, Chinese Center for Disease Control and Prevention, People's Republic of China; and many other high-ranking officials from around the world. This first annual report comes out shortly before the pandemic breaks out in China.

October 2019: Pandemic Predictions and Planning Event. The Johns Hopkins Bloomberg School of Public Health—which two years earlier published *The SPARS Pandemic, 2025-2028*—now holds a much-publicized pandemic simulation. Called Event 201, it physically brings together representatives of governments, corporations, and investors collaborating in wargame style against a pandemic caused by a SARS-CoV virus. Event 201 is sponsored by two of the most influential global predators: Bill Gates and Klaus Schwab. It is also associated with Michael Bloomberg for whom the School of Public Health is named. Participants include the director-general of the Chinese CDC and major global corporations. It deepens the connections between pandemic preparation, Schwab's Great Reset, and Bill Gates' Warp Speed-like program. The top theme is **"Governments, international organizations, and businesses should plan now for how essential corporate capabilities will be utilized during a large-scale pandemic."**[1019]

November 2019: In a direct assault on President Trump's policies and diplomacy, Bloomberg holds his second New Economy Forum in *Beijing, China*[1020] at which China's dictator Xi Jinping makes a smiling appearance. The conference is cosponsored by the Chinese Communists, and the theme is on "recoupling" with China. "Funding Partners" include 3M, Exxon Mobile, FedEx, HSBC, Hyundai, and MasterCard. Shortly after the conference, Bloomberg announces his candidacy for President of the United States, a nation he is consciously sacrificing for the wealth he can generate with the Chinese Communists.

December 1, 2019: Cases of viral pneumonia begin showing up in China, although retrospectively there would be reports of possible cases in November or even earlier. The pandemic will become a boon to all the globalists and the Chinese Communists, whose 400 billionaires will grow their wealth by an *average* of 60% in the first year of COVID-19.

January 2020: Pandemic Predictions and Planning Event. With the same title as the 2017 book by Johns Hopkins, *The SPARS Pandemic* is published as a monograph in the *Journal of International Crisis and Risk*.[1021] It has a new subtitle and focus: "A Futurist Scenario to Facilitate Medical Countermeasure Communication." As in the original book, the star vaccine is Corovax. Emphasis is given to communicating with the public to overcome criticism of the vaccines and to gain compliance with vaccinations. Because of the time it takes to get a journal article published, this scientific publication that originated in 2019 or earlier becomes another amazing prediction and planning event.

January 2020a: The Wuhan Institute is openly taken over by China's People's Liberation Army under General Chen Wei, a leading expert in biochemical weapons and bioterrorism.[1022] However, under the Chinese policy of military-civil fusion (MCF), for several decades all the activities of the Institute were already inseparable from the military.[1023] A State Department memo describes military-civil fusion: [1024]

> As the name suggests, a key part of MCF is the elimination of barriers between China's civilian research and commercial sectors, and its military and defense industrial sectors. The CCP is implementing this strategy, not just through its own research and development efforts, but also by acquiring and diverting the world's cutting-edge technologies—including through theft—in order to achieve military dominance.

January 2020b: Multiple scientific sources within China conclude that SARS-CoV-2 is a chimeric or man-made virus based on, among other things, the artificially tailored spike inserted into the virus which enables it to attack humans and is not found in nature. China begins censoring, punishing, and "disappearing" doctors, scientists, and journalists who describe the pandemic or identify its source as the Wuhan Institute.[1025] One researcher rejects his original conclusions, one physician dies, and another disappears. One courageous scientist flees to America to tell the truth. WHO covers up for the Chinese and repeats their lies. From now on, the entire array of globalists who dominate Western governments, media, educational institutions, medicine, and science will declare that the Wuhan Institute of Virology was not the origin of the virus and that it came from nature. Social media begin deleting or taking down comments calling SARS-CoV-2 the Chinese virus or attributing it to the Wuhan Institute. It is labeled a baseless conspiracy theory.

However, the Communist institute has been creating virulent SARS-Coronaviruses from before 2015, some in collaboration with U.S. scientists and many on its own. Anthony Fauci's NIAID and Francis Collins' NIH fund the collaborative research with the Chinese Communists, and they also fund individual Chinese researchers working in the Wuhan Institute. Every Chinese researcher works under the control of the Chinese Communists and their army, and the Wuhan Institute is essentially a military facility. Beyond that, the Wuhan Institute is notoriously insecure, and Chinese research facilities have had numerous leaks of SARS-CoV and other viruses in the past.

January 21, 2020: The Chinese Communists belatedly admit human-to-human transmission of the new coronavirus, and the U.S. CDC confirms the first case in America.

January 23, 2020: Years of Planning Take Off Faster than the Pandemic.

January 23, 2020a: Pandemic Predictions and Planning Event. *CEPI makes a stunning announcement from its Norway office,[1026] indicating how fully prepared the global predators have become after at least three years preparation for the arrival of the anticipated pandemic. We label this announcement a **Pandemic Predictions and Planning Event** because the virus has not yet been named SARS-CoV-2 and no one is as yet confirming a serious worldwide pandemic. Furthermore, the program would not be able to get its anticipated government funding unless there are no effective treatments. Nonetheless, CEPI barrels ahead with the help of Moderna and Fauci's NIAID:*

CEPI, the Coalition for Epidemic Preparedness Innovations, today announced the initiation of three programmes to develop vaccines against the novel coronavirus, nCoV-2019.

The programmes will leverage rapid response platforms already supported by CEPI as well as a new partnership. The aim is to advance nCoV-2019 vaccine candidates into clinical testing as quickly as possible....

In addition, CEPI today announces a new partnership with Moderna, Inc., (Nasdaq: MRNA) and the U.S. National Institute of Allergy and Infectious Diseases.

So Bill Gates and Klaus Schwab as CEPI founders were now official partners with Anthony Fauci!

January 23, 2020b Moderna announces a collaborative effort with the National Institutes of Health to use its genetic drug development platform to fast-track an experimental vaccine in less than a year.[1027] (A recent approval for an Ebola vaccine took two decades![1028])

Summing up the events of the day, CEPI has new partnerships with Moderna and NIAID. Now Moderna is also in partnership with NIAID and NIH, and NIH owns and profits from a large piece of Moderna's patents for the vaccine.[1029] And Bill Gates more or less tells them all what to do, including CEPI, Moderna, NIH, and Fauci's NIAID.

January 23, 2020c: An article in *JAMA*[1030] coauthored by Anthony Fauci pushes the false narrative of an emergence of a dangerous coronavirus from nature despite his NIAID funding research on creating pathogenic CoVs near the site of the co-called emergence. He also advances the unproven "suspicion" that the original SARS epidemic of 2002-2004 emerged from nature. The article concludes, "The emergence of yet another outbreak of human disease caused by a pathogen from a viral family formerly thought to be relatively benign underscores the perpetual challenge of emerging infectious diseases and the importance of sustained preparedness." Actually, *no SARS-CoV viruses have ever been found in nature or proven to have emerged from nature to cause an epidemic.*

January 30, 2020: WHO reports 7,818 total confirmed SARS-CoV-2 cases worldwide with a risk assessment of *very high* for China and *high* at the global level.

January 31, 2020: President Trump, against Fauci's resistance, bans all traffic from China except returning U.S. citizens.

Early February 2020: Pandemic Predictions and Planning Event. NIH describes how fully prepared the establishment was for the coronavirus pandemic:[1031]

> Years before the COVID-19 pandemic began, experts at the NIH Vaccine Research Center (VRC) were studying coronaviruses to find out how to protect against them. The scientists chose to focus on one "prototype" coronavirus and create a vaccine for it.... The VRC worked with a company called Moderna [*Bill Gates*] to use this information to quickly customize their prototype approach to the SARS-CoV-2 spike protein. *By early February, a COVID-19 vaccine candidate had been designed and manufactured.* This vaccine is called mRNA-1273. By March 16, 2020, this vaccine had entered the first phase of clinical trials. Other vaccines, including a similar one from Pfizer and BioNTech SE [*partnered with China*], entered clinical trials not long after. (Brackets and italics added.)

The "design" and "manufacture" of a specific coronavirus vaccine within days after the identification of the pandemic in early February is either the luckiest guess ever made by mankind, a miracle, or the product of huge investments made with great planning and specific foreknowledge of when the SARS-CoV would be released.

February 4, 2020: CDC sends out its test kits for COVID-19 and independent labs find they do not work. CDC bumbling delays America's response to the pandemic by several weeks as public health officials fly blind, unable to confirm hot spots. It's one of multiple CDC botches to come.[1032]

February 19, 2020: The esteemed British journal *The Lancet* publishes a letter organized by Peter Daszak from more than two dozen professionals fawning over China and rejecting any aspersions cast upon the Communist regime. It rejects the "conspiratorial" accusation that SARS-CoV-2 leaked from the Wuhan Institute. This event epitomizes how global predators in the medical and scientific communities consciously organized to put Communist China's image above scientific truth and the well-being of America and all humanity.

March 2020a: The Trusted News Initiative is launched in December 2019 by the BBC to "protect audiences and users from disinformation, particularly around moments of jeopardy, such as elections." Its goal is to censor anything positive toward President Trump or European populism in the coming elections. However, it soon takes on COVID-19 "disinformation." As of March 2021, the "partners" are "BBC, Facebook, Google/YouTube, Twitter, Microsoft, AFP [a news agency], Reuters, European Broadcasting Union (EBU), Financial Times, The Wall Street Journal, The Hindu, CBC/Radio-Canada, First Draft [leftwing disinformation group], Reuters Institute for the Study of Journalism." Their aim: "An industry collaboration of major news and tech organizations will work together to rapidly identify and stop the spread of harmful Coronavirus disinformation."[1033] In August 2020, to work against Trump's re-election, they will be joined by projects with the *AP*, *The New York Times*, and *Washington Post*.[1034] Their motto should be, "Global predators of the world unite!"

March 2020b: The health bureaucracy worldwide takes action to suppress the use of hydroxychloroquine or any other inexpensive, safe, and effective medications to treat COVID-19 patients, including early, at-home treatment. The suppression is most harsh in wealthy Western nations (who will pay for the development and use of very expensive vaccines).

The active suppression of lifesaving treatments by so many trusted authorities seems preposterous and beyond the imagination. Letting patients die for lack of treatment and murderously overdosing patients to discredit good treatments are such extreme activities that they beg for the full analysis provided in this book. Here is a brief analysis:

The January 2020 Emergency Use Authorization (EUA) act, which enables Operation Warp Speed and the related escalation toward the Great

Reset, has a catch in it. The catch is that the stupendous taxpayer gift to the pharmaceutical industry and its investors under the Emergency Use Authorization cannot go into effect if there are already existing, safe, and effective treatments. *To this day, if hydroxychloroquine and any other array of treatments are recognized by the government as safe and effective for COVID-19, Operation Warp Speed and the Great Reset will lose their legal justification under the Emergency Use Authorization.*

The second reason for stopping all effective treatments for COVID-19 is to make people believe they must take vaccines. *If almost everyone recovers with the available treatments, there will be no rush to take highly experimental, very dangerous, and very expensive vaccines.*

The third reason for making COVID-19 seem untreatable is that the globalists need to keep Americans terrified and docile. The fear of COVID-19 is justifying a totalitarian shutdown or lockdown of America—the one great nation standing in the way of the global predators. This unscrupulous motivation for weakening America is verified by the globalist thought leader, Klaus Schwab, in his book *COVID-19 and the Great Reset* in July 2020. Schwab devotes many pages to explaining that the existence of a strong, patriotic democratic republic like the United States, and especially under President Donald Trump, will thwart the ambitions of the globalists to rule the world through corporate-government fusions.

March 2020c: Family physician Vladimir Zelenko MD, the author of one of the introductions to this book, searches the scientific literature and contacts physicians worldwide, concluding that hydroxychloroquine with additional medications and therapeutics is a very safe and effective treatment for COVID-19. He successfully treats thousands of patients in his office north of New York City. He communicates with President Trump and becomes friends with Rudy Giuliani; however, he meets so much opposition from officials in New York City and fearful people in his community that he is forced to leave his clinic and his hometown.

March 11, 2020a: On this day, WHO declares a pandemic.[1035] On the same day, testifying before Congress, seeking increased funding, Anthony Fauci declares that the lethality of COVID-19 is "ten times" the "seasonable flu."[1036]

The flu has a mortality rate of 0.1 percent. This has a mortality rate of 10 times that. That's the reason I want to emphasize we have to stay ahead of the game in preventing this.

But Fauci had come to much more benign conclusions eleven days earlier in a scientifically documented editorial published online on February 28, 2020, and then republished in hard print on March 26, 2020, in the *New England Journal of Medicine*. Citing specific studies, Fauci declares:[1037]

If one assumes that the number of asymptomatic or minimally symptomatic cases is several times as high as the number of reported cases, the case fatality rate may be considerably less than 1%. This suggests that the overall clinical consequences of COVID-19 *may ultimately be more akin to those of a severe seasonal influenza (which has a case fatality rate of approximately 0.1).*

Which Fauci are we to believe, the one writing a science-based editorial or the one testifying before Congress? The answer is simple—neither one. Fauci always speaks politics to serve his masters and never science to serve the truth.

Actually, COVID-19 is much less dangerous than the flu for people under age 55, and for children, teenagers, and young adults it is a negligible or absent threat. Most deaths occur in patients over 80 who are suffering from other physical disorders. However, with adequate medical treatment—completely opposed by Fauci—deaths in older people can be and are enormously reduced.

March 11, 2020b: "Dr. Anthony Fauci shared a [virtual] platform with a Chinese Communist Party 'health expert,' and it went largely unreported by any Western media outlets. Chinese state media, on the other hand, could not get enough of it, because Fauci delivered a propaganda coup for the communist regime in Beijing." [1038]

March 13, 2020: President Trump, again with resistance from globalists who want to grow international commerce and to weaken America, bans travel from Europe except for U.S. citizens returning home.

March 19, 2020a: California becomes the first state to order a lockdown. This marks the beginning of the tragedy of increasing top-down government control in the United States that weakens America and accustoms its citizens to greater docility and subservience, while the global predators simultaneously bloat themselves with increased wealth, grandiosity, and power.

March 19, 2020b: At a press conference on COVID-19, President Trump reports positive news from other countries, including China, showing the effectiveness of hydroxychloroquine against COVID-19. His support of hydroxychloroquine leads the progressive anti-Trump media to further deride it as "Trump's drug." One commentator on Fox News, Neil Cavuto, almost hysterically accuses President Trump of killing people, warning that anyone who is even taking hydroxychloroquine as a preventive like the President, "it will kill you, I cannot stress enough, it will kill you."[1039]

March 23-24, 2020. Hydroxychloroquine is urgently needed throughout the country as a preventive measure for vulnerable people, to save individual lives of people infected with COVID-19, to reduce hospital admissions and deaths, and to cut short and defeat the pandemic. President Trump's Secretary of HHS Alex Azar directs his deputy, Rick Bright, to make hydroxychloroquine available from the U.S. National Stockpile.[1040] Deputy Bright instead colludes with the FDA's Janet Woodcock to sabotage hydroxychloroquine by restricting its use to hospitalized patients, when its proven worth is in early treatment. Trump is repeatedly stymied by the embedded globalist bureaucracy at the cost of thousands of American lives and a violent lockdown that wrecks individual lives, family life, society, and the economy.

March 25, 2020: New York Governor Cuomo signs an order requiring unprepared and ill-equipped nursing homes to accept COVID-19 patients, turning the facilities into death traps. A year later, in March 2021, Cuomo will be accused by his own administrator of suppressing the number of deaths, raising the actual tragic number from 5,000 to 10,000 or more fatalities.

Cuomo also issues one of the most restrictive executive orders in the country to prevent access to hydroxychloroquine for the treatment of COVID-19 patients.[1041] Cuomo's leadership will influence Democratic governors around the nation to repeat his mistakes, resulting in many thousands of unnecessary deaths.

April 3, 2020: The CDC reverses its earlier policy and recommends wearing face masks in public spaces and indoor gatherings. There is no scientific basis for these requirements, but wearing masks is an effective tool to frighten and demoralize people, isolating them and making them more submissive. Fauci also reverses his original March 8, 2020 statement that "there's no reason to be walking around with a mask." At no time does Fauci or the CDC produce scientific documentation for their stands on face masks, and scientific reviews show they do more harm than good outside of healthcare facilities (Chapter 12).

April 7, 2020: Dr. Deborah Birx confirms that CDC is counting deaths of people *with* coronavirus as *caused* by coronavirus regardless of other comorbid, underlying, or primary causes: "So, I think in this country we've taken a very liberal approach to mortality.... the intent is right now that those if someone dies with COVID-19 we are counting that as a COVID-19 death."[1042]

April 8, 2020: New Jersey becomes the first state to require customers and employees to wear face masks at essential businesses and construction sites.

April 11, 2020: The *Journal of the American Medical Association* (*JAMA*) rushes to put online a study from Brazil purporting to show that chloroquine (the earlier version of hydroxychloroquine) killed many COVID-19 patients. An examination of the prepublication paper and the published paper shows the victims were killed by systematic, substantial overdoses of the drug until an oversight committee stopped the murderous experiment. *JAMA* never retracts the criminal study, which makes American doctors terrified of giving hydroxychloroquine to their patients.

April 14, 2020: President Trump announces the U.S. will withdraw funding from WHO for covering up China's lies early in COVID-19 and for other pro-Chinese actions. Bill Gates criticizes Trump and quickly makes himself the second biggest contributor to WHO after the Chinese Communist Party.

April 14-15, 2020: The Breggins discover the 2015 scientific paper by Menachery showing that Fauci-funded U.S. scientists in 2015 were collaborating with top Wuhan Institute scientists to make lethal SARS-CoVs in American and Wuhan labs. We publish a blog and a video,[1043] and send them to the media and to close allies of President Trump. Two days later, on April 1st, President Trump stops the Fauci-funded collaborative research with the Chinese Communists.

In the blog and video, we point out that the 2015 SARS-CoV is "extremely closely related" to SARS-CoV-2 in its clinical effects. We warn in the video that the 2015 SARS-CoV resisted treatment in mice. We also warn that some of the older vaccinated mice became severely ill and died when exposed to the SARS-CoV virus—a warning yet to heeded by NIAID, FDA, and CDC.

We emphasize that the Wuhan Institute is run by the Chinese military. We already believe that SARS-CoV-2 came from the Wuhan Institute as a pathogen from the line of those being made in 2015 in the U.S./China collaboration.

April 16, 2020: The U.S. Government through BARDA (in HHS), under Rick Bright, awards Moderna $483 million "to Accelerate Development of mRNA Vaccine (mRNA-1273) Against Novel Coronavirus."[1044] The U.S. taxpayer is paying for Moderna's clinical trials, which have not yet begun.

May 15, 2020: President Trump, with Fauci over his shoulder, announces Operation Warp Speed. The program mimics everything Bill Gates announced in January 2017 and is wholly consistent with what Schwab will call the Great Reset with its corporate-government fusions. There is no indication that President Trump knows that the whole idea of Operation Warp Speed was planned and organized years earlier by Bill Gates with Moderna and other investors, confident of a coming vaccine bonanza from the predicted or planned pandemic.

June 3, 2020: Schwab enlists John Kerry and Prince Charles to announce the Great Reset.

July 2020: A team made up of Chinese researchers, with the addition of Baric from North Carolina, publishes a scientific paper about engineering a SARS-CoV-2 into one that causes an unusually high death rate in mice due to encephalitis (see Chapter 3).[1045] Baric continues *making even more dangerous SARS-CoV pathogens with the Chinese Communists.*

July 14, 2020: Schwab releases his book *COVID-19 and The Great Reset*. He touts it as exploring "the far-reaching and dramatic implications of COVID-19 on tomorrow's world, and advocates for a 'Great Reset' in the economy and society."[1046] Note the direct use of COVID-19 to promote the new world order of the Global Reset—something the globalists have been building toward for many decades. Schwab describes how globalism is making strides under COVID-19 but is now deeply threatened by American populism and similar trends in Europe. He warns that the success of globalism depends on stifling America's democratic, patriotic revival under President Trump.

July 15, 2020: Moderna becomes the first drug company to begin clinical trials and announces supposedly successful results in the initial trials. Moderna, remember, was already being promoted by Bill Gates in January through March 2017 as a drug company preparing novel vaccines on rush platforms for the coming great pandemic. The many years of investment, scientific research, and organizational planning by Bill Gates, Klaus Schwab, Michael Bloomberg, Moderna, Fauci, and other globalists are coming to fruition.

July 22, 2020: In a $1.95 billion deal, the U.S. buys 100 million doses of Pfizer's vaccine with an option for 500 million MORE—five months before it is approved.[1047, 1048]

July-August 20, 2020: Attorney Tom Renz asks Dr. Peter Breggin to write a medical-legal report in support of an injunction to stop the oppressive, unending emergency measures in Ohio and other states. Renz succeeds in getting his lawsuit accepted by the court, beginning a nationwide series of legal actions, and the Breggins begin this book. The injunction will cause the governor to halt his emergency edict but not his oppressive activities.

August 7, 2020: Led by Peter A. McCullough MD, a large team of physicians and researchers publish an article in the *American Journal of Medicine* that presents a pathophysiological basis for understanding the existing, safe, and effective early treatment options for COVID-19.[1049] The study carefully outlines a treatment approach that includes hydroxychloroquine in combination with other antiviral medications and zinc, and with prednisone and anticlotting drugs when necessary. *The government and the medical and scientific establishment continue to withhold from doctors and patients any adequate form of lifesaving, early treatment.*

Late August 2020: CDC admits that only 6% of COVID-19 death certificates have only that one cause of death listed and that there are an average of about 2.5 comorbidities per patient, many of which could have been the primary cause of death (Chapter 8).[1050] The inflation of death rates is to frighten people into taking vaccines. The obscure and uniquely inadequate process of counting also makes accurate accounting impossible.

September 2020: Continued warnings about lethality of the COVID-19 vaccines. In the midst of Operation Warp Speed, another major review warns that the vaccines "being expedited through preclinical and clinical development" could "exacerbate COVID-19 through antibody-dependent enhancement (ADE)."[1051] That is, being vaccinated and then exposed to COVID-19 can produce unusually severe respiratory disease related to the cytokine cascades (storms) of severe COVID-19 (Chapter 11). Fauci, Gates, Moderna, Pfizer—everyone knowingly putting any of the vaccines through Operation Warp Speed—are acting recklessly and their actions will lead to thousands of deaths in the U.S. and around the world.

September 14, 2020: An interview with Gates describes him as not a "fan" of President Trump but closer to Fauci.[1052]

September 14, 2020: Li-Meng Yan MD, PhD—who recently escaped from China—and three colleagues publish a paper linking SARS-CoV-2 to a chain of Chinese and American research going back to 2013 and to the Wuhan Institute from which it was released.[1053]

October 2020: Peter A. McCullough MD, MPH, Jane Orient MD, and Elizabeth Lee Vliet MD (the writer of the document) jointly author *A Guide to Home-Based COVID Treatment: Step-By-Step Doctors' Plan That Could Save Your Life.* The free, regularly updated booklet is distributed by the Association of American Physicians and Surgeons (AAPS), but it is suppressed by the mainstream media, social media, the medical and scientific establishment, and every other global predatory institution in America and the world.

October 8, 2020: Dr. Li-Meng Yan and her colleagues put a second paper onto the internet, "SARS-CoV-2 Is an Unrestricted Bioweapon: A Truth Revealed through Uncovering a Large-Scale, Organized Scientific Fraud." [1054]

November 12, 2020: *JAMA Network* issues a preliminary analysis that fluvoxamine vs. placebo is helpful for outpatient early treatment of COVID-19. Fluvoxamine is an SSRI antidepressant with an especially strong profile of adverse effects that include mania, hypomania, and violence [1055] and has gone out of favor in the U.S. Eric Harris, the Columbine High School shooter, was taking fluvoxamine at the time he perpetrated the mass murders. My medical/legal analysis of his medical and related records—made available to me as a medical expert—found that Harris deteriorated while taking the antidepressant for months leading up to the day of the mass murders, with an increase in his dose not long beforehand. On the day of the shootings, he probably continued to take the medication. In the autopsy after his death by suicide at the scene of the murders, Harris had an effective concentration of fluvoxamine in his bloodstream. The coroner ironically described it as a "therapeutic" level.

For those who find it difficult during COVID-19 to believe that the media and pharmaceutical industry work together to deceive the public, the press uniformly ignored the story or falsely reported that Harris was not taking antidepressants at the time of the shootings. That was back in mid-1999.

November 16-19, 2020: Michael Bloomberg's annual New Economy Forum (NEF) is held virtually. The Chinese Communist Party is again a cosponsor. The theme is defiance of President Trump's policies and the necessity of recoupling China with the world's billionaires and giant corporations. Top global corporate executives are lending their support, including McDonald's, IBM, MasterCard, FedEx, Prudential, Goldman Sachs, Honeywell, and numerous huge banks. The Clintons are there and so is Bill Gates.

November 19 and December 8, 2020: The U.S. Senate Oversight Committee for Homeland Security and Government Affairs under Senator Ron Johnson holds two hearings with extremely well-qualified medical experts testifying under oath about the effective, early, home-based treatment of COVID-19 to save lives. Videos of the Senate expert testimony are censored by YouTube, Facebook, and Twitter, and the hearings are not covered in the mainstream media. The Democrats boycott and attack the hearings and the participants. Every effort is made to continue keeping the American people in the dark about inexpensive, safe, and effective treatments for COVID-19.

November 20, 2022: WHO gives "a conditional recommendation against the use of remdesivir. This means that there isn't enough evidence to support its use." This undermining of another treatment is almost certainly aimed at increasing the necessity of the vaccines. The global predators are putting all their emphasis on vaccines for gaining enormous profits and enforcing more fear and greater top-down government, especially in America.

November 24, 2020: Following the election of Joe Biden, which is an absolute necessity for the triumph of the global predators, the Great Reset becomes U.S. policy. At Schwab's 2020 World Economic Forum, John Kerry—Biden's special consultant, cabinet member, and Climate Envoy— makes the final connection. He declares, "In effect, the citizens of the United States have just done a Great Reset. We've done a Great Reset."

December 3, 2020: John Ratcliffe, Director of National Intelligence for the U.S., states:[1056]

> If I could communicate one thing to the American people from this unique vantage point, it is that the People's Republic of China poses the greatest threat to America today, and the greatest threat to democracy and freedom worldwide since World War II.

> The intelligence is clear: Beijing intends to dominate the U.S. and the rest of the planet economically, militarily, and technologically. Many of China's major public initiatives and prominent companies offer only a layer of camouflage to the activities of the Chinese Communist Party.

> I call its approach of economic espionage "rob, replicate, and replace." China robs U.S. companies of their intellectual property, replicates the technology, and then replaces the U.S. firms in the global marketplace.

December 11, 2020: The Pfizer-BioNTech vaccine is authorized by an Emergency Use Authorization by the FDA. The Chinese are also beneficiaries through their corporate partnership with Germany's BioNTech.[1057]

December 18, 2020: Moderna's vaccine is authorized for use by an Emergency Use Authorization by the FDA. NIH claims "joint ownership" of the vaccines,[1058] making it a perfect corporate-government fusion.

Late December 2020: Two scientific articles are posted online and then published in journals indicating the spike protein (S protein) alone can cause inflammation and damage to the endothelial cells in blood vessels in the body and brain in animals and humans.[1059,1060] They document the mRNA vaccines are causing the recipient's body to produce spike proteins that are in themselves potentially very harmful to the body.

December 30, 2020: Peter A. McCullough MD, MPH and 32 other physicians and researchers—including Elizabeth Lee Vliet MD and Vladimir Zelenko MD—publish the most extensive review of the benefits of early treatment of COVID-19.[1061]

January 14, 2021: The Vaccine Credential Initiative (VCI) is unveiled as a coalition of big tech and big healthcare corporations, as well as WHO and others. It claims a "Trustworthy, traceable, verifiable, and universally recognized digital record of vaccination status is urgently needed worldwide to safely enable people to return to work, school, events, and travel."[1062] It is closely involved with the Commons Project,[1063] which is supported by the Rockefeller Foundation and Schwab's WEF, and aims to develop a CommonPass, a vaccine and medical record passport, working with Apple Health for iOS and CommonHealth for Android. Every global predator may end up having a piece of this project and, along with it, a piece of us.

January 19, 2021: "BioNTech-Pfizer and Moderna COVID-19 vaccines sales are expected to generate $14.7bn in revenue by 2023."[1064]

February 2021: Israel experiences an escalating death rate in association with Pfizer vaccine—a controversy that continues to this day. The government remains surprisingly adamant about vaccinating the population.[1065]

February 8, 2021: South Africa suspends use of AstraZeneca's COVID-19 shot after data showing it only gives protection against mild to moderate infection caused by the country's dominant coronavirus variant.[1066]

March 2021: Li-Meng Yan and her colleagues publish a third paper further emphasizing that SARS-CoV-2 was released as a bioweapon in China's unrestricted warfare against the West. They describe the collusion to cover this up and answer globalist critics.[1067]

By now we are in agreement with Dr. Yan. The lengthy, costly, intensive, and worldwide preparation and planning indicates an intention by the global predators, including the Communist regime, to create a coronavirus pandemic if one did not spontaneously arise. The time of the pandemic corresponds with a desperate need for the globalists and for the Communist dictatorship to assault and even depose President Trump, whose policies were undermining their power, control, and wealth.

Furthermore, the global predators, led by China, systematically took advantage of the epidemic and spread it worldwide, especially into America, in their strategy of unrestricted warfare. The pandemic has been greatly to their advantage, vastly enriching their billionaires and hence the Communist Party, helping them reduce their economically burdensome, aged population, enabling them to tighten control over their own population, weakening America, and giving them a biological weapon for current and future use against their enemies. They are especially using the pandemic to strengthen their economy in relation to the rest of the world. Through their principle of "unrestricted warfare," the Chinese Communists spread SARS-CoV-2 throughout the unsuspecting world by continuing flights with hundreds of thousands of potentially infected Chinese flown from Wuhan, Beijing, and other Chinese cities into the United States and the rest of the world. This assault continued until President Trump, on his own and against Fauci's expressed wishes, stopped the traffic from China.

March 8, 2021: We decide to put our book *COVID-19 and the Global Predators* on advance sale with a unique bonus: Everyone who buys the book immediately receives the working manuscript by email. We also periodically email updated versions to the purchasers. Within weeks, many thousands of people have bought the book prepublication and are reading the bill of particulars against Anthony Fauci and other new information in the manuscript. Simultaneously, manuscript copies are sent to the media and critics of COVID-19 policies. Perhaps by coincidence, criticism of Fauci heats up in May and June, surrounding issues we have most fully documented.

March 9, 2021: Fauci meets "side-by-side" with Chinese medical leader Dr. Zhong Nanshan on a virtual panel sponsored by the University of Edinburgh.[1068] Nanshan has been awarded the Medal of the Republic—the highest honor bestowed by the Chinese Community Party. Fauci declares the U.S. has been "hurt" more than any other country by COVID-19, citing falsified CDC data to prove his point. He accuses the past administration of "divisiveness." Both men are critical of President Trump. They praise President Biden and call for global solidarity.[1069] Zhong ends his presentation by "urging other countries not to open their economies just yet, and to perhaps wait until the entire world is vaccinated, which will take a few years." Meantime, China's economy has been largely open for a year with a low vaccination rate.

March 19, 2021: "The PREP Act and COVID-19: Limiting Liability for Medical Countermeasures Updated March 19, 2021. To encourage the expeditious development and deployment of medical countermeasures during a public health emergency, the Public Readiness and Emergency Preparedness Act (PREP Act) authorizes the Secretary of Health and Human Services (HHS) to limit legal liability for losses relating to the administration of medical countermeasures such as diagnostics, treatments, and vaccines."[1070] The pharmaceutical companies were already immune to vaccine lawsuits. This gives nearly complete immunity to healthcare providers. Individual and patient rights—in this instance, the right to sue healthcare providers for injuries—continue to be trampled upon.

March 26, 2021: Former CDC Director Robert Redfield states in an interview that he believes coronavirus escaped from a Wuhan lab.[1071]

April 2021a: As the grip of COVID-19 propaganda on Americans begins to loosen, the Biden Administration and the Chinese Communists are collaborating in turning their attention to ramping up the Global Warming campaign to continue the weakening of America and to increase the spread of totalitarian control worldwide. John Kerry goes on a world tour to convince countries to restrict their economies to ward off global warming.[1072]

April 2021b: Bill Gates is heavily criticized after he spontaneously states that he is against giving India, which is struggling with a resurgence of COVID-19, the legal rights or the know-how to manufacture the Oxford AstraZeneca vaccine. Gates gives reasons, but profit from intellectual property seems key.[1073, 1074]

Behind this conflict between Bill Gates and India, there are three less noticed but ominous facts. First, one man, Bill Gates is deeply invested in and has control over at least three out of four of the world's COVID-19 vaccines—Moderna, Pfizer, and now Oxford AstraZeneca, and he is also invested in BioNTech, the German vaccine manufacturer who is partnered with Pfizer and China. Second, this one man believes his financial investments and corporate influence give him the unrestricted right to make life-and-death decisions for millions and millions of people in the world's second-largest nation. Third, as we have documented, Bill Gates is very aligned with China and WHO. India is an adversary of China. That Gates casually expresses his life-and-death decisions on a TV interview adds a bizarre and unreal aura to the whole horrifying affair.

April 15, 2021: WHO announces it will begin storing and sharing pathogens internationally, declaring "this will pave the way towards a system that will promote the rapid and timely sharing of biological materials with epidemic or pandemic potential, facilitate rapid access to pathogens, and promote equitable access to countermeasures."[1075] *The Scientist* raises the alarm about this.[1076] This is globalism at its most irrational and terrifying. Next, will the UN begin sharing worldwide the design of nuclear weapons, which many experts consider less dangerous than biological warfare with pathogens? And how secure will be the WHO pathogen repositories?

April 24, 2021: The Biden Administration adds to India's COVID-19 crisis by refusing to lift its ban on sending India raw materials for the Oxford AstraZeneca vaccine. We do not accept Biden's claim of putting America's needs first.[1077] China is in constant conflict with India on every level, including militarily, and weakening India suits China's empire building.

April 25, 2021a: The headline in the *Financial Times* lauds the ominous declaration of China's Communist dictator at an international trade conference: "Xi warns against economic decoupling and calls for new world order." [1078] Xi Jinping, Klaus Schwab, Bill Gates, Mike Bloomberg, the tech companies, the big banks, and all the other global predators know that the only obstruction to their ambitions for world domination is the populist, freedom-loving, patriotic, democratic republic of the United States of America and, yes, former President Donald Trump.

April 25, 2021b: On NBC's *Meet the Press*,[1079] NIH Director Francis Collins states that Moderna's and Pfizer's vaccines cannot receive actual permanent FDA approval because more follow-up is needed "to look at any possible late safety signals. There have not been any for Pfizer and Moderna." The deaths reported to VAERS in close association to taking the vaccines have reached 3,486 after only four months.[1080] *These are extraordinarily high numbers that would take any other medicine or vaccine permanently off the market. The CDC's claim that none of these cases are actually related to the vaccines is absurd and unconscionable and can only be explained by tremendous forces impinging on the agency.*

April 30, 2021a: "CDC is transitioning to reporting only patients with COVID-19 vaccine breakthrough infection who were hospitalized or died."[1081] Once again, information is being culled before we can see it to protect the vaccines.

April 30, 2021b: Patrick Coffin, a courageous fighter for liberty, has his YouTube channel taken down shortly before his virtual "Truth Over Fear Summit" is scheduled. He manages nonetheless to have nearly 50,000 signups, but his summit is then shut down by its Australian platform after barely getting started. One week later, with volunteer help from a conservative platform, he resumes his conference. Speakers at the three-day celebration of freedom and truth in science include this writer and the three outstanding physicians who wrote introductions to our book. Patrick stands for all the Americans being cancelled for such things as their opposition to election fraud or criticism of COVID-19 policies, and, more importantly, he stands for resilience as a Re-Founder of America.

May 2021: Six German researchers from Goethe-University of Frankfurt and Ulm University post online a report based on the concept of "Vaccine-Induced COVID-19 Mimicry."[1082] That is, the vaccination by itself can produce similar syndromes to SARS-CoV-2. They focus on vaccine-induced thromboses (clots in blood vessels). One of the frequently clotted veins is in the brain and another in the body. They remark, "This type of adverse event has not been observed in the clinical studies of AstraZeneca, and therefore led immediately to a halt in vaccinations in several European countries. These events were mostly associated with thrombocytopenia," or low blood platelets. These findings confirm that the SARS-CoV-2 spike protein created in the body as a result of vaccination is in itself toxic and can mimic the effects of COVID-19.

May 2, 2021: Michael Bloomberg's newsletter editorial sets the record straight on how much the global predators love the violent Chinese dictatorship and how much they hate the spirit of freedom in America. In one of the most stomach-churning displays of American self-loathing, Bloomberg's editor explains:[1083]

> Only a year ago, the world stood dumbfounded as the richest nation on earth, under the leadership of a COVID-denying president, botched its response to the pandemic, sacrificing hundreds of thousands of lives to a deadly combination of incompetence, hubris, and political partisanship.

> Meanwhile, it was equally obvious that Xi's swift decision to contain the virus with draconian lockdowns, blanket surveillance, and Mao-style social controls delivered stunning results. China flattened the curve and—while Republicans swore off face masks and social distancing as Americans perished...

Contemplate this: A former NYC mayor, now a leading globalist, can unashamedly declare, "Mao good! Trump bad," preferring the Communist nightmare to the American Dream. Everything we Americans hold dear is threatened to extinction by the global predators.

May 7, 2021: Dr. Peter McCullough is interviewed at length by Tucker Carlson on *Fox Nation*, followed by an excerpt that evening on his show on the Fox News Channel. Carlson is openly dismayed to learn about the innumerable deaths caused by the unprecedented and inexcusable suppression of safe and effective treatments for COVID-19. Tucker is seen asking Dr. McCullough "Why?" and "Who's behind it?" multiple times. We send our book manuscript to Tucker in the hope he will use it to answer the essential questions, which would be an enormous media breakthrough in broadcasting the truth.

May 10, 2021: Research that should be required reading for all concerned is published in the *International Journal of Vaccine Theory, Practice, and Research*,[1084] describing the unparalleled circumstances surrounding the Moderna vaccine that make it such a radical, dangerous experiment on humanity:[1085]

Unprecedented

Many aspects of COVID-19 and subsequent vaccine development are unprecedented for a vaccine deployed for use in the general population. Some of these include the following.

1. First to use PEG (polyethylene glycol) in an injection (see text)

2. First to use mRNA vaccine technology against an infectious agent

3. First time Moderna has brought any product to market

4. First to have public health officials telling those receiving the vaccination to expect an adverse reaction

5. First to be implemented publicly with nothing more than preliminary efficacy data

6. First vaccine to make no clear claims about reducing infections, transmissibility, or deaths

7. First coronavirus vaccine ever attempted in humans

8. First injection of genetically modified polynucleotides in the general population

May 11, 2021: Fauci lies to Senator Rand Paul under oath at a Senate hearing by declaring that neither his NIAID nor NIH has funded gain-of-function research with the Wuhan Institute of Virology in China (Chapter 2). Afterward, he continues lying to the press.[1086]

May 12, 2021: *The CDC has now received a total of 4,434 reports of deaths associated with vaccination in the U.S.*[1087] *By any standard, this is catastrophic.* Because underreporting is inherent in these systems, the actual number of deaths could easily be ten times or greater. In the past, a vaccine would have been withdrawn after 100 reports of deaths. Many physicians and countries are now calling for an end to the vaccine slaughter.

A striking pattern emerges. The CDC and other authorities have all along been vastly exaggerating the numbers of deaths caused by COVID-19 and now are ignoring the multitude of deaths being caused by the vaccines. It's all about the billions being made by predatory globalists from vaccines. Simultaneously, they weaken America's resistance to globalism by enforcing submissiveness among citizens and politicians alike.

May 19, 2021: "The COVID-19 pandemic has brought into sharp focus two issues: the human right to access a life-saving drug and a physician's right to exercise his Hippocratic Oath. Accordingly, the Front Line COVID-19 Critical Care Alliance (FLCCC) today called for the 'Immediate and Global Use of Ivermectin to End the COVID-19 Pandemic.'"[1088]

May 21, 2021: An article published in *Clinical Infectious Diseases* demonstrates that patients recently vaccinated with mRNA have SARS-CoV-2 spike protein remnants circulating in their blood stream. This confirms that spike protein distribution is not contained within the injection site, as claimed by advocates, and makes all cells in the body vulnerable to attack.

June 9, 2021: Dr. Tess Lawrie, Director, Evidence-based Medicine Consultancy Ltd., sends an "urgent" report to the British drug monitoring agency (MHRA) based on a preliminary analysis of adverse vaccine reports to the agency from January through May 26, 2021.[1089] She warns:

> We are sharing this preliminary report due to the urgent need
> to communicate information *that should lead to cessation of the*

vaccination rollout while a full investigation is conducted... (Italics added.)

The three vaccines used in Great Britain are made by AstraZeneca, Pfizer, and Moderna (spanning DNA and RNA). The adverse reactions are not limited to any particular vaccine brand or type. The report counts 1,253 deaths and 256,224 individual adverse event reports in what is called the Yellow Card system.

Drawing on Seneff and Nigh,[1090] Lawrie categorizes the types of adverse reactions reported:

- Pathogenic priming, multisystem inflammatory disease, and autoimmunity
- Allergic reactions and anaphylaxis
- Antibody dependent enhancement
- Activation of latent viral infections
- Neurodegeneration and prion diseases
- Emergence of novel variants of SARS-CoV-2
- Integration of the spike protein gene into the human DNA

Lawrie observes that spike proteins which the vaccines cause to be made in the body are "toxic to humans." She summarizes the reports to the British government as including "thromboembolism, multisystem inflammatory disease, immune suppression, autoimmunity, and anaphylaxis, as well as Antibody Dependent Enhancement (ADE)." These bad outcomes have already been documented in animals under laboratory conditions, and that in itself should have prevented the vaccines from being tested in humans as too dangerous.

June 10, 2021: CDC announces upcoming emergency meeting.[1091] "As of May 31, the CDC has received 275 preliminary reports of myocarditis and pericarditis in fully vaccinated 16-to-24-year-olds—a number that's higher than what scientists expected to see, the CDC said Thursday."[1092]

June 11, 2021a: BMJ's *Journal of Medical Ethics* publishes "Voluntary COVID-19 vaccination of children: a social responsibility." The authors endorse vaccinating children to generally benefit society. It also reveals that "Israel, which is a world pioneer in the vaccination of its adult population, signed a contract with Pfizer for sharing data on vaccinated citizens."[1093,1094]

June 11, 2021b: "VAERS data released today showed 329,021 reports of adverse events following COVID vaccines, including 5,888 deaths and 28,441 serious injuries between Dec. 14, 2020, and June 4, 2021."[1095] Giving vaccines to young and younger children is probably the single greatest example of child abuse in human history. A medical blizzard of potential harm like this has never before been unleashed on humanity. The entry to this *Chronology and Overview*, dated July 20, 2021, will summarize the latest reported vaccine-related death totals and comment on the implication of so much mayhem created by a single vaccine in such a short time.

June 22, 2021: Pandemic Predictions and Planning Event. A new report in *Science* led by Ralph Baric's team from the University of North Carolina defines the tragic future for the endless vaccine assault on humanity. It is titled, "Chimeric spike mRNA vaccines protect against Sarbecovirus challenge in mice." Continuing to falsely claim that SARS-CoV-2 emerged from nature, they are now claiming to have developed new mRNA vaccines: "multiplexed-chimeric spikes" that "can prevent SARS-like zoonotic coronavirus infections with pandemic potential." They are perpetuating the myth that SARS-CoV viruses are lurking in nature ready to emerge. This justifies their continuing with humanity-endangering gain-of-function experiments, creating multiple new pandemic CoVs in their labs. This vaccine scheme continues to engorge the wealth of the global predators, while imposing totalitarian control, and will not go away on its own. The research also exposes Anthony Fauci's lies about NIAID and NIH not funding gain-of-function research with Ralph Baric. This creation of "chimeric" pandemic viruses is supported by several grants from NIAID and NIH, as well as by the pharmaceutically based Burroughs Wellcome Fund and by Facebook's Zuckerberg fund.[1096]

July 20, 2021: Reports of vaccine-related deaths in COVID-19 have now exceeded 10,000. A detailed CDC report published in 2015 found that "multiple studies and scientific reviews have found no association between vaccination and deaths except in rare cases."[1097] The study noted that a polio vaccine was withdrawn after five deaths among recipients and five among contact persons. It was called "one of the worst pharmaceutical disasters in U.S. history." The CDC further explained that swine flu vaccine was withdrawn after a Guillain-Barré syndrome (paralysis, often followed by death) resulted in 53 deaths.

In comparison to stopping the use of vaccines after 10 to 53 reported deaths, as of July 9, 2021, there have been *10,991* COVID-19 vaccine-related reported deaths to the CDC.[1098] *Nothing in the history of American medicine has ever come remotely near the size of the COVID-19 vaccine catastrophe with many thousands of deaths from COVID-19 in a vaccination program that seems never-ending.*

Then consider that these reports of vaccine-related deaths to the CDC are but a fraction of the actual number of deaths that have occurred. According to studies we evaluated in the book, only 1% to 10% of serious adverse events are reported to the CDC's reporting system. The actual number of vaccine-related deaths is certainly many, many multiples higher than the already horrifying number of 10,991 deaths reported by early July 2021.

Peter A. McCullough MD, MPH provided us with the following updated information. COVID-19 vaccines are generating record safety reports. In 1990, the Vaccine Adverse Event Reporting Systems ("VAERS") was established as a national early warning system to detect possible safety problems in U.S. licensed vaccines.[1099] The total safety reports in VAERS for all vaccines per year up to 2019 was 16,320. The total safety reports in VAERS for COVID vaccines alone through July 7, 2021, was 438,440.[1100] We note that *compared to previous rate of reporting for adverse vaccine effects to the CDC, the current reporting for COVID-19 is an avalanche.*

Dr. McCullough also observed that people are dying and being hospitalized in record numbers in the days after COVID-19 vaccination. Based on VAERS as of June 25, 2021, there were 9,048 COVID-19 vaccine deaths reported and over 26,818 hospitalizations reported for the COVID-19 vaccines (Pfizer, Moderna, J&J). By historical comparison, from 1999 until December 31, 2019, VAERS received 3,167 (158 per year) adult death reports for all vaccines combined. Thus, the COVID-19 mass vaccination is associated with at least a 57-fold annualized increase in vaccine deaths reported to VAERS. COVID-19 vaccine adverse events account for 99% of all vaccine-related adverse events from December 2020 through the present in VAERS.

We agree with Dr. McCullough *that COVID-19 deaths should have been taken seriously and investigated, and then analyzed for the public, when they reached much less than 150. Then the vaccines should have been withdrawn when the reported deaths exceeded 200. They have now exceeded 10,000. Unlike all other large clinical investigations, the U.S. COVID-19 mass vaccination program has no critical event committee, data safety and monitoring board, and no human ethics committee. In a nutshell, there is nothing in the program to protect Americans from safety problems that arise with the vaccines.*[1101]

We believe that this slaughter of innocents is a mass murder that must be stopped.

August 4, 2021: This is the final entry into the *Chronology and Overview* and also into the book before it is published. It summarizes earlier events that, unfortunately and tragically, confirm that President Trump was aware of considerable medical, scientific, and legal support for any action he would take in saving lives in America and around the world by allowing and promoting preventive care and early home-based treatment of COVID-19 with appropriate medications, including hydroxychloroquine.

According to a private communication to us on August 2, 2021, from Elizabeth Lee Vliet MD, who was involved in these developments, President Trump's adviser, Peter Navarro, and the President himself were aware of the importance and availability of safe, effective treatments and failed to follow through. She also reports that Senator Ron Johnson also communicated directly with the President about the need for allowing American physicians and their patients access to these medications. Her observations are well-documented.

Dr. Vliet was working with Senator Johnson as far back as early March and early April 2020 on organizing a letter from American physicians to President Trump urging him to "give patients the right to try this drug [hydroxychloroquine] early in the course of their COVID-19 diagnosis."[1102] On June 2, 2020, The Association of American Physicians and Surgeons (AAPS) announced a lawsuit against the FDA to release the national stockpile of hydroxychloroquine for use by American citizens in the early and effective treatment of COVID-19.[1103] On June 22, 2020, the AAPS reported, "Today, the Association of American Physicians and Surgeons filed its motion for a preliminary injunction to compel release to the public of hydroxychloroquine by the Food and Drug Administration (FDA) and the Department of Health and Human Services (HHS), in *AAPS v. HHS*, No. 1:20-cv-00493-RJJ-SJB (W.D. Mich.). Nearly 100 million doses of hydroxychloroquine (HCQ) were donated to these agencies, and yet they have not released virtually any of it to the public."[1104] The APA further declared, "AAPS agrees with President Trump's adviser, Peter Navarro, PhD, who decries the obstruction by officials within the FDA to making this medication available to the public. President Trump himself has successfully taken this medication as a preventative measure,[1105] so why can't ordinary Americans?"

As we begin discussing in the Preface to this book, the global predators reversed President Trump's very successful America First program by unleashing COVID-19 on the world. President Trump ultimately supported their goals by shutting down the American economy and society, withdrawing his initial support of early treatment, and strongly endorsing the Gates' vaccine plan called Operation Warp Speed. The President's actions contributed to untold numbers of deaths. His support of the mass vaccinations and the lockdowns continues to demoralize a large portion of our citizenry and have profoundly compromised America's individual and political freedom.

We Americans, as well as others around the world, cannot rely upon heroic leaders to save us. It is time for freedom-loving, patriotic citizens everywhere to recognize and to stand up as a unified resistance against the loss of our freedoms and in support of reviving the spirit of liberty in America and everywhere on Earth.

Is it coming—the end of America as the beacon of liberty for humanity?

Our future is up to us all.

Become a Re-Founder of America!

ACKNOWLEDGMENTS

W e are especially grateful to Peter A. McCullough MD, Elizabeth Lee Vliet MD, and Vladimir "Zev" Zelenko MD, three great doctors and among the most knowledgeable people in the world about the prevention and early treatment of COVID-19. All three have appeared separately on *The Dr. Peter Breggin Hour* radio/TV show and in separate shorter videos on our YouTube Channel.[1106] We are honored that they wrote introductions to our book. Their help is an inestimable contribution to our book, our knowledge, and the public good. They are saving lives with their work in the field of prevention and early treatment of COVID-19. Dr. Zelenko has inspired us beyond words, and Dr. McCullough has educated and supported us beyond anything we could have hoped for. They are Re-Founders of America.

Ginger and I wish to single out Elizabeth Lee Vliet MD for our appreciation. With enormous generosity, she volunteered many hours reading and discussing the manuscript with us, making this a much better book. In early August 2021, Dr. Vliet announced the founding of the nonprofit Truth for Health Foundation. She is the President and CEO, and Dr. McCullough is the Chief Medical Advisor.[1107]

Judy Mikovits MD and Meryl Nass MD were especially welcoming to us early in our investigations surrounding COVID-19, contributing their enormous professional experience and insights into the origins and nature of the pandemic.

We want to thank the experts from many fields we interviewed on videos and the weekly *The Dr. Peter Breggin Hour* radio/TV show: Marilyn Singleton MD JD, Paul Alexander PhD, Eamonn Mathieson MD, Robert Yoho MD, Lee Merritt MD, and Matt Strauss MD. Other outstanding COVID-19 interviews include Dr. Miklos Lukacs, Lt. Frank Moore (FDNY Ret.), attorney Leah Wilson and Stand for Health Freedom, historian Dr. Uwe Alschner, entrepreneur Clay Clark, and journalists Jeffrey Tucker, Leo Hohmann, and Kristina Borjesson.

With special gratitude, we acknowledge the bravery of Li-Meng Yan MD, a scientist who fled China to speak out bravely to the world. We treasure our communications with her and thank her for living in the United States, where she is much-needed and deeply valued.

We want to thank attorney Thomas Renz for asking me to write a medical expert report in support of a legal case aimed at stopping the seemingly unending "emergency declaration" by Ohio's Governor Mike DeWine, who is using COVID-19 to increase his top-down control over the lives of the citizens of Ohio. The huge effort of writing a report for the Ohio court and the success of getting the case accepted by the court galvanized us into further action. More recently, we are pleased to be working on cases of COVID-19 abuse with attorney Robert F. Kennedy, Jr. in the U.S. and attorney Reiner Fuellmich in Germany.

We also want to thank our friend Pam Popper PhD, creator of Make Americans Free Again, for connecting us to Tom Renz. Tom and Pam are both heroes who are devoting themselves to freedom at a time when to do so is risky and all-consuming. C. J. Wheeler is our marvelous publicist, helping to get the book off the ground way before publication. She is a gift to those of us fighting for American values and liberty. Catherine Austin Fitts is a great supporter of our work and has helped the public know about it and helped us understand globalism.

Patrick Coffin has been a great inspiration in his courage, resilience, and deep faith. Brig. Gen. Robert S. Spalding III provided a video interview with outstanding information that confirmed our viewpoint about the existential threat posed by Communist China. Nick Hudson is the courageous Chairman of PANDA (Pandemics - Data & Analytics), an international multidisciplinary group that examines and informs about the human cost of COVID-19 globally.

Diana West's pioneering book *American Betrayal: The Secret Assault on Our Nation's Character* (2014) is about the corruption of our government by Communist influences during World War II and helped inspire our research.

The Association of American Physicians and Surgeons (AAPS) represents thousands of doctors who believe in health freedom and the principle of liberty. I am a member along with many other physicians we have thanked in these acknowledgments.

Ginger's mother, Jean, gets a special thank you for the good humor and insight she brings to the "war room," which is our kitchen/dining room area. She adds immeasurably to our resilience, ability to laugh, and insights.

We have many, many others to thank but, out of respect for their privacy and safety in these threatening times, we must thank them in person.

ABOUT THE AUTHORS

PETER R. BREGGIN MD

Peter R. Breggin MD has been called "The Conscience of Psychiatry" for his many decades of successful efforts to reform the mental health field. His scientific and educational works have provided the foundation for modern criticism of psychiatric drugs, electroshock, and psychosurgery. He is also a leader in promoting more caring and effective therapies. His professional website and his YouTube video channel reach millions annually. However, due to increasing censorship, it is best to follow his work by subscribing to his Free Frequent Alerts on his professional website, www.breggin.com.

From early in his career, Peter has promoted freedom, responsibility, and love in his clinical, educational, professional, and political activities. His values of reason, liberty and love, and his research experience, led him to join others in examining and resisting the oppression behind COVID-19, and in promoting what he calls the Refounding of America.

In the arena of COVID-19, along with his wife Ginger as his co-researcher and consultant, he is currently working with three outstanding attorneys in a variety of state, federal, and international cases to protect and advance individual freedom: Robert F. Kennedy, Jr. and Tom Renz in the United States, and Reiner Fuellmich in Germany and Europe. In these new roles, he draws on many decades of experience as a medical expert in hundreds of legal actions, including landmark cases, on behalf of patient rights in criminal, malpractice, and product liability lawsuits, as well as injunctions to stop abusive medical and psychiatric practices.

Peter is in the private practice of psychiatry in Ithaca, New York. His educational background includes Harvard College, Case-Western Reserve School of Medicine, and psychiatric residency programs at both the State University of NY Upstate Medical Center and the Massachusetts Mental Health Center where he was a Teaching Fellow at Harvard Medical School.

He has authored more than 70 peer-reviewed scientific articles and 24 medical and trade books, including the bestsellers *Toxic Psychiatry* (1991) and

Talking Back to Prozac (with Ginger Breggin, 1994). His most recent three books are (1) *Medication Madness: The Role of Psychiatric Drugs in Cases of Violence, Suicide, and Crime* (2008); (2) *Psychiatric Drug Withdrawal: a Guide for Prescribers, Therapists, Patients and Their Families* (2013); and (3) *Guilt, Shame, and Anxiety: Understanding and Overcoming Our Negative Emotions* (2014).

GINGER ROSS BREGGIN

Ginger Ross Breggin has a background in journalism, book editing, bookmaking, and book publishing. Since 1984, she has partnered with Peter as a coauthor, writer, editor, researcher, organization administrator, advisor, and communicator with the outside world.

When hints of a possible new pandemic reached the U.S. in January 2020, Ginger redirected her attention to researching what would be called SARS-CoV-2. She soon recognized the significance of an obscure reference to a paper published in 2015 in *Nature Medicine*. It was titled, "A SARS-like cluster of circulating bat coronaviruses shows potential for human emergence." It documented that the U.S. had been collaborating with Chinese researchers at the Wuhan Institute of Virology in gain-of-function research—making lethal viruses very similar to SARS-CoV-2. With her husband, they set aside their normal lives and began work on the pandemic, digging deep into the tragedy of the world's response to COVID-19.

Along with her husband Peter, Ginger is a member of several COVID-19 medical and science groups, including the international Doctors for Covid Ethics (D4CE) and the U.S.-based C19 Group which focuses upon early treatments for COVID-19, ongoing research, and the effects of government policies.

Ginger is the coauthor of several books with Peter, including their bestseller *Talking Back to Prozac* (1994) and *The War Against Children of Color: Psychiatry Targets Inner City Youth* (1998). She is a coeditor of *Dimensions of Empathic Therapy*.

Ginger designed and published Peter's book, *Wow, I'm an American: How to Live Like Our Nation's Heroic Founders*. She edited and published *The Conscience of Psychiatry: The Reform Work of Peter R. Breggin, MD.* She has researched, edited, and coauthored many blogs with him on the issues in this book.

Ginger inspired and cofounded with her husband the peer-reviewed scientific journal *Ethical Human Psychology and Psychiatry*, which she managed for many years. As a freshman undergraduate at American University, she was given an annual honors award for the best social sciences paper for all levels of the university.

From 1988-2002, she was the Executive Director of Peter's original nonprofit reform center, the International Center for the Study of Psychiatry and Psychology (ICSPP). In 2010, she cofounded a new reform nonprofit organization with her husband called The Center for the Study of Empathic Therapy, for which she is the executive director. She also works with her husband on his websites and produces his videos and his radio/TV show, *The Dr. Peter Breggin Hour.*

ENDNOTES

These endnotes with live links can be found in digital form on www.breggin. com on the Coronavirus Resource Center. See "Digital Endnotes for COVID-19 and the World Predators" by Peter and Ginger Breggin. Here is the direct link to download the endnotes: https://breggin.com/covid-19-and-the-global-predators-end-notes

1   Peter McCullough, Paul E. Alexander, Robin Armstrong, Cristian Arvinte, Alan F. Bain, Richard P. Bartlett, Robert L. Berkowitz, Andrew C. Berry, Thomas J. Borody, Joseph H. Brewer, Adam M. Brufsky, Teryn Clarke, Roland Derwand13, Alieta Eck14, John Eck, Richard A. Eisner15, George C. Fareed16, Angelina Farella, Silvia N. S. Fonseca, Charles E. Geyer, Jr., Russell S. Gonnering, Karladine E. Graves, Kenneth B. V. Gross, Sabine Hazan, Kristin S. Held , H. Thomas Hight, Stella Immanuel, Michael M. Jacobs, Joseph A. Ladapo, Lionel H. Lee, John Littell, Ivette Lozano, Harpal S. Mangat, Ben Marble, John E. McKinnon , Lee D. Merritt, Jane M. Orient, Ramin Oskou , Donald C. Pompan, Brian C. Procter, Chad Prodromos, Juliana Cepelowicz Rajter, Jean-Jacques Rajter, C. Venkata S. Ram, Salete S. Rios, Harvey A. Risch, Michael J. A. Robb, Molly Rutherford, Martin Scholz , Marilyn M. Singleton, James A. Tumlin49, Brian M. Tyson, Richard G. Urso, Kelly Victory, Elizabeth Lee Vliet, Craig M. Wax, Alexandre G. Wolkoff, Vicki Woolland, Vladimir Zelenko (2020, December). Multifaceted highly targeted sequential multidrug treatment of early ambulatory high-risk SARS-CoV-2 infection (COVID-19). Reviews in Cardiovascular Medicine. 21 (4), 517-530. November 2020. DOI: 10.31083/j.rcm.2020.04.264 https://rcm.imrpress.com/ EN/10.31083/j.rcm.2020.04.264
    The 33 author list and the footnote numbers indicating 56 institutional affiliations are included here to show the scope of support for these protocols for the early treatment of COVID-19 that are being denied, rejected, and suppressed by Anthony Fauci, NIH, the FDA, and the CDC, as well as WHO and many countries.
2   McCullough et al., 2020. See previous endnote.
3   The quote is a personal communication from Dr. McCullough in May 2021. The source research is: Procter BC, Ross C, Pickard V, Smith E, Hanson C, McCullough PA. (2021, March.) Early Ambulatory Multidrug Therapy Reduces Hospitalization and Death in High-Risk Patients with SARS-CoV-2 (COVID-19). ijirms [Internet]. 2021Mar.17 [cited 2021Apr.28];6(03):219 - 221. View of Early Ambulatory Multidrug Therapy Reduces Hospitalization and Death in High-Risk Patients with SARS-CoV-2 (COVID-19) (ijirms.in)
4   The words "we" or "us" refers to Peter and Ginger Breggin. The first person, "I" or "me," refers to Peter Breggin who wrote the text of this book. However, the book has been a joint effort from its initial conception. Ginger has been totally involved in developing our mutual understanding of COVID-19 from the outset. She came up with many insights embedded in the book and has done a large portion of the background research and obtaining of sources for this book. She has also found many of the experts we have consulted and whom I interviewed on the Dr. Peter Breggin Hour, our weekly radio and TV show. https://breggin.com/the-dr-peter-breggin-hour/

5   David R. Martinez et al., including Ralph Baric. (2021, June 22, initially published online). Chimeric spike mRNA vaccines protect against Sarbecovirus challenge in mice. DOI: 10.1126/science.abi4506. https://www.biorxiv.org/content/10.1101/2021.03.11.434872v2.full See June 22, 2021 in the Chronology and Overview for more details.

6   https://www.openvaers.com/covid-data

7   https://www.bloomberg.com/news/features/2021-01-13/china-loves-elon-musk-and-tesla-tsla-how-long-will-that-last

8   https://cepi.net/news_cepi/global-partnership-launched-to-prevent-epidemics-with-new-vac-cines/

9   https://cepi.net/news_cepi/stevanato-group-signs-an-agreement-with-cepi-to-provide-phar-ma-glass-vials-for-2-billion-doses-of-covid-19-vaccines-under-development/; https://cepi.net/wp-content/uploads/2021/03/CEPI-2.0_Strategy-2022-26-Mar21.pdf

10   https://www.pharmaceutical-technology.com/news/clover-biopharmaceuticals-cepi-funds/

11   https://cepi.net/wp-content/uploads/2020/04/CEPI-Annual-Progress-Report-2019_web-site.pdf

12   McCullough, P. and 32 others including Drs. Elizabeth Lee Vliet and Vladimir Zelenko (2020). Multifaceted highly targeted sequential multidrug treatment of early ambulatory high-risk SARS-CoV-2 infection (COVID-19). Reviews in Cardiovascular Medicine. 21 (4), 517-530. November, 2020. DOI: 10.31083/j.rcm.2020.04.264 https://rcm.imrpress.com/EN/10.31083/j.rcm.2020.04.264. See endnote 3 for complete list of authors.

13   https://www.scmp.com/tech/start-ups/article/3098562/china-boasts-two-biggest-tech-unicorns-led-tiktok-owner-bytedance

14   A newspaper, The Epoch Times, provides the best current information on the silent Chinese war against America that we have found. Two of the most revealing recent books, which we had not seen before we reached our own conclusions studying COVID 19, are Gertz, Bill. Deceiving the Sky: Inside Communist China's Drive for Global Supremacy, New York, En-counter Books, 2019 and Schweizer, Peter. Secret Empires, New York, Harpers, 2019. Also see the writings of Gordon Chang @ http://www.gordonchang.com/article.htm.

15   https://sharylattkisson.com/2020/03/watch-confucius-institutes-in-american-schools/·https://duckduckgo.com/?q=China%27s+vast+educational+influence+in+america&ia=web; https://americanannouncement.com/2020/05/chinas-heavy-influence-over-u-s-higher-education-sys-tem/;https://www.washingtonpost.com/opinions/global-opinions/waking-up-to-chinas-infil-tration-of-american-colleges/2018/02/18/99d3bee8-13f7-11e8-9570-29c9830535e5_story.html; https://www.dailysignal.com/2020/02/14/pompeo-warns-of-chinas-hidden-influence-in-america/

16   https://www.aei.org/china-global-investment-tracker/; https://www.cnbc.com/2019/08/21/moodys-chinese-overseas-infrastructure-investment-growth-to-slow.html; https://www.dw.com/en/investing-in-africas-tech-infrastructure-has-china-won-already/a-48540426;

17   NIH also includes the National Cancer Institute (NCI), the National Institute of Aging (NIA), the National Institute of Environmental Health Sciences (NIEHS) and the National Human Genome Research Institute (NHGRI) and many others.

18   Dawsey, J. and Abutaleb, Y. 2020, October 31. "A whole lot of hurt": Fauci warns of covid-19 surge, offers blunt assessment of President Trump's response, Washington Post. https://www.washingtonpost.com/politics/fauci-covid-winter-forecast/2020/10/31/e3970eb0-1b8b-11eb-bb35-2dcfdab0a345_story.html

19   The Hatch Act: Political Activity and the Federal Employee | FDA

20   Can President Trump Fire Dr. Fauci? – FedSmith.com

21   The Hatch Act: Political Activity and the Federal Employee | FDA

22   Dawsey, J. and Abutaleb, Y. 2020, October 31. "A whole lot of hurt": Fauci warns of covid-19 surge, offers blunt assessment of President Trump's response, Washington Post. https://

www.washingtonpost.com/politics/fauci-covid-winter-forecast/2020/10/31/e3970eb0-1b8b-11eb-bb35-2dcfdab0a345_story.html

23 O'Kane, C., 2020, October 23, "We're about to go into a dark winter": Biden says President Trump has no plan for coronavirus, CBS News. https://www.cbsnews.com/news/biden-President Trump-coronavirus-plan-vaccine-winter/

24 http://www.upmc-biosecurity.org/website/events/2001_darkwinter/index.html. For Congressional Testimony about the event, which reads like movie credits for who played the President, etc., see http://www.upmc-biosecurity.org/website/resources/To%20USG/Testimony_Briefings/2001/20010723femarole.html

25 Facebook messages from Dr. Daniele Macchini -- from his FB page:. March 23, 2020 Italian with English translation. Also see, https://www.weforum.org/agenda/2020/03/suddenly-the-er-is-collapsing-a-doctors-stark-warning-from-italys-coronavirus-epicentre

26 https://covid19criticalcare.com/wp-content/uploads/2021/01/FLCCC-PressRelease-NIH-Ivermectin-in-C19-Recommendation-Change-Jan15.2021-final.pdf, https://dominicantoday.com/dr/covid-19/2020/09/29/doctors-cure-6000-patients-with-covid-19-with-ivermectin/; and https://nypost.com/2021/01/04/hair-lice-drug-may-cut-risk-of-covid-19-death-by-80-percent/

27 We tell much of the story in Video and Report: US and China Collaborated to Make a Deadly Coronavirus and in Video and Report: President Trump Cancels US/China Research Making Epidemic Viruses. Breggin, P., 2020, A Special Report, Psychiatric Drug Facts. https://breggin.com/us-chinese-scientists-collaborate-on-coronavirus/

28 https://breggin.com/coronavirus-resource-center/

29 Menachery et al., 2015.

30 Vineet D Menachery, Boyd L Yount Jr, Amy C Sims, Kari Debbink, Sudhakar S Agnihothram, Lisa E Gralinski, Rachel L Graham, Trevor Scobey , Jessica A Plante, Scott R Royal, Jesica Swanstrom, Timothy P Sheahan, Raymond J Pickles, Davide Corti, Scott H Randell, Antonio Lanzavecchia, Wayne A Marasco, Ralph S Baric. (2016) SARS-like WIVl -CoV poised for human emergence. Proc Natl Acad Sci US A 113, 3048-53 (2016). https://pubmed.ncbi.nlm.nih.gov/26976607/ Also obtainable at https://www.pnas.org/content/pnas/113/11/3048.full.pdf

31 CDC, 2020, February 13, Common Human Coronaviruses, Centers for Disease Control. https://www.cdc.gov/coronavirus/general-information.html

32 Because there are so many variations, we are using the spelling of her name found in English-language Chinese scientific papers and newspapers.

33 Menachery et al., 2015.

34 Husseini, S.,2020, May 5, The Long History of Accidental Laboratory Releases of Potential Pandemic Pathogens Is Being Ignored In the COVID-19 Media Coverage, Independent Science News. https://www.independentsciencenews.org/health/the-long-history-of-accidental-laboratory-releases-of-potential-pandemic-pathogens/

35 Thomson, B., 2020, China 'appoints its top military bio-warfare expert to take over secretive virus lab in Wuhan', sparking conspiracy theories that coronavirus outbreak is linked to Beijing's army. Daily Mail. A small-print note states it was published February 2020, https://www.dailymail.co.uk/news/article-8003713/China-appoints-military-bio-weapon-expert-secretive-virus-lab-Wuhan.html

36 https://breggin.com/us-chinese-scientists-collaborate-on-coronavirus/

37 Breggin, P., 2020, 2015 Scientific Paper Proves US & Chinese Scientists Collaborated to Create Coronavirus that Can Infect Humans, Psychiatric Drug Facts. https://breggin.com/us-chinese-scientists-collaborate-on-coronavirus/

38   Owermohle, S., 2020, President Trump cuts U.S. research on bat-human virus transmission over China ties, Politico. https://www.politico.com/news/2020/04/27/President Trump-cuts-research-bat-human-virus-china-213076

39   https://dailycaller.com/2021/04/04/nih-gain-of-function-anthony-fauci-review-board-wuhan-lab/

40   Breggin, P. and Breggin, G. (2020, March April 15 & 16, 2020, Video and Blog: 2015 Scientific Paper Proves US & Chinese Scientists Collaborated to Create Coronavirus that Can Infect Humans. https://breggin.com/us-chinese-scientists-collaborate-on-coronavirus/

41   Breggin, Peter and Breggin, Ginger: Video & Blog: President Trump Cancels Funding of US/China Research Making Epidemic Viruses. https://breggin.com/President Trump-cancels-funding-of-us-china-research-making-epidemic-viruses/

42   Finnegan, C., 2020, President Trump admin pulls NIH grant for coronavirus research over ties to Wuhan lab at heart of conspiracy theories, ABC News. https://abcnews.go.com/Politics/President Trump-admin-pulls-nih-grant-coronavirus-research-ties/story?id=70418101

43   Baric to lead $10 million NIH grant - UNC Gillings School of Global Public Health

44   Donaldson EF, Yount B, Sims AC, Burkett S, Pickles RJ, Baric RS (2008, September). Systematic assembly of a full-length infectious clone of human coronavirus NL63. J Virol. 82(23):11948-57. https://jvi.asm.org/content/jvi/82/23/11948.full.pdf

45   Becker, M., Graham, R., Donaldson, E., Rock, B., Sims, A., Sheahan, T., Pickles, R., Corti, D., Johnson, R. Baric R; and Denison, M. (2008, December) Synthetic recombinant bat SARS-like coronavirus is infectious in cultured cells and in mice. 105 (50), 19944-19949. https://www.pnas.org/content/pnas/105/50/19944.full.pdf

46   Xiao, B. and Xiao, L. The possible origins of 2019-nCoV coronavirus. February 2020. https://www.lifesitenews.com/images/pdfs/The_Possible_Origins_of_the_2019-nCoV_coronavirus.pdf

47   Mishra, A. and Mondal, D. 2020, April 25. Corona leaked likely from Wuhan Institute of Virology: Experts, Sunday Guardian Live. https://www.sundayguardianlive.com/news/corona-leaked-likely-wuhan-institute-virology-experts

48   Breuninger, K. et al., 2020, May 13, China-linked hackers are targeting US coronavirus vaccine research, FBI warns, CNBC. Coronavirus vaccine: China-linked hackers are targeting US research, FBI warns (cnbc.com)

49   FBI National Press Office. 2020, May 13, People's Republic of China (PRC) Targeting of COVID-19 Research Organizations, FBI. https://www.fbi.gov/news/pressrel/press-releases/peoples-republic-of-china-prc-targeting-of-covid-19-research-organizations

50   Conklin, A., 2020, May 13, FBI, CSIA warn of Chinese cyberattacks targeting coronavirus research organizations. Fox Business.com. https://www.foxbusiness.com/technology/fbi-csia-warn-of-chinese-cyberattacks-targeting-coronavirus-research-orgs. Also see https://twitter.com/FBI/status/1121755449890693121 for Christopher Ray's official twitter page.

51   https://www.theepochtimes.com/chinese-army-employs-military-civil-fusion-to-weaponize-industrial-base_3101117.html and https://thediplomat.com/2020/11/us-targets-chinas-quest-for-military-civil-fusion/

52   https://www.state.gov/military-civil-fusion/

53   https://www.airuniversity.af.edu/CASI/Display/Article/2217101/chinas-military-civil-fusion-strategy/. For entire lengthy report, see June 12, 2020 https://www.airuniversity.af.edu/Portals/10/CASI/documents/Research/Other%20topics/CASI%20China's%20Military%20Civil%20Fusion%20Strategy-%20Full%20final.pdf?ver=2020-06-15-152810-733

54   https://www.cnas.org/publications/reports/myths-and-realities-of-chinas-military-civil-fusion-strategy; also see (PDF) China's Military-Civil Fusion Strategy: Building a Strong Nation with a Strong Military (researchgate.net)

55 https://www.ncbi.nlm.nih.gov/pmc/articles/PMC3442247/

56 https://www.cidrap.umn.edu/news-perspective/2013/03/scientists-seek-ethics-review-h5n1-gain-function-research

57 https://www.ncbi.nlm.nih.gov/pmc/articles/PMC3334562/

58 Fauci, A. (2012, September/October, Research on Highly Pathogenic H5N1 Viruses: The Way Forward, mBio, 3(5), e00359-12. https://mbio.asm.org/content/3/5/e00359-12

59 https://www.nap.edu/read/21666/chapter/5, https://www.the-scientist.com/news-opinion/lab-made-coronavirus-triggers-debate-34502

60 https://patents.justia.com/inventor/anthony-s-fauci https://c-vine.com/blog/2020/05/07/follow-the-money-fauci-holds-4-covid-19-related-patents/

61 https://www.nbcnews.com/meet-the-press/meet-press-april-25-2021-n1265222

62 https://www.axios.com/moderna-nih-coronavirus-vaccine-ownership-agreements-2205 1c42-2dee-4b19-938d-099afd71f6a0.html; The NIH Vaccine - Public Citizen and Does Anthony Fauci Own 'Half the Patent' for Moderna's COVID Vaccine? - The Dispatch Fact Check; https://www.commondreams.org/news/2021/04/23/nih-scientist-who-developed-key-vaccine-technology-says-patent-gives-us-leverage

63 Moderna Announces Award from U.S. Government Agency BARDA for up to $483 Million to Accelerate Development of mRNA Vaccine (mRNA-1273) Against Novel Coronavirus | Moderna, Inc. (modernatx.com)

64 https://www.foxbusiness.com/technology/moderna-56m-grant-darpa-to-build-an-espresso-machine-for-medicine

65 https://abcnews.go.com/Technology/bill-melinda-gates-foundation-announces-250-million-covid/story?id=74651890

66 https://endpts.com/biontech-partners-with-bill-and-melinda-gates-foundation-scoring-55m-equity-investment-novartis-sells-china-unit/

67 https://news.yahoo.com/chinas-fosun-seek-approval-biontechs-093837498.html

68 https://news.yahoo.com/chinas-fosun-seek-approval-biontechs-093837498.html and What's Not Being Said About Pfizer Coronavirus Vaccine | New Eastern Outlook (journal-neo.org)

69 https://www.fool.com/investing/2020/09/24/4-coronavirus-vaccine-stocks-the-bill-melinda-gate/

70 https://www.cnn.com/2020/07/22/health/pfizer-covid-19-vaccine-government-contract/index.html

71 Pfizer-BioNTech COVID-19 Vaccine EUA Letter of Authorization (cnbc.com)

72 https://duckduckgo.com/?q=Fauci+with+movie+stars&ia=web

73 West, D. (2013). American Betrayal: The Secret Assault on Our Nation's Character. NY: St. Martin's.

74 Menachery, et al. (2015).

75 Qiu, J., 2020, June 1, How China's 'Bat Woman' Hunted Down Viruses from SARS to the New Coronavirus, Scientific American. A comment attached to the article defends China: "Editor's Note (4/24/20): This article was originally published online on March 11. It has been updated for inclusion in the June 2020 issue of Scientific American and to address rumors that SARS-CoV-2 emerged from Shi Zhengli's lab in China." https://www.scientificamerican.com/article/how-chinas-bat-woman-hunted-down-viruses-from-sars-to-the-new-coronavirus1/

76 Xing-Yi Ge, Jia-Lu Li1, Xing-Lou Yang, Aleksei A. Chmura, Guangjian Zhu, Jonathan H. Epstein, Jonna K. Mazet, Ben Hu, Wei Zhang, Cheng Peng, Yu-Ji Zhang, Chu-Ming Luo, Bing Tan, Ning Wang, Yan Zhu, Gary Crameri, Shu-Yi Zhang, Lin-Fa Wang, Peter Daszak & Zheng-Li Shi. Isolation and characterization of a bat SARS-like coronavirus that uses the ACE2 receptor. Nature, 503 (28), November 2013, pp. 535 ff. https://www.nature.com/articles/nature12711

77   Xing-Lou Yang, Ben Hu, Bo Wang, Mei-Niang Wang, Qian Zhang, Wei Zhanga Li-Jun Wu, Xing-Yi Ge, Yun-Zhi Zhang, Peter Daszak, Lin-Fa Wang, and Zheng-Li Shia. (2016, March) Isolation and Characterization of a Novel Bat Coronavirus Closely Related to the Direct Progenitor of Severe Acute Respiratory Syndrome Coronavirus. Journal of Virology, 90 (6), 3253-56. https://breggin.com/coronavirus/Yang2016-Novel-bat-coronavirus-SARS.pdf

78   EcoHealth Alliance Global Partners Network

79   Johnson & Johnson Announces Its First Phase 3 COVID-19 Vaccine Trial ENSEMBLE is Fully Enrolled | Johnson & Johnson (jnj.com), A Commitment to Caring | Johnson & Johnson (jnj.com)

80   Treating a Virus Threat Like Terrorism - Bloomberg

81   Optimizing Viral Discovery in Bats (plos.org)

82   Charles Calisher, Dennis Carroll, Rita Colwell, Ronald B Corley, Peter Daszak, Christian Drosten, Luis Enjuanes, Jeremy Farrar, Hume Field, Josie Golding, Alexander Gorbalenya, Bart Haagmans, James M Hughes, William B Karesh, Gerald T Keusch, Sai Kit Lam, Juan Lubroth, John S Mackenzie, Larry Madoff, Jonna Mazet, Peter Palese, Stanley Perlman, Leo Poon, Bernard Roizman, Linda Saif, Kanta Subbarao, Mike Turner. Statement in support of the scientists, public health professionals, and medical professionals of China combatting COVID-19. Vol 395 March 7, 2020. https://www.thelancet.com/action/showPdf?pii =S0140-6736%2820%2930418-9

83   Press Release, 2010, Global Health Leaders Launch Decade of Vaccines Collaboration. Bill & Melinda Gates Foundation. https://www.gatesfoundation.org/Media-Center/Press-Releases/2010/12/Global-Health-Leaders-Launch-Decade-of-Vaccines-Collaboration

84   https://www.nature.com/articles/nature11170

85   Fauci denies NIH supported gain-of-function research at Wuhan lab | Washington Examiner

86   https://www.c-span.org/video/?c4962333/senator-paul-dr-fauci-clash-research-funding-wuhan-lab. Also see, https://www.foxnews.com/media/rand-paul-dr-fauci-lied-congress-china-virus-research; https://www.washingtonexaminer.com/news/fauci-denies-nih-supported-gain-of-function-research-wuhan-lab

87   Menachery, et al., 2015.

88   Vineet D Menachery, Boyd L Yount Jr, Amy C Sims, Kari Debbink, Sudhakar S Agnihothram, Lisa E Gralinski, Rachel L Graham, Trevor Scobey , Jessica A Plante, Scott R Royal, Jesica Swanstrom, Timothy P Sheahan, Raymond J Pickles, Davide Corti, Scott H Randell, Antonio Lanzavecchia, Wayne A Marasco, Ralph S Baric. (2016). SARS-like WIVl -CoV poised for human emergence. Proc Natl Acad Sci US A 113, 3048-53 (2016). https://pubmed.ncbi.nlm.nih.gov/26976607/ Also obtainable at https://www.pnas.org/content/pnas/113/11/3048.full.pdf

89   Wang et al. (2018). Serological Evidence of BAT SARS-Related Coronavirus Infections in Humans, China. Virologica Sinica, 33, 104-107. Serological Evidence of Bat SARS-Related Coronavirus Infection in Humans, China (g-scientia.com)

90   Zeng, L.P., Ge, X.Y., Peng, C., Tai, W., Jiang, S., Du, L., and Shi, Z.L. (2017). Cross-neutralization of SARS coronavirus-specific antibodies against bat SARS-like coronaviruses. Sci China Life Sci 60, 1399–1402. https://doi.org/10.1007/s11427-017-9189-3. https://www.ncbi.nlm.nih.gov/pmc/articles/PMC7089274/pdf/11427_2017_Article_9189.pdf

91   Ibid. p. 1400-4001.

92   https://jvi.asm.org/content/jvi/90/6/3253.full.pdf

93   Bat origin of human coronaviruses (nih.gov)

94   Xing-Yi Ge, Jia-Lu Li, Xing-Lou Yang, Aleksei A Chmura, Guangjian Zhu, Jonathan H Epstein, Jonna K Mazet, Ben Hu, Wei Zhang, Cheng Peng, Yu-Ji Zhang, Chu-Ming Luo, Bing Tan, Ning Wang, Yan Zhu, Gary Crameri, Shu-Yi Zhang, Lin-Fa Wang, Peter Daszak, Zheng-

Li Shi. (2013, November). Isolation and characterization of a bat SARS-like coronavirus that uses the ACE2 receptor. 28;503(7477):535-8. doi: 10.1038/nature12711. Epub 2013 Oct 30. The quote is found in the methods section which does not appear in a shortened form of the study but can be found here: Isolation and characterization of a bat SARS-like coronavirus that uses the ACE2 receptor - PubMed (nih.gov)

95   RePORT > RePORTER (nih.gov). We reach this set of data for NIAID funding for potential gain-of-funding research by searching for research study number R01AI079231.

96   Guterl, F., 2020, April 28. Dr. Fauci Backed Controversial Wuhan Lab with U.S. Dollars for Risky Coronavirus Research, Newsweek. https://www.newsweek.com/dr-fauci-backed-controversial-wuhan-lab-millions-us-dollars-risky-coronavirus-research-1500741

97   https://www.newsweek.com/dr-fauci-backed-controversial-wuhan-lab-millions-us-dollars-risky-coronavirus-research-1500741. Also see, https://thenewamerican.com/newsweek-fauci-s-virus-outfit-subsidized-wuhan-virus-lab-famed-virus-fighter-backs-controversial-research/

98   https://projectreporter.nih.gov/project_info_description.cfm?aid=8674931&icde=49750546

99   https://reporter.nih.gov/search/xQW6UJmWfUuOV01ntGvLwQ/project-details/8674931

100   RePORT > RePORTER (nih.gov). Also try putting this into search bar: https://reporter.nih.gov/project-details/8674931

101   Becker, M., Graham, R., Donaldson, E., Rock, B., Sims, A., Sheahan, T., Pickles, R., Corti, D., Johnson, R. Baric R; and Denison, M. (2008, December) Synthetic recombinant bat SARS-like coronavirus is infectious in cultured cells and in mice. 105 (50), 19944-19949. https://www.pnas.org/content/pnas/105/50/19944.full.pdf

102   Ren-Di Jiang, Mei-Qin Liu, Ying Chen, Chao Shan, Yi-Wu Zhou, Xu-Rui Shen, Qian Li, Lei Zhang, Yan Zhu, Hao-RuiSi, Qi Wang, Juan Min, Xi Wang, Wei Zhang, Bei Li, Hua-Jun Zhang, Ralph S.Baric, Peng Zhou, Xing-Lou Yang, Zheng-Li Shi. (2020, July) Pathogenesis of SARS-CoV-2 in Transgenic Mice Expressing Human Angiotensin-Converting Enzyme 2. Cell 182, 50–58 July 9, 2020. Elsevier Inc. https://doi.org/10.1016/j.cell.2020.05.0. Pathogenesis of SARS-CoV-2 in Transgenic Mice Expressing Human Angiotensin-Converting Enzyme 2 - ScienceDirect. Also at Pathogenesis of SARS-CoV-2 in Transgenic Mice Expressing Human Angiotensin-Converting Enzyme 2 - PubMed (nih.gov)

103   https://www.independentsciencenews.org/news/peter-daszaks-ecohealth-alliance-has-hidden-almost-40-million-in-pentagon-funding/; also see, https://childrenshealthdefense.org/defender/ecohealth-alliance-hid-pentagon-funding/

104   Microsoft Word - BLT.17.205005.docx (who.int)

105   Menachery, V. et al. (2015). A SARS-like cluster of circulating bat coronaviruses shows potential for human emergence. Nature Medicine, 21 (12), 1508-1514.

106   Qiu, J., 2020, June 1, How China's 'Bat Woman' Hunted Down Viruses from SARS to the New Coronavirus, Scientific American. A comment attached to the article states, "Editor's Note (4/24/20): This article was originally published online on March 11. It has been updated for inclusion in the June 2020 issue of Scientific American and to address rumors that SARS-CoV-2 emerged from Shi Zhengli's lab in China." https://www.scientificamerican.com/article/how-chinas-bat-woman-hunted-down-viruses-from-sars-to-the-new-coronavirus1/

107   Abbasi, K., 2020, November 13. Covid-19: Politicisation, "corruption," and suppression of science. When good science is suppressed by the medical-political complex, people die, BMJ. https://www.bmj.com/content/371/bmj.m4425

108   https://www.cnn.com/2021/05/27/tech/facebook-covid-19-origin-claims-removal/index.html

109   "Dr. Steven Carl Quay received his MD & PhD from the University of Michigan. He was postdoctorate fellow at the MIT Chemistry Department with Nobel Laureate Gobind Khorana,

resident at the Harvard-Massachusetts General Hospital, and spent a decade on the faculty of Stanford University School of Medicine." bio Steven Carl Quay - Bing.

110 https://zenodo.org/record/4477081#

111 Klotz, L. and Sylvester, E. The Consequences of a Lab Escape of a Potential Pandemic Pathogen Front Public Health. 2014; 2: 116. Published online 2014 Aug 11. doi: 10.3389/fpubh.2014.00116 PMCID: PMC4128296PMID: 25157347; https://www.ncbi.nlm.nih.gov/pmc/articles/PMC4128296/

112 https://www.scientificamerican.com/article/how-chinas-bat-woman-hunted-down-viruses-from-sars-to-the-new-coronavirus1/?amp=true;https://nationalpost.com/news/world/cave-full-of-bats-in-china-identified-as-source-of-virus-almost-identical-to-the-one-killing-hundreds-today;https://www.scientificamerican.com/article/bat-cave-solves-mystery-of-deadly-sars-virus/ [ii]https://www.sciencealert.com/bats-single-cave-china-everything-they-need-make-sars-virus-lethal-strain

113 Zeng, L.P., Ge, X.Y., Peng, C., Tai, W., Jiang, S., Du, L., and Shi, Z.L. (2017). Cross-neutralization of SARS coronavirus-specific antibodies against bat SARS-like coronaviruses. Sci China Life Sci 60, 1399–1402. https://doi.org/10.1007/s11427-017-9189-3.

114 https://www.sciencealert.com/bats-single-cave-china-everything-they-need-make-sars-virus-lethal-strain

115 News. SARS outbreak over, but concerns for lab safety remain. Bulletin of the World Health Organization j June 2004, 82 (6), p. 470. https://apps.who.int/iris/bitstream/handle/10665/269169/PMC2622862.pdf?sequence=1&isAllowed=y

116 https://www.cidrap.umn.edu/news-perspective/2011/09/report-395-mishaps-us-labs-risked-releasing-select-agents

117 Kelly, M. & Cahlan, S., 2020, Was the new coronavirus accidentally released from a Wuhan lab? It's doubtful., Washington Post. https://www.washingtonpost.com/politics/2020/05/01/was-new-coronavirus-accidentally-released-wuhan-lab-its-doubtful/

118 Rogin, J. 2020, April 14, State Department cables warned of safety issues at Wuhan lab studying bat coronaviruses, Washington Post. https://www.washingtonpost.com/opinions/2020/04/14/state-department-cables-warned-safety-issues-wuhan-lab-studying-bat-coronaviruses/

119 News, 2019, December 17, Chinese institutes investigate pathogen outbreaks in lab workers, Nature. Students and staff at two research institutes have tested positive to the Brucella bacterium, which can lead to serious complications. https://www.nature.com/articles/d41586-019-03863-z

120 Young, A., 2017, January 4, CDC keeps secret its mishaps with deadly germs, US Today. https://www.usatoday.com/story/news/2017/01/04/cdc-secret-lab-incidents-select-agents/95972126/

121 Marin, D., 2014, July 11, CDC Botched Handling of Deadly Flu Virus: The third recent mistake in handling of pathogens is a "wake-up call," says Centers for Disease Control head, Scientific America. https://www.scientificamerican.com/article/cdc-botched-handling-of-deadly-flu-virus/

122 McNeil Jr., D., 2017, December 19, A Federal Ban on Making Lethal Viruses Is Lifted, New York Times. https://www.nytimes.com/2017/12/19/health/lethal-viruses-nih.html?_r=0

123 Bender, J., 2014, July 14, Here Are 5 Times Infectious Diseases Escaped from Laboratory Containment. Business Insider. https://www.businessinsider.com/5-terrifying-times-pandemics-e demics-e

124 Martin Furmanski MD Scientist's Working Group on Chemical and Biologic Weapons Center for Arms Control and Nonproliferation February 17, 2014. https://armscontrolcenter.org/wp-content/uploads/2016/02/Escaped-Viruses-final-2-17-14-copy.pdf

125 Bender, J. 2014, There are 5 times infectious diseases have escaped from laboratory containment, Business Insider. https://www.businessinsider.com/5-terrifying-times-pandemics-escaped-from-laboratories-2014-7

126 Young, A. 2017, January 4, CDC keeps secret its mishaps with deadly germs, USA Today. https://www.usatoday.com/story/news/2017/01/04/cdc-secret-lab-incidents-select-agents/95972126/

127 Piper, K., 2019, March 20, How deadly pathogens have escaped the lab—over and over again, Vox. https://www.vox.com/future-perfect/2019/3/20/18260669/deadly-pathogens-escape-lab-smallpox-bird-flu

128 Husseini, S.,2020, May 5, The Long History of Accidental Laboratory Releases of Potential Pandemic Pathogens Is Being Ignored In the COVID-19 Media Coverage. Independent Science News. https://www.independentsciencenews.org/health/the-long-history-of-accidental-laboratory-releases-of-potential-pandemic-pathogens/

129 Grady, D., 2019, April 5, Deadly Germ Research Is Shut Down at Army Lab Over Safety Concerns, The New York Times. https://www.nytimes.com/2019/08/05/health/germs-fort-detrick-biohazard.html

130 Furmanski, M. Threatened pandemics and laboratory escapes: Self-filling prophecies, 2014. March 31. Bulletin of the Atomic Scientists. https://thebulletin.org/2014/03/threatened-pandemics-and-laboratory-escapes-self-fulfilling-prophecies/. In addition, see a similar more detailed in-house report by Furmanski: Scientist's Working Group on Chemical and Biologic Weapons Center for Arms Control and Nonproliferation February 17, 2014. https://armscontrolcenter.org/wp-content/uploads/2016/02/Escaped-Viruses-final-2-17-14-copy.pdf

131 Zhang, F., (2004, July 2), Officials punished for SARS virus leak, China Daily. https://www.chinadaily.com.cn/english/doc/2004-07/02/content_344755.htm

132 Walgate, R., 2004, April 25, SARS escaped Beijing lab twice: Laboratory safety at the Chinese Institute of Virology under close scrutiny, The Scientist. https://www.the-scientist.com/news-analysis/sars-escaped-beijing-lab-twice-50137

133 Mihm, Stephen. (2021, March 27). The History of Lab Leaks Has Lots of Entries. Smallpox, anthrax and influenzas have escaped facilities — sometimes with deadly consequences. https://www.bloomberg.com/opinion/articles/2021-05-27/covid-19-and-lab-leak-history-smallpox-h1n1-sars

134 Bio of Eyal Pinko, director IIMSR.

135 https://web.archive.org/web/20210120090115/https://iimsr.eu/2020/04/16/chinas-biological-warfare/

136 Bio of Eyal Pinko, director IIMSR. https://iimsr.eu/eyal-pinko/

137 Pinko, E., 2020, April 16, China's biological warfare, IIMSR. https://iimsr.eu/2020/04/16/chinas-biological-warfare/

138 https://www.taiwannews.com.tw/en/news/4102619

139 Sørensen, M. D.; Sørensen, B.; Gonzalez-Dosal, R.; Melchjorsen, C. J.; Weibel, J.; Wang, J.; Jun, C. W.; Huanming, Y.; Kristensen, P. (May 2006). Severe acute respiratory syndrome (SARS): development of diagnostics and antivirals. Annals of the New York Academy of Sciences. 1067 (1): 500– 505. https://nyaspubs.onlinelibrary.wiley.com/doi/full/10.1196/annals.1354.072

140 Zhang, F., (2004, July 2), Officials punished for SARS virus leak, China Daily. https://www.chinadaily.com.cn/english/doc/2004-07/02/content_344755.htm

141 CDC, 2017, December 6, SARS Basics Fact Sheet, Centers for Disease Control. https://www.cdc.gov/sars/about/fs-sars.html

142 https://www.cdc.gov/sars/lab/downloads/nucleoseq.pdf complete genome for Urban SARS corona virus April 17, 2003 submitted to the CDC.

143 Zhao Z, Zhang F, Xu M, Huang K, Zhong W, Cai W, Yin Z, Huang S, Deng Z, Wei M, Xiong J, Hawkey PM. 2003. Description and clinical treatment of an early outbreak of severe acute respiratory syndrome (SARS) in Guangzhou, PR China. Journal of Medical Microbiology. 52(Pt

8):715-20. https://citeseerx.ist.psu.edu/viewdoc/download?doi=10.1.1.538.2538&rep=rep1&type=pdf

144   Stacey Knobler, Adel Mahmoud, Stanley Lemon, Alison Mack, Laura Sivitz, and Katherine Oberholtzer. Editors. (2004). Forum on Microbial Threats; Board on Global Health; Institute of Medicine. Learning from SARS: Preparing for the Next Disease Outbreak: Workshop Summary. Washington, DC: The National Academies Press. Learning from SARS book 2004.pdf

145   https://www.state.gov/ensuring-a-transparent-thorough-investigation-of-covid-19s-origin/

146   https://dash.harvard.edu/handle/1/42669767

147   https://www.scmp.com/news/china/science/article/3111314/who-names-line-international-al-team-looking-oronavirus-origins

148   https://www.taiwannews.com.tw/en/news/4104828 https://www.taiwannews.com.tw/en/news/4104828

149   Daszak may have been referring to this gain-of function study: Ren-Di Jiang, Mei-Qin Liu,Ying Chen, Chao Shan, Yi-Wu Zhou, Xu-Rui Shen, Qian Li, Lei Zhang, Yan Zhu, Hao-RuiSi, Qi Wang, Juan Min, Xi Wang, Wei Zhang, Bei Li, Hua-Jun Zhang, Ralph S.Baric, Peng Zhou, Xing-Lou Yang, Zheng-Li Shi. (2020, July) Pathogenesis of SARS-CoV-2 in Transgenic Mice Expressing Human Angiotensin-Converting Enzyme 2. Cell 182, 50–58 July 9, 2020. Elsevier Inc. https://doi.org/10.1016/j.cell.2020.05.0. Pathogenesis of SARS-CoV-2 in Transgenic Mice Expressing Human Angiotensin-Converting Enzyme 2 - ScienceDirect

150   According to the Ren-Di Jiang et al. 2020 (above note), the paper was received 12 March 2020, revised 27 April 2020, and accepted 14 May 2020. That gave him enough time to have second thoughts and to have removed his name from it.

151   https://www.taiwannews.com.tw/en/news/4104828

152   https://www.taiwannews.com.tw/en/news/4104828

153   Ren-Di Jiang, Mei-Qin Liu,Ying Chen, Chao Shan, Yi-Wu Zhou, Xu-Rui Shen, Qian Li, Lei Zhang, Yan Zhu, Hao-RuiSi, Qi Wang, Juan Min, Xi Wang, Wei Zhang, Bei Li, Hua-Jun Zhang, Ralph S.Baric, Peng Zhou, Xing-Lou Yang, Zheng-Li Shi. (2020, July) Pathogenesis of SARS-CoV-2 in Transgenic Mice Expressing Human Angiotensin-Converting Enzyme 2. Cell 182, 50–58 July 9, 2020. Elsevier Inc. https://doi.org/10.1016/j.cell.2020.05.0. Pathogenesis of SARS-CoV-2 in Transgenic Mice Expressing Human Angiotensin-Converting Enzyme 2 - ScienceDirect

154   Vineet D Menachery, Boyd L Yount Jr, Amy C Sims, Kari Debbink, Sudhakar S Agnihothram, Lisa E Gralinski, Rachel L Graham, Trevor Scobey , Jessica A Plante, Scott R Royal, Jesica Swanstrom, Timothy P Sheahan, Raymond J Pickles, Davide Corti, Scott H Randell, Antonio Lanzavecchia, Wayne A Marasco, Ralph S Baric. (2016) SARS-like WIVl -CoV poised for human emergence. Proc Natl Acad Sci US A 113, 3048-53 (2016). https://pubmed.ncbi.nlm.nih.gov/26976607/ Also obtainable at https://www.pnas.org/content/pnas/113/11/3048.full.pdf

155   Botao Xiao Biography, through 2017, School of Biology and Biological Engineering, South China University of Technology. http://www2.scut.edu.cn/biology_en/2017/0614/c5951a169022/page.htm

156   Xiao, B. and Xiao, L., 2020, February, The possible origins of 2019-nCoV coronavirus, Research Gate. [removed from the website] https://chanworld.org/wp-content/uploads/wpforo/default_attachments/1581810860-447056518-Originsof2019-NCoV-XiaoB-

157   Ge XY, Li JL, Yang XL, et al. Isolation and characterization of a bat SARS-like coronavirus that uses the ACE2 receptor. Nature 2013; 503(7477): 535-8. https://www.nature.com/articles/nature12711?fbclid=IwAR1oxB4btiYVmSzncbfTPLtCEORxqfdJygsxayF7cklj3my-1pUF1vC-PUnU

158   Areddy, J. 2020, March 5, Coronavirus epidemic draws scrutiny to labs handling deadly pathogens. Wall Street Journal. https://www.wsj.com/articles/coronavirus-epidemic-draws-scrutiny-to-labs-handling-deadly-pathogens-11583349777

159   https://www.westernjournal.com/wuhan-study-mentioned-tucker-vanishes-claimed-market-didnt-sell-covid-bats-500-miles-infected-bats/

160   Golden, C. D., 2020, April 5, Wuhan Study Mentioned by Tucker Vanishes: Claimed Market Didn't Sell COVID Bats, Was 500 Miles From Infected Bats, The Western Journal. https://www.westernjournal.com/wuhan-study-mentioned-tucker-vanishes-claimed-market-didnt-sell-covid-bats-500-miles-infected-bats/

161   Edmunds, D., 2020, June 7, Former MI6 head claims COVID-19 was made in a Chinese lab, Jerusalem Post. https://www.jpost.com/health-science/former-mi6-head-covid-19-was-made-in-a-chinese-lab-630346

162   Sørensen B, Susrud A, Dalgleish AG (2020). Biovacc-19: A Candidate Vaccine for Covid-19 (SARS-CoV-2) Developed from Analysis of its General Method of Action for Infectivity. QRB Discovery, 1: e6, 1–11 https://doi.org/10.1017/qrd.2020.8 https://www.cambridge.org/core/services/aop-cambridge-core/content/view/DBBC0FA6E3763B0067CAAD8F3363E527/S2633289220000083a.pdf/biovacc19_a_candidate_vaccine_for_covid19_sarscov2_developed_from_analysis_of_its_general_method_of_action_for_infectivity.pdf

163   The Evidence which Suggests that This Is No Naturally Evolved Virus: A Reconstructed Historical Aetiology of the SARS-CoV-2 Spike Birger Sørensen, Angus Dalgleish & Andres Susrud. No indication of publisher. http://rapeutation.com/sorensonnotnatural.pdf

164   https://www.zee5.com/zee5news/covid-19-has-no-credible-natural-ancestor-explosive-study-claims-chinese-scientists-created-virus-in-lab

165   https://usrtk.org/biohazards-blog/new-emails-show-scientists-deliberations-on-how-to-discuss-sars-cov-2-origins/

166   Vineet D Menachery, Rachel L Graham, and Ralph S Baric. Jumping species—a mechanism for coronavirus persistence and survival Curr Opin Virol. 2017 Apr; 23: 1–7.Published online 2017 Mar 31. doi: 10.1016/j.coviro.2017.01.002 https://www.ncbi.nlm.nih.gov/pmc/articles/PMC5474123/

167   Investigative Report, The Coverup of the Century, 2020, June 28, The Epoch Times. Investigative Report: The Coverup of the Century | Epoch Times | CCPVirus | COVID19 | Coronavirus - YouTube

168   Suspected SARS virus and flu germs found in luggage: FBI report describes China's 'biosecurity risk' (yahoo.com)

169   SARS labs unsafe, says WHO | The Scientist Magazine® (the-scientist.com)

170   Opinion | State Department cables warned of safety issues at Wuhan lab studying bat coronaviruses - The Washington Post

171   Walgate, R., 2004, April 25, SARS escaped Beijing lab twice, The Scientist. https://www.the-scientist.com/news-analysis/sars-escaped-beijing-lab-twice-50137

172   Hollingsworth, J., 2020, January 25, A lot has changed since China's SARS outbreak 17 years ago. But some things haven't, CNN. https://edition.cnn.com/2020/01/24/asia/china-sars-coronavirus-intl-hnk/index.html

173   Walgate, R., 2004, June 2, SARS labs unsafe, says WHO, The Scientist. https://www.the-scientist.com/news-analysis/sars-labs-unsafe-says-who-49988

174   Rogin, J., 2020, April 14, State Department cables warned of safety issues at Wuhan lab studying bat coronaviruses, Washington Post. https://www.washingtonpost.com/opinions/2020/04/14/state-department-cables-warned-safety-issues-wuhan-lab-studying-bat-coronaviruses/

175   What's changed between the 2003 SARS outbreak and the current Wuhan coronavirus - CNN

176   SARS escaped Beijing lab twice | The Scientist Magazine® (the-scientist.com)

177   Furmanski, M. Threatened pandemics and laboratory escapes: Self-filling prophecies, 2014. March 31. Bulletin of the Atomic Scientists. https://thebulletin.org/2014/03/threatened-pandemics-and-laboratory-escapes-self-fulfilling-prophecies/

178   SARS escaped Beijing lab twice | The Scientist Magazine® (the-scientist.com)

179   Akst, J., 2015, March 25, CDC Scores Poorly on Biosafety, The Scientist https://www.the-scientist.com/the-nutshell/cdc-scores-poorly-on-biosafety-35739

180   Young, A., et al., 2020, August 17, Here Are Six Accidents UNC Researchers Had With Lab-Created Coronaviruses, ProPublica. https://www.propublica.org/article/here-are-six-accidents-unc-researchers-had-with-lab-created-coronaviruses

181   https://www.taiwannews.com.tw/en/news/4104828\

182   Knox, P., 2020, April 29, Dr Anthony Fauci backed $7.4m funding for controversial Wuhan lab blamed for causing Covid-19 by conspiracy theorists, The Sun (UK). https://www.thesun.co.uk/news/11507231/dr-fauci-7-4-million-funding-wuhan-lab-blamed-covid-19-conspiracy/

183   A SARS-like cluster of circulating bat coronaviruses shows potential for human emergence (breggin.com)

184   2015 Scientific Paper Proves US & Chinese Scientists Collaborated to Create Coronavirus that Can Infect Humans | Psychiatric Drug Facts (breggin.com)

185   Wuhan Study Mentioned by Tucker Vanishes: Claimed Market Didn't Sell COVID Bats, Was 500 Miles From Infected Bats (westernjournal.com)

186   Wuhan Study Mentioned by Tucker Vanishes: Claimed Market Didn't Sell COVID Bats, Was 500 Miles From Infected Bats (westernjournal.com)

187   Yan, Li-Meng Yan ; Kang, Shu; Guan, Jie; Hu, Shanchang. (2020, September 14). Unusual Features of the SARS-CoV-2 Genome Suggesting Sophisticated Laboratory Modification Rather Than Natural Evolution and Delineation of Its Probable Synthetic Route. Prepublication. http://breggin.com/coronavirus/The_Yan_Report.pdf. To confirm the date it was put up and to follow the progress of the paper through publication, go to here: https://zenodo.org/record/4028830#.X2R2T5NKiuV. Also see, Yan, Li-Meng; Kang, Shu; Guan, Jie; Hu, Shanchang. 2020, October 8, SARS-CoV-2 Is an Unrestricted Bioweapon: A Truth Revealed through Uncovering a Large-Scale, Organized Scientific Fraud. Prepublication. https://zenodo.org/record/4073131#.X4OpJOaSk2x "You can cite all versions by using the DOI 10.5281/zenodo.4073130. This DOI represents all versions, and will always resolve to the latest one."

188   Yan, Li-Meng Yan ; Kang, Shu; Guan, Jie; Hu, Shanchang. (2020, September 14). Unusual Features of the SARS-CoV-2 Genome Suggesting Sophisticated Laboratory Modification Rather Than Natural Evolution and Delineation of Its Probable Synthetic Route. Prepublication. http://breggin.com/coronavirus/The_Yan_Report.pdf. To confirm the date it was put up and to follow the progress of the paper through publication, go to here: https://zenodo.org/record/4028830#.X2R2T5NKiuV

189   Yan, Li-Meng; Kang, Shu; Guan, Jie; Hu, Shanchang. (2020, October 8). SARS-CoV-2 Is an Unrestricted Bioweapon: A Truth Revealed through Uncovering a Large-Scale, Organized Scientific Fraud. Prepublication. https://zenodo.org/record/4073131#.X4OpJOaSk2x "You can cite all versions by using the DOI 10.5281/zenodo.4073130. This DOI represents all versions, and will always resolve to the latest one."

190   Microsoft Word - Final Fauci Treachery Report 10.19.2020.docx (breggin.com)

191   Ren-Di Jiang, Mei-Qin Liu,Ying Chen, Chao Shan, Yi-Wu Zhou, Xu-Rui Shen, Qian Li, Lei Zhang, Yan Zhu, Hao-RuiSi, Qi Wang, Juan Min, Xi Wang, Wei Zhang, Bei Li, Hua-Jun Zhang, Ralph S.Baric, Peng Zhou, Xing-Lou Yang, Zheng-Li Shi. (2020) Pathogenesis of SARS-CoV-2 in Transgenic Mice Expressing Human Angiotensin-Converting Enzyme 2. Cell 182, 50–58 July 9, 2020. Elsevier Inc. https://doi.org/10.1016/j.cell.2020.05.0. Patho-

genesis of SARS-CoV-2 in Transgenic Mice Expressing Human Angiotensin-Converting Enzyme 2 - ScienceDirect

192 Former MI6 director says coronavirus is an 'accident' from a Chinese lab – BGR

193 Withers, P. , 2020, May 11, Coronavirus China cover-up: 'Wet market claim doesn't stack up... lab leak is plausible', Express. Coronavirus China cover-up: 'Wet market claim doesn't stack up... lab leak is plausible' | World | News | Express.co.uk

194 Husseini, S.,2020, May 5, The Long History of Accidental Laboratory Releases of Potential Pandemic Pathogens Is Being Ignored In the COVID-19 Media Coverage, Independent Science News. https://www.independentsciencenews.org/health/the-long-history-of-accidental-laboratory-releases-of-potential-pandemic-pathogens/

195 Thomson, B., 2020, China 'appoints its top military bio-warfare expert to take over secretive virus lab in Wuhan', sparking conspiracy theories that coronavirus outbreak is linked to Beijing's army. Daily Mail. A small-print note states it was published February 2020, https://www.dailymail.co.uk/news/article-8003713/China-appoints-military-bio-weapon-expert-secretive-virus-lab-Wuhan.html

196 Sen, S. , 2020, April 30. How China locked down internally for COVID-19, but pushed foreign travel. The Economic Times. https://economictimes.indiatimes.com/blogs/Whathappensif/how-china-locked-down-internally-for-covid-19-but-pushed-foreign-travel/

197 Levenson, M., 2020, Jan. 22, Scale of China's Wuhan Shutdown Is Believed to Be Without Precedent. New York Times. https://www.nytimes.com/2020/01/22/world/asia/coronavirus-quarantines-history.html. January 23, 2020 is often cited in the press as the day of the shutdown of Wuhan, but the actual date, as indicated in this article, was the Thursday before the news came out, or January 16, 2020.

198 BBC.com., 2020, August 27, Coronavirus: Flights within China to 'fully recover' next month, BBC News. https://www.bbc.com/news/business-53927980

199 Nebehay, S., 2020, February 3, WHO chief says widespread travel bans not needed to beat China virus, Reuters. https://www.reuters.com/article/us-china-health-who-idUSKBN1ZX1H3

200 Cheng, E. 2020, February 4. China's aviation authority to allow more foreign flights after the U.S. bans Chinese carriers, CNBC. https://www.cnbc.com/2020/06/04/china-to-allow-more-foreign-flights-after-us-bans-chinese-carriers.html

201 Source of all data: Eder, S. et al., published April 4, 2020 and Updated April 15, 2020; 430,000 People Have Traveled from China to U.S. Since Coronavirus Surfaced, New York Times. https://www.nytimes.com/2020/04/04/us/coronavirus-china-travel-restrictions.html

202 Eder., et al., 2020, April 4, 430,00 People have traveled from China to U.S. since Coronavirus Surfaced, New York Times. https://www.nytimes.com/2020/04/04/us/coronavirus-china-travel-restrictions.html

203 Infographic: How the Global Pandemic Spread From China (theepochtimes.com)

204 https://nypost.com/2020/04/05/President Trump-admin-weighs-legal-action-over-alleged-chinese-hoarding-of-ppe/

205 https://nypost.com/2020/04/05/President Trump-admin-weighs-legal-action-over-alleged-chinese-hoarding-of-ppe/

206 https://www.whitehouse.gov/presidential-actions/memorandum-order-defense-production-act-regarding-3m-company/ and https://nypost.com/2020/04/03/President Trump-slams-3m-after-ordering-company-to-churn-out-masks/

207 https://www.nytimes.com/2020/07/05/business/china-medical-supplies.html

208 OPride Staff, 2017, May 11, he case against WHO director-general candidate Tedros, OPride. https://www.opride.com/2017/05/11/case-director-general-candidate-tedros-adhanom/

209 Ghitis, F., 2017, October 25. Another week, another scandal at the United Nations, Washington Post. https://www.washingtonpost.com/news/democracy-post/wp/2017/10/25/another-week-another-scandal-at-the-united-nations/

210   Chakraborty, B. 2020, March 25, WHO chief's questionable past comes into focus following coronavirus response, Fox News. https://www.foxnews.com/world/who-chief-tedros-questionable-past-coronavirus

211   McNeil Jr., D., 2017, May 13, Candidate to Lead the W.H.O. Accused of Covering Up Epidemics, New York Times. https://www.nytimes.com/2017/05/13/health/candidate-who-director-general-ethiopia-cholera-outbreaks.html

212   Ross, C., 2020, March 24, "Fully Complicit" in the terrible suffering: Health professionals accused him of covering up the previous epidemic to shield two African regimes. https://nationalinterest.org/blog/buzz/fully-complicit-terrible-suffering-tedros-adhanom-ghebreyesus-blamed-2017-cholera

213   McPhillips, D., 2020, May 29, Gates Foundation Donations to WHO Nearly Match Those From U.S. Government, US News and World Report. https://www.usnews.com/news/articles/2020-05-29/gates-foundation-donations-to-who-nearly-match-those-from-us-government

214   Higgins-Dunn, N. 2020, February 26,2020. Travel restrictions 'irrelevant' if coronavirus becomes a pandemic, top US health official says, CNBC.https://www.cnbc.com/2020/02/26/fauci-travel-restrictions-irrelevant-if-coronavirus-becomes-a-pandemic.html

215   White House Briefing, 2020, March 25. Remarks by President Trump, Vice President Pence, and Members of the Coronavirus Task Force in Press Briefing. The White House. https://www.whitehouse.gov/briefings-statements/remarks-president-President Trump-vice-president-pence-members-coronavirus-task-force-press-briefing-11/

216   https://www.cnn.com/2021/05/24/politics/fauci-donald-trump-coronavirus/index.html

217   Morens, D. and Fauci, A. (2020, September 3). Emerging Pandemic Diseases: How We Got to COVID-19. Cell 182, 1099-1091. https://www.cell.com/action/showPdf?pii=S0092-8674%2820%2931012-6

218   Areddy, J., 2020, updated May 26, China Rules Out Animal Market and Lab as Coronavirus Origin The Wallstreet Journal. https://www.wsj.com/articles/china-rules-out-animal-market-and-lab-as-coronavirus-origin-11590517508

219   Cohen, J. Wuhan seafood market may not be source of novel virus spreading globally. 2020, January 26, Science Magazine. https://www.sciencemag.org/news/2020/01/wuhan-seafood-market-may-not-be-source-novel-virus-spreading-globally

220   St. Cavish, C., 2020, March 11. Commentary: No, China's fresh food markets did not cause coronavirus, Los Angeles Times. https://www.latimes.com/food/story/2020-03-11/coronavirus-china-wet-markets

221   Page, J. et. al., 2020, March 6, Missteps, The Wallstreet Journal. https://www.wsj.com/articles/how-it-all-started-chinas-early-coronavirus-missteps-11583508932

222   Xiao, B. and Xiao, L., 2020, February, The possible origins of 2019-nCoV coronavirus. https://web.archive.org/web/20200214144447/https://www.researchgate.net/publication/339070128 The possible origins of 2019-nCoV coronavirus or Microsoft Word - origins of nCoV Lancet (mediaset.it)

223   Mishra, A. and Mondal, D. 2020, April 25. Corona leaked likely from Wuhan Institute of Virology: Experts, Sunday Guardian Live. https://www.sundayguardianlive.com/news/corona-leaked-likely-wuhan-institute-virology-experts

224   Judicial Watch: Fauci Emails Show WHO Entity Pushing for a Press Release 'Especially' Supporting China's Response to the Coronavirus | Judicial Watch

225   Global Preparedness Monitoring Board (who.int)

226   https://www.youtube.com/watch?v=_hpYPMP8tL8

227   Editors of Encyclopaedia Britannica, retrieved December 1, 2020, A Brief Overview of China's Cultural Revolution, Britannica. https://www.britannica.com/story/chinas-cultural-revolution

228   Albawa.Com, 2020, September 2. Is China Locking up Entire Residential Buildings With People Inside to Stop the Coronavirus? Albawa. https://www.msn.com/en-ae/news/other/

is-china-locking-up-entire-residential-buildings-with-people-inside-to-stop-the-coronavirus/ar-BBZOrXP

229 Westcott, B., 2018, August 29, China moves to end two-child limit, finishing decades of family planning. CNN. https://www.cnn.com/2018/08/28/asia/china-family-planning-one-child-intl/index.html

230 Reuters, 2020, November 24, China planning new policies to take on aging population, New York Post. https://nypost.com/2020/11/24/china-planning-new-policies-to-take-on-ageing-population-state-media/

231 https://www.reuters.com/article/china-economy-poll/chinas-economic-growth-seen-hitting-44-year-low-in-2020-bounce-8-4-in-2021-reuters-poll-idUSKBN27C0QB

232 Dunleavey, J., 2020, April 15,Former British spy chief says China and WHO bear responsibility for flawed coronavirus response. Washington Examiner. https://www.washingtonexaminer.com/news/former-british-spy-chief-says-china-and-who-bear-responsibility-for-flawed-coronavirus-response

233 Chakraborty, B. & Diaz, A., 2020, July 10, EXCLUSIVE: Chinese virologist accuses Beijing of coronavirus cover-up, flees Hong Kong:'I know how they treat whistleblowers'. Fox News. https://www.foxnews.com/world/chinese-virologist-coronavirus-cover-up-flee-hong-kong-whistleblower

234 Bowen, E. 2020, July 10. Chinese virologist in hiding after accusing Beijing of coronavirus cover-up, New York Post. https://nypost.com/2020/07/10/chinese-virologist-flees-after-accusing-beijing-of-covid-19-cover-up/

235 Carlson, T., 2020, September 19, TV appearance on Tucker Carlson of Li-Meng Yan, Fox News Channel. https://www.youtube.com/watch?v=qFlqXPl_hZQ

236 Yan, Li-Meng Yan ; Kang, Shu; Guan, Jie; Hu, Shanchang. (2020, September 14). Unusual Features of the SARS-CoV-2 Genome Suggesting Sophisticated Laboratory Modification Rather Than Natural Evolution and Delineation of Its Probable Synthetic Route. Prepublication. http://breggin.com/coronavirus/The_Yan_Report.pdf. To confirm the date it was put up and to follow the progress of the paper through publication, go to here: https://zenodo.org/record/4028830#.X2R2T5NKiuV

237 Sellin, L., 2020, August 4, Refugee Hong Kong Virologist Links COVID-19 to Chinese Military Laboratory, CCNS. https://ccnationalsecurity.org/refugee-hong-kong-virologist-links-covid-19-to-chinese-military-laboratory laboratory /

238 Becker, M.M. et al. Synthetic recombinant bat SARS-like coronavirus is infectious in cultured cells and in mice. Proc Natl Acad Sci U S A 105, 19944-9 (2008). https://scholar.google.com/scholar?hl=en&as_sdt=0%2C33&q=Becker%2C+M.M.+et+al.+Synthetic+recombinant+-bat+SARS-like+coronavirus+is+infectious+in+cultured+cells+and+in+mice.+Proc+Natl+Acad+Sci+U+S+A+105%2C+19944-9+%282008%29.&btnG=

239 Naveira, P., 2020, August 4, Li-Meng Yan: Coronavirus was developed in Chinese military lab, AS English. https://en.as.com/en/2020/08/03/latest_news/1596459547_022260.html

240 Dr. Yan received her MD degree from XiangYa Medical College of Central South University (China), and PhD from Southern Medical University (China). https://virtual.keystonesymposia.org/ks/speakers/view/1097. Also see https://www.cnnindonesia.com/internasional/20200804191248-113-532177/li-meng-yan-pakar-virologi-pengungkap-sumber-corona-di-china.

241 https://www.snopes.com/fact-check/li-meng-yan-covid-19-lab/

242 Liu Y, Yan L-M, Wan L, et al. Viral dynamics in mild and severe cases of COVID-19. https://www.thelancet.com/journals/laninf/article/PIIS1473-3099(20)30232-2/fulltext

243 https://www.nature.com/articles/s41586-020-2342-5?fbclid=IwAR3Kx7Hv3Yv5ppcBY-f1rFvMQfrydRXzoBGdflaeImh1WNW9Eq6MCdCkFFw0

244   https://loop.frontiersin.org/people/556239/overview.

245   Yan, Li-Meng; Kang, Shu; Guan, Jie; Hu, Shanchang. 2020, October 8, SARS-CoV-2 Is an Unrestricted Bioweapon: A Truth Revealed through Uncovering a Large-Scale, Organized Scientific Fraud. Prepublication. https://zenodo.org/record/4073131#.X4OpJOaSk2x "You can cite all versions by using the DOI 10.5281/zenodo.4073130. This DOI represents all versions, and will always resolve to the latest one."

246   Qiao Liang was born to a military family in 1955. He was assistant director of the production office of the air force's political department and held the rank of senior colonel in the air force. He has written books and army manuals and "has repeatedly won national and military awards," according to the bio attached to the report.

247   https://www.c4i.org/unrestricted.pdf Unrestricted Warfare, by Qiao Liang and Wang Xiangsui (Beijing: PLA Literature and Arts Publishing House, February 1999) [FBIS Editor's Note: The following selections are taken from "Unrestricted Warfare," a book published in China in February 1999 which proposes tactics for developing countries, in particular China, to compensate for their military inferiority vis-à-vis the United States during a high-tech war.] Other translations are readily available on the internet. We are using what appears to be a widely relied upon translation from FBIS, The Foreign Broadcast Information Service (FBIS), an open source intelligence component of the Central Intelligence Agency's Directorate of Science and Technology.

248   Yan, Li-Meng, Kang, Shu, Guan, Jie, & Hu, Shanchang. (2021). The Wuhan Laboratory Origin of SARS-CoV-2 and the Validity of the Yan Reports Are Further Proved by the Failure of Two Uninvited "Peer Reviews". http://doi.org/10.5281/zenodo.4650821 The Wuhan Laboratory Origin of SARS-CoV-2 and the Validity of the Yan Reports Are Further Proved by the Failure of Two Uninvited "Peer Reviews" | Zenodo

249   Obama, Barack, 2014, October 17, From the White House, Doing Diligence to Assess the Risks and Benefits of Life Sciences Gain-of-Function Research https://obamawhitehouse.archives.gov/blog/2014/10/17/doing-diligence-assess-risks-and-benefits-life-sciences-gain-function-research

250   https://reporter.nih.gov/search/GCBvbhJJPEOJ8Rp15r1W1A/project-details/9819304#-similar-Projects

251   Lin, C. 2020, April 22. Why US outsourced bat virus research to Wuhan US-funded $3.7 million project approved by President Trump's Covid-19 guru Dr Anthony Fauci in 2015 after US ban imposed on 'monster-germ' research, Asia Times. https://asiatimes.com/2020/04/why-us-outsourced-bat-virus-research-to-wuhan/

252   Owen, G. 2020, April 11, Wuhan lab was performing coronavirus experiments on bats from the caves where the disease is believed to have originated - with a £3m grant, Daily Mail Online.   https://www.dailymail.co.uk/news/article-8211257/Wuhan-lab-performing-experiments-bats-coronavirus-caves.html

253   https://www.independentsentinel.com/report-fauci-funded-gof-research-in-wuhan-due-to-incompetence-at-cdc/

254   Akst, J., 2015, Lab-Made Coronavirus Triggers Debate, The Scientist. https://www.the-scientist.com/news-opinion/lab-made-coronavirus-triggers-debate-34502

255   Also see, Breggin, P. and Breggin, G., 2020, August 3, Why COVID-19 Clinical Trials Cannot Be Trusted: The "Gold Standard" for Science Is Gold for the Drug Companies, www.breggin.com. Find at: https://breggin.com/why-covid-19-clinical-trials-cannot-be-trusted/

256   Branswell, H., 2015, November 9. SARS-like virus in bats shows potential to infect humans, study finds, STAT.https://www.statnews.com/2015/11/09/sars-like-virus-bats-shows-potential-infect-humans-study-finds/

257   Butler, 2015, April 12, engineered bat virus stirs debate over risky research, Nature. https://www.nature.com/news/engineered-bat-virus-stirs-debate-over-risky-research-1.18787

258   https://dailycaller.com/2021/04/04/nih-gain-of-function-anthony-fauci-review-board-wuhan-lab/

259   https://reporter.nih.gov/search/ZBjgKIfaCUyvJmxyKyn6xg/project-details/9819304

260   NIH Director, 2017, December 19, NIH Lifts Funding Pause on Gain-of-Function Research, Office of the Director of NIH. https://www.nih.gov/about-nih/who-we-are/nih-director/statements/nih-lifts-funding-pause-gain-function-research; also, Schnirring, L. 2017, December 19. Feds lift gain-of-function research pause, offer guidance, CIDRAP News. https://www.cidrap.umn.edu/news-perspective/2017/12/feds-lift-gain-function-research-pause-offer-guidance. Akst, J., 2015, Lab-Made Coronavirus Triggers Debate, The Scientist. https://www.the-scientist.com/news-opinion/lab-made-coronavirus-triggers-debate-34502

261   Schnirring, L. 2017, December 19. Feds lift gain-of-function research pause, offer guidance, CIDRAP News. https://www.cidrap.umn.edu/news-perspective/2017/12/feds-lift-gain-function-research-pause-offer-guidance

262   McNeil Jr., D., 2017, December 19, A Federal Ban on Making Lethal Viruses Is Lifted, New York Times. https://www.nytimes.com/2017/12/19/health/lethal-viruses-nih.html?_r=0

263   Burki, T. Ban on gain-of-function studies ends. The US moratorium on gain-of-function experiments has been rescinded, but scientists are split over the benefits—and risks—of such studies, www.thelancet.com/infection Vol 18 February 2018, pp. 148-9. https://www.thelancet.com/action/showPdf?pii=S1473-3099%2818%2930006-9

264   McNeil Jr., D., 2017, December 19, A Federal Ban on Making Lethal Viruses Is Lifted, New York Times. https://www.nytimes.com/2017/12/19/health/lethal-viruses-nih.html

265   Pandey, S., 2019, January 21, Are Biological Weapons More Dangerous Than Nuclear Weapons? What Is the Current State Of Biological Weapons Research? ResearchGate. https://www.researchgate.net/post/Are_Biological_Weapons_More_Dangerous_Than_Nuclear_Weapons_What_Is_The_Current_State_Of_Biological_Weapons_Research

266   Sellin, L. , 2020, September 6, Did Fauci's NIH Institute Financially Assist China's Military? CCNS. https://ccnationalsecurity.org/did-faucis-nih-institute-financially-assist-chinas-military/

267   NIH News Release, 2020, August 27, NIH establishes Centers for Research in Emerging Infectious Diseases. https://www.nih.gov/news-events/news-releases/nih-establishes-centers-research-emerging-infectious-diseases

268   Sellin, L. , 2020, September 6, Did Fauci's NIH Institute Financially Assist China's Military? CCNS. https://ccnationalsecurity.org/did-faucis-nih-institute-financially-assist-chinas-military/

269   https://patents.justia.com/patent/8933106

270   https://patents.justia.com/patent/8309602

271   http://en.cnki.com.cn/Article_en/CJFDTotal-DNGY199902000.htm

272   https://www.ncbi.nlm.nih.gov/pmc/articles/PMC2750777/

273   https://patents.justia.com/patent/9889194

274   https://grantome.com/grant/NIH/R01-AI068002-03

275   https://science.sciencemag.org/content/369/6511/1603.full

276   https://en.cnki.com.cn/Article_en/CJFDTOTAL-JSYX201408012.htm

277   https://en.cnki.com.cn/Article_en/CJFDTOTAL-JSYX201408012.htm

278   https://www.nih.gov/news-events/news-releases/nih-establishes-centers-research-emerging-infectious-diseases

279   Kolata, G., 2019, November 4, Vast Dragnet Targets Theft of Biomedical Secrets for China, New York Times. https://www.nytimes.com/2019/11/04/health/china-nih-scientists.html

280   Weinberger, S., et al., 2020 March 30, Suspected SARS virus and flu samples found in luggage: FBI report describes China's 'biosecurity risk', Yahoo! News. https://news.yahoo.

com/suspected-sars-virus-and-flu-found-in-luggage-fbi-report-describes-chinas-biosecurity-risk-144526820.html

281   In the News, 2020, April 16. The Galveston National Lab and Wuhan Institute of Virology, Galveston National Laboratory, University of Texas Medical Branch. https://www.utmb.edu/gnl/news/2020/04/16/the-galveston-national-lab-and-wuhan-institute-of-virology

282   Lynch, D. and McKay, H., 2020, May 1, Prominent university bio lab urged to reveal extent of relationship with Wuhan lab at center of coronavirus outbreak. https://www.foxnews.com/us/university-texas-biolab-wuhan-connection

283   Rubinstein, R., Principal Deputy General Counsel, 2020, April 24, Letter to James B. Milliken, Chancellor the University of Texas System: Notice of 20 U.S.C. § 1011f Investigation and Record Request/University of Texas System from U.S. Department of Education. https://www2.ed.gov/policy/highered/leg/ut-apr24-2020.pdf

284   Blackwell, T., 2019, August 8, Bio-warfare experts question why Canada was sending lethal viruses to China, National Post. https://nationalpost.com/health/bio-warfare-experts-question-why-canada-was-sending-lethal-viruses-to-china

285   Press Release, 2010, Global Health Leaders Launch Decade of Vaccines Collaboration. Bill & Melinda Gates Foundation. https://www.gatesfoundation.org/Media-Center/Press-Releases/2010/12/Global-Health-Leaders-Launch-Decade-of-Vaccines-Collaboration

286   Morens, D. and Fauci, A. (2020, September 3). Emerging Pandemic Diseases: How We Got to COVID-19. Cell 182, 1099-1091. https://www.cell.com/action/showPdf?pii=S0092-8674%2820%2931012-6

287   Morens, D. and Fauci, A. (2020, September 3). Emerging Pandemic Diseases: How We Got to COVID-19. Cell 182, 1099-1091. https://www.cell.com/action/showPdf?pii=S0092-8674%2820%2931012-6

288   Tucker, J.A. interview, 2020, October 21, Remarkable New Insights On COVID-19, The Dr. Peter Breggin Hour, radio/TV at https://www.youtube.com/user/PeterBreggin

289   Tucker, J. A., 2020, October 1, Lockdowns: The New Totalitarianism, American Institute for Economic Research (AIER). https://www.aier.org/article/lockdown-the-new-totalitarianism/

290   McNeil Jr., G. 2020, December 24, How Much Herd Immunity Is Enough? New York Time. https://www.nytimes.com/2020/12/24/health/herd-immunity-covid-coronavirus.html

291   https://www.cnn.com/asia/live-news/coronavirus-outbreak-hnk-intl-01-24-20/index.html Scroll down the page.

292   Press Release, 2010, Global Health Leaders Launch Decade of Vaccines Collaboration. Bill & Melinda Gates Foundation. https://www.gatesfoundation.org/Media-Center/Press-Releases/2010/12/Global-Health-Leaders-Launch-Decade-of-Vaccines-Collaboration

293   https://www.bing.com/videos/search?q=Gates+says+he+talks+to+Fauci+instead+of+President Trump&docid=608034182707544370&mid=BB6A2E22CCDE06D9B036BB6A2E-22CCDE06D9B036&view=detail&FORM=VIRE

294   CDC Foundation Receives Three Grants Totaling $13.5 Million For Meningitis and Rotavirus Global Advancements | CDC Foundation

295   China CDC - Bill & Melinda Gates Foundation

296   https://www.bloomberg.com/news/articles/2016-06-28/pfizer-bolsters-china-presence-with-350-million-biotech-plant

297   http://www.pfizer.com.cn/(S(3xg00q45fgum2g45uusza0rb))/pfizer-china/index_en.aspx

298   https://www.forbes.com/sites/martaorosz/2020/03/18/german-billionaire-twins-investment-in-biotech-firm-surges-after-deal-with-pfizer-on-coronavirus-vaccine/?sh=3ca58e8057c2

299   Here's the Company That Bill Gates Thinks Is the Clear Coronavirus Vaccine Leader (fool.com)

300   Bill Gates paid WHO $50m to declare a pandemic - YouTube

301 Gates, Bill, 2020, March 25. How We Must Respond to the Coronavirus Pandemic, TED interview with Chris Anderson and Whitney Pennington Rodgers. https://www.youtube. com/watch?v=Xe8fIjxicoo

302 Gates' support for China's claim to have no more COVID-19 cases is at around 24 minutes into the interview.

303 Lockdowns after China's new Covid-19 outbreak impact steel, iron ore (cnbc.com)

304 Pandemic After-Effects: Chinese People Concerned About Food Supply - Bringing you Truth, Inspiration, Hope. (visiontimes.com)

305 China Places Over 22 Million on Lockdown Amid New Covid Wave - The New York Times (nytimes.com)

306 Lockdowns Intensify in Northeastern China as CCP Virus Spreads (theepochtimes.com)

307 "China imposes strict lockdowns on city of 11M amid rise in COVID - The Christian Post

308 https://www.theepochtimes.com/wartime-mode-inside-chinas-current-virus-hotspots_3672911. html

309 https://nypost.com/2020/04/02/chinas-deadly-coronavirus-lie-co-conspira-tor-the-world-health-organization/

310 O'Connor, T., 2020, May 15, China Acknowledges Destroying Early Coronavirus Samples, Confirming U.S. Accusation, Newsweek.https://www.newsweek.com/china-acknowledg-es-destroying-early-coronavirus-samples-confirming-us-accusation-1504484

311 Weissert, W., 2020, May, 4, DHS report: China hid virus' severity to hoard supplies, AP. https://apnews.com/article/bf685dcf52125be54e030834ab7062a8

312 Hydroxychloroquine application is associated with a decreased mortality in critically ill patients with COVID-19 | medRxiv

313 https://economictimes.indiatimes.com/news/international/world-news/how-china-is-using-covid-to-target-people-in-xinjiang/human-rights-violations/slideshow/77992156.cms

314 https://www.breitbart.com/national-security/2020/02/11/famed-dissident-china-weld-ing-people-shut-in-their-homes-to-fight-coronavirus/; https://metro.co.uk/2020/02/02/wuhan-woman-screams-chinese-authorities-barricade-inside-home-12162599/;https://www.cbc.ca/player/play/1703503427818; https://www.thesun.co.uk/news/10925668/coronavirus-patients-welded-homes-china/. I personally spoke with several professional people of Chinese origin with families still in China or who had recently visited China who confirmed these stories of atrocities. One described how the Chinese Communist Party increased its control and surveillance of the nation and made war upon people infected with the COVID-19 as if attacking a foreign enemy. Also, from scientific publications and reports from knowledgeable people, the Chinese government did not take control of patient treatment and doctors used many treatments from traditional Chinese medicine to prednisone and hydroxychloroquine. So two factors may have led to their relative success in controlling COVID-19 (if it is true): totalitarian control and non-interference with early treatment or medical care. While China may have slowed the spread at times, COVID-19 is still being fought with draconian measures wherever it flares up.

315 https://www.mdpi.com/2227-9032/9/1/82/htm

316 Gates' comment on "certificates" is around minute 34 of the interview.

317 Goujon, R., 2019, June 28, Globalists, Nationalists and Patriots, Forbes. https://www.forbes. com/sites/stratfor/2017/06/28/globalists-nationalists-and-patriots/?sh=421dfa8b12ad

318 https://www.forbes.com/sites/randalllane/2021/01/07/a-truth-reckoning-why-were-holding-those-who-lied-for-President-Trump-accountable/?sh=1ac702bb5710; see commentary in https://thefederalistpapers.org/opinion/forbes-editor-threatens-tarnish-company-hires-cer-tain-President-Trump-administration-officials

319 https://nypost.com/2014/07/18/forbes-magazine-sold-to-asian-investor-group/

320   https://thefederalist.com/2020/05/04/has-china-compromised-every-major-mainstream-media-entity/

321   https://www.nbcnews.com/news/us-news/david-koch-billionaire-conservative-activist-philanthropist-dies-79-n1045696

322   https://www.forbes.com/companies/koch-industries/?sh=291fa38674ce

323   https://www.sec.gov/Archives/edgar/data/41077/000119312505225697/dex996.htm

324   https://news.kochind.com/news/2020/how-koch-is-fighting-covid-19

325   https://www.vox.com/2018/7/31/17634070/President Trump-koch-tweet

326   https://www.cnbc.com/2018/04/06/koch-brothers-network-breaks-with-President Trump-over-china-trade-war.html

327   https://www.britannica.com/topic/Cato-Institute

328   https://twitter.com/realDonaldPresident   Trump/status/1024239166429769729?ref_src=twsrc%5Etfw%7Ctwcamp%5Etweetembed%7Ctwterm%5E1024239166429769729%7Ctwgr%5E%7Ctwcon%5Es1_&ref_url=https%3A%2F%2Fwww.vox.com%2F2018%2F7%2F31%2F17634070%2FPresident Trump-koch-tweet

329   https://www.vox.com/2018/7/31/17634070/President Trump-koch-twee

330   https://mises.org/library/children-and-rights

331   Many of our books deal with pharmaceutical fraud and other malfeasances, including Breggin, P. (1991), Toxic Psychiatry; Breggin. P. and Breggin, G. (1994), Talking Back to Prozac; Breggin P. (2008a) Brain Disabling Treatments in Psychiatry, Second Edition; and Breggin, P. (2008), Medication Madness.

332   https://mariettaoh9-12project.com/the-globalists-koch-brothers-and-cato-institute/

333   https://www.cato.org/sites/cato.org/files/pubs/pdf/cato-2019-annual-report-update.pdf?utm_campaign=2019%20Annual%20Report%20PDF%20Clicks&utm_source=PDF-Click&utm_medium=Annual-Report

334   https://theintercept.com/2020/03/26/americans-for-prosperity-cdc-coronavirus/

335   https://www.prwatch.org/news/2018/08/13382/koch-subsidiary-invista-build-1-billion-new-plant-china

336   https://www.bloomberg.com/news/articles/2017-11-07/ruyi-is-said-to-pay-over-2-billion-to-buy-lycra-owner-from-koch

337   https://www.nytimes.com/2019/07/11/business/american-businesses-china.html

338   Yu, Frank, 2020, November 23, US Invested Billions into Companies With Ties to China, Epoch Times. https://www.theepochtimes.com/mkt_app/us-invested-billions-into-companies-with-ties-to-chinese-military_3587923.html?v=ul

339   https://www.theepochtimes.com/President Trump-issues-order-to-end-investments-in-chinese-military-companies_3576876.html

340   Brian T. Kennedy, 2020, September, Facing up to the China Threat, Imprimis.https://www.democraciaparticipativa.net/economia-society/columnistas-invitados/17655-facing-up-to-the-china-threat.html

341   National Center for Health Statistics. 2020, October 10, Weekly Updates by select demographic and geographic characteristics: Provisional death counts for Coronavirus Disease 2019 (COVID-19), CDC, Under "Comorbidities." https://www.cdc.gov/nchs/nvss/vsrr/covid_weekly/index.htm

342   Centers for Disease Control and Prevention (CDC), updated 2020, Sept. 10, COVID-19 Pandemic Planning Scenarios, CDC. https://www.cdc.gov/coronavirus/2019-ncov/hcp/planning-scenarios.html Search for "Table 1. Parameter Values".

343   Brian C. Procter, Casey Ross, Vanessa Pickard, Erica Smith, Cortney Hanson, and Peter A. McCullough (2021, March). Early Ambulatory Multidrug Therapy Reduces Hospitalization and Death in High-Risk Patients with SARS-CoV-2 (COVID-19. International Journal

of Innovative Research in Medical Science (IJIRMS). Volume 06, Issue 03, March 2021, https://doi.org/10.23958/ijirms/vol06. View of Early Ambulatory Multidrug Therapy Reduces Hospitalization and Death in High-Risk Patients with SARS-CoV-2 (COVID-19) (ijirms.in)

344   https://fee.org/articles/physicians-say-hospitals-are-pressuring-er-docs-to-list-covid-19-on-death-certificates-here-s-why/; https://fee.org/articles/physicians-say-hospitals-are-pressuring-er-docs-to-list-covid-19-on-death-certificates-here-s-why/

345   https://www.snopes.com/fact-check/medicare-hospitals-covid-patients/ and https://www.factcheck.org/2020/04/hospital-payments-and-the-covid-19-death-count/

346   https://www.christianpost.com/news/cdc-director-agrees-that-hospitals-have-monetary-incentive-to-inflate-covid-19-data.html; https://www.washingtonexaminer.com/news/cdc-director-acknowledges-hospitals-have-a-monetary-incentive-to-overcount-coronavirus-deaths

347   American Academy of Pediatrics and the Children's Hospital Association, 2020, October 10, Children and COVID-19:State-Level Data Report. https://services.aap.org/en/pages/2019-novel-coronavirus-covid-19-infections/children-and-covid-19-state-level-data-report/

348   Freed, M. et al., 2020, July 24. KFF Coronavirus Stats (based on CDC data up to July 22, 2020). https://www.kff.org/coronavirus-covid-19/issue-brief/what-share-of-people-who-have-died-of-covid-19-are-65-and-older-and-how-does-it-vary-by-state/ CDC data at https://www.cdc.gov/coronavirus/2019-ncov/need-extra-precautions/older-adults.html

349   Freed, M. et al., 2020, July 24. KFF Coronavirus Stats (based on CDC data up to July 22, 2020). https://www.kff.org/coronavirus-covid-19/issue-brief/what-share-of-people-who-have-died-of-covid-19-are-65-and-older-and-how-does-it-vary-by-state/ CDC data at https://www.cdc.gov/coronavirus/2019-ncov/need-extra-precautions/older-adults.html

350   Berenson, A., 2020, June 27, Getting realistic about the coronavirus death rate. New York Post. https://nypost.com/2020/06/25/getting-realistic-about-the-coronavirus-death-rate/

351   Richardson, I. 2020, June 5, Fact check: CDC's estimates COVID-19 death rate around 0.26%, doesn't confirm it, USA Today. https://www.usatoday.com/story/news/factcheck/2020/06/05/fact-check-cdc-estimates-covid-19-death-rate-0-26/5269331002/

352   National Center for Health Statistics. Weekly Updates by select demographic and geographic characteristics: Provisional death counts for Coronavirus Disease 2019 (COVID-19). See the small print under "Comorbidities" near bottom of the page. https://www.cdc.gov/nchs/nvss/vsrr/covid_weekly/index.htm

353   Collin County, 2020, May 18, 2020 Commissioners Court, Collin County. https://collincountytx.new.swagit.com/videos/62477

354   Collin County, 2020, COVID-19 Case Definition, Collin County E Agenda. https://eagenda.collincountytx.gov/docs/2020/CC/20200518_2481/48410_Explanation.pdf

355   Hellerstedt, J., 2020, DSHS Surveillance Case Definitions for 2019 Novel Coronavirus Disease (COVID-19)- Revised: May 11, 2020. This is a download link to the PDF referenced: https://www.google.com/url?sa=t&rct=j&q=&esrc=s&source=web&cd=&ved=2ahUKEwjmqfe85bTqAhUBhXIEHaWjCzAQFjAAegQIARAB&url=https%3A%2F%2Fwww.dshs.state.tx.us%2Fcoronavirus%2Fdocs%2FDSHS-COVID19CaseDefinitionandInvestigationPrioritizationGuidance.pdf&usg=AOvVaw3w1gx6jwpHw_wfmZcpO9Q5

356   CDC, 2020, FAQ: COVID-19 Data and Surveillance, CDC. https://www.cdc.gov/coronavirus/2019-ncov/covid-data/faq-surveillance.html

357   Kulldorff, M., Gupta, S. and Bhattacharya, J. 2020, October 4, Great Barrington Pledge. https://gbdeclaration.org/

358   In addition to other sources we have cited, see the recently obtained excellent book, Reiss, Karina and Bhakta, Sucharit (2020). Corona False Alarm? Facts and Figures. White River Junction, VT: Chelsea Green Publications.

359 Stand for Health Freedom Campaign Inspired by Peter Breggin MD, Ginger Breggin and Jean Ross, Ages 84, 69 and 94, Stand for Health Freedom. https://standforhealthfreedom. com/action/declaration-of-older-folks/

360 Private Communication to Peter Breggin, February 26, 2021, originally sent to the National Center for Public Policy Research in Washington, DC.

361 Stokkermans, T. J., Goyal, A., Bansal, P. Trichonas, G. Chloroquine And Hydroxychloroquine Toxicity. StatPearls [Internet]. Treasure Island (FL): StatPearls. Last Update: July 4, 2020 https://www.ncbi.nlm.nih.gov/books/NBK537086/?report=printable

362 Breggin, P., 2020, Scientific Study or Megadose Mass Murder. The Breggins' Coronavirus Resource Center. https://breggin.com/scientific-study-or-megadose-mass-murder/

363 Borba, M., Val, F., and Sampaio, V. et al. (2020 April 24). Effect of High vs Low Doses of Chloroquine Diphosphate as Adjunctive Therapy for Patients Hospitalized with Severe Acute Respiratory Syndrome Coronavirus (SARS-CoV-2) Infection: A Randomized Clinical Trial. JAMA Network Open. https://jamanetwork.com/journals/jamanetworkopen/fullarticle/2765499

364 Goodman & Gilman's The Pharmacological Basis of Therapeutics (2011, p. 1405). New York: McGraw Hill.

365 Goodman & Gilman's The Pharmacological Basis of Therapeutics (2011, p. 1404).

366 Borba, S. et al, 2020. Chloroquine diphosphate in two different dosages as adjunctive therapy of hospitalized patients with severe respiratory syndrome in the context of coronavirus (SARS-CoV-2) infection: Preliminary safety results of a randomized, double-blinded, phase IIb clinical trial (CloroCovid-19 Study) Unpublished at the time. See section on "Ethical Aspects." https://www.medrxiv.org/content/10.1101/2020.04.07.20056424v1.full.pdf

367 Borba, S. et al, 2020.

368 Thomas, K. and Knvul, S. (Published April 12, 2020. Updated June 15, 2020). Chloroquine Study Halted Over Risk of Fatal Heart Complications: A research trial of coronavirus patients in Brazil ended after patients taking a higher dose of chloroquine, one of the drugs President Trump has promoted, developed irregular heart rates. https://www.nytimes.com/2020/04/12/health/chloroquine-coronavirus-President Trump.html We are no longer emphasize that they are attacking "President Trump's drug," because as mentioned earlier, that is a diversion. The medication has the support of many countries and untold numbers of doctors, so it is hardly "President Trump's drug." In addition, the medication is being attacked around the world, not just in America.

369 Borba, M., Val, F., and Sampaio, V. et al. (2020 April 24). Effect of High vs Low Doses of Chloroquine Diphosphate as Adjunctive Therapy for Patients Hospitalized with Severe Acute Respiratory Syndrome Coronavirus (SARS-CoV-2) Infection: A Randomized Clinical Trial. JAMA Network Open. https://jamanetwork.com/journals/jamanetworkopen/fullarticle/2765499

370 AMA Ed Hub. CME credits for reading Borba et al. Ibid. https://edhub.ama-assn.org/jn-learning/module/2765499

371 Wessel, L. (2020, June 22). Science, 'It's a nightmare.' How Brazilian scientists became ensnared in chloroquine politics. https://www.sciencemag.org/news/2020/06/it-s-nightmare-how-brazilian-scientists-became-ensnared-chloroquine-politics

372 For published comments by Meryl Nass and others on hydroxychloroquine overdoses in clinical trials, see Svab, P., 2020, August 5, COVID-19 Hydroxychloroquine Studies Flawed, Experts Say, Epoch Times. https://www.theepochtimes.com/covid-19-hydroxychloroquine-studies-flawed-experts-say_3450045.html

373 The Zelenko Protocol, undated general information about Dr. Zelenko's work. https://faculty.utrgv.edu/eleftherios.gkioulekas/zelenko/ZelenkoProtocol.pdf and also Zelenko, V. (2020,

April 28). Letter to the Physicians of the World. https://docs.google.com/document/d/1p-jgHlqI-ZuKOziN3txQsN5zz62v3K043pR3DdhEmcos/edit

374   Derwand, R., Scholz, M., and Zelenko, V. (2020, December). COVID-19 outpatients: early risk-stratified treatment with zinc plus low-dose hydroxychloroquine and azithromycin: a retrospective case series study. International Journal of Antimicrobial Agents, 56 (6) and McCullough, P. and 32 others including Vladimir Zelenko (2020). Multifaceted highly targeted sequential multidrug treatment of early ambulatory high-risk SARS-CoV-2 infection (COVID-19). Reviews in Cardiovascular Medicine. Reviews in Cardiovascular Medicine. 21 (4), 517-530. DOI: 10.31083/j.rcm.2020.04.264 https://rcm.imrpress.com/EN/10.31083/j.rcm.2020.04.264

375   https://radiopatriot.net/2020/04/24/an-american-hero-dr-zelenko/

376   https://www.youtube.com/watch?v=uJ4ibn3ehnw

377   https://aapsonline.org/heroic-testimony-from-drs-mccullough-risch-and-fareed/ and https://breggin.com/covid-19-home-treatment/

378   US Senate Hearing Explores Ivermectin as "Miracle Drug" for COVID-19 While Mainstream Media Outlets Ignore (trialsitenews.com)

379   https://www.educationviews.org/six-courageous-doctors-testify-before-u-s-senate-committee-on-covid-treatments/ and https://www.fox7austin.com/news/doctor-pleads-for-review-of-data-on-ivermectin-as-covid-19-treatment-during-senate-hearing

380   Peter McCullough, Paul E. Alexander, et al. (2020, December). Multifaceted highly targeted sequential multidrug treatment of early ambulatory high-risk SARS-CoV-2 infection (COVID-19). Reviews in Cardiovascular Medicine. 21 (4), 517-530. November 2020. DOI: 10.31083/j.rcm.2020.04.264 https://rcm.imrpress.com/EN/10.31083/j.rcm.2020.04.264

381   Procter BC, Ross C, Pickard V, Smith E, Hanson C, McCullough PA. (2021, March.) Early Ambulatory Multidrug Therapy Reduces Hospitalization and Death in High-Risk Patients with SARS-CoV-2 (COVID-19). ijirms [Internet]. 2021Mar.17 [cited 2021Apr.28];6(03):219 - 221. View of Early Ambulatory Multidrug Therapy Reduces Hospitalization and Death in High-Risk Patients with SARS-CoV-2 (COVID-19) (ijirms.in)

382   https://www.youtube.com/user/PeterBreggin. My coverage of their work is continuous with several one-hour TV interviews and shorter videos. Check the two sections of the YouTube channel, "Dr. Breggin's Weekly TV Show" and "Dr. Breggin's Coronavirus Commentary." For a partial backup on our YouTube Channel, see go to Brighteon.com and search Dr. Peter Breggin.

383   The Truth for Health Foundation, 2021, https://www.truthforhealthfoundation.org/patient-guide/patient-treatment-guide/, and the Association of American Physicians and Surgeons, 2020, available without charge, and updated when necessary. https://aapsonline.org/covid-patientguide/

384   McCullough, P. et al. (2020). Multifaceted highly targeted sequential multidrug treatment of early ambulatory high-risk SARS-CoV-2 infection (COVID-19). Reviews in Cardiovascular Medicine. 21 (4), 517-530. November, 2020. DOI: 10.31083/j.rcm.2020.04.264 https://rcm.imrpress.com/EN/10.31083/j.rcm.2020.04.264, and Procter BC, Ross C, Pickard V, Smith E, Hanson C, McCullough PA. (2021, March.) Multidrug Therapy Reduces Hospitalization and Death in High-Risk Patients with SARS-CoV-2 (COVID-19). ijirms [Internet]. 2021Mar.17 [cited 2021Apr.28];6(03):219 - 221. https://www.ijirms.in/index.php/ijirms/article/view/1100,)

385   Zelenko, Vladimir; Risch, Harvey and Fareed, George, 2020. Medical Studies Support MDs Prescribing Hydroxychloroquine for Early Stage COVID-19 and for Prophylaxis. link to the eBook: https://drive.google.com/file/d/1l6y3L_KGb1ilMW0FaP4VZsd7WvX2IU3z/view

386   http://breggin.com/coronavirus/HCQ/Medical-Studies-Support-MDs-Prescribing-Hydroxychloroquine-for-Early-Stage-COVID.pdf

387   Risch, H., 2020, The Key to Defeating COVID-19 Already Exists. We Need to Start Using it | Opinion, Newsweek. https://www.newsweek.com/key-defeating-covid-19-already-exists-we-need-start-using-it-opinion-1519535

388   Peiffer-Smadja And Costagliola' S. (2020, July 20). Early Outpatient Treatment Of Symptomatic, High-Risk Covid-19 Patients That Should Be Ramped-Up Immediately As Key To The Pandemic Crisis. American Journal Of Epidemiology. Doi: 10.1093/Aje/Kwaa151 Pmcid: Pmc 7454270 Pmid: 32685975 https://www.ncbi.nlm.nih.gov/pmc/articles/PMC7454270/

389   Yu, B., et al., 2020, Low dose of hydroxychloroquine reduces fatality of critically ill patients with COVID-19, Science China. http://breggin.com/coronavirus/HCQ/HCQ-reduces-fatality-China-study-May-15-2020.pdf

390   Kim, M., et al., 2020, Treatment Response to Hydroxychloroquine, Lopinavir-Ritonavir, and Antibiotics for Moderate COVID-19: A First Report on the Pharmacological Outcomes from South Korea. doi: https://doi.org/10.1101/2020.05.13.20094193 http://breggin.com/coronavirus/HCQ/Kim_HCQ-Retrospective-SKorea-study-May-2020.pdf

391   Kupferschmidt, K., 2020, Three big studies dim hopes that hydroxychloroquine can treat or prevent COVID-19, AAAS Science. https://www.sciencemag.org/news/2020/06/three-big-studies-dim-hopes-hydroxychloroquine-can-treat-or-prevent-covid-19

392   https://www.americasfrontlinedoctors.com/

393   Most Banned Videos, 2020, See the DC Doctor Press Conference Big Tech Is Fighting to Suppress, Banned Video. https://banned.video/watch?id=5f1fc7a468370e02f29f34cf

394   Andrew Torba, 2020, "A group of doctors who are on the frontlines…", gab. https://gab.com/a/posts/104591612298957345

395   PatriotDude, 2020, BREAKING: American Doctors Address COVID-19 Misinformation. Hydroxychloroquine Works!!, Brighteon. https://www.brighteon.com/3571f9ae-ec43-4254-8a56-1a931c250888

396   The Emergency Use Authorization is part of Federal Food, Drug, and Cosmetic Act, Pub. L. No. 75-717 (as amended at 21 U.S.C. §§ 301-399f). It is interpreted and applied by the FDA. Cites to specific FDA guidelines follow.

397   FDA, 2017, January, Emergency Use Authorization of Medical Use Products and Related Authorities. Food and Drug Administration, https://www.fda.gov/regulatory-information/search-fda-guidance-documents/emergency-use-authorization-medical-products-and-related-authorities#declaration Part III. Emergency Use Authorization, section B, 1., d. Part d. is titled "No Alternatives." Another key section says it only has to rise to the standard of "may" be effective.

398   Statement from NIH and BARDA on the FDA Emergency Use Authorization of the Moderna COVID-19 Vaccine | National Institutes of Health (NIH). The statement describes the EUA approval but without mentioning that according to the statute, it only means "may" be effective, which is below FDA standards.

399   https://www.hhs.gov/coronavirus/explaining-operation-warp-speed/index.html

400   https://www.cbc.ca/news/world/us-bright-whistleblower-hydroxychloroquine-1.5542131

401   Head of vaccine agency says he was ousted for resisting hydroxychloroquine - Axios

402   Homeland Security, 2017, January, Biological Incident Annex to the Response and Recovery Federal Interagency Operation Plans. Final-January 2017. https://web.archive.org/web/20201104004301/https://www.fema.gov/media-library-data/1511178017324-92a7a7f808b-3f03e5fa2f8495bdfe335/BIA_Annex_Final_1-23-17_(508_Compliant_6-28-17).pdf

403   FDA, 2017, Plaquenil. link to pdf: http://breggin.com/coronavirus/HCQ/HCQ-label.pdf

404 WHO, 2015, WHO Model List of Essential Medicines. http://breggin.com/coronavirus/HCQ/WHO_NOV2015.pdf

405 WHO, 2017, The cardiotoxicity of antimalarials, WHO. link to pdf: http://breggin.com/coronavirus/HCQ/WHO-cardiotoxicity-of-antimalarials.pdf#page=36

406 Credible Meds, 2020, Combined List of Drugs That Prolong QT And/Or Cause Torsades De Pointes (TPD), Credible Meds. link to pdf: http://breggin.com/coronavirus/HCQ/CombinedList-QT-interval-affecting-drugs.pdf

407 FDA, 2018, Understanding Unapproved Use of Approved Drugs "Off Label", FDA. https://www.fda.gov/patients/learn-about-expanded-access-and-other-treatment-options/understanding-unapproved-use-approved-drugs-label

408 Singer, J., 2020, Doctors, Not Politicians, Ought to Decide Whether Off-Label Drug Use of Hydroxychloroquine Is Appropriate for COVID-19 Patients, Cato Institute. https://www.cato.org/publications/commentary/doctors-not-politicians-ought-decide-whether-label-drug-use

409 Wittich, C., et al, 2012, Ten Common Questions (and Their Answers) About Off-label Drug Use, Mayo Clinic. https://www.ncbi.nlm.nih.gov/pmc/articles/PMC3538391/pdf/main.pdf

410 FDA, 2020, Hydroxychloroquine of Chloroquine for COVID-19: Drug Safety Communication – FDA Cautions Against Use Outside of the Hospital Setting of a Clinical Trial Due to Risk of Heart Rhythm Problems, FDA. https://www.fda.gov/safety/medical-product-safety-information/hydroxychloroquine-or-chloroquine-covid-19-drug-safety-communication-fda-cautions-against-use

411 FDA, 2020, Frequently Asked Questions on the Emergency Use Authorization (EUA) for Chloroquine Phosphate and Hydroxychloroquine Sulfate for Certain Hospitalized COVID-19 Patients, FDA. https://www.fda.gov/media/136784/download

412 https://clinicaltrials.gov/

413 FDA, 2020, Coronavirus (COVID-19) Update: FDA Revokes Emergency Use Authorization for Chloroquine and Hydroxychloroquine, FDA. https://www.fda.gov/news-events/press-announcements/coronavirus-covid-19-update-fda-revokes-emergency-use-authorization-chloroquine-and

414 Hixenbaugh, M., 2020, Scientists were close to a coronavirus vaccine years ago. Then the money dried up., NBC News. https://www.nbcnews.com/health/health-care/scientists-were-close-coronavirus-vaccine-years-ago-then-money-dried-n1150091

415 NIH, Last Updated: 2020, October 9, Therapeutic Management of Patients with COVID-19, National Institutes of Health. https://www.covid19treatmentguidelines.nih.gov/therapeutic-management/

416 Superville, D. and Seitz, A., 2020, President Trump's coronavirus retweets spark claims of censorship, The Globe and Mail. https://www.theglobeandmail.com/world/us-politics/article-President Trump-again-pushes-hydro

417 Glasspiegel, R. (2020, July 25, date of one censorship), Outkick. https://www.outkick.com/twitter-censors-donald-President Trump-and-puts-don-jr-in-penalty-box-over-hydroxychloroquine/ This blog contains copies of censored materials and confirmation of the censorships.

418 Pease, H. , 2020, August 10, America's Frontline Doctors Censored by Big Tech, Dr. Harold Pease, Liberty Under Fire, Org. http://libertyunderfire.org/?utm_source=Newspaper%20Mailing%20List&utm_campaign=de763dd1e4-EMAIL_CAMPAIGN_2020_08_05_04_32&utm_medium=email&utm_term=0_95cfc15da5-de763dd1e4-389273633

419 Breggin, P., Breggin, G., 2020, The Elite Strikes Back Against President Trump's Populist Drug, Psychiatric Drug Facts. https://breggin.com/the-elite-strikes-back-against-President Trumps-populist-drug/

420 Richardson, V., 2020, Michigan Gov. Gretchen Whitmer hit for hydroxychloroquine crackdown as debate escalates, Washington Times. https://www.washingtontimes.com/news/2020/mar/29/gretchen-whitmer-michigan-governor-hydroxychloroqu/ and News Release, NIH, July 27, NIH to invest $58M to catalyze data science and health research innovation in Africa, National Institutes of Health.

421 The COVID-19 Treatment Guidelines Panel recommends against the use of ivermectin for the treatment of COVID-19, except in a clinical trial

422 https://www.covid19treatmentguidelines.nih.gov/antiviral-therapy/ivermectin/

423 https://www.covid19treatmentguidelines.nih.gov/statement-on-ivermectin/

424 S. C. Atkinson, M. D. Audsley, K. G. Lieu, G. A. Marsh, D. R. Thomas, S. M. Heaton, J. J. Paxman, K. M. Wagstaff, A. M. Buckle, G. W. Moseley, D. A. Jans and N. A. Borg. Recognition by host nuclear transport proteins drives disorder-to-order transition in Hendra virus V. Scientific Reports. 8, 358 (2018); S. N. Y Yang, S. C. Atkinson, C. Wang, A. Lee, M. A. Bogoyevitch, N. A. Borg and D. A. Jans. The broad spectrum antiviral ivermectin targets the host nuclear transport importin a/b1 heterodimer. Antiviral Research. 177, 104760 (2020); V. Götz, L. Magar, D. Dornfeld, S. Giese, A. Pohlmann, D. Höper, B.-W. Kong, D. A. Jans, M. Beer, O. Haller and M. Schwemmle. Influenza A viruses escape from MxA restriction at the expense of efficient nuclear vRNP import. Scientific Reports. 6, 23138 (2016); C. Lv, W. Liu, B. Wang, R. Dang, L. Qiu, J. Ren, C. Yan, Z. Yang and X. Wang. Ivermectin inhibits DNA polymerase UL42 of pseudorabies virus entrance into the nucleus and proliferation of the virus in vitro and vivo. Antiviral Research. 177, 104760 (2020); E. Mastrangelo, M. Pezzullo, T. De Burghgraeve, S. Kaptein, B. Pastorino, K. Dallmeier, X. de Lamballerie, J. Neyts, A. M. Hanson, D. N. Frick, M. Bolognesi and M. Milani. Ivermectin is a potent inhibitor of flavivirus replication specifically targeting NS3 helicase activity: new prospects for an old drug. Journal of Antimicrobial Chemotherapy. 67, 1884-1894 (2012); M. Y. F. Tay, J. E. Fraser, W. K. K. Chan, N. J. Moreland, A. P. Rathore, C. Wang, S. G. Vasudevan and D. A. Jans. Nuclear localization of dengue virus (DENV) 1–4 nonstructural protein 5; protection against all 4 DENV serotypes by the inhibitor Ivermectin. Antiviral Research. 99, 301-306 (2013); F. S. Varghese, P. Kaukinen, S. Gläsker, M. Bespalov, L. Hanski, K. Wennerberg, B. M. Kümmerer and T. Ahola. Discovery of berberine, abamectin and ivermectin as antivirals against chikungunya and other alphaviruses. Antiviral Research. 126, 117-124 (2016); K. M. Wagstaff, H. Sivakumaran, S. M. Heaton, D. Harrich, D. A. Jans. Ivermectin is a specific inhibitor of importina/b-mediated nuclear import able to inhibit replication of HIV-1 and dengue virus. Biochemical Journal. 443, 851-856 (2012); C. R. King, T. M. Tessier, M. J. Dodge, J. B. Weinberg, J. S. Mymryk, Inhibition of Human Adenovirus Replication by the Importin a/b1 Nuclear Import Inhibitor Ivermectin. Journal of Virology. 94, e00710-20 (2020).

425 L. Caly, J. D. Druce, M. G. Catton, D. A. Jans, K. M. Wagstaff, The FDA-approved drug ivermectin inhibits the replication of SARS-CoV-2 in vitro. Antiviral Res. 178, 104787 (2020).

426 X. Zhang et al., Inhibitory effects of ivermectin on nitric oxide and prostaglandin E2 production in LPS-stimulated RAW 264.7 macrophages. Int Immunopharmacol. 9, 354- 359 (2009); X. Ci et al., Avermectin exerts anti-inflammatory effect by downregulating the nuclear transcription factor kappa-B and mitogen-activated protein kinase activation pathway. Fundam Clin Pharmacol. 23, 449-455 (2009); X. Zhang, Y. Song, X. Ci, N. An, Y. Ju, H. Li, X. Wang, C. Han, J. Cui and X. Deng. Ivermectin inhibits LPS-induced production of inflammatory cytokines and improves LPS-induced survival in mice. Inflamm Res. 57, 524-529 (2008).

427 A. P. Arévalo et al., https://www.biorxiv.org/content/10.1101/2020.11.02.363242v1.full and G. D. de Melo et al., https://www.biorxiv.org/content/10.1101/2020.11.21.392639v1.full.

Also see Pierre Kory's Congressional testimony: https://www.pandemicdebate.com/post/these-data-show-that-ivermectin-is-effectively-a-miracle-drug-against-covid-19-pierre-kory-md.

428 Carvallo H. https://clinicaltrials.gov/ct2/show/NCT04425850 (2020). 35. Shouman W. https://clinicaltrials.gov/ct2/show/NCT04422561 (2020). 36. P. Behera et al., https://www.medrxiv.org/content/10.1101/2020.10.29.20222661v1.full (2020); A. Elgazzar et al., https://www.researchsquare.com/article/rs-100956/v2 (2020); 88. Bernigaud, D. Guillemot, A. Ahmed-Belkacem, L. Grimaldi-Bensouda, A. Lespine, F. Berry, L. Softic, C. Chenost, G. Do-Pham, B. Giraudeau, S. Fourati, O. Chosidow. Bénéfice de l'ivermectine : de la gale à la COVID-19, un exemple de sérendipité. Annales de Dermatologie et de Vénéréologie. 147, Issue 12, Supplement, Page A194 (2020).

429 Robin RC, Alam RF, Saber S, Bhiuyan E, Murshed R, Alam MT. A case series of 100 COVID-19 positive patients treated with combination of ivermectin and doxycycline. Journal of Bangladesh College of Physicians and Surgeons. 38, Supp 10-15 (2020); Carvallo HE et al., https://www.medrxiv.org/content/10.1101/2020.09.10.20191619v1 (2020); Mahmud R. https://clinicaltrials.gov/ct2/show/NCT04523831 (2020); M. S. I. Khan, C. R. Debnath, P. N. Nath, M. A. Mahtab, H. Nabeka, S. Matsuda and S. M. F. Akbar. Ivermectin treatment may improve the prognosis of patients with COVID19. Archivos de Bronconeumología. 10.1016/j.arbres.2020.08.007 (2020); F. I. Gorial et al., https://www.medrxiv.org/content/10.1101/2020.07.07.20145979v1 (2020); J. Morgenstern et al., https://www.medrxiv.org/content/10.1101/2020.10.29.20222505v1.full (2020); A. Elgazzar et al., https://www.researchsquare.com/article/rs-100956/v2 (2020).

430 M. S. I. Khan, C. R. Debnath, P. N. Nath, M. A. Mahtab, H. Nabeka, S. Matsuda and S. M. F. Akbar. Ivermectin treatment may improve the prognosis of patients with COVID19. Archivos de Bronconeumología. 10.1016/j.arbres.2020.08.007 (2020); J. C. Rajter, M. S. Sherman, N. Fatteh, F. Vogel, J. Sacks, J. J. Rajter. Use of ivermectin is associated with lower mortality in hospitalized patients with COVID-19 (ICON study). Chest. 10.1016/j.chest.2020.10.009 (2020); Hashim HA et al., https://www.medrxiv.org/content/10.1101/2020.10.26.20219345v1 (2020); A. Elgazzar et al., https://www.researchsquare.com/article/rs-100956/v2 (2020); M. S. Niaee et al., https://www.researchsquare.com/article/rs-109670/v1 (2020); A. Portmann-Baracco, M. Bryce-Alberti, R. A. Accinelli. Antiviral and antiinflammatory properties of ivermectin and its potential use in Covid-19. Arch Broncopneumol. July 7, doi: 10.1016/j.arbres.2020.06.011 (2020).

431 J. C. Rajter, M. S. Sherman, N. Fatteh, F. Vogel, J. Sacks, J. J. Rajter. Use of ivermectin is associated with lower mortality in hospitalized patients with COVID-19 (ICON study). Chest. 10.1016/j.chest.2020.10.009 (2020); Hashim HA et al., https://www.medrxiv.org/content/10.1101/2020.10.26.20219345v1 (2020); A. Elgazzar et al., https://www.researchsquare.com/article/rs-100956/v2 (2020).

432 J. J. Chamie. https://www.researchgate.net/publication/344469305 (2020); A Connel. An old drug tackles new tricks: ivermectin treatment in three brazilain towns. TrialSiteNews.com. https://www.trialsitenews.com/an-old-drug-tackles-new-tricksivermectin-treatment-in-three-brazilian-towns/ (2020). J. J. Chamie. COVID-19 en Paraguay - departamentos mas afectados. twitter.com. https://twitter.com/jjchamie/status/1322014560551841794/photo/1 (2020).

433 L.H. Kircik, J. Q. Del Rosso, A. M. Layton, J. Schauber . Over 25 years of clinical experience with ivermectin: an overview of safety for an increasing number of indications. J Drugs Dermatol. 15, 325-332 (2016).

434 World Health Organization. Model list of essential medicines - 21st list. https://www.who.int/publications/i/item/WHOMVPEMPIAU2019.06 (2019).

435 https://apnews.com/article/fact-checking-afs:Content:9768999400#:~:text=No%20evidence%20ivermectin%20is%20a%20miracle%20drug%20against,COVID-19%20and%20will%20prevent%20people%20from%20getting%20sick.

436 Richardson, V. (2020, August 29). Hydroxychloroquine 'very safe,' says Dr. Scott Atlas; blasts 'garbage' medical studies. Washington Times. https://www.washingtontimes.com/news/2020/aug/29/hydroxychloroquine-uproar-shows-objective-science-/

437 https://www.theepochtimes.com/scott-atlas-resigns-as-covid-19-adviser-to-President Trump_3599657.html

438 Fareed, G. et al., August 22, 2020, Open letter to Dr. Anthony Fauci regarding the use of hydroxychloroquine for treating COVID-19. The Desert Review. https://www.thedesertreview.com/opinion/columnists/open-letter-to-dr-anthony-fauci-regarding-the-use-of-hydroxychloroquine-for-treating-covid-19/article_31d37842-dd8f-11ea-80b5-bf80983bc072.html

439 Cohen, A., Bruer, W., 2020, US stockpile stuck with 63 million doses of hydroxychloroquine, CNN Health. https://www.cnn.com/2020/06/17/health/hydroxychloroquine-national-stockpile/index.html

440 Mulangu, S., et al, 2019, A Randomized, Controlled Trial of Ebola Virus Disease Therapeutics, The New England Journal of Medicine. N Engl J Med 2019; 381:2293-2303. DOI: 10.1056/NEJMoa1910993 https://www.nejm.org/doi/full/10.1056/NEJMoa1910993

441 MedicineNet, 2020, Remdesivir (RDV): Experimental Antiviral for Coronavirus (COVID-19), MedicineNet. https://www.medicinenet.com/remdesivir_rdv_ebola_covid-19_coronavirus_trial/article.htm#what_is_remdesivir_rdv

442 Wang, Y., et al, 2020, Remdesivir in adults with severe COVID-19: a randomized, double-blind, placebo-controlled, multicentre trial. DOI: https://doi.org/10.1016/S0140-6736(20)31022-9 https://www.thelancet.com/action/showPdf?pii=S0140-6736%2820%2931022-9

443 The authors found that "More patients in the remdesivir group than the placebo group discontinued the study drug because of adverse events or serious adverse events (18 [12%] in the remdesivir group vs four [5%] in the placebo group), among whom seven (5%) were due to respiratory failure or acute respiratory distress syndrome in the remdesivir group. (p. 10)

444 Dunn, A., 2020, There are already 72 drugs in human trials for coronavirus in the US. With hundreds more on the way, a top drug regulator warns we could run out of researches to test them all, Business Insider. https://www.businessinsider.com/fda-woodcock-overwhelming-amount-of-coronavirus-drugs-in-the-works-2020-4

445 NIH, 2019-2020, Appendix A, Table 2. COVID-19 Treatment Guidelines Panel Financial Disclosure for Companies Related to COVID-19 Treatment or Diagnostics, NIH. https://www.covid19treatmentguidelines.nih.gov/panel-financial-disclosure/

446 NIH, 2020, What's New in the Guidelines, NIH. https://www.covid19treatmentguidelines.nih.gov/whats-new/

447 NIH, 2020, NIH clinical trial of remdesivir to treat COVID-19 begins, NIH. https://www.nih.gov/news-events/news-releases/nih-clinical-trial-remdesivir-treat-covid-19-begins

448 Strauss, D., 2020, FULL INTERVIEW: Houston-area doctor in viral video touting hydroxychloroquine as virus cure doubles down on claims, Click2Houston. https://www.click2houston.com/news/local/2020/07/31/full-interview-houston-doctor-in-viral-video-touting-hydroxychloroquine-as-virus-cure-doubles-down-on-claims/

449 Brennan, Z., 2020, Remdesivir helps coronavirus patient – but at what cost?, Politico. https://www.politico.com/news/2020/05/06/remdesivir-helps-coronavirus-patients-but-at-what-cost-240230

450 Rowland, C., 2020, Government researchers changed metric to measure coronavirus drug remdesivir during clinical trial, Washington Post. https://www.washingtonpost.com/business/2020/05/01/government-researchers-changed-metric-measure-coronavirus-drug-remdesivir-during-clinical-trial/

451 Schwitzer, G., 2020, What the public didn't hear about the NIH remdesivir trial, Health News Review. https://www.healthnewsreview.org/2020/04/what-the-public-didnt-hear-about-the-nih-remdesivir-trial/

452 Evans S (2007) When and How Can Endpoints Be Changed after Initiation of a Randomized Clinical Trial. PLOS Clin Trial 2(4): e18. https://doi.org/10.1371/journal.pctr.0020018

453 Clinical Trials, 2020, History of Changes for Study, Clinical Trials. https://clinicaltrials.gov/ct2/history/NCT04280705?A=10&B=16&C=Side-by-Side#StudyPageTop

454 Clinical Trials, 2020, Adaptive COVID-19 Treatment Trial, Clinical Trials. https://clinicaltrials.gov/ct2/show/NCT04280705

455 NIH, 2020, NIH Clinical Trial Shows Remdesivir Accelerates Recovery from Advanced COVID-19, NIH. http://breggin.com/coronavirus/remdesivir/NIH-release-Remdesivir-Accelerates-Recovery-4-29-20.pdf

456 Edwards, E., 2020, Remdesivir shows promising results for coronavirus, Fauci says, NBC News. https://www.nbcnews.com/health/health-news/coronavirus-drug-remdesivir-shows-promise-large-trial-n1195171

457 Clancy, C.J., 2020, Fauci on remdesivir for COVID-19: 'This will be the standard of care', Healio. http://breggin.com/coronavirus/remdesivir/Fauci-on-remdesivir-for-COVID-19-This-will-be-the-standard-of-care.pdf

458 https://www.fda.gov/news-events/press-announcements/fda-approves-first-treatment-covid-19

459 Remdesivir for the Treatment of Covid-19 — Final Report | NEJM

460 World's Top 5 Medical Journals and How Much They Cost - The Practo Blog for Doctors

461 NIH. (last reviewed February 5, 2020). About NIAID: Director, Anthony Fauci, M.D. https://www.niaid.nih.gov/about/anthony-s-fauci-md-bio

462 Schwitzer, G., 2020, What the public didn't hear about the NIH remdesivir trial, Health News Review. https://www.healthnewsreview.org/2020/04/what-the-public-didnt-hear-about-the-nih-remdesivir-trial/

463 Centers for Disease Control and Prevention (CDC), May 12, 2021, Selected Adverse Events Reported after COVID-19 Vaccination | CDC

464 Paul Elias Alexander, PhD; Parvez Dara, MD, MBA; Howard Tenenbaum, DDS, PhD. (2021, June 1). COVID-19 spike protein may be a potentially unsafe toxic endothelial pathogen. TrialSiteNews. https://trialsitenews.com/the-covid-19-spike-protein-may-be-a-potentially-unsafe-toxic-endothelial-pathogen/

465 https://www.sciencedirect.com/science/article/pii/S0264410X03004894

466 Deming D., et al. 2006. Vaccine efficacy in senescent mice challenged with recombinant SARS-CoV bearing epidemic and zoonotic spike variants. PLoS Med. 3:e525. [PMC free article] [PubMed] [Google Scholar]. Vaccine Efficacy in Senescent Mice Challenged with Recombinant SARS-CoV Bearing Epidemic and Zoonotic Spike Variants (plos.org)

467 Yasui, F. et al. (2008). Prior Immunization with Severe Acute Respiratory Syndrome (SARS)-Associated Coronavirus (SARS-CoV) Nucleocapsid Protein Causes Severe Pneumonia in Mice Infected with SARS-CoV1. DOI:10.4049/jimmunol.181.9.6337 [PDF] Prior Immunization with Severe Acute Respiratory Syndrome (SARS)-Associated Coronavirus (SARS-CoV) Nucleocapsid Protein Causes Severe Pneumonia in Mice Infected with SARS-CoV1 | Semantic Scholar

468 Wen Shi Lee, Adam K. Wheatley, Stephen J. Kent, and Brandon J. DeKosky.(2020, October). Antibody-dependent enhancement and SARS-CoV-2 vaccines and therapies. Nature Microbiology, 5, 1185-1199. https://www.nature.com/articles/s41564-020-00789-5.pdf

469 https://duckduckgo.com/?q=Antibody-dependent+enhancement&ia=web&iai=r1-0&page=1&sexp=%7B%22minusexp%22%3A%22a%22%2C%22biaexp%22%3A%22b%22%2C%22msvrtexp%22%3A%22b%22%2C%22mliexp%22%3A%22b%22%7D

470 Alana F Ogata, Chi-An Cheng, Michaël Desjardins, Yasmeen Senussi, Amy C Sherman, Megan Powell, Lewis Novack, Salena Von, Xiaofang Li, Lindsey R Baden (2021, May 20). Circulating SARS-CoV-2 Vaccine Antigen Detected in the Plasma of mRNA-1273 Vaccine Recipients. Clinical Infectious Diseases, ciab465, https://doi.org/10.1093/cid/ciab465. https://academic.oup.com/cid/advance-article/doi/10.1093/cid/ciab465/6279075. Also see new data on organ concentrations of the spike protein in vaccinated humans, ahead in this chapter.

471 https://www.hcs.harvard.edu/hghr/online/dna-vaccines/

472 https://www.sciencemag.org/news/2021/05/further-evidence-offered-claim-genes-pandemic-coronavirus-can-integrate-human-dna

473 Liguo Zhang, Alexsia Richards, M. Inmaculada Barrasa, Stephen H. Hughes, Richard A. Young, Rudolf Jaenisch. (2021). Reverse-transcribed SARS-CoV-2 RNA can integrate into the genome of cultured human cells and can be expressed in patient-derived tissues. Proceedings of the National Academy of Sciences May 2021, 118 (21) e2105968118; DOI : 10.1073/pnas.2105968118. https://www.pnas.org/content/118/21/e2105968118. Also see, SARS-CoV-2 RNA can be reverse-transcribed (news-medical.net) and

474 https://extranet.who.int/pqweb/sites/default/files/documents/Status_COVID_VAX_20Jan2021_v2.pdf

475 https://www.sciencemag.org/news/2020/12/suspicions-grow-nanoparticles-pfizer-s-covid-19-vaccine-trigger-rare-allergic-reactions

476 https://www.livemint.com/science/health/deaths-after-covid-19-vaccination-in-europe-not-directly-tied-to-shots-experts-11611485520372.html

477 https://www.reuters.com/article/us-health-coronavirus-vaccine-idUSKBN28E1G3

478 Deaths after Covid-19 vaccination in Europe not directly tied to shots: Experts (livemint.com)

479 Vaccination in Israel: Challenging mortality figures? - Israel National News.

480 https://rxview.adverahealth.com/calculating-the-reporting-rate-of-adverse-events-in-faers-a-new-methodology

481 Testimony Before the Subcommittees on Health and Environment, and Oversight and Investigations, Committee on Commerce, and Subcommittee on Health, Committee on Veterans' Affairs, House of Representatives United States General Accounting Office. (GAO). February 9, 2000. ADVERSE EVENTS Surveillance Systems for Adverse Events and Medical Errors Statement of Janet Heinrich, Associate Director Health Financing and Public Health Issues Health, Education, and Human Services Division. T-HEHS-00-61 Adverse Events: Surveillance Systems for Adverse Events and Medical Errors (gao.gov)

482 Grant Final Report Grant ID: R18 HS 017045 Electronic Support for Public Health—Vaccine Adverse Event Reporting System (ESP:VAERS) Inclusive dates: 12/01/07 - 09/30/10 Principal Investigator: Lazarus, Ross, MBBS, MPH, MMed, GDCompSci Team members: Michael Klompas, MD, MPH Performing Organization: Harvard Pilgrim Health Care, Inc. Project Officer: Steve Bernstein Submitted to: The Agency for Healthcare Research and Quality (AHRQ) U.S. Department of Health and Human Services 540 Gaither Road Rockville, MD 20850 www.ahrq.gov Electronic Support for Public Health—Vaccine Adverse Event

483 https://www.cdc.gov/vaccinesafety/concerns/concerns-history.html

484 https://www.cdc.gov/coronavirus/2019-ncov/vaccines/safety/adverse-events.html

485 https://nypost.com/2021/02/10/greenwald-blasts-facebook-for-censoring-covid-vaccine-opinions/t Reporting System (ESP:VAERS) (nvic.org); https://www.businessinsider.com/tuckercarlson-attacked-covid-19-vaccine-again-on-show-2021-5?op=1; https://www.christianpost.com/news/facebook-expands-censorship-of-covid-vaccine-conspiracy-theories.html; https://www.overtoncountynews.com/news/local_state/court-asked-to-prohibit-facebook-from-censoring-vaccine-critics/article_4193379e-b292-11eb-8d8a-fbcb8d28d195.html

486   Tseng CT, Sbrana E, Iwata-Yoshikawa N, Newman PC, Garron T, et al. (2012, April 20) Immunization with SARS Coronavirus Vaccines Leads to Pulmonary Immunopathology on Challenge with the SARS Virus. PLOS ONE 7(8): 10.1371/annotation/2965cfae-b77d-4014-8b7b-236e01a35492. https://journals.plos.org/plosone/article?id=10.1371/journal.pone.0035421

487   Yuyang Lei, Jiao Zhang, Cara R. Schiavon, Ming He , Lili Chen, Hui Shen, Yichi Zhang, Qian Yin, Yoshitake Cho, Leonardo Andrade, Gerry S. Shadel, Mark Hepokoski, Ting Lei, Hongliang Wang, Jin Zhang, Jason X.-J. Yuan, Atul Malhotra, Uri Manor, Shengpeng Wang, Zu-Yi Yuan, and John Y-J. Shyy. (2020, December) SARS-CoV-2 Spike Protein Impairs Endothelial Function via Downregulation of ACE2. bioRxiv preprint doi: https://doi.org/10.1101/2020.12.04.409144; this version posted December 4, 2020 https://www.biorxiv.org/content/biorxiv/early/2020/12/04/2020.12.04.409144.full.pdf

488   Gerard J. Nuovo, Cynthia Magro, Toni Shaffer, Hamdy Awad, David Suster, Sheridan Mikhail , Bing He, Jean-Jacques Michaille,Benjamin Liechty, Esmerina Tili. (2021) Endothelial cell damage is the central part of COVID-19 and a mouse model induced by injection of the S1 subunit of the spike protein. Annals of Diagnostic Pathology 51 (2021) 151682 https://www.ncbi.nlm.nih.gov/pmc/articles/PMC7758180; A. P. Arévalo et al., https://www.biorxiv.org/content/10.1101/2020.11.02.363242v1.full and G. D. de Melo et al., https://www.biorxiv.org/content/10.1101/2020.11.21.392639v1.full. Also see Pierre Kory's Congressional testimony: https://www.pandemicdebate.com/post/these-data-show-that-ivermectin-is-effectively-a-miracle-drug-against-covid-19-pierre-kory-md

489   https://foreignaffairsintelligencecouncil.wordpress.com/2021/02/04/doctors-now-warn-about-permanent-damage-and-cardiovascular-events-following-covid-19-vaccination/. It cites Hooman Noorchashm, MD, PhD.

490   COVID-19 Autopsies Put Endothelial Damage, Angiogenesis in the Spotlight | tctmd.com

491   Endothelial dysfunction in COVID-19: Current findings and therapeutic implications - PubMed (nih.gov)

492   Endothelial cell damage is the central part of COVID-19 and a mouse model induced by injection of the S1 subunit of the spike protein (nih.gov)

493   https://www.forbes.com/sites/victoriaforster/2021/01/11/covid-19-vaccines-cant-alter-your-dna-heres-why/

494   https://alschner-klartext.de/2021/05/31/spike-protein-in-der-muttermilch-ist-toxisch/ in German, Google translator available.

495   https://abc7.com/menstrual-cycles-and-covid-vaccine-side-effects-women-coronavirus-period/10552668/

496   Paul Elias Alexander, PhD; Parvez Dara, MD, MBA; Howard Tenenbaum, DDS, PhD. (2021, June 1). The COVID-19 spike protein may be a potentially unsafe toxic endothelial pathogen. TrialSiteNews. https://trialsitenews.com/the-covid-19-spike-protein-may-be-a-potentially-unsafe-toxic-endothelial-pathogen/

497   https://www.wsj.com/articles/BL-WHB-1322; Population control: Is it a tool of the rich? - BBC News

498   https://www.sandiegouniontribune.com/news/science/story/2020-12-15/fda-review-moderna-covid-19-vaccine-safe-effective and Experimental vaccine death rate for Israel's elderly 40 times higher than COVID-19 deaths: researchers | News | LifeSite (lifesitenews.com). Original article here: The uncovering of the vaccination data in Israel reveals a frightening picture (nakim.org)

499   https://abcnews.go.com/Politics/fda-raises-concerns-safety-efficacy-moderna-covid-19/story?id=74735461

500   The discussion of attempted vaccinations begins a little after 8 minutes in the video. https://breggin.com/trump-cancels-funding-of-us-china-research-making-epidemic-viruses/

501 Over 900 Died After Receiving COVID-19 Vaccine, But Experts Say Data Is Misinterpreted (ibtimes.com)

502 Yuyang et al., 2020. https://www.biorxiv.org/content/biorxiv/early/2020/12/04/2020.12.04.409144. full.pdf

503 Nuovo et al., 2021. https://www.ncbi.nlm.nih.gov/pmc/articles/PMC7758180

504 Dr. Gøtzsche, P., 2020, February, Vaccines: Truth, Lies, and Controversy. Denmark: People's Press. Kindle only. https://www.amazon.com/Vaccines-controversy-Peter-C-G%C3%B8tzsche-ebook/dp/B0848FPKNP

505 Gates Foundation.org. (2016, September). Cochrane Collaboration. https://www.gatesfoundation.org/How-We-Work/Quick-Links/Grants-Database/Grants/2016/09/OPP1158795

506 Gates Foundation (2020, June 4). Bill & Melinda Gates Foundation pledges $1.6 billion to Gavi, the Vaccine Alliance, to protect the next generation with lifesaving vaccines. https://www.gatesfoundation.org/Media-Center/Press-Releases/2020/06/Bill-and-Melinda-Gates-Foundation-pledges-to-Gavi-the-Vaccine-Alliance

507 For coverage, see Hart, E.M. (2018). HPV vaccine safety: Cochrane launches urgent investigation into review after criticisms. BMJ 2018; 362 doi: https://doi.org/10.1136/bmj.k3472. (Published 09 August 2018)). Cite this as BMJ 2018;362;k3472; the 2018 article in BMJ by Lars Jørgensen, Peter C Gøtzsche, and Tom Jefferson, which criticizes Cocherane and the HIV study, can be found here: https://ebm.bmj.com/content/ebmed/23/5/165.full. pdf. For additional commentary, see https://www.cochrane.org/sites/default/files/public/uploads/cochrane_hpv_response_3sep18.pdf; https://thewire.in/health/gotzsche-affair-spotlights-challenges-of-producing-good-evidence-in-medicine and https://www.sciencemag.org/news/2018/09/evidence-based-medicine-group-turmoil-after-expulsion-co-founder

508 https://breggin.com/alert-73-meet-me-in-denmark-march-9-2019/

509 Breggin, Peter. Moving past the vaccine/autism controversy - to examine potential vaccine neurological harms. International Journal of Risk & Safety in Medicine, 1 (2020) 1–15 1 DOI 10.3233/JRS-200052 IOS https://breggin.com/wp-content/uploads/Breggin-P.-2020-Moving-past-the-vaccine-1.pdf

510 Branswell, H., 2019, August 29, With new grants, Gates Foundation takes an early step toward a universal flu vaccine https://www.statnews.com/2019/08/29/gates-foundation-grants-universal-flu-vaccine/

511 New, W. and Saez, C., Bill Gates Calls For "Vaccine Decade;" Explains How Patent System Drives Public Health Aid, Intellectual Property Rights. https://www.ip-watch.org/2011/05/17/bill-gates-calls-for-vaccine-decade-explains-how-patent-system-drives-public-health-aid/

512 Calfas, J. (2020, April 5) Bill Gates to help fund coronavirus-vaccine development. The Wall Street Journal. Retrieved June 19, 2020 at https://www.wsj.com/articles/bill-gates-to-spend-billions-on-coronavirus-vaccine-development-11586124716 Schwab, T. (2020, March 17). Bill Gates' Charity Paradox. A Nation investigation illustrates the moral hazards surrounding the Gates Foundation's $50 billion charitable enterprise. The Nation. Retrieved June 16, 2020 at https://www.thenation.com/article/society/bill-gates-foundation-philanthropy/

513 Schwab, T. (2020, March 17). Bill Gates' Charity Paradox. A Nation investigation illustrates the moral hazards surrounding the Gates Foundation's $50 billion charitable enterprise. The Nation. Retrieved June 16, 2020 at https://www.thenation.com/article/society/bill-gates-foundation-philanthropy/

514 https://www.aier.org/article/who-deletes-naturally-acquired-immunity-from-its-website/

515 https://www.aier.org/article/who-deletes-naturally-acquired-immunity-from-its-website/

516 Boyle, P., 2020, March 31, Here's why we can't rush a COVID-19 vaccine, American Association of Medical Colleges (AAMC). https://www.aamc.org/news-insights/here-s-why-we-can-t-rush-covid-19-vaccine

517   Dr. McCullough on the Dr. Peter Breggin Hour, December 9, 2020. Peter A. McCullough, MD, MPH--Save Yourself, Your Family, the World From COVID-19 - YouTube

518   https://timesofindia.indiatimes.com/world/us/gates-foundation-now-whos-biggest-funder/articleshow/76145069.cms

519   https://www.americasfrontlinedoctors.com/mission-statement/

520   https://blogs.bmj.com/bmj/2020/11/27/covid-19-vaccines-where-are-the-data/

521   https://www.webmd.com/lung/news/20200928/doctors-wary-of-rushed-covid-vaccine#1

522   https://www.jpost.com/health-science/could-an-mrna-vaccine-be-dangerous-in-the-long-term-649253

523   Covid vaccines: Will drug companies make bumper profits? - BBC News

524   Huge Profits For Vaccine Manufacturers 2021 - Bing News. This reference is to a collection of articles about huge vaccine profits.

525   The material in brackets is elaborated by P. Breggin.

526   Rancourt, D. (2020 April). Masks Don't Work A review of science relevant to COVID-19 social policy. https://www.algora.com/Algora_blog/2020/05/23/masks-dont-work-a-review-of-science-relevant-to-covid-19-socia-l-policy

527   Berenson, A., 2020, November. Unreported Truths About Covid-19 and Lockdowns: Part 3: Masks. No publisher listed.

528   Forgey, Q. 2020, May 27, Fauci says he wears mask as 'symbol' of good behavior, Politico. https://www.politico.com/news/2020/05/27/fauci-wears-mask-as-symbol-of-good-behavior-283847

529   https://www.snopes.com/fact-check/medicare-hospitals-covid-patients/ and https://www.factcheck.org/2020/04/hospital-payments-and-the-covid-19-death-count/

530   https://foreignaffairsintelligencecouncil.wordpress.com/2021/02/04/doctors-now-warn-about-permanent-damage-and-cardiovascular-events-following-covid-19-vaccination/; https://noorchashm.medium.com/a-letter-of-warning-to-fda-and-pfizer-on-the-immunological-danger-of-covid-19-vaccination-in-the-7d17d037982d and https://www.realclearpolitics.com/video/2021/03/24/hooman_noorchashm_not_everyone_should_get_the_covid_vaccine.html

531   https://www.columbian.com/news/2021/mar/02/amid-covid-19-pandemic-flu-has-virtually-disappeared-in-the-u-s/. Also see: https://apnews.com/article/flu-has-disappeared-us-pandemic-2145d999319b53d8a32a829a324f398d and many others.

532   Decreased Influenza Activity During the COVID-19 Pandemic — United States, Australia, Chile, and South Africa, 2020 | MMWR (cdc.gov)

533   Rosen, J., 2020, EXCLUSIVE: Internal HHS investigation finds CDC's early test kits were 'contaminated', ABC News. https://abc3340.com/news/nation-world/exclusive-internal-hhs-investigation-finds-cdcs-early-test-kits-were-contaminated

534   Madrigal, A., Meyer, R., 2020, 'How Could the CDC Make That Mistake?', The Atlantic. https://www.theatlantic.com/health/archive/2020/05/cdc-and-states-are-misreporting-covid-19-test-data-pennsylvania-georgia-texas/611935/

535   Collin County, 2020, May 18, 2020 Commissioners Court, Collin County. https://collincountytx.new.swagit.com/videos/62477

536   Collin County, 2020, COVID-19 Case Definition, Collin County E Agenda. https://eagenda.collincountytx.gov/docs/2020/CC/20200518_2481/48410_Explanation.pdf

537   Hellerstedt, J., 2020, DSHS Surveillance Case Definitions for 2019 Novel Coronavirus Disease (COVID-19)- Revised: May 11, 2020. This is a download link to the PDF referenced: https://www.google.com/url?sa=t&rct=j&q=&esrc=s&source=web&cd=&ved=2ahUKEwjmqfe85bTqAhUBhXIEHaWjCzAQFjAAegQIARAB&url=https%3A%2F%2Fwww.dshs.state.tx.us%2Fcoronavirus%2Fdocs%2FDSHS-COVID19CaseDefinitionandInvestigationPrioritizationGuidance.pdf&usg=AOvVaw3w1gx6jwpHw_wfmZcpO9Q5

538 CDC, 2020, FAQ: COVID-19 Data and Surveillance, CDC. https://www.cdc.gov/coronavirus/2019-ncov/covid-data/faq-surveillance.html

539 Reinhard, B., et al, 2020, CDC wants states to count 'probable' coronavirus cases and deaths, but most aren't doing it, The Washington Post. http://breggin.com/coronavirus/CDC-wants-states-to-count-probable-covid-19-cases-WAPO.pdf

540 Mandavilli, A., 2020, August 29, Your coronavirus test is positive. Maybe it shouldn't be, The New York Times. https://www.nytimes.com/2020/08/29/health/coronavirus-testing.html

541 Shaw, A., 2020, Cuomo, facing criticism for COVID handling, blames President Trump for virus coming to New York, Fox News. https://www.foxnews.com/politics/cuomo-covid-handling-blames-President-Trump-ny

542 Goodman, J., 2020, How Delays and Unheeded Warnings Hindered New York's Virus Fight, The New York Times. https://www.nytimes.com/2020/04/08/nyregion/new-york-coronavirus-response-delays.html

543 https://isabelcastillo.com/states-coronavirus-deaths

544 Singer, J., 2020, Doctors, Not Politicians, Out To Decide Whether Off-Label Drug Use of Hydroxychloroquine Is Appropriate for COVID-19 Patients, Reason. https://reason.com/2020/04/07/doctors-not-politicians-ought-to-decide-whether-off-label-drug-use-of-hydroxychloroquine-is-appropriate-for-covid-19-patients/

545 Richardson, V., 2020, Cuomo pressured to reverse restrictions on hydroxychloroquine, The Washington Times. https://www.washingtontimes.com/news/2020/apr/5/rudy-giuliani-urges-andrew-cuomo-lift-hydroxychlor/

546 New York nursing home deaths topped 12K, Cuomo health chief reveals | Fox News

547 New York State Department of Health, 2020, Advisory: Hospital Discharges and Admissions to Nursing Homes, Web Archive. (the document has since been removed from the New York State) [https://coronavirus.health.ny.gov/system/files/documents/2020/03/doh_covid19_nhadmissionsreadmissions_-032520.pdf] and can currently only be found through a web archive) https://web.archive.org/web/20200407103413/https://coronavirus.health.ny.gov/system/files/documents/2020/03/doh_covid19-_nhadmissionsreadmissions_-032520.pdf

548 Civiletti, D., 2020, As coronavirus began to escalate, state officials ordered nursing homes to accept COVID patients, Riverhead Local. https://riverheadlocal.com/2020/04/17/as-coronavirus-began-to-escalate-state-officials-ordered-nursing-homes-to-accept-covid-patients/

549 Fonrouge, G., et al, 2020, Coronavirus patients admitted to Queens nursing home – with body bags, New York Post. https://nypost.com/2020/04/23/coronavirus-patients-admitted-to-queens-nursing-home-with-body-bags/

550 Reisman, N., 2020, State Ends Policy Allowing COVID-Positive Nursing Home Staffers to Work, Spectrum News. https://spectrumlocalnews.com/nys/central-ny/ny-state-of-politics/2020/04/29/new-york-allows-asymptomatic-nursing-home-staffers-to-work-with-covid-residents

551 CBS New York, 2020, Coronavirus Toll: 'Absolutely Horrifying' Surge Of 98 Dead At NYC Nursing Home, CBS New York. https://newyork.cbslocal.com/2020/05/02/coronavirus-toll-absolutely-horrifying-surge-of-98-dead-at-nyc-nursing-home/

552 The New York Times, 2020, Covid in the U.S.: Latest Map and Case Count, The New York Times. https://www.nytimes.com/interactive/2020/us/coronavirus-us-cases.html

553 Campanile, C., 2020, Cuomo's nursing home policies amid coronavirus 'a disaster,' says ex-Gov.Pataki, New York Post. https://nypost.com/2020/05/01/cuomos-coronavirus-nursing-home-policies-a-disaster-ex-gov-pataki/

554 New York attorney general accuses Gov. Cuomo of undercounting nursing home deaths by as much as 50% (msn.com)

555 Donahue, A., 2014, 'That silly little dress': the story behind Marilyn Monroe's iconic scene, The Guardian. https://www.theguardian.com/film/filmblog/2014/sep/15/marilyn-monroe-seven-year-itch-dress

556 MTA Service for the Coronavirus Pandemic. Updated January 25, 2021.

557 Guse, C., 2020, MTA workers dying from coronavirus at triple the rate of agencies that employ NYC first responders, Daily News. https://www.nydailynews.com/coronavirus/ny-corona-virus-mta-nyc-transit-workers-deaths-20200408-f37damz5tjdmnc4pahurf3cjay-story.html

558 COVID NYC Update: MTA launches living memorial to honor transit workers lost to coronavirus - ABC7 New York (abc7ny.com)

559 Loria, K., 2015, The Army tested 'germ warfare' on the NYC subway by smashing lightbulbs full of bacteria, Business Insider. https://www.businessinsider.com/biological-agents-were-tested-on-the-new-york-city-subway-2015-11

560 link to purchase Clouds of Secrecy: https://www.amazon.com/Clouds-Secrecy-Armys-War-fare-Populated/dp/082263001X?tag=TIsafetynet-20

561 Department of the Army, 1977, U.S. Army Activity in the U.S. Biological Warfare Programs. https://nsarchive2.gwu.edu//NSAEBB/NSAEBB58/RNCBW_USABWP.pdf

562 Daly, M., 1998, The Day Subway Got Dusted, Daily News. https://www.nydailynews.com/archives/news/day-subway-dusted-article-1.794791

563 Vaughan, A., 2020, Are you more likely to die of covid-19 if you live in a polluted area?, New Scientist. https://www.newscientist.com/article/2241778-are-you-more-likely-to-die-of-covid-19-if-you-live-in-a-polluted-area/

564 Rice, A., 2020, How far does coronavirus travel in the air? Preliminary study finds virus on tiny particles of air pollution, USA Today. https://www.usatoday.com/story/news/health/2020/04/27/coronavirus-found-air-pollution-particles-preliminary-study-finds/3033646001/

565 Tamman, M., 2020, Special Report: A night on the New York subway – Homeless find shelter underground during pandemic, Reuters. https://www.reuters.com/article/us-health-coronavirus-new-york-homeless/special-report-a-night-on-the-new-york-subway-homeless-find-shelter-underground-during-pandemic-idUSKCN2241IJ

566 Moore, T. and Meyer, D., 2020, NYPD deploying 1K cops to remove homeless from subways. New York Post. https://nypost.com/2020/05/05/over-one-thousand-nypd-cops-set-to-re-move-subway-homeless/

567 Associated Press, 2020, "Surreal": New York Funeral Homes Struggle as COVID-19 Deaths Surge (Video), The National Herald. https://www.thenationalherald.com/archive_corona-virus/arthro/surreal_new_york_funeral_homes_struggle_as_covid_19_deaths_surge_vid-eo-266298/

568 Winsor, M., Shapiro, E., 2020, US coronavirus death toll surpasses 60,000 and 100 bodies found in trucks outside NYC funeral home, ABC News. https://abcnews.go.com/Health/coronavirus-updates-us-federal-inmate-dies-covid-19/story?id=70399771

569 Wallace, D., Givas, N., 2020, New York state rescinds DNR order for cardiac patients amid coronavirus crisis, Fox News. https://www.foxnews.com/us/new-york-state-instructs-para-medics-not-to-revive-cardiac-patients-amid-coronavirus-crisis-report

570 Post Editorial Board, 2020, Daniel Vargas is whom NOT to release from prison amid coro-navirus crisis, New York Post. https://nypost.com/2020/04/12/daniel-vargas-is-who-not-to-release-from-prison-amid-coronavirus/

571 Binder, J., 2020, Murder, Burglary Soars in New York City During Coronavirus Lockdown, Breitbart. https://www.breitbart.com/politics/2020/04/26/murder-burglary-soars-new-york-city-during-coronavirus-lockdown/

572 Eder, S., et al., 2020, 430,000 People Have Traveled From China to U.S. Since Coronavirus Surfaced, The New York Times. https://www.nytimes.com/2020/04/04/us/coronavirus-chi-na-travel-restrictions.html

573 Murphy, H., 2020, All I Wanted Was to Be Interrogated, The New York Times. https://www.nytimes.com/2020/04/04/us/airline-travel-coronavirus.html

574 Steinbach, Y., 2020, Airline passenger describes packed flight to NYC surrounded by people not wearing masks, New York Post. https://nypost.com/2020/04/23/airline-passenger-describes-flight-with-many-not-wearing-masks/

575 US travel restrictions state by state | CNN Travel

576 Barnello, J., WHAM staff, 2020, Contact tracing 'army' being put together in NY by Bloomberg, Johns Hopkins, 13 WHAM. https://13wham.com/news/local/watch-gov-cuomo-holds-daily-covid-19-briefing-from-albany-04-30-2020

577 Prestigiacomo, A., 2020, WHO Official: Lockdowns Driving Infections From Street To 'Household,' Suggests Removal Of Likely Infected Family Members To 'Dignified' Isolation, The Daily Wire. https://www.dailywire.com/news/who-official-lockdowns-driving-infections-from-street-to-household-suggests-removal-of-likely-infected-family-members-to-dignified-isolation

578 World Health Organization (WHO), 2020, Live from WHO Headquarters – coronavirus – COVID-19 daily press briefing 30 March 2020, YouTube. https://www.youtube.com/watch?time_continue=3004&v=2v3vlw14NbM&feature=emb_logo

579 New York Governor Under Fire For Data On Nursing Home Infections : NPR

580 https://www.washingtonpost.com/politics/cuomo-toxic-workplace/2021/03/06/7f7c5b9c-7dd3-11eb-b3d1-9e5aa3d5220c_story.html

581 https://www.msn.com/en-us/news/us/2-attorneys-named-to-lead-investigation-into-cuomo-sexual-harassment-allegations/ar-BB1enRpv

582 https://renzlaw.files.wordpress.com/2020/09/a01495_50958838b9ee4115b7800a12d3ee8652.pdf; https://breggin.com/alert-166-covid-19-lawsuit-update/ and https://spectator.org/dem-governors-face-dozens-of-lockdown-lawsuits/

583 https://www.who.int/medicines/ebola-treatment/TheCoalitionEpidemicPreparednessInnovations-an-overview.pdf

584 https://cepi.net/news_cepi/global-partnership-launched-to-prevent-epidemics-with-new-vaccines/

585 Improving coordination and fostering an enabling environment (who.int)

586 https://cepi.net/get_involved/support-cepi/

587 "The Salesforce Conversations with Peter Schwartz and Klaus Schwab. Join Peter Schwartz, SVP, Strategic Planning, Salesforce as he welcomes Klaus Schwab, World Economic Forum Executive Chairman and Founder, into the Salesforce LIVE Studio for a chat about the future of global governance." 2014. https://sfdc.hubs.vidyard.com/watch/lemzpqnyZA5yQfedOpoDTQ

588 CEPI conceived 2015 - Bing and https://wellcomeopenresearch.org/articles/5-284/v1

589 https://www.federalregister.gov/documents/2017/01/13/2017-00721/emergency-use-authorization-of-medical-products-and-related-authorities-guidance-for-industry-and

590 https://ywww.fda.gov/media/97321/download

591 CEPI officially launched – CEPI

592 https://cepi.net/about/whoweare/

593 https://www.who.int/director-general/speeches/detail/who-director-general-s-opening-remarks-at-148th-session-of-the-executive-board

594 https://www.who.int/medicines/ebola-treatment/TheCoalitionEpidemicPreparednessInnovations-an-overview.pdf

595 Vaccine Testing and Approval Process | CDC

596 CEPI: preparing for the worst - The Lancet Infectious Diseases

597 https://www.who.int/publications/m/item/a-coordinated-global-research-roadmap

598 P. 8.
599 WHO (undated). Improving coordination and fostering an enabling environment. Improving coordination and fostering an enabling environment (who.int)
600 Improving coordination and fostering an enabling environment (who.int)
601 https://www.britannica.com/topic/transnationalism
602 https://www.youtube.com/watch?v=6Af6b_wyiwI
603 Newshttps://duckduckgo.com/?q=Bill+Gates+at+Davos+2017&iax=videos&ia=videos&pn=1&iai=https%3A%2F%2Fwww.youtube.com%2Fwatch%3Fv%3Dldn1bzTOXPQ
604 https://CEPI.net/
605 Bill Gates warned of a deadly pandemic for years — and said we wouldn't be ready to handle it - CBS
606 https://CEPI.net/
607 https://www.cbsnews.com/news/coronavirus-bill-gates-epidemic-warning-readiness/
608 https://www.cbsnews.com/news/coronavirus-bill-gates-epidemic-warning-readiness/
609 https://www.cbsnews.com/news/davos-world-economic-forum-bill-gates-outsmart-global-epidemics-cepi-coalition-for-epidemic-preparedness/
610 Federal Register :: Emergency Use Authorization of Medical Products and Related Authorities; Guidance for Industry and Other Stakeholders; Availability
611 Bill Gates says Covid-19 vaccine tech should not be shared with India, now there is a vaccine shortage - Technology News (indiatoday.in); COVID-19 vaccine formulas shouldn't be shared with India: Bill Gates (yahoo.com)
612 Gates promotes Johnson & Johnson's vaccine in media interviews but the degree of his investment or control remains unclear to us.
613 Joe Biden's last major speech as Vice President in full | World Economic Forum (weforum.org)
614 https://www.youtube.com/watch?v=3DNCamgK-GQ
615 President Xi's speech to Davos in full | World Economic Forum (weforum.org)
616 Jack Ma: America has wasted its wealth | World Economic Forum (weforum.org)
617 Alibaba's market cap as of April 2021 - Bing
618 Fauci, A., 2017, January 10, Speech on Pandemic Preparedness in Washington, DC, at Georgetown University. Hosted by The Center for Global Health Science and Security at Georgetown University Medical Center, the Harvard Global Health Institute, and Health Affairs. The following link is to a video of the introduction and the complete speech and discussion, almost 60 minutes. https://ghss.georgetown.edu/pandemicprep2017/
619 https://www.washingtonpost.com/wp-dyn/articles/A2401-2005Jan11.html
620 Fauci talks death threats, gay bathhouses with Terry Gross - WHYY
621 https://aep.lib.rochester.edu/node/49111
622 https://www.washingtonpost.com/history/2020/05/20/fauci-aids-nih-coronavirus/
623 Farber, Celia, 2015, October 5, AIDS and the AZT Scandal: SPIN's 1989 Feature, 'Sins of Omission.' The story of AZT, one of the most toxic, expensive, and controversial drugs in the history of medicine. Spin. https://www.spin.com/featured/aids-and-the-azt-scandal-spin-1989-feature-sins-of-omission/
624 https://www.independent.co.uk/arts-entertainment/rise-and-fall-azt-it-was-drug-had-work-it-brought-hope-people-hiv-and-aids-and-millions-company-developed-it-it-had-work-there-was-nothing-else-many-who-used-azt-it-didn-t-2320491.html
625 https://www.spin.com/featured/aids-and-the-azt-scandal-spin-1989-feature-sins-of-omission/
626 https://www.thebody.com/article/pneumonias-hiv#hiv
627 https://www.thebody.com/article/pneumonias-hiv#hiv Engel,(2006). The Epidemic:A Global History of AIDS. New York, NY: SmithsonianBooks/HarperCollins. Cited in

628   https://www.nejm.org/doi/full/10.1056/NEJM199212243272604

629   https://www.huffpost.com/entry/whitewashing-aids-history_b_4762295

630   https://aapsonline.org/americans-need-covid-treatment-now-states-aaps/

631   The long road to PCP prophylaxis in AIDS. An early history. - POZ

632   Schachtel, Jordan, 2021, February 19, Flashback: Fauci describes Ebola quarantines as 'draconian,' warns about 'unintended consequences,' From the Dossier. Flashback: Fauci describes Ebol. quarantines as 'draconian,' warns of 'unintended consequences' - The Dossier (substack. com); for additional coverage, also see Several US states break with CDC and enact stricter Ebola guidelines | Ebola | The Guardian and Fauci warned of 'unintended consequences' of 'draconian' quarantines during 2014 Ebola outbreak | Fox News and 5 Quotes From Sunday's Talk Shows - WSJ

633   Fauci, A. (2017, February 9). What Three Decades Of Pandemic Threats Can Teach Us About The Future, Health Affairs. https://www.healthaffairs.org/do/10.1377/hblog20170209.058678/full/

634   Vliet, Elizabeth Lee, 2020, December 17, Senate Hearings Offer Holiday Hope for Early COVID Treatment, Association of American Physicians and Surgeons (AAPS). https://aapsonline.org/senate-hearings-offer-holiday-hope-for-early-covid-treatment/ Also see https://www.c-span.org/video/?478159-1/senate-hearing-covid-19-outpatient-treatment and https://www.cnsnews.com/article/national/susan-jones/medical-expert-failure-treat-covid-patients-early-home-creating

635   https://www.hsgac.senate.gov/imo/media/doc/Testimony-Jha-2020-11-19.pdf

636   https://www.brown.edu/news/2020-02-26/jha

637   https://sites.sph.harvard.edu/china-health-partnership/2020/12/18/hchp-newsletter-1/; https://sites.sph.harvard.edu/china-health-partnership/2020/12/18/hchp-newsletter-1/

638   https://sites.sph.harvard.edu/china-health-partnership/about-us-and-contact/

639   Jha, A., 2020, January 10, Moderator introducing "Panel I-domestic" at 9:30 am on domestic issues in pandemic preparation, Georgetown University, Center for Global Health Sciences and Security. His presentation begins at 6:14 minutes of the panel. Pandemic Preparedness in the Next Administration | Center for Global Health Science and Security | Georgetown University

640   spars-pandemic-scenario.pdf (jhsphcenterforhealthsecurity.s3.amazonaws.com)

641   Johns Hopkins Center for Health Security, 2020, Event 201: A Global Pandemic Exercise. https://centerforhealthsecurity.org/event201/

642   Event 201: A Global Pandemic Response, Johns Hopkins Bloomberg School of Public Health. Undated. https://www.scmp.com/news/china/society/article/3074991/coronavirus-chinas-first-confirmed-covid-19-case-traced-back

643   Chapman, G. 2015, May 18, Bill Gates Calls for "Germ Games" Instead of War Games, Business Insider. https://www.businessinsider.com/afp-bill-gates-calls-for-germ-games-instead-of-war-games-2015-3?op=1

644   Event 201: A Global Pandemic Response, Johns Hopkins Bloomberg School of Public Health. Undated. https://www.scmp.com/news/china/society/article/3074991/coronavirus-chinas-first-confirmed-covid-19-case-traced-back

645   Johns Hopkins Center for Health Security, 2020, January 24 (dated from link), Statement about nCoV and our pandemic exercise. https://www.centerforhealthsecurity.org/news/center-news/2020-01-24-Statement-of-Clarification-Event201.html

646   Ma, J. Pearce, 2020, March, 13, South China Morning Post, Exclusive | Coronavirus: China's first confirmed Covid-19 case traced back to November 17, Coronavirus: China's first confirmed Covid-19 case traced back to November 17 | South China Morning Post (scmp. com)

647   https://www.centerforhealthsecurity.org/event201/

648   Brunson, E. K., Chandler, H., Gronvall, G. K., Ravi, S., Sell, T. K., Shearer, M. P., & Schoch-Spana, M. (2020). The SPARS pandemic 2025–2028: A futuristic scenario to facilitate medical countermeasure communication. Journal of International Crisis and Risk Communication Research, 3(1), 71–102. https://doi.org/10.30658/jicrcr.3.1.4 https://stars.library.ucf.edu/jicrcr/vol3/iss1/4/

649   Norbert Pardi, Michael J. Hogan, Frederick W. Porter and Drew Weissman. (2018, April). mRNA vaccines — a new era in vaccinology. Nature Reviews, 17, 261-274. Online January 2018. https://www.nature.com/articles/nrd.2017.243.pdf?origin=ppub This is a detailed review with many citations to earlier research.

650   https://content.iospress.com/articles/international-journal-of-risk-and-safety-in-medicine/jrs574; https://content.iospress.com/articles/international-journal-of-risk-and-safety-in-medicine/jrs574 and https://www.ncbi.nlm.nih.gov/pmc/articles/PMC2872821/

651   NIH, 2020, NIH Clinical trial of remdesivir to treat COVID-19 begins, National Institutes of Health (NIH). https://www.nih.gov/news-events/news-releases/nih-clinical-trial-remdesivir-treat-covid-19-begins

652   AAPS, 2020, Where's the Evidence on COVID-19 Treatment?, Association of American Physicians and Surgeons (AAPS). https://aapsonline.org/evidence-hydroxychloroquine/

653   Orient, J., 2020, Association of American Physicians and Surgeons (AAPS) Suggests Different Approach to COVID-19 Surge, GlobeNewswire. https://www.globenewswire.com/news-release/2020/07/13/2061509/0/en/Association-of-American-Physicians-and-Surgeons-AAPS-Suggests-Different-Approach-to-COVID-19-Surge.html

654   AAPS, 2020, How many COVID deaths are preventable?, Association of American Physicians and Surgeons. https://aapsonline.org/how-many-covid-deaths-are-preventable/

655   Zelenko, V., et al, 2020, Medical Studies Support MDs Prescribing Hydroxychloroquine for Early Stage COVID-19 and for Prophylaxis. a link to the eBook: https://files.internetprotocol.co/ebook-covid-19.pdf

656   Breggin, P., Breggin, G., 2020, America: Time to Save the World Again!, Psychiatric Drug Facts. https://breggin.com/america-time-to-save-the-world-again/

657   Sharp, R., Gordon, J., 2020, 'It did no good and in some extreme cases killed them': Fox News host Neil Cavuto renews attack on President Donald Trump for recommending hydroxychloroquine, Daily Mail. https://www.dailymail.co.uk/news/article-8337773/Fox-News-host-Neil-Cavuto-renews-attack-Donald-President-Trump-recommending-hydroxychloroquine.html

658   Breggin, P., Breggin, G., 2020, Study Attacking "President Trumps Miracle Drug" hydroxychloroquine Actually Shows It Works, Psychiatric Drug Facts. https://breggin.com/negative-study-of-President Trump-miracle-drug-actually-shows-it-works/

659   Strauss, D., Eisenbaum, J., 2020, FULL INTERVIEW: Houston-area doctor in viral video touting hydroxychloroquine as virus cure doubles down on claims, Click 2 Houston. https://www.click2houston.com/news/local/2020/07/31/full-interview-houston-doctor-in-viral-video-touting-hydroxychloroquine-as-virus-cure-doubles-down-on-claims/

660   Brennan, Z., 2020, Remdesivir helps coronavirus patients – but at what cost?, Politico. https://www.politico.com/news/2020/05/06/remdesivir-helps-coronavirus-patients-but-at-what-cost-240230

661   Breggin, P., webpage relating to the book Brain Disabling Treatments in Psychiatry, Psychiatric Drug Facts. https://breggin.com/brain-disabling-treatments-in-psychiatry/

662   Breggin, P., Dr. Peter Breggin's Antidepressant Drug Resource & Information Center, Psychiatric Drug Facts. https://breggin.com/antidepressant-drugs-resource-center/

663   Eli Lilly document PDF:

664   Eli Lilly document PDF: http://breggin.com/wp-content/uploads/1997/01/elilillydocspart4.pdf

665 Breggin, P., 2009, Medication Madness: How Psychiatric Drugs Cause Violence, Suicide, and Crime, Psychiatric Drug Facts. https://breggin.com/medication-madness-how-psychiatric-drugs-cause-violence-suicide-and-crime/

666 Kirsch, I., 2014, Antidepressants and the Placebo Effect. http://breggin.com/antidepressant-drugs-resources/Kirsch2014.pdf

667 My first book about psychiatric drugs and industry corruption was a medical book Psychiatric Drugs: Hazards to the Brain (New York: Springer Publishing Company, 1983—38 years ago.

668 Breggin, P. DECLARATION OF PETER R. BREGGIN, MD. SUPERIOR COURT OF THE STATE OF CALIFORNIA COUNTY OF SANTA CLARA. Case No.: CV 773623. Lacuzong v. Davidson. July 21, 2001. http://breggin.com/wp-content/uploads/2004/01/Breggin%20Paxil%20Lacuzong%20Report%20Filed%20with%20Court.pdf

669 Breggin, P. Court filing makes public my previously suppressed analysis of Paxil's effects. Ethical Human Psychology and Psychiatry, 8, 77-84, 2006. 57-58 http://breggin.com/wp-content/uploads/2003/01/courtfiling.pbreggin.2006.pdf; Breggin, Peter. Drug company suppressed data on paroxetine-induced stimulation: Implications for violence and suicide. Ethical Human Psychology and Psychiatry, 8, 255-263 http://breggin.com/wp-content/uploads/2001/01/Peter%20Breggins%20Paxil%20Special%20Report%20III.pdf ; Breggin, Peter. How GlaxoSmithKline suppressed data on Paxil-induced akathisia: Implications for suicide and violence. Ethical Human Psychology and Psychiatry, 8, 91-100, 2006. http://breggin.com/wp-content/uploads/2002/01/Peter%20Breggins%20Paxil%20special%20report%20II.pdf

670 Yaqub, F. Reforming Big Pharma. Lancet, V. 383, Issue 9915, p. 402, February 1, 2014. https://www.thelancet.com/journals/lancet/article/PIIS0140-6736(14)60139-2/fulltext

671 This summary is taken from the Penguin Random House offering of the book. https://www.penguinrandomhouse.com/books/3901/the-truth-about-the-drug-companies-by-marcia-angell-md/

672 The Lancet, 2020, Reviving the US CDC, The Lancet. https://www.thelancet.com/journals/lancet/article/PIIS0140-6736(20)31140-5/fulltext

673 Breggin, P., Psychosurgery for political purposes (1975). webpage containing PDF: http://breggin.com/psychosurgery-for-political-purposes-1975/.

674 Breggin, P. (2008). Brain-Disabling Treatments in Psychiatry: Drugs, Electroshock and the Psychopharmaceutical Complex, Second Edition. NY: Springer; Breggin, P. (2008). Medication Madness: The Role of Psychiatric Drugs in Cases of Violence, Suicide, and Crime. NY: St. Martin's Press.

675 https://www.nytimes.com/2008/11/25/health/25psych.html

676 https://www.medpagetoday.com/psychiatry/bipolardisorder/11915

677 Breggin, P., Breggin, G., 2020, Negative Study of "President Trump Miracle Drug" Actually Shows It Works, Psychiatric Drug Facts. https://breggin.com/negative-study-of-President Trump-miracle-drug-actually-shows-it-works/

678 Elfrink, T., 2020, WHO pauses trial of hydroxychloroquine, once touted by President Trump as 'game changer,' over safety concerns, The Washington Post. http://breggin.com/coronavirus/HCQ/HCQ-WHO-stops-arm-of-study-WAPO.pdf

679 https://www.webmd.com/lung/news/20200605/lancet-retracts-hydroxychloroquine-study; https://www.nbcnews.com/health/health-news/lancet-retracts-large-study-hydroxychloroquine-n1225091; https://www.theguardian.com/world/2020/jun/04/covid-19-lancet-retracts-paper-that-halted-hydroxychloroquine-trials

680 Sachs, J., Horton, R., Bagenal, J., Amor, Y., Ozge, C., LaFortune, G. 2020, July 9. The Lancet COVID-19 Commission, The Lancet. https://www.thelancet.com/journals/lancet/article/PIIS0140-6736%2820%2931494-X/fulltext?rss=yes

681 The Lancet COVID-19 Commission, 2020, November 23, Members of the Lancet COVID Commission Task Force on the Origins of SARS-CoV-2 Named, The Lancet. https://covid-19commission.org/media-blog/members-of-the-lancet-covid-commission-task-force-on-the-origins-of-sars-cov-2-named

682 Breggin, G. and Breggin, P. 2020, November 25, Lancet COVID-19 Commission"—A Mortal Blow to Science.www.breggin.com. The"Lancet COVID-19 Commission"—A Mortal Blow to Science | Psychiatric Drug Facts (breggin.com)

683 https://articles.mercola.com/sites/articles/archive/2020/12/16/covid-conspiracy.aspx

684 https://usrtk.org/biohazards-blog/ecohealth-alliance-orchestrated-key-scientists-statement-on-natural-origin-of-sars-cov-2/

685 https://usrtk.org/biohazards-blog/ecohealth-alliance-orchestrated-key-scientists-statement-on-natural-origin-of-sars-cov-2/

686 https://www.scmp.com/news/china/science/article/3111314/who-names-line-international-al-team-looking-coronavirus-origins

687 Transforming our world: thec2030 Agenda for Sustainable Development, United Nations. https://sustainabledevelopment.un.org/content/documents/21252030%20Agenda%20for%20Sustainable%20Development%20web.pdf

688 p. 107, Schwab, COVID-19: The Great Reset

689 The Davos Agenda | World Economic Forum (weforum.org)

690 Full text: Xi Jinping's speech at the virtual Davos Agenda event - CGTN

691 https://www.msn.com/en-xl/news/other/china-leads-world-in-terms-of-fighting-pandemic-klaus-schwab/ar-BB1bHl3s

692 https://foreignpolicy.com/2018/09/10/google-is-handing-the-future-of-the-internet-to-china/ and https://www.cjr.org/the_new_gatekeepers/china-facebook-google.php

693 Schwab, K. 2019, December 2, What Kind of Capitalism Do We Want?, Time. https://time.com/5742066/klaus-schwab-stakeholder-capitalism-davos/

694 Davos Manifesto

695 TIME, 2020, undated on-line issue, circa late October, The Great Reset. The Great Reset: How to Build a Better World Post-COVID-19 | TIME

696 Schwab, K., 2020, June 3, Now is the time for a 'great reset,' World Economic Forum. https://www.weforum.org/agenda/2020/06/now-is-the-time-for-a-great-reset/

697 Di Caro, B., 2020, November 17, The Great Reset: Building Future Resilience to Global Risks, World Economic Forum. https://www.weforum.org/agenda/2020/11/the-great-reset-building-future-resilience-to-global-risks/

698 The Great Reset Initiative, circa 2020, September, World Economic Forum. https://www.weforum.org/great-reset

699 https://www.bloomberg.com/news/newsletters/2021-02-06/bloomberg-new-economy-u-s-insurrection-clears-the-way-for-coups

700 Smith, W. 2020, November 24, Beware the "Great Rest" New World Order, The Epoch Times. https://www.theepochtimes.com/beware-the-great-reset-new-world-order_3584620.html

701 https://www.breitbart.com/tag/vatican-china-deal/

702 YouTube presentation by K. Schwab on transhumanism, https://www.youtube.com/watch?v=3VkjA2ag-0M.

703 Breggin, P. (1975). The Politics of Psychosurgery. Duquesne Law Review 13, 841-861. Psychosurgery for political purposes (1975) | Psychiatric Drug Facts (breggin.com). Also see, Dr. Breggin's Psychosurgery Page at PSYCHOSURGERY PAGE | Psychiatric Drug Facts (breggin.com)

704 https://breggin.com/ect-resources-center/deep-brain-stimulation-dbs/

705   https://breggin.com/psychosurgery-page/
706   https://breggin.com/alert-85-the-film-on-mind-control-the-minds-of-men-passes-1-3-million-viewers-on-youtube/
707   http://truthstreammedia.com/2019/10/25/breggin-the-most-important-video-we-have-ever-watched/ http://truthstreammedia.com/2019/10/25/breggin-the-most-important-video-we-have-ever-watched/
708   https://www.theepochtimes.com/world-economic-forum-retracts-statement-suggesting-lockdowns-improved-cities-worldwide_3714613.html; https://www.breitbart.com/europe/2021/02/27/great-reset-wef-hails-quieter-lockdown-cities-as-businesses-collapse/
709   Shafarevich, Igor. (2019). The Socialist Phenomenon. With a Forward by Aleksandr Solzhenitsyn. Gideon House. www.gideonhouse.com.
710   https://www.wealthypersons.com/john-kerry-net-worth-2020-2021/
711   https://www.law.cornell.edu/uscode/text/18/953
712   https://pjmedia.com/news-and-politics/tyler-o-neil/2018/03/21/as-biden-and-kerry-went-soft-on-china-sons-made-nuclear-military-business-deals-with-chinese-govt-n56848,
713   Kerry, John, transcript of 1971 testimony before Congress, NPR. https://www.npr.org/templates/story/story.php?storyId=3875422
714   https://www.foxbusiness.com/politics/obama-biden-cronies-made-billions-off-china-trade-deals-and-regulatory-policies-report; https://pjmedia.com/news-and-politics/tyler-o-neil/2018/03/21/as-biden-and-kerry-went-soft-on-china-sons-made-nuclear-military-business-deals-with-chinese-govt-n56848; https://canadafreepress.com/article/joe-bidens-and-john-kerrys-sons-made-nuclear-and-military-deals-with-china; https://www.breitbart.com/politics/2018/03/15/secret-empire-joe-biden-john-kerry-billion-dollar-deal-chinese-government/; https://nypost.com/2018/03/15/inside-the-shady-private-equity-firm-run-by-kerry-and-bidens-kids/
715   https://nypost.com/2018/03/15/inside-the-shady-private-equity-firm-run-by-kerry-and-bidens-kids/
716   https://www.washingtonpost.com/politics/2020/12/04/energy-202-biden-calls-climate-change-an-emergency-now-he-under-pressure-officially-declare-it-one/
717   https://www.washingtonexaminer.com/news/john-kerry-on-climate-change-weve-got-to-treat-this-like-a-war
718   https://www.vanityfair.com/style/2020/05/prince-charles-pandemic-great-reset
719   The World Economic Forum COVID Action Platform, 2020, June 24, 'Normal wasn't working' - John Kerry, Phillip Atiba Goff and others on the new social contract post-COVID, World Economic Forum. https://www.weforum.org/agenda/2020/06/great-reset-social-contract-john-kerry-phillip-goff/
720   https://www.msn.com/en-us/news/politics/biden-says-our-darkest-days-in-battling-covid-19-are-ahead-of-us/ar-BB1c9dtf?rt=0&referrerID=InAppShare+%22+target&c=13444059561558240975&mkt=en-us
721   https://www.theguardian.com/uk-news/2020/jun/03/pandemic-is-chance-to-reset-global-economy-says-prince-charles
722   https://www.ecohealthalliance.org/2020/04/regarding-nih-termination-of-coronavirus-research-funding
723   Biden Announces John Kerry as Climate Envoy, Washington Post. https://news.yahoo.com/biden-announces-john-kerry-climate-205601020.html; also see https://edition.cnn.com/2020/11/23/politics/john-kerry-biden-climate-envoy/index.html
724   Kerry, John. 2020, November 24, Video of speech at World Economic Forum COVID Action Platform committee. https://www.youtube.com/watch?v=eeQsGVfbD6E&feature=emb_imp_woyt Also see, https://www.weforum.org/agenda/2020/11/the-great-reset-building-future-resilience-to-global-risks

725　The Heartland Institute, 2020, November 30, John Kerry: 'Great Reset' Will Happen, Red-State. https://redstate.com/heartlandinstitute/2020/11/30/john-kerry-great-reset-will-happen-n286949 This section is modified from RedState. We have listened to Kerry's speech and checked the quotes.

726　Gates, Bill, 2020, February 10, Innovating to Zero, TED Talk, https://www.youtube.com/watch?v=JaF-fq2Zn7I. The anxious laughter and comments implying we will need a massive reduction in human life occur around 5 minutes into the video.

727　Gates, B. (2012, February 18). Questions On Population Growth. A Conversation with Bill Gates: Population Growth. On www.gatesnotes.com. Direct to video @: A Conversation with Bill Gates: Population Growth | Bill Gates (gatesnotes.com)

728　https://www.tribuneindia.com/news/world/declare-climate-emergency-un-chief-guterres-urges-world-leaders-183725

729　https://www.theguardian.com/environment/2020/dec/12/un-secretary-general-all-countries-declare-climate-emergencies-antonio-guterres-climate-ambition-summit

730　https://www.sandiegouniontribune.com/communities/north-county/encinitas/story/2020-12-21/global-climate-condition-gains-emergency-status-in-encinitas

731　https://www.bloomberg.com/news/articles/2020-12-12/global-leaders-meet-to-discuss-cutting-emissions-climate-update

732　https://weather.com/news/climate/news/ice-age-climate-change-earth-glacial-interglacial-period

733　Crew, B., 2015, July 13. Professor Claims That a 'Mini Ice Age' Is Coming in The Next 15 Years, ScienceAlert. https://www.sciencealert.com/a-mini-ice-age-is-coming-in-the-next-15-years

734　Breggin, P. (1992). Beyond Conflict: From Self-Help and Psychotherapy to Peacemaking. St. Martin's Press: New York. The discussion of the perpetrator syndrome begins on p. 89.

735　https://www.insidermonkey.com/blog/10-biggest-companies-that-do-not-support-trump-572558/

736　https://www.forbes.com/sites/jonathanponciano/2021/02/10/twitter-market-value-surges-to-50-billion-despite-trump-ban-and-shares-could-run-another-25/?sh=76bb804757f5

737　https://companiesmarketcap.com/

738　https://www.forbes.com/sites/hayleycuccinello/2020/04/07/the-20-richest-american-billionaires-2020/?sh=5e1cf65a1aeb

739　https://www.foxnews.com/opinion/warren-buffett-has-given-1-2-billion-to-abortion-groups; https://www.lifenews.com/2020/04/10/warren-buffet-has-spent-4-billion-funding-the-abortion-industry-enough-to-kill-8-million-babies-in-abortions/; https://www.lifenews.com/2014/03/18/warren-buffet-foundation-hires-abortion-activist-to-spend-millions-on-population-control/;

740　https://www.zdnet.com/article/oracle-said-to-close-chinas-r-d-center-axing-nearly-1000-employees/

741　Is Larry Ellison a Republican? The Oracle CEO Just Acquired TikTok (distractify.com)

742　https://www.youtube.com/watch?v=rxFouplAFDY

743　https://www.youtube.com/watch?v=aM52EOtmTwY

744　https://www.forbes.com/sites/walterloeb/2020/10/07/walmarts-political-dichotomy-is-no-surprise/?sh=142df6f72981

745　https://www.politico.com/news/2021/01/12/cato-fellow-election-conspiracy-theories-458248

746　https://www.rollingstone.com/politics/politics-news/nike-phil-knight-republican-donor-746122/

747　Ross, C. 2020, November 23, The Wall Street Journal, Washington Post Among Newspapers Paid Millions by Beijing-Controlled News Outlet to Publish Propaganda this Year, The Tennessee Star. https://tennesseestar.com/2020/11/23/the-wall-street-journal-washington-post-among-newspapers-paid-millions-by-beijing-controlled-news-outlet-to-publish-propaganda-this-year/

748 Fitton, T. (2020). A Republic Under Assault: The Left's Ongoing Attack on American Freedom. New York: Threshold Editions. There are many good examinations of American media economic ties to the Chinese Communist run nation, including Clark, C., 2020, May 4, A Rundown Of Major U.S. Corporate Media's Business Ties To China, The Federalist. https://thefederalist.com/2020/05/04/has-china-compromised-every-major-mainstream-media-entity/

749 Foundation Fact Sheet, Bill & Melinda Gates Foundation. https://www.gatesfoundation.org/Who-We-Are/General-Information/Foundation-Factsheet

750 Center News, dated from link 2020, January 17, The Johns Hopkins Center for Health Security, World Economic Forum, and Bill & Melinda Gates Foundation Call for Public-Private Cooperation for Pandemic Preparedness and Response. Johns Hopkins Center for Health Security. https://www.centerforhealthsecurity.org/news/center-news/2020-01-17-Event201-recommendations.html

751 https://www.forbes.com/sites/moneyshow/2020/12/15/buffett-and-gates-the-foundation-of-a-blue-chip-portfolio/?sh=129312f02681

752 Speights, K., 2020, February 19, Warren Buffett's Big Biotech Bet Is 1 of His Riskiest Wagers Ever, Motley Fool. https://www.fool.com/investing/2020/02/19/warren-buffetts-big-biotech-bet-1-of-his-riskiest-.aspx

753 Schwab, T. (2020, March 17). Bill Gates' Charity Paradox. A Nation investigation illustrates the moral hazards surrounding the Gates Foundation's $50 billion charitable enterprise. The Nation. Retrieved June 16, 2020 at https://www.thenation.com/article/society/bill-gates-foundation-philanthropy/

754 https://www.usatoday.com/story/news/factcheck/2020/06/11/fact-check-bill-gates-has-given-over-50-billion-charitable-causes/3169864001/

755 https://www.forbes.com/profile/bill-gates/?sh=6ae5a6f6689f

756 https://www.independent.co.uk/news/world/politics/gates-foundation-accused-dangerously-skewing-aid-priorities-promoting-big-business-a6822036.html

757 https://www.globaljustice.org.uk/resources/gated-development-gates-foundation-always-force-good https://www.globaljustice.org.uk/sites/default/files/files/resources/gjn_gates_report_june_2016_web_final_version_2.pdf

758 For example, https://www.ibtimes.com/colombian-oil-money-flowed-clintons-state-department-took-no-action-prevent-labor-1874464 and Scandal Without End: Is The Clinton Foundation A Fraud? | Investor's Business Daily

759 https://www.clintonfoundation.org/contributors

760 the clinton global initiative fraud scam - Bing video

761 https://www.npr.org/templates/story/story.php?storyId=98467642

762 The Epoch Times has one of the first investigations of the fraud. https://www.theepochtimes.com/2020-election-investigation-who-is-stealing-america_3617562.html?utm_source=newsnoe&utm_medium=email&utm_campaign=breaking-2020-12-15-3

763 https://www.clintonfoundation.org/clinton-global-initiative/commitments/delian-project-democracy-through-technology

764 https://www.washingtontimes.com/news/2020/jan/25/george-soros-89-still-quest-destroy-america/

765 The video of the Soros October 2009 interview was found at https://duckduckgo.com/?q=Soros+financial+times+interview&iax=videos&ia=videos&iai=https%3A%2F%2Fwww.youtube.com%2Fwatch%3Fv%3DYlG_zYgG05o and at https://www.youtube.com/watch?v=TOjckJWqb0A. The official transcript can be found at https://www.ft.com/content/6e2d-

fb82-c018-11de-aed2-00144feab49a. Backups are in our "COVID-19 and the Global Predator Resources" at https://breggin.com/covid-19-and-the-global-predators-resources/

766 https://www.forbes.com/sites/daviddawkins/2021/07/19/george-soros-and-bill-gates-backed-consortium-to-buy-uk-maker-of-covid-lateral-flow-tests-for-41-million/ and https://www.worldtribune.com/big-deal-gates-soros-group-acquires-for-profit-covid-testing-firm/

767 The quote begins at 1:55 of the video. https://www.bing.com/videos/search?q=Chris+Hedges+and+George+Soros&docid=607990790830492646&mid=7FE96CBD92B0BA64C-BA97FE96CBD92B0BA64CBA9&view=detail&FORM=VIRE

768 https://www.investopedia.com/articles/investing/090815/3-best-investments-george-soros-ever-made.asp

769 https://www.washingtontimes.com/news/2020/aug/20/george-soros-funded-das-oversee-big-cities-skyrock/; https://theduran.com/george-soros-revealed-to-have-funded-das-who-now-oppose-police-in-american-cities/; https://pjmedia.com/news-and-politics/jeff-reynolds/2020/06/03/yes-george-soros-sent-money-to-fund-the-riots-and-so-did-taxpayers-n487939; https://www.thegatewaypundit.com/2020/07/exclusive-open-society-foundation-funded-george-soros-supported-groups-individuals-behind-us-riots-since-2014/; https://www.thegatewaypundit.com/2020/07/exclusive-open-society-foundation-funded-george-soros-supported-groups-individuals-behind-us-riots-since-2014/; https://www.msn.com/en-us/news/crime/newly-elected-soros-funded-los-angeles-da-announces-sweeping-reforms-on-first-day-in-office-including-end-to-cash-bail/ar-BB1bJNV4

770 https://www.washingtonexaminer.com/news/george-soros-contributes-record-50m-to-back-2020-democratic-efforts

771 Brown becomes president of Open Society Forum - Bing

772 Gates, Bill, 2020, February 10, Innovating to Zero, TED Talk, https://www.youtube.com/watch?v=JaF-fq2Zn7I.

773 Durden, T., 2020, October 24, Blockbuster Report Reveals How Biden Family Was Compromised By China, ZeroHedge, https://www.zerohedge.com/geopolitical/blockbuster-report-reveals-how-biden-family-was-compromised-china. The original report, https://www.baldingsworld.com/wp-content/uploads/2020/10/KVBJHB.pdf, which is extremely detailed about the Biden's intimate connections with China, was published by Typhoon Investigations which describes itself as "an activist research firm protecting companies from foreign influence and threats. © Droits d'auteur 2020 Typhoon Investigations." It has inevitably come under attack, but the information has not to our knowledge been discredited and there are now multiple confirmatory resources.

774 Secret Empires

775 Schweizer, P. and Bruner, S. 2020, October 24, Longstanding claims of Biden corruption all but confirmed with Hunter's emails, New York Post, https://nypost.com/2020/10/24/biden-corruption-claims-all-but-confirmed-with-hunter-emails/. First of several articles in the newspaper.

776 foreign business entanglements

777 Tony Bobulinski

778 Tucker Carlson Tonight, 2020, October 28, Tucker Carlson: What Tony Bobulinski told me and why it matters. Millions of dollars linked directly to the Communist Party of China went to Joe Biden's family https://www.foxnews.com/opinion/tucker-carlson-tony-bobulinski-joe-biden-hunter-biden-china

779 Rodgers, H., 2020, October 28, EXCLUSIVE: Senate Committee Successfully Verifies All Bobulinski Materials Reviewed To Date, Daily Caller. https://dailycaller.com/2020/10/28/senate-committee-successfully-verifies-bobulinski-materials-to-date/

780 Akan, E., 2020, November 14, Beijing, Wall Street Could Deepen Ties Under Potential Biden Presidency. Truly U.S.A. News. https://trulytimes.com/beijing-and-wall-street-could-deep-

en-ties-under-biden-presidency.html. Also appeared with the same title in the Epoch Times on November 11, 2020, p. A-11.

781   memorandum of understanding

782   Just the News

783   https://www.uscc.gov/research/chinese-companies-listed-major-us-stock-exchanges

784   Zillgitt, J. and Medina, M., 2019. October 17,, LeBron James' controversial comments "furthers his brand power in China, USA TODAY. https://www.usatoday.com/story/sports/nba/2019/10/17/lebron-james-nike-china-revenue/3989915002//

785   Manfred, T.,2013, January 16, Behind-The-Scenes Video Of LeBron James Having Dinner With Warren Buffett And Bill Gates In Las Vegas, Business Insider. https://www.businessinsider.com.au/lebron-james-dinner-with-warren-buffett-video-2013-1

786   Whitlock calls out LeBron James for selective outrage | Fox News

787   Kranish, M., 2020, January 1, Bloomberg's business in China has grown. That could create unprecedented entanglements if he is elected president, The Washington Post. Bloomberg?s business in China has grown. That could create unprecedented entanglements if he is elected president. - The Washington Post

788   https://www.cbsnews.com/news/jeff-bezos-climate-change-grant-first-round-announced/

789   Bezos Asked Bloomberg to Run for President (politicalwire.com)

790   Bloomberg New Economy Forum Announces New Global Co-Hosts and Preliminary Speaker Line-up for 2020 Virtual Event - Bloomberg

791   CCIEE, undated, Brief Introduction China Center for International Economic Exchanges http://images.mofcom.gov.cn/fi2/accessory/200906/1243930413765.pdf. Also see http://www.zhejianginvestment.com/Organizations/ShowArticle.asp?ArticleID=262

792   https://en.wikipedia.org/wiki/China_Center_for_International_Economic_Exchanges

793   https://www.neweconomyforum.com/2019-new-economy-forum-beijing/

794   https://www.bloomberg.com/news/live-blog/2019-11-04/bloomberg-s-new-economy-forum-in-beijing-day-one

795   https://www.bloomberg.com/news/live-blog/2019-11-04/bloomberg-s-new-economy-forum-in-beijing-day-two

796   https://www.usatoday.com/story/news/politics/elections/2019/11/24/michael-bloomberg-presidential-campaign-announcement/4290486002/

797   Trump Impeachment: Censure or Senate Trial Would Come Out the Same - Bloomberg

798   Bloomberg News, 2020, November 16, BlackRock's Fink Calls on Investors to Embrace ESG: NEF Update, yahoo!finance. https://finance.yahoo.com/news/blackrock-fink-eskom-ruyterset-001626898.html

799   Leap: $50M PROGRAM IN Human Organs, Physiology, and Engineering. Wellcome Leap: Unconventional Projects. Funded at Scale. Scroll down the page to find to find the descriptor.

800   Breggin, P. and Breggin, G. 2019, October 25, Dr. Breggin: The Most Important Video We Have Ever Watched!, Truthstream Media by Aaron and Melissa Dykes. https://truthstreammedia.com/2019/10/25/breggin-the-most-important-video-we-have-ever-watched/

801   Mok, A. 3 key takeaways from Bloomberg New Economy Forum - CGTN

802   Laura Ingraham. 2020, November 20, Ingram Angle, Fox News. https://www.foxnews.com/media/laura-ingraham-globalists-bloomberg-forum-world-health-organization-bill-gates

803   Creitz, C. Ingraham: Globalists are already planning to "subvert your rights and take your prosperity." https://www.foxnews.com/media/laura-ingraham-globalists-bloomberg-forum-world-health-organization-bill-gates

804   https://www.aier.org/article/we-hadnt-really-thought-through-the-economic-impacts-melinda-gates/

805   Bill Gates is trying to dim the sun (washingtonexaminer.com)

806   Planned Harvard balloon test stirs solar geoengineering unease (trust.org)

807   https://www.theepochtimes.com/mkt_morningbrief/bill-gates-funded-company-releases-genetically-modified-mosquitoes-in-us_3806432.html?utm_source=morningbrief-noe&utm_medium=email&utm_campaign=mb-2021-05-08&mktids=ecb6f3c427c9c6980d-9d0e22fae7a5ae&est=BHkLWDdBxK3aqlP43ZNUgcXuFKGfy17Sjn1eRJxvOoNMu-OKElCNtGZz7uREiCvM%3D; https://www.cdc.gov/mosquitoes/mosquito-control/community/sit/genetically-modified-mosquitoes.html.

808   https://www.mosquitomagnet.com/articles/gmo-mosquitoes-pros-cons

809   https://www.regulations.gov/document/EPA-HQ-OPP-2019-0274-0359

810   https://landreport.com/2021/01/bill-gates-americas-top-farmland-owner/ and https://www.marketwatch.com/story/bill-gates-is-now-the-largest-farmland-owner-in-america-11610818582?mod=hp_minor_pos21

811   Ted Turner Talks Politics, Fidel Castro and Jane Fonda With O'Reilly | Fox News

812   America's Biggest Owner Of Farmland Is Now Bill Gates (forbes.com)

813   Elkins, K. 2019, Februay14, Bill Gates suggests higher taxes on the rich—the current system is 'not progressive enough,' he says. CNBC. https://www.cnbc.com/2019/02/13/bill-gates-suggests-higher-taxes-on-those-with-great-wealth.html

814   The Global Progressive Form. 2020. Founded in 2003. https://www.globalprogressiveforum.org/

815   Lux, M., 2017, January 25, A Progressive Vision of Globalization. HuffPost. https://www.huffpost.com/entry/a-progressive-vision-of-g_b_9071000?guccounter=1&guce_referrer=aHR0cHM6Ly93d3cuZ29vZ2xlLmNvbS8&guce_referrer_sig=AQAAALhN6vn-vDxm6kkQNYpXEAexsuz8FSUXeioYJ312ds2ud4nMZ7dWFlO7EyY_rfrOVOnJQ9f-G5Thxdh6G7Cn2oPrhHjCaYlWY7rox3_kVlYTLbXx4Ix4ZMsG63Q3bJ_bDqxEXcS-L4y-csYFDRRhmlYoZ6YzYsk0kaY-jolDCQGpIOV

816   Gangitano, A. 2019, December 19, Tech industry cash flows to Democrats despite 2020 scrutiny. The Hill. https://thehill.com/policy/technology/475221-tech-industry-cash-flows-to-democrats-despite-2020-scrutiny

817   Freedland, C., January/February 2011, The Rise of the New Global Elite, The Atlantic. http://globalization.qwriting.qc.cuny.edu/2011/02/20/files/2010/09/The-Rise-of-the-New-Global-Elite-Magazine-The-Atlantic.pdf

818   See this essay by Boris Johnson: https://www.boris-johnson.com/2007/10/25/global-population-control/ or this from WorldAtlas: https://www.worldatlas.com/articles/population-control-methods.html

819   Lerer, Lisa (March 19, 2020). "Suddenly, a New Normal." https://www.nytimes.com/2020/03/19/us/politics/mitt-romney-coronavirus-checks.html

820   Roth, D. (2020, April 11), The new normal, according to Bill Gates (and when it arrives), LinkedIn. https://www.linkedin.com/pulse/new-normal-according-bill-gates-when-arrives-daniel-roth/ For Gates' concepts of the changes, also see Bariso, J., 2020, April 3, Bill Gates Says the Coronavirus Will Change Life Forever, Here's How to Adapt, Inc. https://www.inc.com/justin-bariso/bill-gates-says-coronavirus-will-change-life-forever-heres-how-to-adapt.html

821   International, 2020, August 18, The covid-19 pandemic will be over by the end of 2021, says Bill Gates, But millions of deaths are yet to come in poor countries. The Economist. https://www.economist.com/international/2020/08/18/the-covid-19-pandemic-will-be-over-by-the-end-of-2021-says-bill-gates

822   Chapman, G. 2015, May 18, Bill Gates Calls for "Germ Games" Instead of War Games, Business Insider. https://www.businessinsider.com/afp-bill-gates-calls-for-germ-games-instead-of-war-games-2015-3?op=1

823   Max, S., 2015, March 12, From the Gates Foundation, Direct Investment, Not Just Grants, New York Times. https://www.nytimes.com/2015/03/13/business/from-the-gates-foundation-direct-investment-not-just-grants.html

824 https://www.brusselstimes.com/news/eu-affairs/168606/eu-vaccines-millions-of-doses-exported-to-rich-countries-less-to-poor-countries/; https://abcnews.go.com/Health/wireStory/stalled-jab-vaccine-shortages-hit-poor-countries-76990671

825 Schwab, T. (2020, March 17). Bill Gates' Charity Paradox. A Nation investigation illustrates the moral hazards surrounding the Gates Foundation's $50 billion charitable enterprise. The Nation. Retrieved June 16, 2020 at https://www.thenation.com/article/society/bill-gates-foundation-philanthropy/

826 Gates Foundation (2020, June 4). Bill & Melinda Gates Foundation pledges $1.6 billion to Gavi, the Vaccine Alliance, to protect the next generation with lifesaving vaccines. Retrieved on June 16, 2020 from https://www.gatesfoundation.org/Media-Center/Press-Releases/2020/06/Bill-and-Melinda-Gates-Foundation-pledges-to-Gavi-the-Vaccine-Alliance

827 Calfas, J. (2020, April 5) Bill Gates to help fund coronavirus-vaccine development. The Wall Street Journal. Retrieved June 19, 2020 at https://www.wsj.com/articles/bill-gates-to-spend-billions-on-coronavirus-vaccine-development-11586124716

828 Schwab, T. (2020, March 17). Bill Gates' Charity Paradox. A Nation investigation illustrates the moral hazards surrounding the Gates Foundation's $50 billion charitable enterprise. The Nation. Retrieved June 16, 2020 at https://www.thenation.com/article/society/bill-gates-foundation-philanthropy/

829 Bill Gates: Testimony before the Committee on Science and Technology, U.S. House Representatives, 2008. Microsoft.com. https://news.microsoft.com/2008/03/12/bill-gates-testimony-before-the-committee-on-science-and-technology-u-s-house-of-representatives/

830 Bill Gates on the new normal. (A collection of his videos on the subject.) https://www.google.com/search?q=fauci+on+the+new+normal&oq=fauci+on+the+new+normal&aqs=chrome..69i57.7284j0j4&sourceid=chrome&ie=UTF-8

831 Funke, D. 2020, April 15. Facebook posts falsely claim Dr. Fauci has millions invested in a coronavirus vaccine, PolitiFact. https://www.politifact.com/factchecks/2020/apr/15/facebook-posts/facebook-posts-falsely-claim-dr-fauci-has-millions/

832 Branswell, H. 2020, September 14, Bill Gates slams 'shocking' U.S. response to Covid-19 pandemic, STAT. https://www.statnews.com/2020/09/14/bill-gates-slams-mismanaged-u-s-response-to-covid-19-pandemic/

833 Morens, D. and Fauci, A. (2020, September 3). Emerging Pandemic Diseases: How We Got to COVID-19. Cell 182, 1099-1091. https://www.cell.com/action/showPdf?pii=S0092-8674%2820%2931012-6

834 https://www.bloomberg.com/news/newsletters/2020-12-19/bloomberg-new-economy-china-doubles-down-on-state-control

835 Dawsey, J. and Abutaleb, Y. 2020, October 31. "A whole lot of hurt": Fauci warns of covid-19 surge, offers blunt assessment of President Trump's response, Washington Post. https://www.washingtonpost.com/politics/fauci-covid-winter-forecast/2020/10/31/e3970eb0-1b8b-11eb-bb35-2dcfdab0a345_story.html

836 resume.pdf (breggin.com)

837 http://breggin.com/psychosurgery-for-political-purposes-1975/. Also see http://breggin.com/psychosurgery-for-political-purposes-1975/ and http://breggin.com/wp-content/uploads/2008/01/campaignsagainst.pbreggin.1995.pdf

838 https://breggin.com/judge-applies-nuremberg-code-in-kaimowitz-psychosurgery-case/. Also see http://breggin.com/psychosurgery-for-political-purposes-1975/

839 https://www.youtube.com/user/PeterBreggin. Also see https://breggin.com/?s=psychosurgery. My YouTube Channel came under heavy assault by YouTube in May and June 2021 and its future is uncertain at the time of this writing. The items taken down seem almost random covering both psychiatry and COVID-19. We are shifting our videos elsewhere in preparation for a permanent shutdown.

840   Breggin, P. (1975). The Politics of Psychosurgery. Duquesne Law Review 13, 841-861. Psychosurgery for political purposes (1975) | Psychiatric Drug Facts (breggin.com). Also see, Dr. Breggin's Psychosurgery Page at PSYCHOSURGERY PAGE | Psychiatric Drug Facts (breggin.com)

841   https://breggin.com/ect-resources/FINAL-FINAL-AFFIDAVIT-12.1.17.pdf

842   Alert 67: MORE HUGE ECT NEWS | Psychiatric Drug Facts (breg). gin.com)

843   DK Law Group Announces ECT Settlement! | ECT Justice!

844   ECT Shock Treatment Class Action – Case Update April 2018 - Mad In America

845   My book critical of ECT can be downloaded as a PDF from my free ECT Resource Center at https://breggin.com/ect-resources-center/. The book itself is at http://breggin.com/ect-resources/Breggin_1979_AAA_Complete_ECT_Book_Overview_244_pages_Brain_Damage_Memory_Loss_Abuse_Etc_.pdf

846   ECT Resources Center | Psychiatric Drug Facts (breggin.com). More specifically, see my many scientific articles at the Resource Center, especially

847   Breggin, P. and Breggin, R. (1998). The War Against Children of Color: Psychiatry Targets Inner City Children. Monroe, Main: Common Courage Press. Updated version of The War Against Children.

848   https://www.judicialwatch.org/press-releases/fauci-emails/; https://www.judicialwatch.org/tom-fittons-weekly-update/new-fauci-emails-on-china-who/; https://www.judicialwatch.org/press-releases/emails-who-terms/

849   US Grant To Wuhan Lab To Enhance Bat-Based Coronaviruses Was Never Scrutinized By HHS Review Board, NIH Says | The Daily Caller

850   Informed, heroic physicians have been reporting this kind of success since March, beginning with Vladimir Zelenko. It was then confirmed by later publications. The precise numbers cited are from: Brian C. Procter, Casey Ross, Vanessa Pickard, Erica Smith, Cortney Hanson, and Peter A. McCullough (2021, March). Early Ambulatory Multidrug Therapy Reduces Hospitalization and Death in High-Risk Patients with SARS-CoV-2 (COVID-19. International Journal of Innovative Research in Medical Science (IJIRMS). Volume 06, Issue 03, March 2021, https://doi.org/10.23958/ijirms/vol06. View of Early Ambulatory Multidrug Therapy Reduces Hospitalization and Death in High-Risk Patients with SARS-CoV-2 (COVID-19) (ijirms.in)

851   https://www.cnn.com/2020/06/17/health/hydroxychloroquine-national-stockpile/index.html and https://thetexan.news/ted-cruz-letter-to-fda-says-hydroxychloroquine-restrictions-may-be-directly-costing-lives/

852   https://www.washingtonpost.com/politics/fauci-covid-winter-forecast/2020/10/31/e3970eb0-1b8b-11eb-bb35-2dcfdab0a345_story and htmlhttps://www.cbsnews.com/news/fauci-washington-post-white-house-coronavirus-pandemic/

853   https://www.washingtonpost.com/politics/fauci-covid-winter-forecast/2020/10/31/e3970eb0-1b8b-11eb-bb35-2dcfdab0a345_story and htmlhttps://www.cbsnews.com/news/fauci-washington-post-white-house-coronavirus-pandemic/

854   https://www.youtube.com/watch?v=NUHsEmlIoE4

855   Press Release, 2010, Global Health Leaders Launch Decade of Vaccines Collaboration. Bill & Melinda Gates Foundation. https://www.gatesfoundation.org/Media-Center/Press-Releases/2010/12/Global-Health-Leaders-Launch-Decade-of-Vaccines-Collaboration

856   China the only major economy to see positive growth in 2020: official - CGTN

857   Dr. Fauci's COVID-19 Treachery | Psychiatric Drug Facts (breggin.com)

858   Dr. Breggin's COVID-19 Totalitarianism Legal Report | Psychiatric Drug Facts

859   https://www.rollingstone.com/politics/politics-news/trump-knew-coronavirus-deadly-bob-woodward-book-1057051/

860   Halpern, D. (2020). Coping With COVID-19: Learning from Past Pandemics to Avoid Pitfalls and Panic. Global Health: Science and Practice, 8 (2), 155-165. https://www.ghspjournal.org/content/ghsp/8/2/155.full.pdf

861   Wu., C., 2020, August 26. C.D.C. Now Says People Without Covid-19 Symptoms Do Not Need Testing, New York Times. https://www.nytimes.com/2020/08/06/health/coronavirus-asymptomatic-transmission.html The quote was taking from the Google search page for the article from the NYT.

862   Weise, E. and Rodriquez, A., 2020, August 27, CDC clarifies surprise guidelines that people without COVID-19 symptoms don't need testing. USA Today. https://www.usatoday.com/story/news/2020/08/27/cdc-walks-back-surprise-coronavirus-asymptomatic-testing-guidelines/5645630002/ Also, see van Bruden, I. 2020, August 27, Asymptomatic people no longer require CCP virus test: New CDC guidance. https://www.theepochtimes.com/asymptomatic-people-no-longer-require-ccp-virus-test-new-cdc-guidance_3477773.html

863   Papenfuss, M., 2020, September 3, Fauci Warns 7 States To Take Extra Holiday Precautions Against COVID-19 Surge. Huffington Post. https://www.huffpost.com/entry/anthony-fauci-labor-day-covid-7-central-states_n_5f51a149c5b62b3add3e3e0d?ri18n=true&ncid=newsltushpmgnews

864   Morens, D. and Fauci, A. (2020, September 3). Emerging Pandemic Diseases: How We Got to COVID-19. Cell 182, 1099-1091. https://www.cell.com/action/showPdf?pii=S0092-8674%2820%2931012-6

865   See the constantly revised chart of rates of death in several countries including the US at the top of our Coronavirus Resource Center, Chart from Our World In Data: Daily confirmed COVID-19 deaths, rolling 3-day average. https://breggin.com/coronavirus-resource-center/

866   Breggin, P. and Breggin, G., 2020, CDC Surges Covid-19 Stats: Death Rates Remain Low. More Cases Needed for Herd Immunity. https://breggin.com/cdc-surge-stats-exaggerated/

867   American Psychological Association, 2015, October 22, Fear-Based Appeals Effective at Changing Attitudes, Behaviors After All. Press Release. https://www.apa.org/news/press/releases/2015/10/fear-based-appeals. The article described is Tannenbaum M. et al. (2015). Appealing to Fear: A Meta-Analysis of Fear Appeal Effectiveness and Theories. Psychological Bulletin, 141, 1178 –1204.

868   Tannenbaum M. et al. (2015). Appealing to Fear: A Meta-Analysis of Fear Appeal Effectiveness and Theories. Psychological Bulletin, 141, 1178 –1204.

869   Ruiter, R., Kessels; L., ., , G-J & Kok, G. (2014). Sixty years of fear appeal research: current state of the evidence. Int J Psychol. 2014 Apr;49(2):63-70. doi: 10.1002/ijop.12042. Epub 2014 Feb 24. https://pubmed.ncbi.nlm.nih.gov/24811876/

870   Jeni A. Stolow, Lina M. Moses, Alyssa M. Lederer and Rebecca Carter, 2020, June 11, How Fear Appeal Approaches in COVID-19 Health Communication May Be Harming the Global Community. Health Education. Health Education & Behavior, 47(4) 531–535 https://journals.sagepub.com/doi/pdf/10.1177/1090198120935073

871   Merrett, L. (2020, Winter). Planet Covid. Journal of American Physicians and Surgeons, 25 (4), 97.

872   Breggin, P. . "Psychiatry's Role in the Holocaust." International Journal of Risk and Safety in Medicine, 4:133-148, 1993. Adapted from a paper delivered at "Medical Science Without Compassion" in Cologne, Germany and originally published in the conference proceedings. http://breggin.com/wp-content/uploads/2008/01/psychiatrysrole.pbreggin.1993.pdf

873   Evans, N. and Inglesby, T. (2019). Biosecurity and public health ethics issues raised by biological threats. Chapter 66, pp. 774-785 in The Oxford Handbook of Public Health Ethics. Eds. Mastroianni, A, Kahn, J., & Kass. N. New York: Oxford University Press.

874 Faden, R., Shebaya, S., & Siegel, A. (2019). Distinctive Challenges of Public Health Ethics. Chapter 2, pp. 12-20 in The Oxford Handbook of Public Health Ethics. Eds. Mastroianni, A, Kahn, J., & Kass. N. New York: Oxford University Press.

875 Faden, R &, Shebaya, S. (2019). Public Health Programs and Policies: Ethical justifications. Chapter 3, pp. 12-20 in The Oxford Handbook of Public Health Ethics. Eds. Mastroianni, A, Kahn, J., & Kass. N. New York: Oxford University Press.

876 For a specific critic of the handling of COVID-19 from a Constitutional perspective, see: Jackie McDermott and Lana Ulrich. (2020, April 15). COVID-19 and the Constitution — Key Takeaways. From the National Constitution Center. https://constitutioncenter.org/interactive-constitution/blog/covid-19-and-the-constitution-key-takeaways. Also, see Heath, I. (2017). The missing person: The outcome of the rule-based totalitarianism of too much contemporary healthcare. Patient Educ Couns. 2017 Nov;100(11):1969-1974. doi: 10.1016/j.pec.2017.03.030. Epub 2017 Apr 3; Fleming, K. (2004, December. Rapid Response: National Health Care and Totalitarianism. BMJ 2004;329:1424; A General Surgeon. (2018). As We Continue to Drift Into a Totalitarian Medical System: A View of a Country Boy. Scand. J. Surg. https://doi.org/10.1177/1457496918757579

877 Buchanan, D. (2019). Public Health Programs and Policies: Ethical justifications. Chapter 8, pp. 77-80 in The Oxford Handbook of Public Health Ethics. Eds. Mastroianni, A, Kahn, J., & Kass. N. New York: Oxford University Press.

878 Pavlov, Ivan (2010, originally published in 1927) Conditioned reflexes: An investigation of the physiological activity of the cerebral cortex. Ann Neurosci. 2010 Jul; 17(3): 136–141. doi: 10.5214/ans.0972-7531.1017309. The pages for the excerpt are not paginated. https://www.ncbi.nlm.nih.gov/pmc/articles/PMC4116985/

879 https://www.britannica.com/science/Stockholm-syndrome

880 Masson sent me this around August 25, 2020.

881 Breggin, P. Guilt, Shame and Anxiety: Understanding and Overcoming Negative Emotions. (Prometheus Books, Amherst, NY, 2014). Also see, Breggin, P. The biological evolution of guilt, shame and anxiety: A new theory of negative legacy emotions. Medical Hypotheses, 85, 17-24, 2015 and Breggin, P. Understanding and Overcoming Guilt, Shame, and Anxiety: Based on the Theory of Negative Legacy Emotions. In Kirkcaldy, B. (Ed). Promoting Psychological Well-Being in Children and Families. New York, NY: Palgrave Macmillan. Chapter 5, pp. 68-80, 2015.

882 Breggin, P. Wow, I'm an American! How to Live Like Our Nation's Heroic Founders. (Lake Edge Press, Ithaca, NY, 2009).

883 Breggin, P. Liberty and Love in the Former USSR, pp. 180-189. In Chapter 9 in Breggin, P. Beyond Conflict: From Self-Help and Psychotherapy to Peacemaking. (St. Martin's, NY, 1992).

884 The following discussion of coercion is modified from Breggin (1992), p. 75 ff.

885 Breggin, P. (2015). Understanding and Overcoming Guilt, Shame, and Anxiety: Based on the Theory of Negative Legacy Emotions. In Kirkcaldy, B. (Ed). Promoting Psychological Well-Being in Children and Families. New York, NY: Palgrave Macmillan. Chapter 5, pp. 68-80, 2015. http://breggin.com/wp-content/uploads/2015/09/Breggin%2C%20P.%20%282015%29%20Understanding%20%26%20overcoming%20GSA.pdf ;

886 Breggin, P. and Stolzer, J. Psychological Helplessness and Feeling Undeserving of Love: Windows into Suffering and Healing. The Humanistic Psychologist, June 2020, 48 (2), 113-122. https://breggin.com/psychological-helplessness-and-feeling-undeserving-of-love-windows-into-suffering-and-healing/

887 Breggin, P. The biological evolution of guilt, shame and anxiety: A new theory of negative legacy emotions. Medical Hypotheses, 85, 17-24, 2015. https://breggin.com/understanding-and-overcoming-guilt-shame-and-anxiety-based-on-theory-of-negative-legacy-emotions/

888   Breggin, P. and Stolzer, J. Psychological Helplessness and Feeling Undeserving of Love: Windows into Suffering and Healing. The Humanistic Psychologist, June 2020, 48 (2), 113-122. https://breggin.com/psychological-helplessness-and-feeling-undeserving-of-love-windows-into-suffering-and-healing/

889   Breggin, P. (1991). Toxic Psychiatry. New York, St. Martins, Chapter 1.

890   Goffman, G. (1961). Asylums: Essays on the Social Situation of Mental Patients and Other Inmates, New York, Anchor Books.

891   Breggin, P. (1964). Coercion of Voluntary Patients in an Open Hospital. AMA Archives of General Psychiatry. http://breggin.com/wp-content/uploads/2005/05/coercionof.pbreggin.1982.pdf

892   Altheide, D. T(2006, November) Terrorism and the Politics of Fear. Cultural Studies <—> Critical Methodologies. November, 2006. 415-439. https://journals.sagepub.com/doi/pdf/10.1177/1532708605285733

893   Muller-Hill B. Murderous science: Elimination by scientific selection of Jews, gypsies, and others, Germany, 1933- 1945. New York: Oxford University Press, 1988.

894   Nuremberg Code Against Nazi Experiments Should Stop Further COVID-19 Vaccinations | Psychiatric Drug Facts (breggin.com)

895   Leonardi, A. (2020). Bill de Blasio asks New Yorkers to take photos of people violating social distancing rules. https://www.washingtonexaminer.com/news/bill-de-blasio-asks-new-yorkers-to-take-photos-of-people-violating-social-distancing-rules Also, see FoxNews.Com (2020). https://www.foxnews.com/us/de-blasio-new-yorkers-report-social-distancing-violations-text

896   Garcetti, E. (2020). LA mayor, threatens to cut power to noncomplying businesses: 'We will shut you down.' - Washington Times https://www.washingtontimes.com/news/2020/mar/25/eric-garcetti-la-mayor-threatens-to-cut-power-to-n/

897   Stickings, T. (2020). Best-case scenario, new study predicts. Daily Mail, London. https://www.dailymail.co.uk/news/article-8082327/15-MILLION-people-die-best-case-coronavirus-scenario.html

898   McKibbin, W. and Roshen, F. (2010).The Global Macroeconomic Impacts of COVID-19: Seven Scenarios. University of Australia. McKibbin, W. and Roshen, F. (2010). https://anu.prezly.com/coronavirus-is-highly-uncertain-and-the-costs-could-be-high?utm_source=email&utm_medium=campaign&utm_id=campaign_Ejmb_lbzJ.contact_RBhn_IDAN&asset_type=attachment&asset_id=171409#attachment-171409

899   Frieden, T. (2020). Could coronavirus kill a million Americans. Think Global Health. https://www.thinkglobalhealth.org/article/could-coronavirus-kill-million-americans

900   Fink, S. (2020) White House Takes New Line After Dire Report on Death Toll. https://www.nytimes.com/2020/03/16/us/coronavirus-fatality-rate-white-house.html

901   Faust, J. Lin, Z., & del Rio, C. (2020, August 13, 2020). Comparison of Estimated Excess Deaths in New York City During the COVID-19 and 1918 Influenza Pandemics. JAMA Network Open. 2020;3(8):e2017527. doi:10.1001/jamanetworkopen.2020.17527 https://jamanetwork.com/journals/jamanetworkopen/fullarticle/2769236

902   Epstein, R. (2020, August 17). COVID-19 Confusion. Hoover Institute. https://www.hoover.org/research/covid-19-confusion

903   Papazian (2020, June 1). Fauci's Flip Flops. Great America. https://amgreatness.com/2020/06/01/faucis-flip-flops/

904   Snyder, T. (2017(). On Tyranny: Twenty Lessons from the Twentieth Century. New York: Tim Duggan Books. He is the Levin Professor of History at Yale. s

905   https://www.silverdoctors.com/headlines/world-news/President Trump-vs-the-fed-america-will-be-sacrificed-at-the-nwo-altar-through-the-fed-and-by-the-globalists/

906 Roose, K., 2021, February 2, How the Biden Administration Can Help Solve Our Reality Crisis, NYT. How the Biden Administration Can Help Solve Our Reality Crisis - The New York Times (nytimes.com)

907 https://www.washingtonexaminer.com/tag/donald-President Trump?source=%2Fnews%2F-maxine-waters-calls-on-individuals-to-block-President Trump-officials-from-retailers

908 https://www.the-sun.com/news/1380148/nancy-pelosi-President Trump-enemy-state-defend-constitution/

909 https://www.nytimes.com/2021/01/09/technology/apple-google-parler.html

910 https://www.nytimes.com/2021/01/08/technology/twitter-President Trump-suspended.html

911 https://nypost.com/article/what-is-cancel-culture-breaking-down-the-toxic-online-trend/

912 https://www.msn.com/en-us/news/us/universities-face-pressure-to-vet-ex-President Trump-officials-before-hiring-them/ar-BB1d1zDP

913 https://www.theguardian.com/books/2021/jan/19/open-letter-calls-for-publishing-boycott-of-President Trump-administration-memoirs and https://www.foxnews.com/politics/publishing-professionals-black-list-President Trump-officials

914 https://www.judicialwatch.org/press-releases/judicial-watch-new-irs-documents-used-donor-lists-to-target-audits/

915 https://www.npr.org/2017/10/27/560308997/irs-apologizes-for-aggressive-scrutiny-of-conservative-groups, https://newspunch.com/judicial-watch-obama-irs/ and https://aclj.org/free-speech/another-victory-over-the-abusive-irs-targeting-of-conservatives

916 Pelosi Threatens Impeachment If Pence Doesn't Invoke 25th Amendment (thefederalist.com)

917 https://www.fox5ny.com/news/new-york-area-national-guard-units-assisting-with-inauguration-security

918 https://thehill.com/policy/national-security/534669-fear-of-insider-attack-prompts-additional-fbi-screening-of-national

919 https://thehill.com/homenews/state-watch/534791-texas-governor-calls-fbi-vetting-of-national-guard-troops-offensive

920 https://www.baltimoresun.com/news/nation-world/ct-nw-my-pillow-ceo-mike-lindell-twitter-ban-20210126-7wtatyxgznbzzhljfkv4f2bgbe-story.html

921 https://www.foxnews.com/tech/live-updates-big-tech-1-12-2021

922 https://www.nbcnews.com/politics/congress/aoc-demands-ted-cruz-resign-over-capitol-riot-you-almost-n1256052

923 https://www.nytimes.com/2021/01/27/opinion/how-to-defeat-americas-homegrown-insurgency.html

924 https://thefederalist.com/2020/10/06/breaking-dni-declassifies-handwritten-notes-from-john-brennan-2016-cia-referral-on-clinton-campaigns-collusion-operation/

925 https://twitter.com/tomselliott/status/1352007118392582148 and https://news.yahoo.com/john-brennan-says-biden-admin-152005224.html

926 https://time.com/4486502/hillary-clinton-basket-of-deplorables-transcript/

927 https://www.theguardian.com/world/2008/apr/14/barackobama.uselections2008

928 President of Soros-Linked Voting Software Firm on Biden Transition Team – President Trump Lawyers « Aletho News

929 The Empire Doubles Down: Open Society Foundations Will Now Be Run by Lord Malloch Brown « Aletho News

930 https://medium.com/age-of-awareness/yes-all-white-people-are-racist-eefa97cc5605 Also see, https://www.forbes.com/sites/johnrex/2020/07/15/ive-been-complicit-in-systemic-racism-heres-what-im-doing-to-change/?sh=2f80d74b5078

931 https://apnews.com/article/pfizer-vaccine-effective-early-data-4f4ae2e3bad122d-17742be22a2240ae8

932 https://www.msn.com/en-us/news/us/why-did-amazon-wait-until-biden-e2-80-99s-inau-guration-to-offer-help-with-vaccine-distribution/ar-BB1cWxQV

933 https://www.who.int/news/item/20-01-2021-who-information-notice-for-ivd-users-2020-05

934 https://www.washingtonpost.com/news/post-politics/wp/2017/01/20/the-campaign-to-impeach-president-trump-has-begun/

935 Epoch Times Statement on YouTube Demonetization (theepochtimes.com)

936 https://www.newsweek.com/biden-china-virus-ban-may-foretell-national-security-threat-ening-policies-opinion-1565208

937 https://nypost.com/2021/02/19/china-lab-probed-for-covid-to-get-us-funds-until-2024-report/

938 https://www.americanrhetoric.com/speeches/alexandersolzhenitsynharvard.htm

939 https://www.cnbc.com/2021/01/25/davos-watch-chinas-xi-jinping-speaking-with-wef-founder-klaus-schwab.html

940 https://www.theepochtimes.com/beijing-substantially-involved-in-us-2020-election-chi-na-analyst_3626810.html An interview with China expert Gordon Chang.

941 Browne, A., Editorial Director, 2021, January 2, Bloomberg New Economy: How the Pandemic Shoved Us Into the Future, Turning Points, New Economy Forum. https://www.bloomberg.com/news/newsletters/2021-01-02/bloomberg-new-economy-how-the-pandemic-shoved-us-into-the-future

942 https://www.marketwatch.com/story/heres-the-case-for-elon-musk-warren-buffet-and-the-rest-of-americas-billionaires-sending-3-000-stimulus-checks-to-everybody-11607613932

943 https://www.toddstarnes.com/uncategorized/big-tech-strikes-back-as-president-trump-is-indefinitely-banned-from-facebook-and-twitter/; https://miningawareness.wordpress.com/2021/01/10/free-speech-crackdown-repressive-big-tech-gangs-up-against-competi-tor-parler-founder-ceo-responds-start-boycotting-big-tech-asap/; https://www.toddstarnes.com/uncategorized/big-tech-strikes-back-as-president-trump-is-indefinitely-banned-from-facebook-and-twitter/

944 https://www.westernjournal.com/another-big-tech-platform-just-permanently-banned-don-ald-trump/; https://www.cbsnews.com/news/snapchat-trump-ban/

945 The hidden Covid-19 health crisis: Elderly people are dying from isolation (nbcnews.com)

946 More people are dying during the pandemic – and not just from COVID-19 | American Heart Association. The July 1, 2020 JAMA report can be found here: https://jamanetwork.com/journals/jama/fullarticle/2768086

947 https://www.marketwatch.com/story/heres-the-case-for-elon-musk-warren-buffet-and-the-rest-of-americas-billionaires-sending-3-000-stimulus-checks-to-everybody-11607613932

948 https://www.counterpunch.org/2004/02/24/kerry-s-china-connection/;https://www.theat-lantic.com/ideas/archive/2020/12/risk-john-kerry-following-his-own-china-policy/617459/

949 https://www.theepochtimes.com/mkt_app/georgia-mired-in-election-disputes-fosters-cozy-ties-with-china_3614801.html

950 A stunning analysis of the history of human enslavement and its inherent ties to progressiv-ism is found in Shafarevich, Igor. (2019). The Socialist Phenomenon. With a Forward by Aleksandr Solzhenitsyn. Gideon House. www.gideonhouse.com.

951 Spalding's interview with Breggin is on https://www.youtube.com/watch?v=_3DWfqnMFXA and on https://www.brighteon.com/aaa1122f-a50d-411f-bd6a-da62285eb7ec

952 Black History Month | The National WWII Museum | New Orleans (nationalww2museum.org)

953 https://www.historynet.com/robert-byrd-consorts-kkk-grand-dragon.htm

954 The Trust Imperative (theepochtimes.com)

955 https://nypost.com/2020/08/06/bill-gates-climate-change-could-cause-more-misery-than-covid-19/

956 https://www.weforum.org/agenda/2018/06/the-world-will-be-carbon-neutral-by-2050-but-at-what-cost/

957 For a video and transcript of his speech, go here: https://reason.com/volokh/2020/11/12/video-and-transcript-of-justice-alitos-keynote-address-to-the-federalist-society/ For more commentary, https://nypost.com/2020/11/13/justice-alito-warns-of-threats-to-religious-liberty-in-speech/and Justice Alito: COVID restrictions 'previously unimaginable' | The Spokesman-Review

958 https://www.americanrhetoric.com/speeches/alexandersolzhenitsynharvard.htm

959 https://www.forbes.com/sites/daviddawkins/2021/07/19/george-soros-and-bill-gates-backed-consortium-to-buy-uk-maker-of-covid-lateral-flow-tests-for-41-million/ and https://www.worldtribune.com/big-deal-gates-soros-group-acquires-for-profit-covid-testing-firm/

960 https://www.battlefields.org/learn/articles/civil-war-casualties

961 This emphasis on trust was inspired by thoughts I generated in an interview with to journalist Conan Milner of The Epoch Times which he drew on extensively for his article: The Trust Imperative: Human beings depend on trusting relationships, and suffer deeply when trust is broken, Epoch Times published on February 3, 2021.. The quote is from the subhead of his article and reflects my thoughts. The Trust Imperative (theepochtimes.com) Also see Breggin, Peter. (2009). Wow, I'm an American! How to Live Like Our Heroic Founders. Wow, I'm an American: Breggin, Peter R, Breggin, Ginger R, Breggin, Ginger R: 9780982456019: Amazon.com: Books

962 The Weekly "Dr. Peter Breggin Hour" on PRN.fm | Psychiatric Drug Facts

963 All the observations in the Chronology and Overview are documented in the text of the book, except for a few which have endnotes.

964 Baric, Ralph S.; Yount, Boyd; Hensley, Lisa; Peel, Sheila; and Chen, Wan. (1997, March). Episodic Evolution Mediates Interspecies Transfer of a Murine Coronavirus. Journal of Virology, 71 (3), 1945-1955 https://jvi.asm.org/content/jvi/71/3/1946.full.pdf

965 https://jvi.asm.org/content/74/3/1393.short

966 Kuo, Lili; Godeke, Gert-Jan; Raamsman, Martin; Masters, Paul; and Rottier, Peter. (2000, February). Journal of Virology 74 (3), 1393–1406. 1393.full.pdf (asm.org)

967 O'Kane, C., 2020, October 23, "We're about to go into a dark winter": Biden says President Trump has no plan for coronavirus, CBS News. https://www.cbsnews.com/news/biden-President-Trump-coronavirus-plan-vaccine-winter/

968 Zhang, F., (2004, July 2), Officials punished for SARS virus leak, China Daily. https://www.chinadaily.com.cn/english/doc/2004-07/02/content_344755.htm

969 https://www.cdc.gov/sars/lab/downloads/nucleoseq.pdf complete genome for Urban SARS corona virus April 17, 2003 submitted to the CDC.

970 Andrea Savarino, Johan R Boelaert, Antonio Cassone, Giancarlo Majori, and Roberto Cauda. (2003, November). Effects of chloroquine on viral infections: an old drug against today's diseases? Lancet Infect Dis 2003; 3: 722–27. https://www.thelancet.com/action/showPdf?pii=S1473-3099%2803%2900806-5. Also many other articles including https://www.ncbi.nlm.nih.gov/pmc/articles/PMC1232869/pdf/1743-422X-2-69.pdf. Also see: Martin J Vincent, Eric Bergeron, Suzanne Benjannet, Bobbie R Erickson, Pierre E Rollin, Thomas G Ksiazek, Nabil G Seidah and Stuart T. Nichol. (2005) Chloroquine is a potent inhibitor of SARS coronavirus infection and spread. Virology Journal, 2, 69. doi:10.1186/1743-422X-2-69. https://www.ncbi.nlm.nih.gov/pmc/articles/PMC1232869/pdf/1743-422X-2-69.pdf

971 Deming D., et al. 2006. Vaccine efficacy in senescent mice challenged with recombinant SARS-CoV bearing epidemic and zoonotic spike variants. PLoS Med. 3:e525. [PMC free article] [PubMed] [Google Scholar]. Vaccine Efficacy in Senescent Mice Challenged with Recombinant SARS-CoV Bearing Epidemic and Zoonotic Spike Variants (plos.org)

972   https://www.gavi.org/investing-gavi/innovative-financing/iffim
973   https://www.pnas.org/content/pnas/105/50/19944.full.pdf
974   Immigrant-Founded Moderna Leading The Way In Covid-19 Response (forbes.com)
975   Tseng CT, Sbrana E, Iwata-Yoshikawa N, Newman PC, Garron T, et al. (2012) Immunization with SARS Coronavirus Vaccines Leads to Pulmonary Immunopathology on Challenge with the SARS Virus. PLOS ONE 7(8): 10.1371/annotation/2965cfae-b77d-4014-8b7b-236e 01a35492. https://journals.plos.org/plosone/article?id=10.1371/journal.pone.0035421
976   Top 10 vaccine manufacturers in the world (yahoo.com)
977   Press Release, 2010, Global Health Leaders Launch Decade of Vaccines Collaboration. Bill & Melinda Gates Foundation. https://www.gatesfoundation.org/Media-Center/Press-Releases/2010/12/Global-Health-Leaders-Launch-Decade-of-Vaccines-Collaboration. The January date is cited in this document.
978   The booklet has been expunged from the Rockefeller website but remains available here: https://issuu.com/dueprocesstv/docs/scenario-for_the-future
979   https://www.gatesfoundation.org/Ideas/Media-Center/Press-Releases/2010/12/Global-Health-Leaders-Launch-Decade-of-Vaccines-Collaboration
980   This is goal 3.8, located under "Goal 3. Ensure healthy lives and promote well-being for all at all ages." https://sdgs.un.org/2030agendagrant/https://sustainabledevelopment.un.org/content/documents/21252030%20Agenda%20for%20Sustainable%20Development%20web.pdf
981   Bolles, M. et al. (2011). A double-inactivated severe acute respiratory syndrome coronavirus vaccine provides incomplete protection in mice and induces increased eosinophilic proinflammatory pulmonary response upon challenge. J. Virol. 85, 12201–12215 (2011). https://www.ncbi.nlm.nih.gov/pmc/articles/PMC3209347/ The Acknowledgments in toto: "This work was supported by grants from the National Institutes of Health, Division of Allergy and Infectious Diseases (AI075297;, U54AI080680;, and U54AI081680-01), and the UNC-CH Medical Science Training Program (T32GM008719 [M.B.]). We thank Kanta Subbarao at the NIH for generously providing the icMA15 virus. We recognize the generous gift of the adjuvanted and unadjuvanted DIV from BEI. We also thank Vineet Menachery for help in manuscript preparation."
982   https://sph.unc.edu/sph-news/baric-to-lead-10-million-nih-grant/S
983   https://reporter.nih.gov/search/GCBvbhJJPEOJ8Rp15r1W1A/project-details/9819304#-similar-Projects
984   Bill & Melinda Gates Foundation announces $250 million COVID vaccine commitment - ABC News (go.com). "Gates, who had invested in the vaccine technology in 2015, emphasized the importance of messenger RNA (mRNA) vaccines ending this pandemic and potentially ones to come in the future."
985   https://www.business-standard.com/article/current-affairs/scientists-in-china-discussed-weaponising-coronavirus-in-2015-report-121050900916_1.html; https://www.news.com.au/world/coronavirus/leaked-chinese-document-reveals-a-sinister-plan-to-unleash-coronaviruses/news-story/53674e8108ad5a655e07e990daa85465; https://www.dailymail.co.uk/news/article-9556415/China-preparing-WW3-biological-weapons-six-years-investigators-say.html; https://www.independent.co.uk/asia/china/china-covid-leak-genetic-weapons-b1844805.html
986   CEPI conceived 2015 - Bing
987   The Bill & Melinda Gates Foundation | Gavi, the Vaccine Alliance. Also see, https://www.gavi.org/investing-gavi/resource-mobilisation-process/gavi-pledging-conference-january-2015. Gavi is the "Global Alliance for Vaccines and Immunisation."
988   https://www.scidev.net/global/features/under-fire-critics-challenge-gavi-s-vaccine-spending-practices/

989 https://www.youtube.com/watch?v=6Af6b_wyiwI

990 Vineet D Menachery, Boyd L Yount Jr, Kari Debbink1, Sudhakar Agnihothram, Lisa E Gralinski, Jessica A Plante, Rachel L Graham, Trevor Scobey, Xing-Yi Ge, Eric F Donaldson, Scott H Randell, Antonio Lanzavecchia, Wayne A Marasco, Zhengli-Li Shi & Ralph S Baric. (2015) A SARS-like cluster of circulating bat coronaviruses shows potential for human emergence. Nature Medicine, 21 (12), 1508-1514. December 2015. With follow-up letter included: https://www.nature.com/articles/nm.3985

991 https://web.archive.org/web/20160630124045/http://www.nature.com/natureconferences/viir2016/sponsors.html. Also see: https://thenationalpulse.com/exclusive/fauci-scientists-attended-wuhan-lab-conference/

992 History----Wuhan Institute of Virology (cas.cn)

993 http://english.whiov.cas.cn/About_Us2016/Directors2016/

994 Homeland Security, 2017, January, Biological Incident Annex to the Response and Recovery Federal Interagency Operation Plans. Final-January 2017. https://web.archive.org/web/20201104004301/https://www.fema.gov/media-library-data/1511178017324-92a7a7f808b-3f03e5fa2f8495bdfe335/BIA_Annex_Final_1-23-17_(508_Compliant_6-28-17).pdf

995 Homeland Security, 2017, January, Biological Incident Annex to the Response and Recovery Federal Interagency Operation Plans. Final-January 2017. https://web.archive.org/web/20201104004301/https://www.fema.gov/media-library-data/1511178017324-92a7a7f808b-3f03e5fa2f8495bdfe335/BIA_Annex_Final_1-23-17_(508_Compliant_6-28-17).pdf. For a color chart of the agencies, see: Pandemic (hstoday.us)

996 President Barack Obama. 2019, January 9. The White House. Recommended Policy Guidance for Potential Pandemic Pathogen Care and Oversight | whitehouse.gov (archives.gov). Recommended Policy Guidance for Potential Pandemic Pathogen Care and Oversight | whitehouse.gov (archives.gov)

997 Eleven Days Before Trump Took Office, Obama-Biden White House'Depopulation' Science Czar Holdren Advised Lifting Gain-Of-Function Research Ban; Trump Kept In The Dark - The American Report

998 Secretary Sylvia Mathews Burwell | whitehouse.gov (archives.gov)

999 Eleven Days Before Trump Took Office, Obama-Biden White House'Depopulation' Science Czar Holdren Advised Lifting Gain-Of-Function Research Ban; Trump Kept In The Dark - The American Report

1000 NOT-OD-17-071: Notice Announcing the Removal of the Funding Pause for Gain-of-Function Research Projects (nih.gov)

1001 https://www.federalregister.gov/documents/2017/01/13/2017-00721/emergency-use-authorization-of-medical-products-and-related-authorities-guidance-for-industry-and

1002 Monaco, Lisa, 2020, March 3, Pandemic Disease Is a Threat to National Security, Foreign AffairsLisa Monaco served as Assistant to the President for Homeland Security and Counterterrorism from 2013 to 2017. https://www.foreignaffairs.com/articles/2020-03-03/pandemic-disease-threat-national-security

1003 CEPI officially launched – CEPI

1004 Bill Gates warns tens of millions could be killed by bio-terrorism | Bill Gates | The Guardian

1005 https://www.who.int/medicines/ebola-treatment/TheCoalitionEpidemicPreparednessInnovations-an-overview.pdf

1006 Rewriting the Genetic Code: A Cancer Cure In the Making | Tal Zaks | TEDxBeaconStreet - Bing video. For commentary, see: https://medicalkidnap.com/2021/03/09/modernas-top-scientist-on-mrna-technology-in-covid-shots-we-are-actually-hacking-the-software-of-life/#:~:text=Tal%20Zaks%2C%20the%20chief%20medical%20officer%20at%20Moderna,we%20think%20about%20prevention%20and%20treatment%20of%20disease.%E2%80%9D and Artificial Spike Proteins And The End of Human Health (drsircus.com)

1007   Norbert Pardi, Michael J. Hogan, Frederick W. Porter and Drew Weissman. (2018, April). mRNA vaccines — a new era in vaccinology. Nature Reviews 17, 261-279. https://www. nature.com/articles/nrd.2017.243.pdf?origin=ppub

1008   "TÜBINGEN, Germany / BOSTON – March 15, 2020. CureVac AG, a clinical stage biopharmaceutical company pioneering mRNA-based drugs for vaccines and therapeutics, confirmed today that internal efforts are focused on the development of a coronavirus vaccine with the goal to reach, help and to protect people and patients worldwide." https://www. curevac.com/en/2020/03/15/curevac-focuses-on-the-development-of-mrna-based-coro-navirus-vaccine-to-protect-people-worldwide/ It is impossible at this time to entangle all the various interlocking investments in vaccines, many of which include or lead back to Bill Gates.

1009   The next epidemic is coming. Here's how we can make sure we're ready. | Bill Gates gatesnotes. com)

1010   Crystal Watson, Tara Kirk Sell, Matthew Watson, Caitlin Rivers, Christopher Hurtado, Matthew P. Shearer, Ashley Geleta, and Thomas Inglesby ("Posted October 09, 2018"). Technologies to Address Global Catastrophic Biological Risks. Johns Hopkins University, Bloomberg School of Public Health, Center for Health Security. Copyright 2018." No sponsors are listed. Technologies to Address Global Catastrophic Biological Risks (centerforhealthse-curity.org) and 181009-gcbr-tech-report.pdf (jhsphcenterforhealthsecurity.s3.amazonaws. com)

1011   P. 6 of the JHU report.

1012   P. 47 of the JHU report.

1013   https://reporter.nih.gov/search/cVFEMwgD-UyuwQkDAvSL8Q/project-details/8674931

1014   WEF_HGHI_Outbreak_Readiness_Business_Impact.pdf (weforum.org)

1015   https://health.news/2020-07-08-pandemic-bill-gates-negotiated-contact-tracing-deal.html#

1016   https://www.congress.gov/bill/116th-congress/house-bill/6666/text. With sponsors: Text - H.R.6666 - 116th Congress (2019-2020): COVID-19 Testing, Reaching, And Contacting Everyone (TRACE) Act | Congress.gov | Library of Congress

1017   https://truepundit.com/exclusive-bill-gates-negotiated-100-billion-contact-tracing-deal-with-dem-ocratic-congressman-sponsor-of-bill-six-months-before-coronavirus-pandemic/

1018   GPMB_annualreport_2019.pdf (who.int)

1019   https://www.centerforhealthsecurity.org/news/center-news/2020/2020-01-17-Event201-recommendations.html

1020   https://www.neweconomyforum.com/2019-new-economy-forum-beijing/

1021   Brunson, E. K., Chandler, H., Gronvall, G. K., Ravi, S., Sell, T. K., Shearer, M. P., & Schoch-Spana, M. (2020). The SPARS pandemic 2025–2028: A futuristic scenario to facilitate medical countermeasure communication. Journal of International Crisis and Risk Communication Research, 3(1), 71–102. https://doi.org/10.30658/jicrcr.3.1.4; https://stars.library.ucf.edu/jicrcr/vol3/iss1/4/

1022   https://www.dailymail.co.uk/news/article-8003713/China-appoints-military-bio-weap-on-expert-secretive-virus-lab-Wuhan.html and http://www.asianews.it/news-en/Prof-Tritto:-COVID-19-was-created-in-the-Wuhan-laboratory-and-is-now-in-the-hands-of-the-Chinese-military-50719.html

1023   https://www.cnas.org/publications/reports/myths-and-realities-of-chinas-military-civil-fu-sion-strategy

1024   https://www.state.gov/wp-content/uploads/2020/05/What-is-MCF-One-Pager.pdf

1025   In addition to Chapter 6, here are some additional reports and an Indian scientific article: https://www.hindustantimes.com/india-news/journals-retract-two-major-covid-studies/sto-ry-EVzDlWf9bUoTgBmVvRUgVK.html; https://www.theweek.in/news/sci-tech/2020/02/01/

research-paper-by-indian-scientists-on-coronavirus-fuels-bioweapon-theories.htm; Prashant Pradhan, Ashutosh Kumar Pandey, Akhilesh Mishra, Parul Gupta , Praveen Kumar Tripathi, Manoj Balakrishnan Menon, James Gomes, Perumal Vivekanandan and Bishwajit Kundu (2020, January 31) Uncanny similarity of unique inserts in the 2019-nCoV spike protein to HIV-1 gp120 and Gag. bioRxiv. https://www.biorxiv.org/content/10.1101/2020.01.30.92 7871v1.full.pdf

1026 https://cepi.net/news_cepi/cepi-to-fund-three-programmes-to-develop-vaccines-against-the-novel-coronavirus-ncov-2019/

1027 Headline: "A coalition backed by Bill Gates is funding biotechs that are scrambling to develop vaccines for the deadly Wuhan coronavirus." https://markets.businessinsider.com/news/stocks/vaccines-for-wuhan-china-cornonavirus-moderna-inovio-cepi-2020-1-1028840542?op=1

1028 The inside story of how scientists produced an Ebola vaccine (statnews.com) and FDA approves an Ebola vaccine, long in development, for the first time - STAT (statnews.com)

1029 https://www.axios.com/moderna-nih-coronavirus-vaccine-ownership-agreements-22051c 42-2dee-4b19-938d-099afd71f6a0.html

1030 Paules CI, Marston HD, Fauci AS. "Coronavirus Infections - More than Just the Common Cold." JAMA. 2020;323(8):707-708. https://pubmed.ncbi.nlm.nih.gov/31971553/

1031 National Institutes of Health. Undated. Retrieved May 14, 2021. COVID-19 Vaccine Development: Behind the Scenes. Years of research enable a COVID-19 vaccine to be developed in record time. https://covid19.nih.gov/news-and-stories/vaccine-development

1032 The CDC Is Still Botching the Coronavirus Testing Process – Reason.com; Early CDC Coronavirus Test Came with Inconsistent Instructions And Cost The U.S. Weeks | WBUR News

1033 https://www.bbc.co.uk/mediacentre/latestnews/2020/coronavirus-trusted-news

1034 Trusted News Initiative (TNI) steps up global fight against disinformation with new focus on US presidential election - Media Centre (bbc.co.uk)

1035 https://www.who.int/director-general/speeches/detail/who-director-general-s-opening-remarks-at-the-media-briefing-on-covid-19---11-march-2020

1036 Ronald Bailey. COVID-19 Mortality Rate 'Ten Times Worse' Than Seasonal Flu, Says Dr. Anthony Fauci. Reason Mar 11 2020 https:// reason.com /2020 /03/11/covid-19-mortality-rate-ten-times-worse than-seasonal-flu-says-dr-anthony-fauci/; and Joseph Guzman Coronavirus 10 times more lethal than seasonal flu, top health official says. The Hill https:// thehil l.com/ changing-america/well-being/prevention-cures/487086-coronavi rus-10-times-more-lethal-than-seasonal accessed 9.1.2020

1037 Covid-19 — Navigating the Uncharted | NEJM

1038 https://newlevellers.blogspot.com/2021/03/fauci-shares-stage-with-top-ccp.html. For more about D. Zhong Nanshan: https://www.npr.org/sections/goatsandsoda/2020/04/02/825957192/dr-zhong-is-the-supreme-commander-in-china-s-war-against-coronavirus

1039 Fox News' Neil Cavuto Shocked by Trump Hydroxychloroquine Announcement (businessinsider.com)

1040 Steven Hatfield. How a Single Point Failure Destroyed the National Pandemic Plan - Dr. Steven Hatfill (drstevenhatfill.com) and https://www.kmblegal.com/sites/default/files/NEW%20R.%20Bright%20OSC%20Complaint_Redacted.pdf

1041 No. 202.11: Continuing Temporary Suspension and Modification of Laws Relating to the Disaster Emergency (ny.gov)

1042 For the Brix video and transcription: https://www.realclearpolitics.com/video/2020/04/08/dr_birx_unlike_some_countries_if_someone_dies_with_covid-19_we_are_counting_that_as_a_covid-19_death.html

1043 Find the blog and the Youtube presentation by Dr. Breggin that quickly reached 40,000 viewers and exceeded 59,000 on May 24, 2020: https://breggin.com/us-chinese-scientists-collaborate-on-coronavirus/

1044   https://investors.modernatx.com/news-releases/news-release-details/moderna-announc-es-award-us-government-agency-barda-483-million

1045   Ren-Di Jiang, Mei-Qin Liu, Ying Chen, Chao Shan, Yi-Wu Zhou, Xu-Rui Shen, Qian Li, Lei Zhang, Yan Zhu, Hao-RuiSi, Qi Wang, Juan Min, Xi Wang, Wei Zhang, Bei Li, Hua-Jun Zhang, Ralph S.Baric, Peng Zhou, Xing-Lou Yang, Zheng-Li Shi. (2020, July) Pathogenesis of SARS-CoV-2 in Transgenic Mice Expressing Human Angiotensin-Converting Enzyme 2. Cell 182, 50–58 July 9, 2020. https://doi.org/10.1016/j.cell.2020.05.0. Pathogenesis of SARS-CoV-2 in Transgenic Mice Expressing Human Angiotensin-Converting Enzyme 2 - ScienceDirect

1046   https://www.weforum.org/press/2020/07/klaus-schwab-and-thierry-malleret-release-covid-19-the-great-reset-the-first-policy-book-on-the-covid-crisis-globally

1047   https://www.cnn.com/2020/07/22/health/pfizer-covid-19-vaccine-government-contract/index.html

1048   U.S. Government Engages Pfizer to Produce Millions of Doses of COVID-19 Vaccine | HHS.gov

1049   The paper was originally made available on August 7, 2020 online and then appeared in the printed journal in January 2021: McCullough, P., et al. (2021, January). Pathophysiological Basis and Rationale for Early Outpatient Treatment of SARS-CoV-2 (COVID-19) Infection. The American Journal of Medicine, Vol 134 (1), 16-22. https://reader.elsevier.com/reader/sd/pii/S0002934320306732?token=0999741F25467F59E95ACBE97F83721EE81555E-A4437C92130EAC544BEB80A2868218CFDDDB04C043CE32B7C5402D20A&origin-Region=us-east-1&originCreation=20210426124409

1050   https://www.cdc.gov/nchs/data/health_policy/covid19-comorbidity-expanded-12092020-508.pdf

1051   Wen Shi Lee, Adam K. Wheatley, Stephen J. Kent, and Brandon J. DeKosky.(2020, October). Antibody-dependent enhancement and SARS-CoV-2 vaccines and therapies. Nature Microbiology, 5, 1185-1199. https://www.nature.com/articles/s41564-020-00789-5.pdf

1052   Branswell, H. 2020, September 14, Bill Gates slams 'shocking' U.S. response to Covid-19 pandemic, STAT. https://www.statnews.com/2020/09/14/bill-gates-slams-mismanaged-u-s-response-to-covid-19-pandemic/

1053   Yan, Li-Meng Yan ; Kang, Shu; Guan, Jie; Hu, Shanchang. (2020, September 14). Unusual Features of the SARS-CoV-2 Genome Suggesting Sophisticated Laboratory Modification Rather Than Natural Evolution and Delineation of Its Probable Synthetic Route. Prepubli-cation. http://breggin.com/coronavirus/The_Yan_Report.pdf.

1054   Yan, Li-Meng; Kang, Shu; Guan, Jie; Hu, Shanchang. 2020, October 8, SARS-CoV-2 Is an Unrestricted Bioweapon: A Truth Revealed through Uncovering a Large-Scale, Organized Scientific Fraud. Prepublication. https://zenodo.org/record/4073131#.X4OpJOaSk2x "You can cite all versions by using the DOI 10.5281/zenodo.4073130. This DOI represents all versions, and will always resolve to the latest one."

1055   http://breggin.com/wp-content/uploads/1990/01/fluvoxamine.pdf

1056   https://www.wsj.com/articles/china-is-national-security-threat-no-1-11607019599?mod=ar-ticle_inline

1057   BioNTech in China alliance with Fosun over potential coronavirus vaccine | Reuters

1058   The NIH claims joint ownership of Moderna's coronavirus vaccine - Axios

1059   Yuyang,etal.,2020.https://www.biorxiv.org/content/biorxiv/early/2020/12/04/2020.12.04.409144.full.pdf

1060   Nuovo et al., 2021. https://www.bloomberg.com/news/newsletters/2021-05-01/the-new-economy-saturday-maybe-china-won-t-beat-the-u-s-after-all

1061   McCullough, P. et al. (2020, December). Multifaceted highly targeted sequential multidrug treatment of early ambulatory high-risk SARS-CoV-2 infection (COVID-19). Reviews in

Cardiovascular Medicine. 21 (4), 517-530. November 2020. DOI: 10.31083/j.rcm.2020.04.264 https://rcm.imrpress.com/EN/10.31083/j.rcm.2020.04.264. See the graph at https://breggin. com/coronavirus-resource-center/. Keep in mind that death rates are vastly over-inflated.

1062 https://www.mitre.org/news/press-releases/coalition-of-health-technology-industry-leaders-announce-vaccination-credential-initiative

1063 https://thecommonsproject.org/commonpass

1064 https://www.busiweek.com/pfizer-biontech-and-moderna-to-earn-14-7b-from-covid-19-vaccines-sales-by-2023/

1065 Controversy as Israeli Stats Report Appears to Suggest 40 times Higher Deaths in Elderly After 1st But Before 2nd Pfizer Dose Than From COVID - NewsRescue.com; Controversy as Israeli Stats Report Appears to Suggest 40 times Higher Deaths in Elderly After 1st But Before 2nd Pfizer Dose Than From COVID - NewsRescue.com; As Israel Reopens,'Whoever Does Not Get Vaccinated Will Be Left Behind' - The New York Times (nytimes.com)

1066 South Africa suspends use of AstraZeneca's COVID-19 vaccine after it fails to clearly stop virus variant | Science | AAAS (sciencemag.org)

1067 The Wuhan Laboratory Origin of SARS-CoV-2 and the Validity of the Yan Reports Are Further Proved by the Failure of Two Uninvited "Peer Reviews" | Zenodo

1068 https://www.theepochtimes.com/beijing-seizes-upon-faucis-call-for-solidarity-to-push-covid-19-propaganda_3731620.html; Fauci and the Communists – AIER

1069 https://www.worldtribune.com/fauci-makes-nice-with-chinese-communists-at-future-of-health-forum/

1070 Congressional Research Service. The PREP Act and COVID-19: Limiting Liability for Medical Countermeasures (congress.gov)

1071 https://www.mediaite.com/tv/breaking-former-cdc-director-says-he-believes-coronavirus-escaped-from-wuhan-lab/

1072 https://www.msn.com/en-us/news/politics/bidens-climate-duo-puts-us-back-in-global-warming-fight/ar-BB1fJzxM

1073 Bill Gates says Covid-19 vaccine tech should not be shared with India, now there is a vaccine shortage - Technology News (indiatoday.in); COVID-19 vaccine formulas shouldn't be shared with India: Bill Gates (yahoo.com)

1074 Gates promotes Johnson & Johnson's vaccine in media interviews but the degree of his investment or control remains unclear to us.

1075 https://www.who.int/director-general/speeches/detail/who-director-general-s-opening-remarks-at-the-member-state-information-session-on-who-biohub---15-april-2021

1076 https://www.the-scientist.com/news-opinion/biohub-network-aims-to-advance-sharing-of-pathogens-for-research-68849

1077 US denies India's request to lift ban on export of COVID-19 vaccine raw materials (jagranjosh.com)

1078 https://www.ft.com/content/096dd554-499b-468c-b5fa-38b0352941a0?segmentid=a-cee4131-99c2-09d3-a635-873e61754ec6

1079 https://www.nbcnews.com/meet-the-press/meet-press-april-25-2021-n1265222

1080 https://www.cdc.gov/coronavirus/2019-ncov/vaccines/safety/adverse-events.html . On the second page. These numbers are both hidden and ignored by the CDC which continues to claim there are no proven cases to report, but the best use of CDC's VAERS or the FDA's FAERS is precisely to identify gross patterns like these, whereupon the marketing of the vaccine or drug would, in the past, be stopped, and a massive investigation begun.

1081 https://www.cdc.gov/vaccines/covid-19/health-departments/breakthrough-cases.html

1082 Kowarz E, Krutzke L, Reis J, Bracharz S, Kochanek S, Marschalek R. "Vaccine-induced covid-19 mimicry" Syndrome: Splanchnic reactions withing the SARS-CoV-2 spike open reading frame result in spike protein variant that may cause thromboembolic events in patients immunized with vector based vaccines. The article has been accepted by Clinical Infectious Diseases. https://www.researchsquare.com/article/rs-558954/v1

1083 https://outlook.live.com/mail/deeplink?popoutv2=1&version=20210426004.07

1084 Ogata AF, Cheng C-A, Desjardins M, Senussi Y, Sherman AC, Powell M, et al. Circulating SARS-CoV2 vaccine antigen detected in the plasma of mRNA-1273 vaccine recipients. Clinical Infectious Diseases. https://doi.org/10.1093/cid/ciab465. https://www.researchsquare.com/article/rs-558954/v1

1085 Seneff S, Nigh G. Worse Than the Disease? Reviewing Some Possible Unintended Consequences of the mRNA Vaccines Against COVID-19. International Journal of Vaccine Theory, Practice, and Research. 2021;2(1):402-43. Circulating SARS-CoV-2 Vaccine Antigen Detected in the Plasma of mRNA-1273 Vaccine Recipients | Clinical Infectious Diseases | Oxford Academic (oup.com)

1086 Hours after clashing with Sen. Rand Paul, R-Ky., at a Senate hearing Tuesday, Dr. Anthony Fauci called Paul's claim that the National Institutes of Health had funded "gain-of-function research" in Wuhan "preposterous" in an interview with PolitiFact. https://www.politifact.com/article/2021/may/11/united-facts-america-fauci-disputes-sen-rand-paul-/

1087 Centers for Disease Control and Prevention (CDC), May 12, 2021, Selected Adverse Events Reported after COVID-19 Vaccination | CDC

1088 https://www.thedesertreview.com/opinion/letters_to_editor/ivermectin-for-the-world/article_2aa653ca-b1ae-11eb-903e-a78b2b053ba4.html

1089 http://medisolve.org/yellowcard_urgentprelimreport.pdf

1090 Seneff S, Nigh G. Worse Than the Disease? Reviewing Some Possible Unintended Consequences of the mRNA Vaccines Against COVID-19. International Journal of Vaccine Theory, Practice, and Research. 2021;2(1):402-43. Circulating SARS-CoV-2 Vaccine Antigen Detected in the Plasma of mRNA-1273 Vaccine Recipients | Clinical Infectious Diseases | Oxford Academic (oup.com)

1091 ACIP Meeting Agenda June 18 2021 (cdc.gov)

1092 'We clearly have an imbalance': CDC holding emergency meeting over COVID vax concern (nypost.com)

1093 Voluntary COVID-19 vaccination of children: a social responsibility (bmj.com)

1094 11221-moh-pfizer-collaboration-agreement-redacted.pdf (govextra.gov.il)

1095 https://childrenshealthdefense.org/defender/vaers-data-injuries-deaths-vaccinating-5-year-olds/

1096 David R. Martinez, Alexandra Schäfer, Sarah R. Leist, Gabriela De La Cruz, Ande West, Elena N. Atochina-Vasserman, Lisa C. Lindesmith, Norbert Pardi, Robert Parks, Maggie Barr, Dapeng Li, Boyd Yount, Kevin O. Saunders, Drew Weissman, Barton F. Haynes, Stephanie A. Montgomery, Ralph S. Baric. (2021, June 22, published online). Chimeric spike mRNA vaccines protect against Sarbecovirus challenge in mice. DOI: 10.1126/science.abi4506. Chimeric spike mRNA vaccines protect against Sarbecovirus challenge in mice | Science (sciencemag.org)."Acknowledgments. David R. Martinez is currently supported by a Burroughs Wellcome Fund Postdoctoral Enrichment Program Award and a Hanna H. Gray Fellowship from the Howard Hugues Medical Institute and was supported by an NIH NIAID T32 AI007151 and an NIAID F32 AI152296. This research was also supported by funding from the Chan Zuckerberg Initiative awarded to R.S.B... This project was funded in part by the National Institute of Allergy and Infectious Diseases, NIH, U.S. Department of Health and Human Services award U01 AI149644, U54 CA260543, AI157155 and AI110700 to R.S.B., AI124429 and a BioNTech SRA to D.W., and E.A.V., as well as an animal models contract from the NIH... Competing interests: The University of North Carolina at Chapel Hill has filed provisional patents for which D.R.M. and R.S.B are co-inventors (U.S. Provisional Application No. 63/106,247 filed on October 27th, 2020) for the chimeric vaccine constructs and their applications described in this study. (HHSN272201700036I)."

1097 Elaine R. Miller, Pedro L. Moro, Maria Cano, and Tom Shimabukuro. (2015, June 26). Deaths following vaccination: What does the evidence show? Vaccine, 33, 3288-3292. The article is from the Immunization Safety Office (ISO) Centers for Disease Control and Prevention (CDC). https://www.sciencedirect.com/science/article/pii/S0264410X15006556?casa_token=dU-jB6I2SHDQAAAAA:KD0i_GcZ74p5wYMAubCcUd-17RSvTUleAnB_afIofV_xNnG6I-e3RRfAPrw8mj4DQ4bKylilBgA

1098 https://www.openvaers.com/covid-data.

1099 https://www.openvaers.com/covid-data.

1100 VAERS may be publicly accessed at https://www.openvaers.com/covid-data. Accessed July 9, 2021, by Peter McCullough.

1101 https://www.authorea.com/users/414448/articles/522499-sars-cov-2-mass-vaccination-urgent-questions-on-vaccine-safety-that-demand-answers-from-international-health-agencies-regulatory-authorities-governments-and-vaccine-developers

1102 https://www.ronjohnson.senate.gov/2020/4/open-letter-to-president-trump-urging-action-on-hydroxychloroquine

1103 https://aapsonline.org/hcqsuit and https://www.ronjohnson.senate.gov/2020/4/open-letter-to-president-trump-urging-action-on-hydroxychloroquine

1104 https://aapsonline.org/preliminary-injunction-sought-to-release-hydroxychloroquine-to-the-public/

1105 https://abcnews.go.com/Politics/trump-taking-hydroxychloroquine-unproven-drug-touted-covid-19/story?id=70751728

1106 The Dr. Peter Breggin YouTube Channel, https://www.youtube.com/user/PeterBreggin https://www.nbcnews.com/meet-the-press/meet-press-april-25-2021-n1265222. If our YouTube Channel is taken down, the alternative will be http://www.brighteon.com/. To keep track of our work and our platforms, sign up for our Free Frequent Alerts on http://www.breggin.com/.

1107 www.truthforhealth.org

INDEX

[Created with **TExtract** / www.TExtract.com]